BIOGENIC STRUCTURES: THEIR USE IN INTERPRETING DEPOSITIONAL ENVIRONMENTS

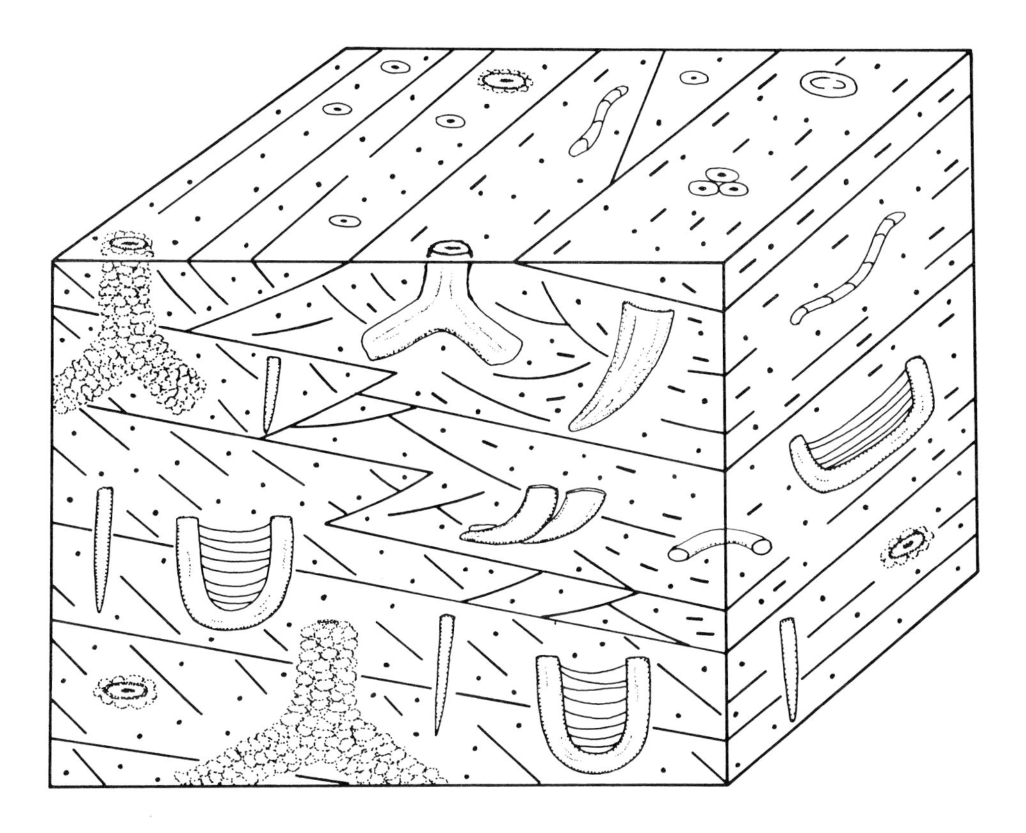

Edited by

H. Allen Curran

Department of Geology, Smith College,

Northampton, Massachusetts

SOCIETY OF ECONOMIC PALEONTOLOGISTS AND MINERALOGISTS

R.L. Ethington, Editor of Special Publications

Special Publication No. 35

Tulsa, Oklahoma, U.S.A. ISBN: 0-918985-15-3 *March 1985*

PREFACE

Organisms of one sort or another today inhabit virtually every sedimentary environment on Earth, and the rock record tells us that this has been the case through the greater part of our planet's history. Furthermore, organisms leave their mark in most sedimentary settings, either directly in the form of body fossils or indirectly as biogenic structures. Such structures include trace fossils which record the bioturbation activity of animals (e.g. burrows, tracks, trails, grazing patterns) and the biostratification activity (e.g. biogenic graded bedding and mottling, stromatolites) of animals and plants in sediment. Also included are the bioerosion structures (e.g. borings, scrapings) of plants and animals and other evidences of animal activity such as fecal pellets and coprolites.

In addition to their often profound modifying effects on substrates, ancient biogenic structures preserve a record of organism behavioral activity in response to substrate and other paleoenvironmental controls. Thus, biogenic structures can be highly useful as facies indicators and can provide valuable clues to the interpretation of paleodepositional environments. Nonetheless, as the studies of this volume well illustrate, knowledge gained from biogenic structures always should be used in concert with other available sedimentologic, paleontologic, and stratigraphic information in order to obtain a thorough understanding of an ancient environmental setting.

The purpose of this volume is to present a broad spectrum of case-book examples of the use of biogenic structures in the interpretation of depositional environments. The papers herein discuss a wide range of environments from Phanerozoic non-marine to deep-sea settings and include studies of bioerosion and biostratification structures generated by plants as well as trace fossils produced by invertebrates and vertebrates. Although the breadth of this volume is great, as is always true with a large topic and a collection of symposium papers, coverage is not all-inclusive in terms of possible environments and/or geologic ages. This collection is no exception in that it contains examples exclusively from the Phanerozoic and it has a bias toward siliciclastic sedimentary settings and toward marginal marine to shallow subtidal environments. A good, recently published, companion for this volume would be the SEPM-sponsored March, 1984 special issue of the *Journal of Paleontology* (v. 58, no. 2) that contains more papers on trace fossils in marine carbonate and non-marine terrigenous settings.

The papers of this volume were initially presented at a poster session symposium on biogenic structures and depositional environments sponsored by the Society of Economic Paleontologists and Mineralogists at the joint annual meeting of the SEPM and American Association of Petroleum Geologists in Denver in June, 1980. Some papers were added from a trace fossils session held at the 26th International Geological Congress in Paris in July, 1980, and several other papers subsequently were added to fill gaps.

I am most grateful to the many individuals who have helped to make this volume possible. In particular, I thank the contributing authors who have given their full cooperation and assistance and have displayed patience and good humor throughout the editorial and production process. Many colleagues graciously assisted as article reviewers; they are acknowledged within the volume. Raymond L. Ethington, SEPM Special Publications Editor, provided much guidance and advice, as well as editorial assistance. Smith College students Dillon Scott and Jennifer Sakurai aided greatly with copy editing, and Nancy Mosher drafted the cover logos. Finally, I thank Pam Garrish, Composition Department at Edwards Brothers, Inc., for her assistance during the final production stages of the volume.

H. Allen Curran
Editor

A Publication of
The Society of Economic Paleontologists and Mineralogists
a division of
The American Association of Petroleum Geologists

BIOGENIC STRUCTURES: THEIR USE IN INTERPRETING DEPOSITIONAL ENVIRONMENTS

Edited by H. Allen Curran

CONTENTS

PART I
ASSEMBLAGES OF BIOGENIC STRUCTURES IN NON-MARINE AND MARGINAL MARINE ENVIRONMENTS

INTRODUCTION.—The papers in this first part of the volume cover trace fossil assemblages in a range of environments from dunes to marginal marine deltaic settings. The lead paper by A. A. Ekdale and M. Dane Picard deals with trace fossils and bioturbation textures in Jurassic eolianites from Utah. Most eolianites are thought to possess a sparse fossil record, but Ekdale and Picard report several new and distinctive trace fossil forms that presumably were produced by insects. Their Table 1 summarizes well the known distribution of animal traces, trace fossils, and body fossils in dune and interdune deposits.

Large escape burrows attributed to the activity of Devonian bivalves in a fluvial environment are analyzed by Richard Thoms and Thomas Berg. They suggest a modern bivalve as a tracemaker analog for the fossil escape burrows. The study demonstrates well how knowledge of the structures produced by a modern organism under known living conditions can be used to interpret with greater clarity the significance of ancient biogenic sedimentary structures and their paleodepositional setting.

William Hakes reports on a trace fossil assemblage from an Upper Pennsylvanian sequence in Kansas that is thought to be indicative of a lowered salinity environment. The unique aspects of the assemblage are the small size of the individual trace fossils along with high diversity, moderate abundance, and excellent preservation. Small-sized trace fossils easily can be overlooked, but assemblages of such forms may be a real key to the recognition of brackish-marine paleoenvironments in the rock record.

These findings are reinforced and confirmed by Grzegorz Pieńkowski's study of Jurassic trace fossil assemblages from Poland. Again assemblages of smaller than normal size trace fossils are thought to characterize brackish-marine settings. Pieńkowski also describes trace fossil assemblages from fluvial settings, and he gives useful criteria for differentiating in core samples deposits of shore-prograding, barrier-lagoon, and deltaic environments.

Tracks and trackway areas made by Triassic vertebrates are the topic of the paper by Georges Demathieu. Through careful analysis of tracks, Demathieu demonstrates that it often is possible to determine what the tracemaker vertebrate was, at least to its proper higher taxonomic level. Through the use of statistical methods, Demathieu also is able to estimate population sizes and to reconstruct biological relationships, as summarized in his Figure 4. Furthermore, the trackways appear to characterize sandy areas situated between a sea or lagoon and a vegetated, terrestrial environment.

The study by Molly Miller and Larry Knox on Pennsylvanian coal-bearing sequences in Tennessee should serve as a model for the integration of lithologic, sedimentologic, and trace fossil data towards interpretation of depositional environments. Miller and Knox show that trace fossil assemblages can be highly useful in the recognition of subenvironments within ancient barrier island to marine-dominated deltaic systems. Figures 9 and 10 are particularly valuable because they summarize in detail information on the known onshore to offshore distribution of a number of commonly recognized ichnogenera.

Michael Eagar and his five co-authors present a most comprehensive study of the trace fossil assemblages of Mid-Carboniferous deltaic sequences in the Central Pennine Basin, England. This paper should be highly useful to many ichnologists, particularly those working with Paleozoic trace fossils, because it is profusely illustrated and contains descriptions and discussions of many trace fossil forms. The analysis of the evolution of bivalve escape structures is particularly interesting. Throughout, the authors demonstrate how trace fossil assemblages can be used to recognize and define subenvironments and varying modes of sedimentation within ancient deltaic sequences.

The study of a Cambrian deltaic sequence in Spain by Iain Legg shows that other sedimentologic factors beyond simple increase in water depth are crucial in determining the distribution of trace fossils. Figure 2 presents a useful summary of the distribution of common Cambrian trace fossils across a range of environments from tidal channels to open shelf. Again, changes in trace fossil size and depth of burrowing, in this case with forms of *Arenicolites* and *Diplocraterion*, are shown to be indicative of onshore to offshore transition. Three new ichnospecies also are described in this paper.

The final paper of this section by Larry Knox and Molly Miller analyzes and compares traces produced by modern gastropods with similar trace fossils found in Pennsylvanian sandstones in Tennessee. This study demonstrates how trace morphology can vary with substrate consistency. Furthermore, the paper illustrates well how neoichnologic investigation can be coupled with trace fossil studies to enhance our ability to interpret ancient environments.

TRACE FOSSILS IN A JURASSIC EOLIANITE, ENTRADA SANDSTONE, UTAH, U.S.A.

A.A. EKDALE AND M.DANE PICARD
Department of Geology and Geophysics, University of Utah, Salt Lake City, Utah 84112

ABSTRACT

Most ancient eolianites possess a meager fossil record. Trace fossils and bioturbate textures are present, however, in eolian beds of the Entrada Sandstone (Jurassic) southeast of Moab, Utah. Three new ichnogenera and ichnospecies are described in this report.

The most noticeable of the trace fossils are trails (*Entradichnus meniscus* n. ichnogen. and ichnosp.) which parallel bedding planes in cross-stratified sandstone with well-developed parting lineation. Trails are long, unbranched, and gently curved. Many specimens contain an internal structure of meniscate backfill. The trails are oriented parallel to the depositional dip of cross-strata, suggesting that their creators moved down the lee sides of dunes, pushing sediment back up behind them. Similar meniscate trails are produced in modern sand dunes by the larvae of tipulid insects ("crane flies").

A second trace fossil type consists of small, vertical burrows (*Pustulichnus gregarious* n. ichnogen. and ichnosp.) preserved as bumps in convex epirelief on cross-strata surfaces. These bumps may represent upward extensions of the meniscate trails described above, or they may represent shallow burrows made by sphecid insects ("sand wasps").

Larger, plug-shaped, vertical burrows containing laminated fill (*Digitichnus laminatus* n. ichnogen. and ichnosp.) are rare in the Entrada. Origin of these burrows is unknown. Moderate to thorough bioturbation of sandstone lenses also is present.

Sedimentary structures that together indicate an eolian origin for the sandstone are: (1) large-scale, high angle (mean of 22°), sweeping cross-stratification; (2) large-scale soft sediment deformation, including small-scale soft sediment faulting; (3) eolian ripple marks (large ripple index and high ripple symmetry index) parallel to the dip of foreset slopes of cross-strata; and (4) multiple parallel-truncation bedding planes. Paleocurrent measurements are unimodal and suggest that winds blew to the south and southeast. Eolian petrographic characteristics of the sandstone are: (1) bimodal textures; (2) frosting of grains; (3) rounded or well-rounded coarser grains; (4) minor matrix; (5) high quartz content; (6) dominantly calcite cement; and (7) moderately to well-sorted grains.

These ancient dunes are believed to have been formed in a sand sea where deposition persisted for a long period of time. Deposition was probably within 30° of the paleoequator, and the climate was semiarid or arid and hot.

INTRODUCTION

Trace fossils in the Entrada Sandstone were studied at two localities in east-central Utah (Fig. 1). One site is located at Muleshoe Canyon, about 28 km southeast of the city of Moab, and the other is located at Dry Valley, about 25 km farther south.

The first occurrence was discovered by Picard in 1975; the second occurrence was found by Picard and Ekdale in 1980. Trails, burrows, and bioturbate textures similar to those described here have been observed elsewhere in the Entrada by Picard, but they are not as numerous and well-preserved at other sites. Additional good trace fossil localities in Entrada Sandstone probably will be found with further field work.

STRATIGRAPHY

This paper follows essentially the same stratigraphic nomenclature for the Entrada Sandstone and bounding formations as that employed by O'Sullivan (1980) in a recent study of the Middle Jurassic San Rafael Group. O'Sullivan's rock units are, from bottom to top, a reworked zone of the Carmel Formation (about 4 m thick), a lower red zone of the Carmel Formation (about 10 m thick), Entrada Sandstone (about 80 m thick), and the Moab Tongue, referred to as the Moab Member in this paper (about 25 m thick). This nomenclature coincides with that used in unpublished subsurface studies by Picard.

The Carmel Formation lies unconformably on the Navajo Sandstone (Jurassic). Basal beds are 1 to 5 m thick and consist of reddish-brown, earthy siltstone or white, horizontally stratified sandstone that seems to have been reworked from the Navajo. The Carmel Formation is 20 to 25 m thick in the study area and is characterized by siltstone, sandy siltstone, and clayey siltstone (nomenclature of Picard, 1971). Silt-sized grains and minor very fine sand grains are dominant in nearly all rocks of the formation. Erosional remnants are sculptured into "hoodoos", "goblins" and "stone-babies".

The Entrada Sandstone, which contains the trace fossils reported in this paper, ranges in thickness from 35 to 65 m in the area southeast of Moab. The cross-stratified sandstone of the formation is light brown, reddish brown, reddish orange, grayish red, or moderate red. It is chiefly very fine- to fine-grained sand and commonly is bimodal with medium to coarse, well-rounded, frosted grains present.

The white and pinkish gray Moab Member, which is 25 to 30 m thick, characteristically forms cliffs marked by large-scale, sweeping, trough cross-stratification. The sandstone is a subarkose or quartzarenite that is very fine- to fine-grained (dominant), well-sorted and carbonate-cemented. Between Arches National Park and Gateway, Colorado, and sporadically south of there, one or more thin purplish-red shale beds mark the base of the Moab Member (Wright *et al.*, 1962).

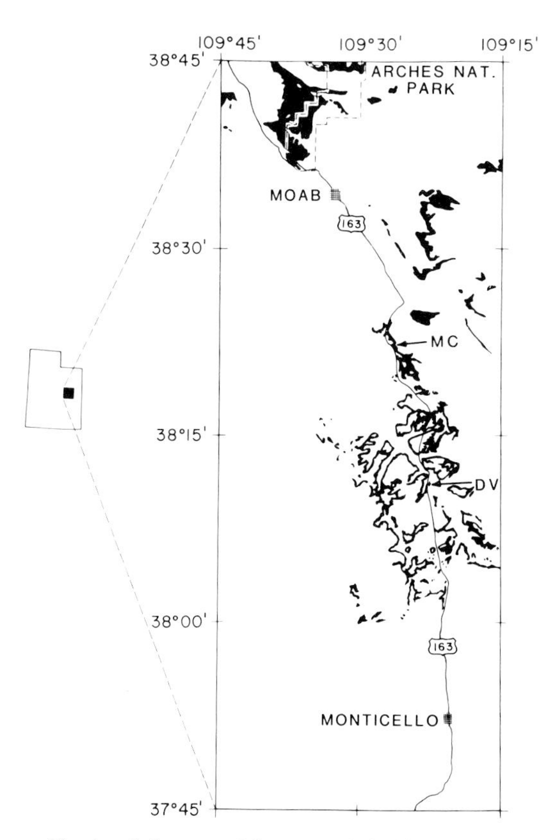

Fig. 1.—Index map of the outcrop belt (black) of the Entrada Sandstone (Jurassic) in east-central Utah. Trace fossiliferous localities at Muleshoe Canyon (MC) and Dry Valley (DV) indicated by arrows.

SEDIMENTOLOGY

Petrography

Thin sections were made of sandstone from the medium- and large-scale, high-angle, cross-stratified facies with trails and of totally bioturbated sandstone. All of the sandstones are reddish brown with hematite staining and pigmentation. They are dominantly subarkose, i.e., with K-feldspar content greater than that of plagioclase. Quartzarenite is rare. Grain size ranges from very fine- to coarse-grained, but is mainly fine- to medium-grained. Sorting ranges from poorly sorted to very well-sorted in the sandstone; most samples are bimodal (Pl. A-D). Matrix is minor, ranging from 3 to 5 percent of the total sandstone. Trace amounts of coarse mica are present. Very fine to fine grains are subangular to subrounded; medium to coarse grains are rounded or well-rounded. Quartz grains have a frosted surface. The general petrographic characteristics of the sandstone are typical of eolian sand (Picard, 1977b).

The rocks can be divided into three groups: (I) sandstone with small amounts (1–10%) of authigenic carbonate cement and slight bimodality of grains; (II) sandstone with large amounts (10 to 20%) of authigenic carbonate cement and marked textural bimodality; and (III) sandstone with very large amounts (20–30%) of authigenic carbonate cement and slight bimodality of grains. Clay minerals are noticeable (1–2%) in the type-I sandstone but not in the other two types. The pronounced bimodality of the type-II sandstone reflects a large population of medium and coarse grains and of very fine grains, but few fine grains. Slightly bimodal sandstone (types I and II) shows a smaller gap in the grain sizes present but still is clearly bimodal. Porosity is greatest (5–10%) in the type-II sandstone, but it is less than in probable eolian beds (average of 16%) of the Entrada Sandstone in northeastern Utah (Otto and Picard, 1976). Calcite is the dominant carbonate cement; dolomite is more abundant than siderite, but both are present only in trace amounts. The calcite cement is almost entirely sparite; micrite is rare. Silica and chert cement are present in all three sandstone types, but only in small amounts (about 1%). Trails and bioturbated facies without distinct trails are found in all three types of sandstone in east-central Utah.

Paleocurrent Analysis

Paleocurrent directions were determined from medium- and large-scale cross-stratification in sandstone at both localities. Forty-seven measurements were made from wedge-planar, tabular-planar (Pl. 1B), and trough cross-stratification. High inclination angles (average of 22°), bimodal grain size and their large-scale cross-stratification sets indicate that the beds are mostly of eolian origin. The eolian cross-stratification frequently forms festoon cosets. Some paleocurrent measurements are from beds containing trace fossils; others are from beds which lie within 3 m stratigraphically of trace fossiliferous horizons.

The paleocurrent direction at Dry Valley is unimodal with the wind directed to the south (Fig. 2). There is little scatter in the measurements. At Muleshoe Canyon, where the paleocurrent direction also is unimodal with moderate scatter, the wind was directed to the southeast (Fig. 2). Two-thirds of the measurements at Dry Valley are between 170° and 189°; at Muleshoe Canyon more than two-thirds of the measurements are between 110° and 130°.

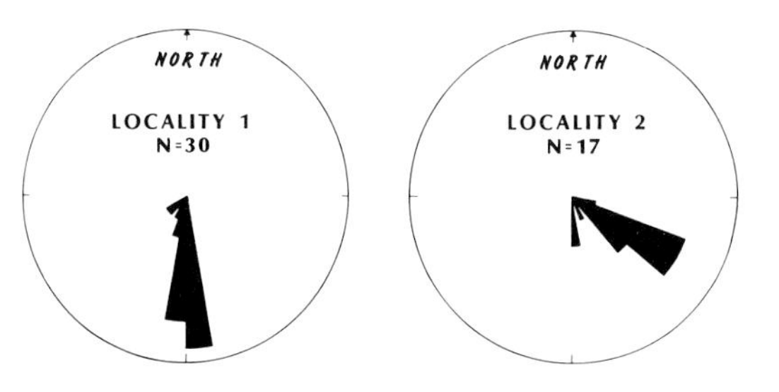

Fig. 2.—Paleocurrent directions determined from cross-strata in Entrada Sandstone in outcrop at Dry Valley (locality 1) and Muleshoe Canyon (locality 2).

Otto and Picard (1975) found that dominant paleocurrent directions for the Entrada Sandstone in northeastern Utah (based on 275 measurements) are unimodal, and 79 percent of them are between 150° and 299°. The scatter in measurements for northeastern Utah for the whole formation is much greater than it is in the measurements for east-central Utah reported here. This is to be expected because of the larger area and greater stratigraphic thickness studied by Otto and Picard (1975). Based on 360 paleocurrent measurements in the Entrada Sandstone in northwestern New Mexico, Tanner (1965) determined that the dominant wind transport direction was southwest. Measurements made by Poole (1963) in the Colorado Plateau yield paleocurrent directions similar to those in the other studies summarized here for the Entrada Sandstone.

Depositional Environment

Several characteristics indicate that the trace fossils described in this paper occur in eolian beds. A major feature of these strata is large-scale, high-angle cross-stratification. Long, sweeping, asymptotic topsets and bottomsets are common in this part of the Entrada Sandstone. Wedge-planar and trough cross- stratification are the most abundant types. The paleocurrent measurements indicate that the pattern is unimodal with small scatter in directions.

Beds approximately correlative with the trace fossil horizons at Dry Valley, which crop out about a mile to the west of the interval, display large-scale soft sediment deformation. These beds contain large sedimentary folds and faults as well as small-scale soft sediment faulting. Soft sediment deformation is present in other deposits considered to be eolian, notably in the Coconino, DeChelly, Lyons, Wingate, Navajo, and Nugget Formations.

Eolian ripple marks (large ripple index and high ripple symmetry index) occur in the Entrada Sandstone

TABLE 1.—SUMMARY OF ANIMAL TRACES, TRACE FOSSILS AND BODY FOSSILS IN DUNE AND INTERDUNE DEPOSITS

Animals	Modern Traces	Trace Fossils	Body Fossils
Phylum ARTHROPODA			
Class INSECTA			
Order Hymenoptera (wasps and ants)	common-abundant[1]		
Order Orthoptera (camel crickets)	common[1]		
Order Diptera (crane flies)	rare-common[1]	present?[4]	
Order Isoptera (termites)	rare[1]		
Order Coleoptera (beetles)	rare-common[1]		
Unidentified Insects			present[2]
Class ARACHNIDA			
Order Araneida (wolf spiders)	common[1]		
Order Scorpionida (scorpions)		present[2,3]	
Class CRUSTACEA			
Subclass Ostracoda (ostracods)			present[3]
Unidentified Crustaceans			present[2]
Class MYRIAPODA			
Subclass Diplopoda (millipedes)	rare?[2]	present[2]	
Phylum MOLLUSCA			
Class GASTROPODA			
Subclass Pulmonata (land snails)	rare-common[1]		
Class BIVALVIA			
Order Unioida (fresh-water clams)	rare-common[1]		present[3]
Phylum CHORDATA			
Class OSTEICHTHYES			
Various Orders (fish)			present[2]
Class AMPHIBIA			
Order Anura (toads)	rare[1]		
Class REPTILIA			
Order Crocodilia (crocodiles)			present[2,3]
Order Squamata (lizards)	common?[2]	present[2,3]	present[2]
Order Pterosauria (pterodactyls)		present[3]	present[2]
Order Ornithischia and/or Saurischia (dinosaurs)		present[2,3]	present[2,3]
Class MAMMALIA			
Order Rodentia (rats, mice and gophers)	rare-common[1]		present[2]
Unidentified Burrows, Trails and Tracks		present[2,3,4]	

[1]Ahlbrandt *et. al.* (1978)
[2]McKee (1979)
[3]Picard (1977a)
[4]Picard and Ekdale (this paper)

close to the fossil localities. Their crests are approximately parallel to the dip of the foreset slopes of the cross-strata. They record ripple mark formation across the lee sides of dunes.

The paleontological evidence in the Entrada Sandstone also supports an eolian origin. No marine fossils have been found within the large-scale cross-stratified beds. In fact, the trace fossils described here are the most significant finds in the formation. Stokes (oral communication, 1981) has observed dinosaur tracks in beds at the very top of the Moab Member on the south flank of the Salt Valley anticline.

PALEONTOLOGY

Fossils in Eolian Deposits

Most eolianites typically possess a meager fossil record. Primary sedimentary structures dominate; biogenic sedimentary structures (trace fossils) are uncommon; body fossils are rare.

Ahlbrandt *et al.* (1978) reviewed the major groups of burrowing organisms that inhabit modern dune fields and described their burrows. Most of the present-day burrowers are insects (e.g., wasps, flies, beetles, crickets, ants, and termites), although burrowing arachnids, molluscs, toads and rodents also are common (Table 1). In order for organism traces to be preserved in eolian deposits, the dune sand apparently must be moist and cohesive, organically reinforced (e.g., by a spider web), or buried rapidly.

Although ancient eolian deposits are common throughout the geologic column from the Precambrian to the present, fossils in eolianites are reported infrequently (Table 1). Body fossils include remains of such animals as dinosaurs (both bones and eggs), other reptiles, fish, insects and crustaceans (McKee, 1979), which occur in dune and (or) wet interdune facies. Reptilian trackways and possible millipede trails occur in Permian eolianites of the Coconino Sandstone in northwestern Arizona (McKee, 1944, 1947) and in the De Chelly Sandstone in northeastern Arizona (McKee, 1934; Peirce, 1963). Trackways of reptiles and scorpionids are present in eolian beds of the Permian Lyons Sandstone in Colorado (Henderson, 1924) and in the Jurassic Nugget Sandstone in northeastern Utah (Albers, 1975; Picard, 1977a; Stokes, 1978). A variety of surface trails and infaunal burrows, presumably produced by dune-dwelling invertebrates, occur in the Pennsylvanian-Permian Casper Formation in southeastern Wyoming (Hanley *et al.*, 1971) and in the Jurassic Navajo-Nugget Sandstone in Utah, Colorado, and Wyoming (Picard, 1977a; Stokes, 1978).

Trace Fossils in Entrada Sandstone

Bioturbation and trace fossils occur in the Entrada Sandstone, but they are rare. In fact, this report appears to be the first published account of such structures in the formation.

Biogenic structures, including distinct trace fossils as well as non-descript bioturbate textures, are known from two localities in eastern Utah (Dry Valley and Muleshoe Canyon) out of a total of eight examined during this study for trace fossils. They occur in eolian sandstone beds which crop out east of U.S. Highway 163 between Moab and Monticello (Fig. 1). No body fossils of animals, vertebrate trackways, organic remains of plants or root casts were found at any of the outcrops investigated.

Meniscate trails.—The most obvious and abundant of the trace fossils in the Entrada Sandstone are meniscate trails, herein named *Entradichnus meniscus* (see SYSTEMATIC ICHNOLOGY), which occur on bedding planes in cross-stratified units that exhibit well-developed parting lineation (Pl. 2A). The trails typically occur in convex epirelief in localized patches, where densities may be as high as 10 trails per 1000cm^2. The trails are long, unbranched and gently curved. Some individual trails can be followed for more than half a meter. All trails maintain a constant width of 5 mm over their entire length, and a circular cross-section is evident wherever the rock is broken oblique to bedding. Many specimens reveal a well-developed internal structure of meniscate backfill when slightly weathered (Pl. 2B), but no burrow lining is apparent.

At Dry Valley the meniscate trails exhibit an obvious preferred orientation parallel to depositional dip of the cross-strata (Figs. 2 and 3). The crescentic internal meniscae are almost always concave down-dip, indicat-

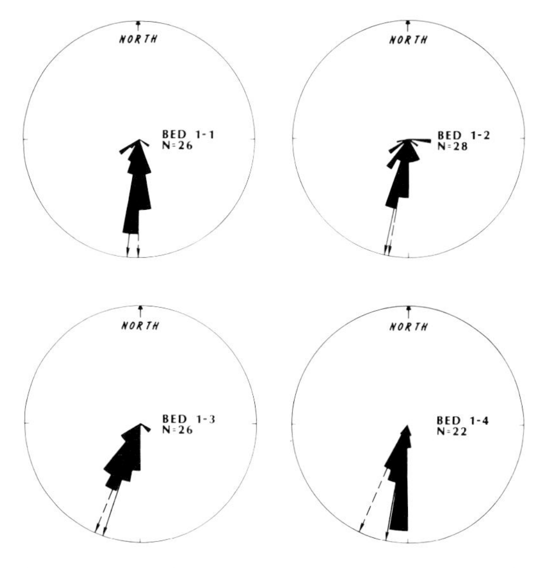

Fig. 3.—Orientation of trails on four different cross-bed surfaces at Dry Valley. Solid arrows indicate average orientation of trails; dashed arrows indicate dip direction of cross-beds.

ing that the creators of the trails usually moved down the lee side of the sand dune, pushing sediment back up behind them. Cross-overs are common, but no branching of the trails was observed.

Simple, unbranched, horizontal burrows in the fossil record commonly are assigned to the ichnogenus *Planolites* (Häntzschel, 1975). *Planolites* appears to be restricted to marine facies, however, and generally is regarded as the feces-filled burrow of a deposit-feeding worm (e.g., see Chamberlain, 1971; Alpert, 1975; Ekdale, 1977). The characteristic meniscate backfill structure of the Entrada trails is a feature not observed in *Planolites* (Alpert, 1975). The Entrada trace fossils also resemble the ichnogenus *Scoyenia*, which commonly exhibits backfilling and, in fact, typifies many nonmarine (although probably subaqueous) clastic deposits (Häntzschel, 1975; Bromley and Asgaard, 1979). However, *Scoyenia* characteristically possesses a wrinkled or rope-like external sculpture that is missing from the Entrada specimens. Howard (1966, p. 40–41) describes "chevron trails" with meniscate backfilling in Upper Cretaceous sandstone of north-central Utah, but they are more than twice the diameter of the Entrada specimens and apparently represent grazing trails at the water-sediment interface in shallow marine environments. The meniscate trails in the Entrada Sandstone also differ from *Muensteria* and *Taenidium*, which are both systems of cylindrical, back-filled tunnels (Häntzschel, 1975), because the Entrada trails are much smaller, do not branch, and are always oriented parallel to bedding planes.

The meniscate trails in the Entrada Sandstone apparently are identical to infaunal trails produced in modern sand dunes by the larvae of tipulid dipteran insects ("crane flies"), which burrow just beneath the sand surface and create backfilled burrows in convex epirelief (Ahlbrandt et al., 1978). Crane fly body fossils have been described in rocks as old as Triassic, although none have been reported in the Entrada Sandstone.

Small, vertical burrows.—A second common trace fossil in the Entrada Sandstone, herein named *Pustulichnus gregarious* (see SYSTEMATIC ICHNOLOGY), consists of numerous small, vertical burrows which typically appear in convex epirelief on cross-bed surfaces (Pl. 2D). These bump-like structures usually are 5 mm in diameter and are raised 2 or 3 mm in relief. Their distribution is patchy, but densities of about 170 bumps per 1000 cm^2 are not uncommon.

These small burrows are somewhat reminiscent of *Pelecypodichnus* (=*Lockeia*), which is a small, almond-shaped, dwelling/resting burrow of a bivalved organism (Häntzschel, 1975; Bromley and Asgaard, 1979), but *Pelecypodichnus* is not circular in cross-section like the Entrada traces. *Mamillichnis*, a similar bump-like trace fossil (Chamberlain, 1971), possesses a granulated, ring-like "flange" surrounding the raised central portion of the burrow; this flange is not present in the Entrada specimens. While the short, vertical burrows in the Entrada Sandstone are always seen in convex epirelief in eolian sandstone, *Pelecypodichnus* and *Mamillichnis* generally occur in convex hyporelief in sandstone beds believed to have been deposited under water.

These short, vertical burrows commonly occur in the same layers as the meniscate trails and exhibit the same dimensions, so it is possible that they simple represent vertical extensions of the trails, perhaps produced as the creators of the trails entered or exited from the sediment. No direct connection has been observed, however, between the vertical burrows and the meniscate trails.

The small, vertical burrows may not be genetically related to the meniscate trails. Sphecid hymenopteran insects ("sand wasps"), for example, build burrows displaying a variety of sizes and shapes, including shallow "trial burrows" and "sleeping burrows" (Ahlbrandt *et al.*, 1978, p. 842), in modern sand dunes. Eroded casts of such burrows possibly would result in structures resembling the bump-like structures in the Entrada Sandstone. Although sand wasp body fossils are known only from the Eocene to the present, other members of the Order Hymenoptera have existed since the Triassic. No hymenopteran body fossils have been reported in the Entrada Sandstone, however.

Plug-shaped burrows.—Another trace fossil, herein named *Digitichnus laminatus* (see SYSTEMATIC ICHNOLOGY), is larger but much less common in the Entrada Sandstone than the two aforementioned types. This trace fossil is a plug-shaped, vertical burrow containing laminated (not meniscate) fill (Pl. 1C). The diameter varies from 18 to 22 mm and tapers downward slightly to a rounded distal end. Most specimens are 3 to 5 cm long and appear to be truly vertical (i.e., inclined about 30° from the cross-bed surface).

The plug-shaped burrows superficially resemble the common Paleozoic trace fossils *Bergaueria* (Alpert, 1973; Häntzschel, 1975), a cylindrical vertical burrow with a rounded bottom; and *Conostichus* (Chamberlain, 1971), a sharply tapering conical burrow. *Bergaueria* typically exhibits a very shallow concave-up form for the internal laminae, and *Conostichus* (including also the "plug-shaped burrows" of Howard, 1966, p. 48–49) possesses a sort of cone-in-cone internal structure. Both trace fossils are virtually restricted to sandy marine facies and usually are interpreted as burrows of soft-bodied anthozoan coelenterates (burrowing sea anemones). In contrast, the plug-shaped burrows in the Entrada Sandstone possess flat internal laminae and presumably had a much different origin from that of anemone burrows.

The plug-shaped burrows also apparently resemble laminated vertical burrows which occur in Tertiary lacustrine deposits and have been referred to "*Taenidium*" by Toots (1967). Unlike the straight, tapering, plug-shaped burrows in the Entrada Sandstone, however, the "*Taenidium*" specimens often curve irregularly and do not taper. Moreover, the internal laminae, which are flat in the Entrada burrows, are "saucer-shaped" in the "*Taenidium*" of Toots (1967).

Trace fossil distribution.—The meniscate trails (*Entradichnus*) and short, vertical burrows (*Pustulichnus*) are abundant at Dry Valley and sparse at Muleshoe Canyon; the plug-shaped burrows (*Digitichnus*) occur only at Dry Valley, where they are sparse. In addition, moderate to thorough bioturbation of localized sand lenses occurs at both Dry Valley and Muleshoe Canyon and at two other nearby localities. Bioturbate textures, consisting of a clotted sedimentary fabric with the original stratification either partially disrupted or totally obscured by abundant indistinct burrows (Pl. 2E and 2F), occur in trough-shaped zones within the larger, trough cross-stratified units. The intensely bioturbated beds possibly represent moist or subaqueous interdune deposits in which certain arthropods laid eggs which subsequently hatched into teeming populations of burrowing worm-like larvae.

SYSTEMATIC ICHNOLOGY

Ichnogenus DIGITICHNUS new ichnogenus

Diagnosis.— Unlined, unbranched, vertical burrow with planar laminated internal fill; diameter uniform throughout most of length, but tapers downward slightly to a rounded distal end; differs from *Bergaueria* Prantl in lacking curved, concave- up shape of laminae; differs from *Conostichus* Lesquereux in lacking sharply conical form and cone-in-cone shape of laminae; differs from *Taenidium* Heer in being strictly straight, vertical and unbranched.

Derivation.—*Digit-*, from finger-like shape and size; *-ichnus*, from Greek *iknos*, meaning trace.

Type Species.—*Digitichnus laminatus* new ichnospecies.

DIGITICHNUS LAMINATUS new ichnospecies

Pl. 1C

Diagnosis.—Type species of *Digitichnus*; elongate, rounded, finger-like burrow, circular in cross-section, 18 to 22 mm in diameter, 30 to 50 mm in length; typically occurs in full relief in vertical orientation (i.e., oriented at acute angle to cross-stratification plane) within cross-stratified eolian sand.

Derivation.—*laminatus*, referring to planar internal laminae.

Type specimen.—Holotype, UUIP Specimen J-225; Department of Geology and Geophysics, Paleontological Collections, University of Utah.

Type locality.—Outcrop of Entrada Sandstone (Jurassic) at Dry Valley, Utah (see Appendix, Locality 1).

Remarks.—Probably represents passively filled dwelling and/or resting burrows of unidentified eolian dune-dwelling animals.

Ichnogenus ENTRADICHNUS new ichnogenus

Diagnosis.—Unbranched, unornamented, unlined, backfilled burrow, gently but irregularly sinuous; oriented parallel to bedding; differs from *Planolites* Nicholson in possessing internal meniscate backfill structure which is not of fecal origin; differs from *Scoyenia* White in lacking an external sculpture; differs from *Muensteria* von Sternberg and *Taenidium* Heer in being smaller, unbranched and oriented along a single plane.

Derivation.—*Entrad-*, from Entrada Sandstone, in which the type specimens occur; *-ichnus*, from Greek *iknos*, meaning trace.

Type species.—*Entradichnus meniscus* new ichnospecies.

ENTRADICHNUS MENISCUS new ichnospecies

Pl. 2A, B, D

Diagnosis.—Type species of *Entradichnus*; infaunal trail, circular in cross-section, uniformly 5 mm in diameter throughout entire length; typically occurs in great profusion in convex epirelief on eolian cross-stratification planes, oriented parallel to depositional dip with internal meniscae concave down-dip.

Derivation.—*meniscus*, from characteristic meniscate internal structure.

Type specimen.—Holotype, UUIP Specimen J-223; Paratypes, UUIP Specimens J-222, J-224, J-226; Department of Geology and Geophysics, Paleontological Collections, University of Utah.

Type locality.—Outcrop of Entrada Sandstone (Jurassic) at Dry Valley, Utah (see Appendix, Locality 1).

Remarks.—Apparently represents back-filled locomotion burrows created by eolian dune-dwelling crane fly larvae.

Ichnogenus PUSTULICHNUS new ichnogenus

Diagnosis.—Short, small-diameter, unlined, vertical burrow lacking external sculpture and internal

---

EXPLANATION OF PLATE 1

FIG. A.— Trace fossiliferous outcrop of Entrada Sandstone at Dry Valley, east-central Utah. The trace fossils occur in the upper part of the section in beds exhibiting well-developed parting lineation. Snow-capped LaSal Mountains in background.

B. — Tabular-planar cross-bed set from eolian sandstone at Dry Valley outcrop. (Held in palm of hand.)

C. — Plug-shaped burrow (*Digitichnus laminatus*) containing planar, horizontal laminae.

D. — Thin section of cross-stratified sandstone which contains trails. Note the laminated, bimodal texture.

A
5cm
B
C
5cm
D
0.5mm

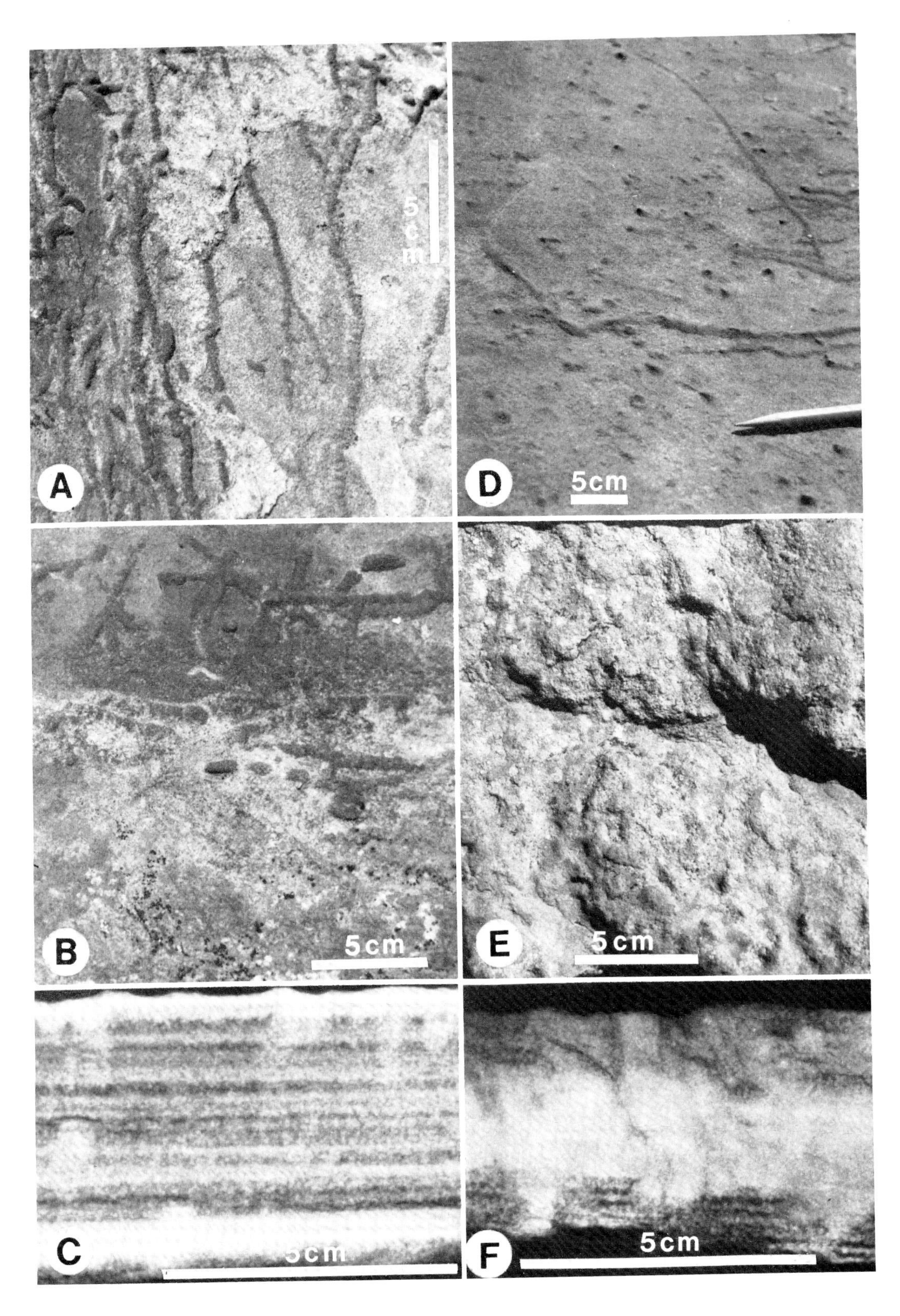
A
5 cm
D
5cm
B
5cm
E
5cm
C
5cm
F
5cm

EXPLANATION OF PLATE 2

FIG. A.— Trails (*Entradichnus meniscus*) exposed in convex epirelief on a cross-bed surface. Note alignment of the trails, which parallel the depositional dip of the sand cross-beds.
B. — Trails (*E. meniscus*) exposed in convex epirelief on a cross-bed surface. Note the meniscate internal structure of the trails.
C. — X-ray radiograph (negative) of cross-stratified sandstone which contains trails and short vertical burrows.
D. — Short vertical burrows (*Pustulichnus gregarious*) and trails (*E. meniscus*) exposed in convex epirelief on a cross-bed surface.
E. — Burrowed sandstone exhibiting a clotted, bioturbate texture but no distinct trace fossils.
F. — X-ray radiograph (negative) of burrowed sandstone with clotted, bioturbated texture.

structure; diameter usually exceeds length; oriented normal to bedding; differs from *Pelecypodichnus* Seilacher (=*Lockeia* James) in possessing a rounded, circular, mound-like shape exposed on tops of beds; differs from *Mammillichnis* Chamberlain in lacking a flange around the raised center of the trace.

Derivation.—Pustul-, from pustular shape; - *ichnus*, from Greek *iknos*, meaning trace.

Type species.—Pustulichnus gregarious new ichnospecies.

PUSTULICHNUS GREGARIOUS new ichnospecies
Pl. 2D

Diagnosis.—Type species of *Pustulichnus*; small, rounded, bumplike trace, circular in cross-section, 5 mm in diameter; typically occurs in great profusion in convex epirelief on eolian cross-stratification planes.

Derivation.—gregarious, referring to typically high density of occurrence in localized patches.

Type specimen.—Holotype, UUIP Specimen J-222; Paratypes, UUIP Specimen J-224; Department of Geology and Geophysics, Paleontological Collections, University of Utah.

Type locality.—Outcrop of Entrada Sandstone (Jurassic) at Dry Valley, Utah (see Appendix, Locality 1).

Remarks.—Possibly represent eroded casts of shallow resting burrows produced by eolian dune-dwelling sand wasps.

SUMMARY AND CONCLUSIONS

Trace fossils created by invertebrate animals have seldom been reported in ancient sand dune deposits. Eolian beds of the Entrada Sandstone in eastern Utah contain three types of trace fossils (infaunal trails with meniscate backfill structure; short, vertical burrows; and plug-shaped, vertical burrows) in cross-stratified units and contain bioturbate textures in other zones. Animal trackways and body fossils are absent from these rock units, as are plant remains and root casts. The burrows are thought to have been produced in moist sand of active dunes by invertebrate organisms, most likely insects. The bioturbated zones may have been produced in very localized, subaqueous, interdune environments.

ACKNOWLEDGMENTS

This work was supported by a University of Utah Mineral Leasing Funds grant to both authors. An early version of the manuscript was read by D. W. Boyd, H. A. Curran, and E. D. McKee, who offered valuable suggestions for its improvement.

INDEX TO LOCALITIES

Locality 1.—Dry Valley. Outcrop on east side of U.S. Highway 163 about 52 km south of Moab, Utah. Sec. II, T30S, R23E, San Juan County, Utah.

Locality 2.—Muleshoe Canyon. Outcrop on east side of U.S. Highway 163 about 28 km south of Moab, Utah. Sec. 18, T28S, R23E, San Juan County, Utah.

REFERENCES

AHLBRANDT, T.S., ANDREWS, S., AND GWYNNE, D.T., 1978, Bioturbation of eolian deposits: Jour. Sed. Petrology, v. 48, p. 839–848.

ALPERT, S.P., *Bergaueria* Prantl (Cambrian and Ordovician), a probable actinian trace fossil: Jour. Paleontology, v. 47, p. 919–924.

______, 1975, *Planolites* and *Skolithos* from the Upper Precambrian-Lower Cambrian, White-Inyo Mountains, California: Jour. Paleontology, v. 49, p. 508–521.

ALBERS, S.H., 1975, Paleoenvironment of the Upper Triassic-Lower Jurassic(?) Nugget(?) Sandstone near Heber, Utah; [unpub. M.S. thesis]: Salt Lake City, Univ. of Utah, 94 p.

BROMLEY, R.G., AND ASGAARD, U., 1979, Triassic freshwater ichnocoenoses from Carlsberg Fjord, East Greenland: Palaeogeogr., Palaeoclimatol., Palaeoecol, v. 28, p. 39–80.

CHAMBERLAIN, C.K., 1971, Morphology and ethology of trace fossils from the Ouachita Mountains, southeast Oklahoma: Jour. Paleontology, v. 45, p. 212–246.

EKDALE, A.A., 1977, Abyssal trace fossils in worldwide Deep Sea Drilling Project cores, *in* Crimes, T.P., and Harper, J.C. (eds.), Trace Fossils 2: Geol. Jour. Spec. Issue 9, Liverpool, Seel House Press, p. 163–182.

HÄNTZSCHEL, W., 1975, Trace fossils and problematica, *in* Teichert, C. (ed.), Treatise on Invertebrate Paleontology, Pt. W. (Supp. 1): Lawrence, Kansas, Geological Soc. America and Univ. Kansas Press, 269 p.

HANLEY, J.H., STEIDTMANN, J.R., AND TOOTS, H., 1971, Trace fossils from the Casper Sandstone (Permian), southern Laramie Basin, Wyoming and Colorado: Jour. Sed. Petrology, v. 41, p. 1065–1068.

HENDERSON, J., 1924, Footprints in Pennsylvanian sandstones of Colorado: Jour. Geology, v. 32, p. 226–229.

HOWARD, J.D., 1966, Characteristic trace fossils in Upper Cretaceous sandstones of the Book Cliffs and Wasatch Plateau: Utah Geological Mineralogical Survey Bull., No. 80, p. 35–53.

MCKEE, E.D., 1934, An investigation of the light-colored, cross-bedded sandstones of Canyon De Chelly, Arizona: Am. Jour. Sci., v. 28, p. 219–233.

______, 1944, Trails that go uphill: Plateau, v. 16, p. 61–72.

______, 1947, Experiments on the development of tracks in fine cross-bedded sand: Jour. Sed. Petrology, v. 17, p. 23–28.

______, 1979, A study of global sand seas: U.S. Geological Survey Prof. Paper 1052, 429 p.

O'SULLIVAN, R.B., 1980, Stratigraphic sections of Middle Jurassic San Rafael Group from Wilson Arch to Bluff in southeastern Utah: U.S. Geological Survey, Oil and Gas Invest. Chart 102.

OTTO, E.P. AND PICARD, M.D., 1975, Stratigraphy and oil and gas potential of Entrada Sandstone (Jurassic), northeastern Utah, *in* Bolyard, D.W., (ed.), Deep Drilling Frontiers of the Central Rocky Mountains: Denver, Rocky Mtn. Assoc. Geologists, p. 129–139.

______, AND ______, 1976, Petrology of Entrada Sandtone (Jurassic), northeastern Utah, *in* Hill, J.G. (ed.), Geology of the Cordilleran Hingeline: Denver, Rocky Mtn. Assoc. Geologists, p. 231–245.

PICARD, M.D., 1971, Classification of fine-grained sedimentary rocks: Jour. Sed. Petrology, v. 41, p. 179–195.

______, 1977a, Stratigraphic analysis of the Navajo Sandstone: A discussion: Jour. Sed. Petrology, v. 47, p. 475–483.

______, 1977b, Petrology of the Jurassic Nugget Sandstone, northeast Utah and southwest Wyoming, *in* Rocky Mountain thrust belt: geology and resources: Wyoming Geol. Assoc., p. 239–258.

PEIRCE, H.W., 1963, Stratigraphy of the DeChelly Sandstone of Arizona and Utah: [unpub. Ph.D. thesis]: Tucson, Univ. of Arizona, 206p.

POOLE, F.G., 1963, Palaeowinds in the western United States, *in* Nairn, A.E. M., (ed.), Problems in Paleoclimatology: London, Interscience Publishers, p. 394–405.

STOKES, W.L., 1978, Animal tracks in the Navajo-Nugget Sandstone: Contribs. to Geology, Univ. Wyoming, v. 16, p. 103–107.

TANNER, W.F., 1965, Upper Jurassic paleogeography of the Four Corners region: Jour. Sed. Petrology, v. 35, p. 564–574.

TOOTS, H., 1967, Invertebrate burrows in the non-marine Miocene of Wyoming: Contribs. to Geology, Univ. Wyoming, v. 6, p. 93–96.

WRIGHT, J.C., SHAWE, D.R., AND LOHMAN, S.W., 1962, Definition of members of Jurassic Entrada Sandstone in east- central Utah and west-central Colorado: Am. Assoc. Petroleum Geologist Bull., v. 46, p. 2057–2070.

INTERPRETATION OF BIVALVE TRACE FOSSILS IN FLUVIAL BEDS OF THE BASAL CATSKILL FORMATION (LATE DEVONIAN), EASTERN U.S.A.

RICHARD E. THOMS AND THOMAS M.BERG
Geology Department, Portland State University, Portland, Oregon 97207, and Pennsylvania Geological Survey, Department of Environmental Resources, Harrisburg, Pennsylvania 17120

ABSTRACT

Structures attributed to the upward escape from anastrophic burial of specimens of *Archanodon catskillensis* (Vanuxem) (Archanodontidae) in the basal sandstone member of the Late Devonian Catskill Formation in northeastern Pennsylvania and equivalent beds in southern New York and northern New Jersey exhibit the following vectorial features: (1) preferential curvature (in vertical section), (2) ellipse parallelism (in bedding plane cross-sections), and (3) internal crescent asymmetry (also seen in vertical section). The utility of these vectorial features in the reconstruction of past sedimentary environments depends upon the discovery and understanding of a suitable Holocene analogue. Populations of *Margaritifera margaritifera* (Linné) have been observed from the lower Siletz River, Oregon in both the field and the laboratory. These observations indicate preferential orientation by *M. margaritifera* in response to unidirectional current flow and ability to burrow quickly upward following anastrophic burial. Morphologic features of *M. margaritifera* point to its capability of producing primary biogenic sedimentary structures similar to those in the Catskill Formation. Comparison with similar, though smaller, burrows in the British Upper Carboniferous reinforces the interpretation that the Catskill burrows are the product of the upward escape activities of *Archanodon catskillensis*, living in an environment of unidirectional flow regime with rapidly accreting sediments.

INTRODUCTION

Numerous biogenic sedimentary structures, interpreted herein as fossilized bivalve burrows, occur in the basal sandstone member of the Devonian Catskill Formation in northeastern Pennsylvania and portions of adjacent states. The initial investigations on this topic were reported by Berg (1973), who described the occurrences and characteristics of the burrows, pointing out their probable utility in vectorial analysis of the enclosing rock. Thoms and Berg (1974), proposed a mechanism for the origin of the burrows in an analogy with the Holocene freshwater pearl-mussel, *Margaritifera margaritifera* (Linné). Later, it was discovered that Eagar (1948) had proposed *M. margaritifera* as an analogue for the Anthracosiidae, relative to his studies on bivalve faunas of the Coal Measures and their potential for burrowing. His proposal was further developed experimentally by Trueman (1968), whose studies of infaunal burial, myography, locomotion, and pedal activity of this species are part of his considerable researches on the burrowing activities of Holocene bivalves. Eagar (1978) summarized evidence for the burrowing activities of certain Carboniferous Anthracosiidae as compared with those of *M. margaritifera*.

It is obvious that the discovery of a suitable analogue for the large Devonian escape burrows will be of immense value in assisting in the interpretation of several particulars in the sedimentary history of the Catskill Formation and possibly in similar units where structures of this type occur. Chief among these are the distributary patterns and rates of sedimentation in the basal sandstone. The following discussion is an attempt to establish the suitability of *M. margaritifera* as a Holocene analogue for the bivalve responsible for the Catskill burrows.

OCCURRENCE OF THE BURROWS

The distribution and stratigraphic position of the basal sandstone of the Catskill Formation in eastern Pennsylvania and portions of adjacent states are shown in Figure 1. The basal sandstone was named Towamensing Member in the Lehigh River area (Epstein *et al.* 1974), and has been mapped from this area to the Delaware River north of Port Jervis, New York. According to Berg (1973), the Towamensing is fine to medium grained and displays parallel laminated planar bedding, small to medium scale trough cross-bedding, few ripple marks, and some apparent lag gravels. It thickens toward the northeast, exhibiting non-red fining upward cycles, abundant plant remains, and slightly calcareous shale chip intraformational conglomerates. This sandstone is interpreted at present to be a delta front or possible lower delta plain deposit. Specifically, river mouth bars appear to be the best interpretation. Numerous sedimentary structures, interpreted herein as fossil burrows, occur in the Towamensing, being particularly abundant at the eighteen localities indicated in Figure 1. Molds of the valves (Fig. 1 and Pl. 1) of the bivalve *Archanodon catskillensis* (Vanuxem) are abundant at Harrity and New Milford, Pennsylvania, and at Oxford, New York (Clarke, 1901), and are found in proximity to, but rarely in, the burrows. One bivalve was found in a burrow at Ressaca, Pennsylvania. *Archanodon* is apparently the oldest known freshwater mollusk, occurring not only in the Catskill Formation but also in the Old Red Sandstone in southern Ireland and southern

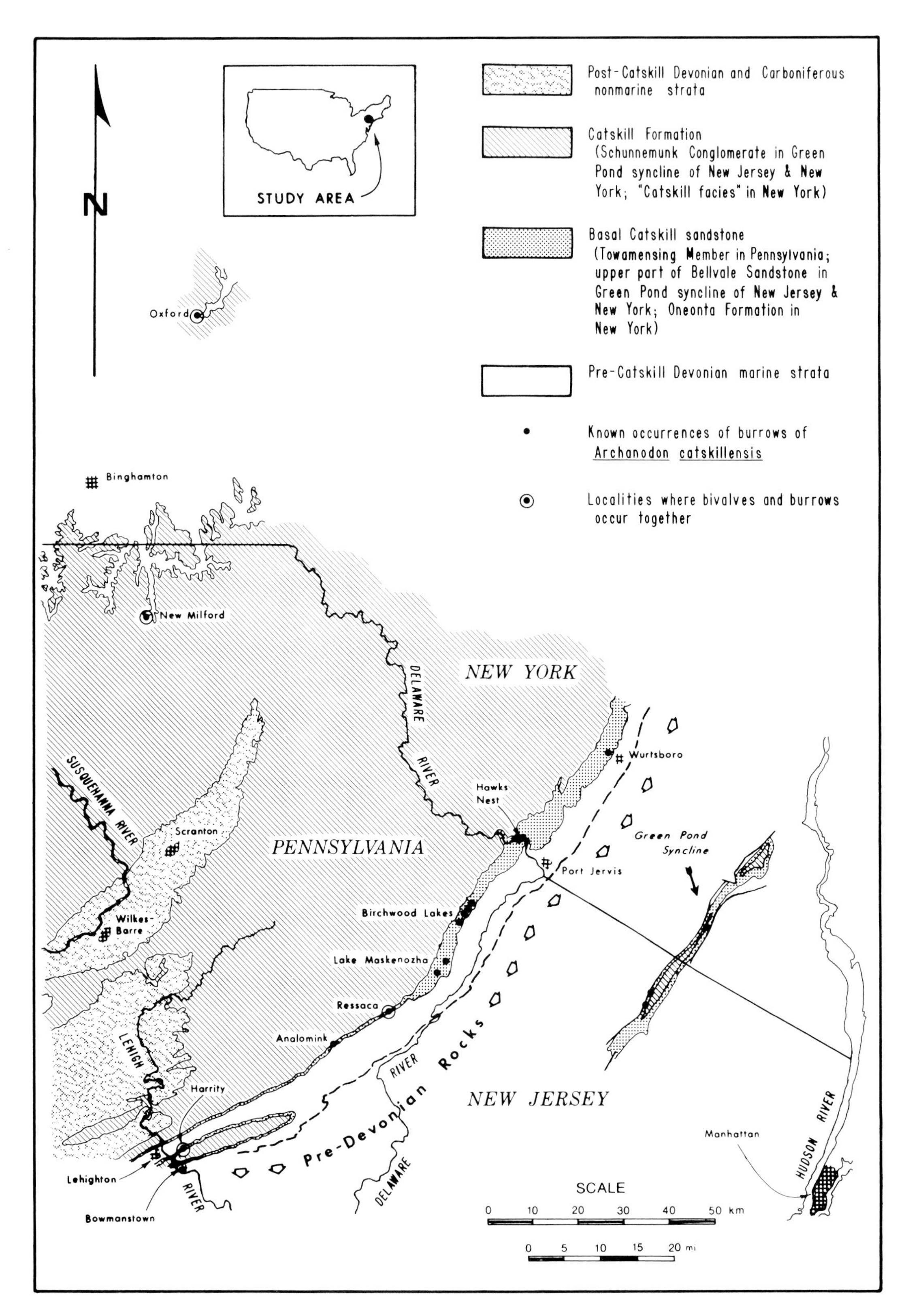

STUDY AREA
N
Post-Catskill Devonian and Carboniferous nonmarine strata
Catskill Formation (Schunnemunk Conglomerate in Green Pond syncline of New Jersey & New York; "Catskill facies" in New York)
Basal Catskill sandstone (Towamensing Member in Pennsylvania; upper part of Bellvale Sandstone in Green Pond syncline of New Jersey & New York; Oneonta Formation in New York)
Pre-Catskill Devonian marine strata
Known occurrences of burrows of Archanodon catskillensis
Localities where bivalves and burrows occur together
Oxford
Binghamton
New Milford
NEW YORK
DELAWARE RIVER
Wurtsboro
Hawks Nest
Green Pond Syncline
Port Jervis
SUSQUEHANNA RIVER
Scranton
PENNSYLVANIA
Wilkes-Barre
Birchwood Lakes
Lake Maskenozha
Ressaca
Analomink
LEHIGH RIVER
Harrity
Lehighton
Bowmanstown
Pre-Devonian Rocks
NEW JERSEY
DELAWARE RIVER
Manhattan
HUDSON RIVER
SCALE
0 10 20 30 40 50 km
0 5 10 15 20 mi

Fig. 1.—Stratigraphic and geographic occurrence of the burrows, showing principal localities.

Wales (*Archanodon jukesi*) and in the Devonian of the Rhineland (*Archanodon rhenana*). The burrows have been found in homotaxially correlative upper fluvial beds of the Bellvale Sandstone (Middle Devonian) of New Jersey and New York (Berg, 1977), and in the Oneonta Formation in New York.

CHARACTERISTICS OF THE BURROWS

The burrows exhibit the following features (Fig. 2): (1) preferential basal curvature, seen in axial sections; (2) elliptical outlines with parallelism of ellipses, seen in transverse sections; and (3) internal asymmetric crescentic structures, quite regular in shape and size, also seen in axial sections. The burrows vary from 3 to 10 cm in diameter and are generally no longer than 1.3 m, although at least one burrow at Hawks Nest, New York has a length of nearly 2 m. The majority average about 0.5 m in length. The sediment inside the burrow structures is the same as that of the surrounding rock. The margins of many of the burrows are marked by a zone of weakness, which results in ready fracture around them. This zone is marked by reorientation of mica flakes from bedding-parallel to burrow-parallel, evidently caused by the movement of material from its original position during the production of the burrows (R.M.C. Eagar, personal communication, 1981).

INTERPRETATION OF THE BURROWS

The inferred relationships of the features of the structures, both one to the other and to the bedding plane, are shown in Figure 2. The structures can most likely be attributed to the activities of *Archanodon catskillensis* because this species is the only invertebrate fossil in the basal sandstone and because it possesses the appropriate morphology and size (Plate 1). Berg (1973) first hypothesized that the basal curvature might be explained by comparison with the burrowing activities of *Siliqua patula*. When *S. patula* burrows into the sand, it tends both to curve its burrow and orient its hinge line toward the sea. If this analogy were correct, then the basal curvature of the Catskill burrows might represent seaward curvature. However, this hypothesis has been dismissed for the following reasons: (1) Some of the burrows are quite long, in one case about 2 m (the downward escape activities of *S. patula* do not involve such deep burrowing). (2) Examination of x-ray radiographs of axial sections of *S. patula* burrows reveals none of the structures seen in the Catskill burrows (H.E. Clifton, personal communication, 1973). (3) Only one specimen of *A. catskillensis* (or any other body fossil) has ever been found in a burrow; if they were the product of continuous occupation or downward escape more specimens could be expected. (4) The features of the burrows are well preserved, as are the bedding features, indicating no reworking, either by animal activity or by wave action. Hence, the Catskill burrows probably represent continued bivalve occupation with upward movement only, rather than both downward and upward movement. If some burrows do represent downward movement, then others could logically be expected to exhibit features different from those characteristic of upward movement. But only one type of burrow, always with the same internal features, has been found.

Another analogue must therefore be proposed. The burrows of the common Holocene freshwater mussel *Margaritifera margaritifera* (Linné) are interpreted as the best analogue for upward escape from anastrophic burial of individuals of *A. catskillensis* that live in a unidirectional flow system and are subjected to rapid rates of sedimentation. The basal curvature of the *Archanodon* burrows is apparently related to rapid vertical and lateral accretion of sets of trough cross strata. (Fig. 3), during which the bivalve attempted to regain its normal geotropic orientation following emplacement on its side by a rapid current. In the case of *M. margaritifera*, an erect position, with the commissural plane perpendicular to the substratum, precedes burrowing. (R.M.C. Eagar, personal communication, 1981). The shape and internal features of the *Archanodon* burrows also can be related to the orientation and morphology of a bivalve similar to *M. margaritifera* (Fig. 4). The internal asymmetric crescents can be accounted for, both in shape as well as position in the burrow, by attributing them to extension and dilation of the foot in the course of upward movement (Fig. 4). That the burrows were formed contemporaneously with the deposition of the basal sandstone is indicated by the identical composition of the rock both inside and outside the burrows. Incremental, bed-by-bed deposition of the sandstone is indicated by parting laminations which define individual beds. Downwarped laminae within burrows match closely with laminae outside the burrows.

MORPHOLOGY AND BEHAVIOR OF *MARGARITIFERA MARGARITIFERA*

Living populations of *Margaritifera margaritifera* in the Siletz River, a northern Oregon coastal stream, were examined to determine variation in size, orientation, and burrowing behavior. A definite preferential orientation was observed, wherein individuals were buried to about one-half of their length, with the hinge line and excurrent opening oriented downstream and the incurrent opening oriented upstream. A slight inclination of the long axis of the bivalves in the upstream direction was also noted. This pattern of common orientation was disrupted in more turbulent

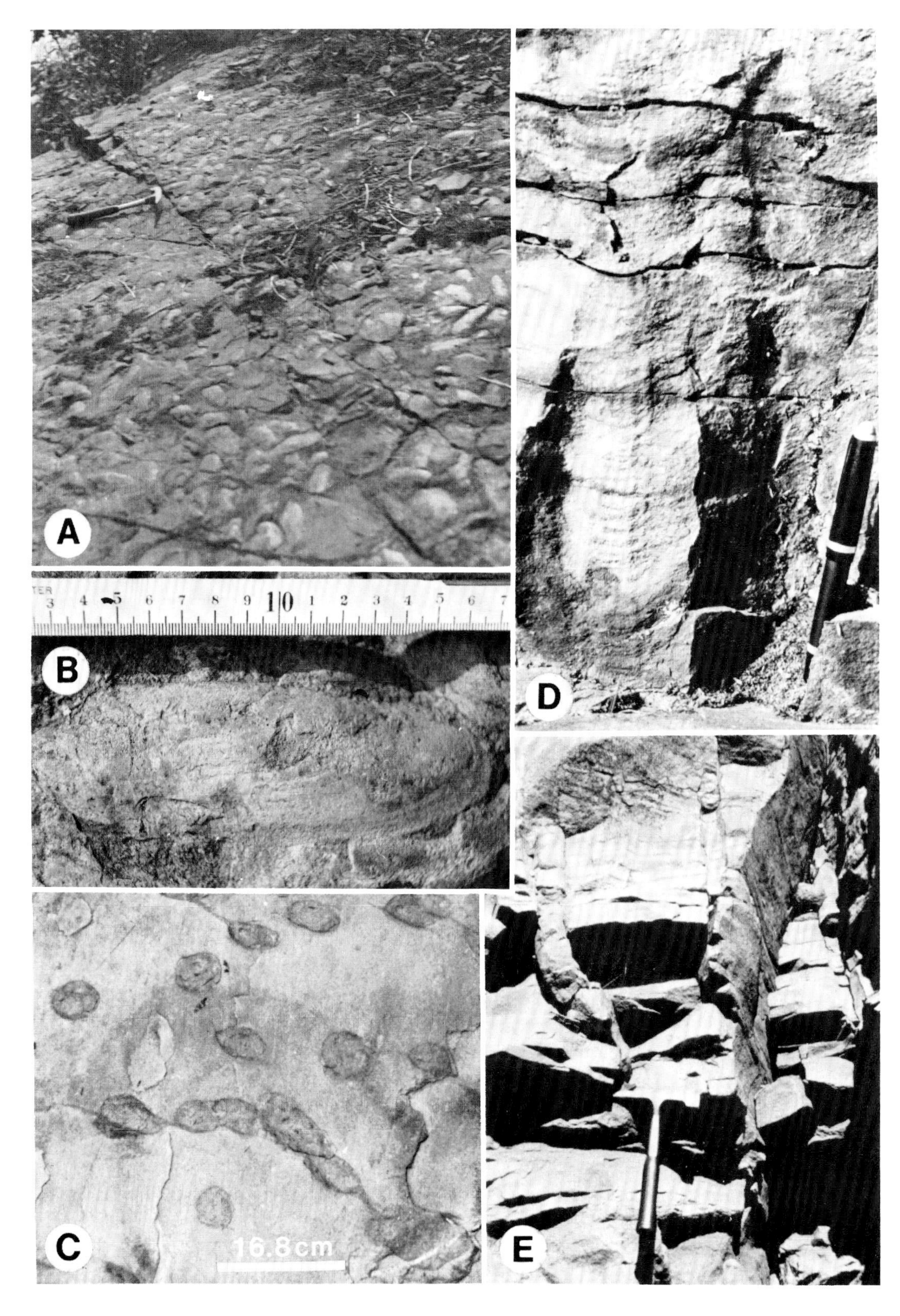
A
B
C
D
E
16.8cm

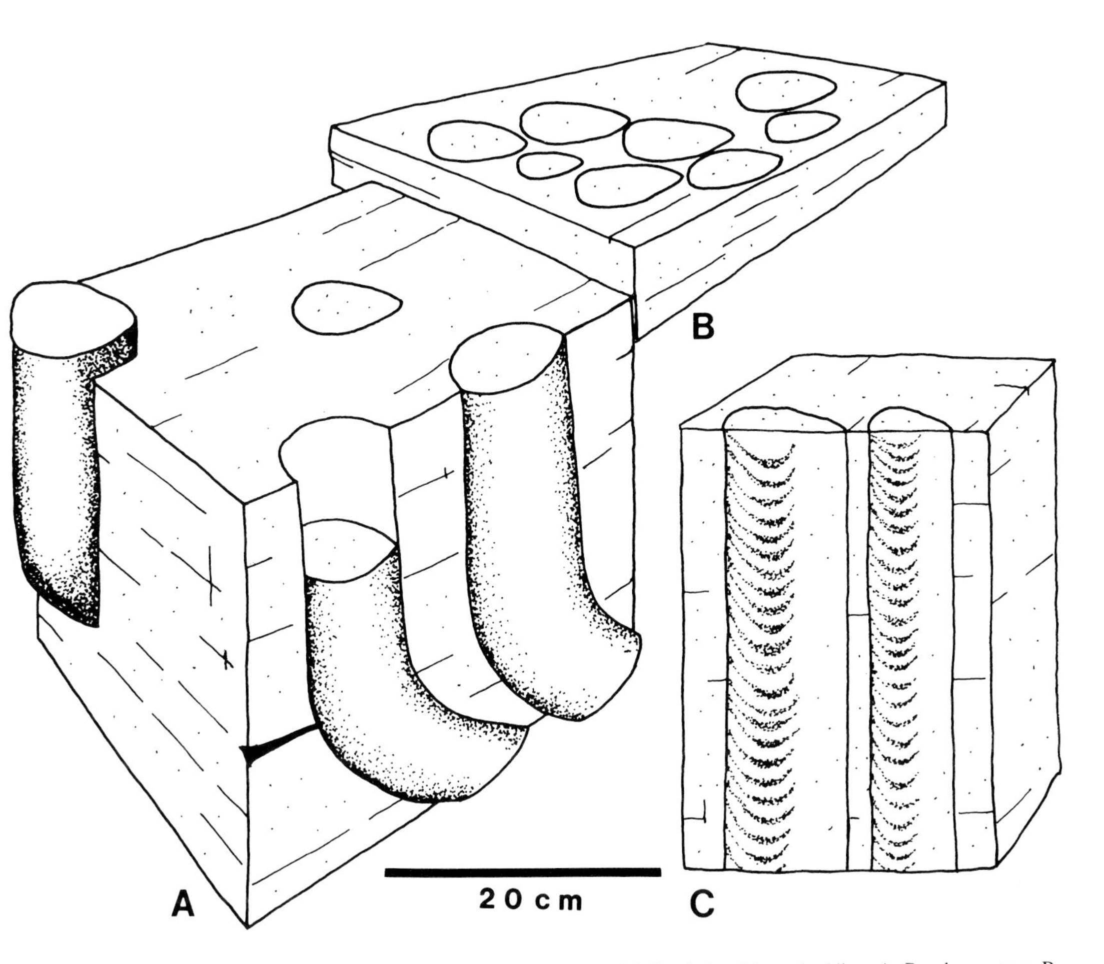

Fig. 2.—Block diagrams showing vectorial features of the burrows and their relationships to bedding. A. Basal curvature. B. Parallel orientation of ellipses. C. Internal crescentic structures.

water, such as in eddies behind boulders. Similar observations were recorded by Zhadin (1939). Swifter currents produced by seasonal flooding dislodge the mussels and move them as part of the normal bedload to new positions downstream. Following the work of Kranz (1974), an attempt was made to force several specimens into upward escape from anastrophic burial by covering them with about 3 cm of sand. They were found to burrow upward to the surface in a barely emergent position in about one hour. Unfortunately, very little published information exists on the burrowing activities of this species, in spite of its former commercial importance. Trueman (1966, p. 73) has recorded a speed of 20 cm/hr. for this species in over-the-sand locomotion. He also did myographic studies (1968) on its infaunal burial and noted a major pulsation of the foot at about 3 second intervals. Although essentially an epifaunal species, *M. margaritifera* will burrow downward to whatever depth required to prevent dislodgement by the current. It survives well under a variety of water velocities, both in seasonal variations in a single stream, as well as from stream to stream.

Explanation of Plate 1

Fig. A.— External molds of *Archanodon catskillensis* on bedding plane. Near Harrity, Pennsylvania.

B. — External mold of a left valve of *A. catskillensis* with scale in centimeters.

C. — Bedding-plane cross-sections of burrows near Harrity, Pennsylvania, showing parallel orientation of ellipses.

D. — Vertical section of burrow near Harrity, Pennsylvania, showing internal crescentic structures. Fountain pen is 14 cm long.

E. — Vertical section of burrow at Hawks Nest, New York, showing basal curvature. Note other burrows in outcrop.

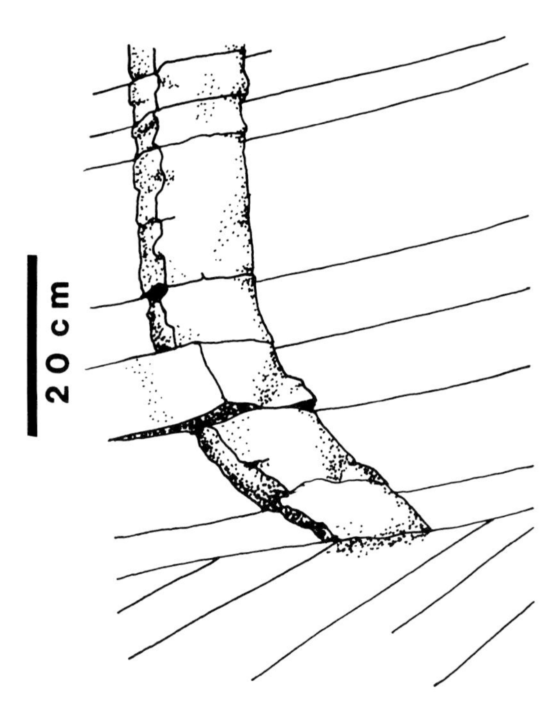

Fig. 3.—Relationship of basal curvature of the burrow to bedding.

Trueman (1968, p. 402) noted that specimens which display a distinct inflection of the ventral margin appear to be characteristic of populations from swiftly moving streams. This feature is not seen in individuals from sluggish streams or lakes, thereby confirming observations made by Eagar (1947, p. 145) of evolutionary parallelism between *M. margaritifera* and the Anthracosiidae. The inflected ventral margin, as well as lack of a pallial sinus in this non-siphonate bivalve, are two key features also possessed by *A. catskillensis* (Pl. 1).

If this analogy is suitable, then in addition to the vectorial analysis concerning current direction, the following ideas can be entertained: (1) Each burrow may represent, at the most, the activity of an individual during one year. This conclusion is based on the observation of the movement, caused by seasonal flooding, of extant populations of *M. margaritifera* in swift water streams. In actuality, each burrow may well represent even less time if bed load scouring occurred more than annually. (2) Whatever the rate of accumulation of the basal Catskill sandstone, it occurred at intervals and in amounts which allowed for the recovery from anastrophic burial of many individuals. The single known case of bivalve remains found in a burrow attests to the success of this species in escaping from burial. (3) Experimental observations of upward-burrowing activity of Holocene unionoids may allow the attachment of some average or approximate time value to the incremental patterns of asymmetric crescents seen in burrow cross sections.

COMPARISON WITH SIMILAR STRUCTURES

A number of sedimentary structures, both non-biogenic and biogenic, bear some degree of similarity to the Catskill burrows. Of the several kinds of non-biogenic cylindrical structures recorded in the literature, three are worth noting here.

Gabelman (1955) described such structures in a possible Permian calcareous siltstone in the Bush Creek area of Colorado. They are circular in bedding-plane cross section, average about 4 cm in diameter (some are as large as 12 cm), are about 0.5 to 1.0 m in length, and ascend through the siltstone nearly perpendicular to the bedding. Although nearly all of the structures are simple and uniform in direction, some branch upwards, some downwards, some curve, and some are constricted in places. The walls of the cylinders are sharply defined, exhibiting concentric stylolitic cracks. The material in the cylinder is the same as that enclosing it, except in a few cases where a fine grained breccia of the siltsone occurs in the cylinder matrix. Gabelman did not record either basal curvature or internal crescents in these structures. He has concluded that they originated from the action of fresh water rising through salt water-saturated, partly consolidated silt.

Allen (1961) has recorded the occurrence of "sandstone-plugged pipes" in the Lower Old Red Sandstone in Shropshire. These structures measure as much as 1 m in length and average about 10 cm in diameter. They are composed of the same material as the surrounding sandstone, although sandstone and siltstone debris often fills some of them. They tend to rise normal to the bedding, but some follow the bedding plane before rising. Some cylinders interlock, while others exhibit lateral structures which fan outward into the surrounding rock. Allen attributed these structures to the later plugging of ducts eroded by the migration of ground water rising through unconsolidated sediments.

Dionne and Laverdière (1972) have described cylindrical structures in Quaternary sediments in Quebec. The cylinders contain essentially the same matrix as the enclosing rock, but exhibit concentric rings of medium sand, fine sand, and silt in bedding-plane cross section. When seen in transverse section, the rings appear as elongated, irregular, cone-in-cone structures. Although the length of these structures approximates that of the Catskill burrows (about 1.5 m), their diameters are significantly greater (about 30 cm). Dionne and Laverdiere interpret the cylinders as either the infilling of whirlpool excavations or the result of springs rising through unconsolidated sediment. In the cases just cited, although the structures are similar to the Catskill burrows in general geometry and size, as well as in containing matrices identical (or nearly so) to the enclosing rock, none exhibits the regularity of pattern of orientation or internal crescentic structures as seen in the Catskill burrows.

Among biogenic sedimentary structures, those produced by the escape burrowing of anemones and bivalves, particularly nonsiphonate freshwater forms, also are important to this study. Curran and Frey (1977) have reviewed the escape structures of living actinians, as well as fossil traces attributed to their activities. Aside from their generally smaller size (4–15 cm long and 2–3 cm wide), the acutely conical and symmetrical internal structures possessed by these traces readily distinguish them from the Catskill burrows.

Structures produced by bivalve escape burrowing have been mentioned by a number of authors, but prior to 1977 none had been sufficiently illustrated or described in the published literature to allow ready comparison with the Catskill burrows. Nonetheless, the manner of disruption of the beds at the margins of bivalve escape burrows and the production of concentric features in the central part of the structures seemed well established in the literature and are certainly comparable with the features seen in the Catskill burrows. In 1978, following some discussion of their occurrence and significance by Eagar (1973), small bivalve escape burrows of the British Upper Carboniferous were described in detail by Hardy and Broadhurst. The burrows occur in the Lower Westphalian A interval of the British Pennines, principally in flagstones and finer grained sedimentary rocks of fluvial origin. They range in length from about 5 to 10 cm, up to 1 m. As seen in bedding-plane cross section, they are elliptical in outline, with the long axis about 1 to 2 cm in length and the short axis about 0.75 to 1 cm. On occasion, they are found in great abundance, with recorded densities of about 150 burrows per m^2.

In such cases, a common orientation of the ellipses is notably pronounced. In transverse section, the burrows exhibit regularly spaced, downwardly convex laminae. Eagar, and Hardy and Broadhurst have all

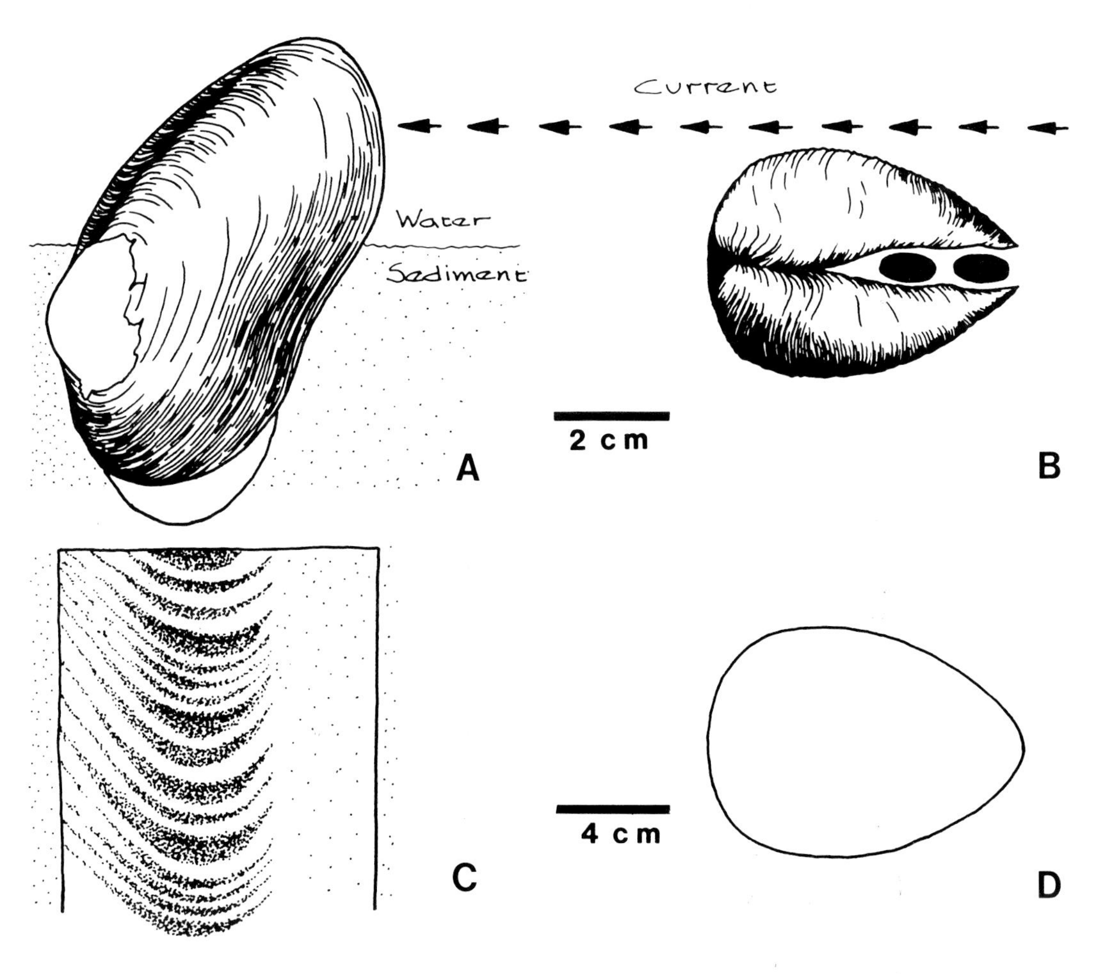

Fig. 4.—Morphology and orientation of *M. margaritifera* in relation to the fossil burrows. *A*, Lateral view of *M. margaritifera*, seen in normal living position. *B*, Posterior view of *M. margaritifera*. *C*, Vertical cross-section of fossil burrow. *D*, Bedding-plane cross-section of fossil burrow.

interpreted the burrows as the product of upward escape burrowing of individuals of the genus *Carbonicola* that lived in a unidirectional flow regime and were subjected to periodic anastrophic burial. Infrequently, specimens of *Carbonicola* are found in the burrows, in the position appropriate to upward burrowing. Except for their much smaller size and the somewhat different internal structure, the burrows are strikingly similar to those in the Catskill Formation, especially in bedding plane cross section and in common orientation of numerous specimens. It is interesting to note that in neither study (British Pennines and Catskill Formation) did the researchers learn of the conclusions of the others until the basic interpretation had been made and further inquiry had been directed to the literature.

CONCLUSIONS

Use of *Margaritifera margaritifera* as an analogue for *Archanodon catskillensis* provides a reasonable hypothesis for the origin of the Catskill burrows. In particular, it allows for further understanding of the following features, all suitable for vectorial analysis: (1) preferential basal curvature, apparently produced by initial parallel emplacement of bivalves by a rapid current; (2) ellipse parallelism interpreted, as in the case of *Carbonicola* burrows in the British Upper Carboniferous, as the effect of a common response of numerous bivalves living in a unidirectional flow regime; and (3) internal crescent asymmetry and the position of the crescents within the burrow, inferred to be related to the morphology and orientation of *A. catskillensis*. Present and future research on this topic entails search for similar burrows at various stratigraphic levels, as well as thorough testing of the present hypothesis in both field and aquarium studies.

ACKNOWLEDGMENTS

Thanks are due to the Oregon Fish Commission; to Arden Callender and Paul Smith for their assistance in the field and laboratory; to Joseph J. Kohut and J. Wyatt Durham for helpful comments and criticisms; to E. R. Trueman and R. M. C. Eagar for their comments and suggestions; and to Chris Thoms for the illustrations accompanying this paper. We also wish to acknowledge the help of W. D. Sevon of the Pennsylvania Geological Survey; he was the first to point out the existence of burrows at Harrity, Pennsylvania.

REFERENCES

ALLEN, J.R.L., 1961, Sandstone-plugged pipes in the Lower Old Red Sandstone of Shropshire, England: Jour. Sed. Petrology, 31, p. 325–335.

BERG, T.M., 1973, Pelecypod burrows in the Basal Sandstone Member of the Catskill Formation, Northeastern Pennsylvania: Geol. Soc. America Abstracts with Programs, v. 5, p. 137.

______, 1977, Bivalve burrow structures in the Bellvale Sandstone, New Jersey and New York: Bull. New Jersey Acad. Sci., 22, no. 2, p. 1–5.

CLARKE, J.M., 1901, *Amnigenia* as an indicator of fresh-water deposits during the Devonic of New York, Ireland, and the Rhineland: New York State Museum, Paleontologic Paper 2, Bulletin 49, p. 199–203.

CURRAN, H.A. AND FREY, R.W., 1977, Pleistocene trace fossils from North Carolina (U.S.A.) and their Holocene analogues, *in* T.P. Crimes and J.C. Harper, eds., Trace Fossils 2: Geol. Jour. Spec. Issue No. 79, Liverpool, Seel House Press, p. 139–162.

DIONNE, J.C. AND LAVERDIÈRE, C., 1972, Structure cylindrique verticale dans un depot meuble Quaternaire, au Nord de Montreal, Quebec: Canadian Jour. Earth Sci., v. 9, p. 528–543.

EAGAR, R.M.C., 1947, A study of a non-marine lamellibranch succession in the *Anthraconaia lenisulcata* zone of the Yorkshire Coal Measures: Philos. Trans. Roy. Soc. London, ser. B, v. 223, p. 1–54.

______, 1948, Variation in shape of shell with respect to ecological station. A review dealing with recent Unionidae and certain species of the Anthracosiidae in Upper Carboniferous times: Proc. Roy. Soc. Edin., ser. B, v. 63, p. 130–148.

______, 1978, Shape and function of the shell: A comparison of some living and fossil bivalve molluscs: Biol. Rev., v. 53, p. 169–210.

EPSTEIN, J.B., SEVON, W.D., AND GLAESER, J.D., 1974, Geology and mineral resources of the Lehighton and Palmerton quadrangles, Carbon and Northampton Counties, Pennsylvania: Pennsylvania Geological Survey, 4th Series, Atlas 195 cd, 460 p.

GABELMAN, J.W., 1955, Cylindrical structures in Permian (?) siltstone, Eagle County, Colorado: Jour. Geology, v. 63, p. 214–227.

HARDY, P.G. AND BROADHURST, F.M., 1978, Refugee communities of *Carbonicola*, Lethaia, v. II, p. 175–178.

KRANZ, P.M., 1974, The anastrophic burial of bivalves and its paleoecological significance: Jour. Geology, v. 82, p. 237–265.

THOMS, R.E. AND BERG, T.M., 1974, Comparison of the burrowing habits of a Devonian pelecypod with those of a Recent analogue: Geol. Soc. America Abstracts with Programs, v. 6, p. 267.

TRUEMAN, E.R., 1966, The fluid dynamics of the bivalve molluscs, *Mya* and *Margaritifera*: Jour. Exp. Biol., v. 45, p. 369–382.

______, 1968, The locomotion of the freshwater clam *Margaritifera margaritifera* (Unionacea: Margaritanidae): Malacologia, v. 6, p. 401–410.

ZHADIN, V.I., 1939, Contribution to the ecology of the pearl shell (*Margaritana margaritifera*) [in Russian]: Izvestiya vsesoyuznogo nauchno-issledovateshogo. Instituta ozerno-rechnogo rynogo khozyaisna, v. 23, p. 351–358.

TRACE FOSSILS FROM BRACKISH-MARINE SHALES, UPPER PENNSYLVANIAN OF KANSAS, U.S.A.

WILLIAM G. HAKES
Phillips Petroleum Company, The Adelphi, John Adam Street, London WC2N 6BW, England

ABSTRACT

Several trace fossil assemblages from the clastic members of three Upper Pennsylvanian cyclothems in the state of Kansas are considered to represent marginal marine (brackish-marine) environments. Trace fossils were collected from several lenticular- or flaser-bedded siltstone and fine-grained sandstone horizons within the Rock Lake Shale, Timberhill Siltstone, and Lawrence Shale. These units were commonly thought to be nonmarine (freshwater to subaerial exposure) due to the general lack of marine body fossils. Common trace fossils are *Asteriacites, Lingulichnus, Isopodichnus, Chondrites, Didymaulichnus, Pelecypodichnus,* and *Planolites*. The trace fossil assemblages are characterized by the following: (1) small-sized members; (2) excellent preservation; (3) high to moderate diversity; (4) moderate abundance; (5) no preferential preservation; and (6) all ethological groups except grazing trails. The small size of these trace fossils is related to a lowering of salinity as a result of freshwater influx into a shallow marine environment accompanying the deposition of the silts and sands upon which the ichnofauna is preserved. These trace fossils may have been overlooked because of their small size, and it is proposed that they could be found elsewhere within similar shallow water stratigraphic sequences.

INTRODUCTION

The role salinity plays in the distribution of trace fossils is at present inadequately understood. The nonmarine *Scoyenia* ichnofacies was established by Seilacher (1967) to distinguish freshwater and terrestrial assemblages from those of the marine environment. The members of this ichnofacies are not well defined and neither is the influence of salinity on the behavioral patterns which determine trace fossil morphology (Frey and Seilacher, 1980).

Trace fossils collected from several shale and siltstone units in Late Pennsylvanian strata in eastern Kansas comprise an assemblage which is representative of marginal marine (reduced marine salinity) environments. This interpretation is based upon the distribution of body fossils in conjunction with an understanding of the regional geologic setting. The ichnofauna is composed of predominantly marine trace fossils whose individuals are reduced in size compared with examples of the same ichnogenera from marine, shallow water paleoenvironments. It is proposed that the identification of these assemblages elsewhere in shallow marine environments could be used to recognize similar conditions of reduced salinity.

UPPER PENNSYLVANIAN—KANSAS

The Upper Pennsylvanian of eastern Kansas has received detailed attention since the recognition by Moore (1929, 1936) of its cyclic nature. The stratigraphic sequence consists of alternating clastic and carbonate units which were laid down under shallow water (probably not greater than 100 to 150 m) marine to nonmarine conditions (Heckel, 1977).

STRATIGRAPHY

Trace fossils discussed in detail here were collected from three Late Pennsylvanian (Missourian to Virgilian) clastic units in eastern Kansas (Figs. 1 and 2).

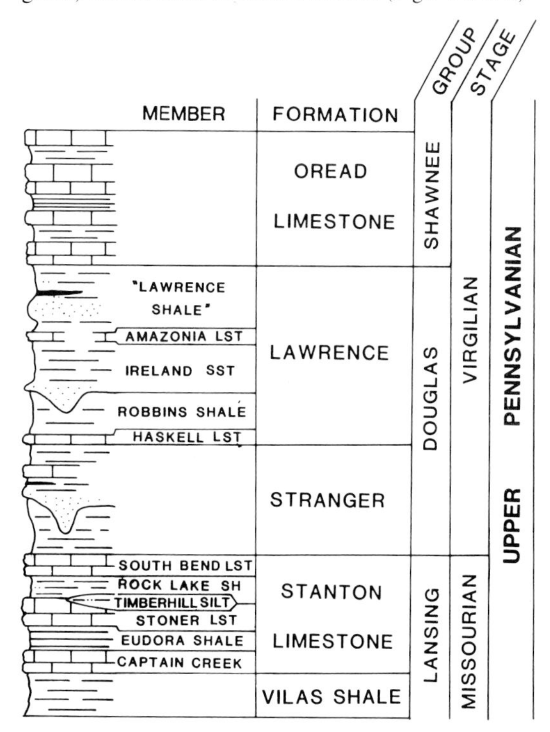

Fig. 1.—Stratigraphic position of the three Upper Pennsylvanian clastic units studied (modified from Zeller, 1968).

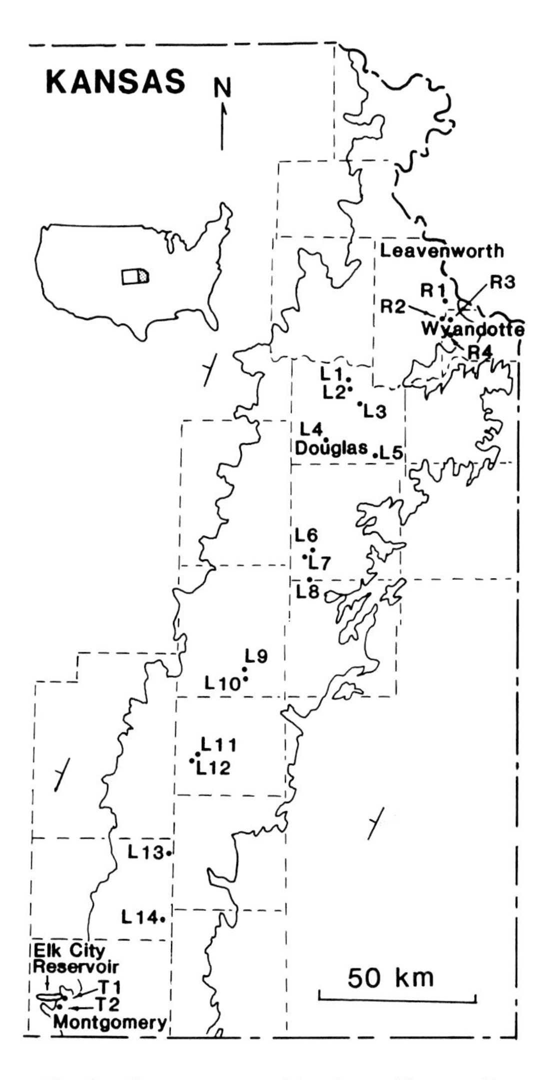

Fig. 2.—Outcrop pattern of Lansing to Shawnee Group (Missourian to Virgilian, Upper Pennsylvanian) strata in eastern Kansas. The Rock Lake Shale localities are in Leavenworth, Wyandotte, and Douglas Counties. The Timberhill Siltstone localities are in Montgomery County. The Lawrence Shale was sampled over the entire outcrop belt.

The Rock Lake Shale and the Timberhill Siltstone belong to the upper part of the Missourian Lansing Group (Fig. 1). The Rock Lake Shale ranges in thickness from approximately 1 to 21 m and is locally absent along the outcrop belt in the southeastern part of the state. It lies stratigraphically between the South Bend Limestone and Stoner Limestone Members. In southeastern Kansas, the Stoner Limestone is absent, and the Rock Lake rests directly upon the Eudora Shale Member. There, the Timberhill Siltstone occurs, locally separating the two (Heckel, 1975). The Timberhill Siltstone is a light gray siltstone unit approximately 1 m thick. Two localities were sampled in Montgomery County, southeastern Kansas, near the Elk City Reservoir (Fig. 2). Trace fossils from the Rock Lake Shale were collected from northeastern Kansas in Leavenworth, Wyandotte, and Douglas Counties where the unit is a medium gray, clayey shale with discrete beds of siltstone containing macerated plant material. These beds are commonly flaser- or lenticular-bedded with the thickness of individual lentils being 1 to 5 mm. Trace fossils were collected from this horizon (Fig. 3). At these localities, the Rock Lake Shale ranges from 1.4 to 2.6 m thick (Hakes, 1976).

The Lawrence Shale is Virgilian age and is the uppermost "unnamed" member of the Lawrence Formation lying directly beneath the Oread Limestone Formation (Fig. 1). The shale ranges from 45 to 69 m in thickness from northeastern to southeastern Kansas. The unit is principally a medium gray, laminated, occasionally very friable shale with locally developed beds of siltstone and fine-grained sandstone. The siltstone units are commonly 1 m thick and consist of wavy-bedded to lenticular- to flaser-bedded lentils with macerated plant debris. Individual bed thickness is rarely greater than 10 mm. Sandstone units may exceed 5 m in thickness. Individual beds are typically several centimeters thick and locally exhibit wave ripples. Coals are developed. Trace fossils were collected from northeastern to southeastern Kansas (Hakes, 1977).

DEPOSITIONAL SETTING

During the Late Pennsylvanian, epeiric seas spread over Kansas, southeastern Nebraska, southwestern Iowa, and western Missouri (Moore, 1966). As a result of fluctuating water levels, together with varying amounts of detrital influx from either the Ouachita Fold Belt to the southeast or the Canadian Shield to the north, a depositional pattern of alternating clastics and carbonates developed (Wanless *et al.*, 1970; Heckel, 1977; Cubitt, 1979). The clastic units consisted of gray, laminated to black fissile shales, with the local appearance of channel sandstones. The gray shale units exhibit pronounced lateral facies variations (Moore, 1966; Heckel, 1977). The carbonates are fossiliferous marine limestones which exhibit lateral homogeneity in outcrop and subsurface (Moore, 1966; Toomey, 1969; Troell, 1969).

The units dip gently, in general, a few degrees to the west, and the outcrop pattern of these units approximates strike (Zeller, 1968). Because of this, the shoreline deposits along the eastern and northern basin margins largely have been eroded away. Fagerstrom and Burchett (1972) described evidence for subaerial exposure within the Shawnee Group of southeastern Nebraska, but such occurrences are rare. On the whole, the truly nonmarine (subaerial to freshwater) facies are absent, and the most common evidence for the existence of nonmarine conditions remaining is recorded in the clastic units which were influenced by this land-derived detritus. The expected result would be increased turbidity plus reduced and fluctuating salinities (Heckel, 1972, 1977).

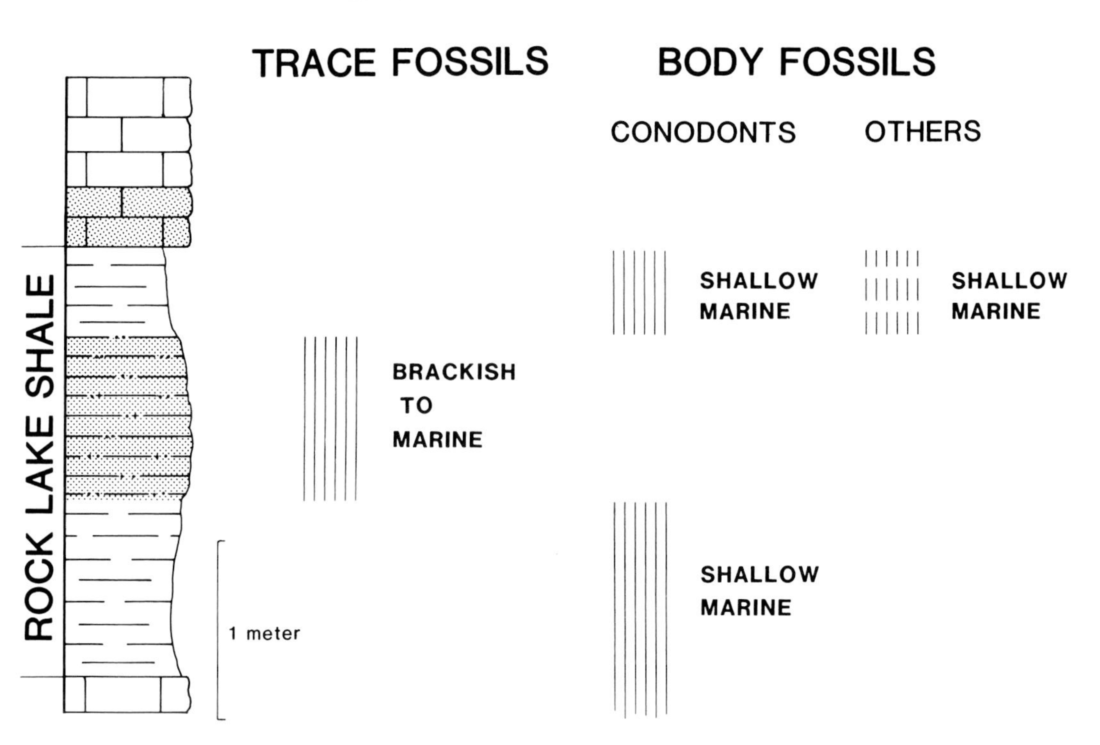

Fig. 3.—Idealized stratigraphic section of Rock Lake Shale showing relationship of body fossil distribution to trace fossil distribution. Note that body fossils and trace fossils are mutually exclusive. Trace fossils in Timberhill Siltstone and Lawrence Shale occur in similar discrete units. Conodont distribution from Heckel and Baesemann (1975); body fossil distribution from Moore (1966).

KANSAS TRACE FOSSILS

The trace fossils discussed below are from the Rock Lake Shale, Timberhill Siltsone, and Lawrence Shale (Fig. 1). These assemblages are believed to represent marginal marine (brackish to marine) conditions. There are also other types of trace fossil assemblages within the Upper Pennsylvanian in Kansas, and in keeping with the shallow water regional setting, the marine forms are generally indicative of the *Skolithos* ichnofacies (Hakes, 1976). Limulid trails are also known from channel sandstones (Bandel, 1967).

Ichnofauna

The distribution of trace fossils within the Rock Lake Shale, Timberhill Siltstone, and Lawrence Shale is shown in Table 1. Common individuals are figured in Plates 1 and 2. All three of these assemblages are preserved in a similar manner on small lentils of siltstone or fine-grained sandstone. In all cases, the lentils freely weather out and are easily collected in the scree slope. In fact, preservation or definition is enhanced by the weathering process. These lentils possess sharp erosive bases and grade upward into the intercalated clays. Preservation is best on the soles of beds, but because the sediments are so fine-grained, preservation on the tops of and within beds is also very good.

The use here of the ichnogenus *Isopodichnus* deserves mention. *Isopodichnus* has traditionally been considered to contain two basic forms. These are the band-shaped (Bandform) sets of two parallel flattened ridges plus the buckle-shaped (Buckleform) of discrete bilobed or coffee-bean pods (Schindewolf, 1928, fig. 6–10; Seilacher, 1960, Pl. 1). Both forms are typically preserved in positive relief on the soles of beds. The band-shaped traces were interpreted to be the locomotion trails of small nonmarine arthropods such as phyllopods, notostrachans, and branchiopods plus possibly annelids (Seilacher, 1960, 1963; Trewin 1976; Bromley and Asgaard, 1979). The buckle-shaped traces were thought to be the resting impressions of the same or very similar animals. The band-shaped and buckle-shaped forms were placed together in *Isopodichnus problematicus* Schindewolf, 1928. *I. eutendorfensis* Linck, 1942, proposed for buckle-shaped forms, has not been popularly accepted (Häntzschel, 1975), probably due to its relatively large size (Trewin, 1976). The taxonomy for *Isopodichnus* was therefore left in the state which was complicated by Seilacher's (1970) observation that large forms of *Isopodichnus* may be indistinguishable from small forms of *Cruziana*.

Bromley and Asgaard (1979) attempted to lend stability by extending the use of the ichnogenera

Cruziana and *Rusophycus* to these small traces. The result of their amended taxonomy is that *Isopodichnus eutendorfensis* was placed in synonymy with *Rusophycus eutendorfensis* and *Isopodichnus problematicus* in *Cruziana problematica*.

Only the band-shaped forms of *Isopodichnus* were recognized in the Rock Lake Shale, Lawrence Shale (Hakes, 1976, 1977), and the Timberhill Siltstone (Pl. 1B). For the purposes of this paper, they are included within the ichnospecies *I. problematicus*. *Cruziana problematica* is not used because the transverse ridges characteristic of *Cruziana* are commonly considered to occur in pairs (Häntzschel, 1975), which they do not always on *C. problematica* or *I. problematicus*. This distinction may not, with further detailed taxonomic work, prove significant, but for the present, Bromley and Asgaard's suggestion is acknowledged but not followed.

The salinity ranges of trace fossils are discussed in this paper and are based upon reported occurrences in the literature. Therefore, if a certain trace fossil is known only from nonmarine environments, then it is considered to be nonmarine. The same applies for marine traces. Because of this, a potential for circular reasoning exists, and it is hoped that the subsequent discussion will explain other occurrences due to facies variation.

Rock Lake Shale.—Trace fossils in the Rock Lake Shale were described by Hakes (1976) (Table 1). This unit possesses a very diverse ichnofauna containing nearly thirty ichnogenera. Several of these forms are paleoenvironmentally significant. *Asteriacites lumbricalis* (12 to 15 mm in diameter) is interpreted to have been produced by an ophiuroid (Seilacher, 1953); *Bergaueria* (commonly 13 mm in diameter) and *Conostichus* (commonly 20 mm in diameter) by sea anemones (Alpert, 1973; Chamberlain, 1971). *Lingulichnus* is considered to have been produced by linguloid brachiopods (Hakes, 1976). *Didymaulichnus lyelli* (4 to 7 mm wide) is typically a molluscan trace. *Isopodichnus* (1.2 to 2.8 mm wide) is predominantly a freshwater to ?brackish indicator (Seilacher, 1963; Hänztschel, 1975); *Scalarituba* a marine indicator (Conkin and Conkin, 1968), as is *Chondrites*. *Chondrites* may occur on the tops and bottoms of beds and does not develop more than a few branches, in contrast to the intricate patterns normally developed.

TABLE 1.—TRACE FOSSILS FROM THE ROCK LAKE SHALE, TIMBERHILL SILTSTONE, AND LAWRENCE SHALE, UPPER PENNSYLVANIAN, KANSAS.

Ichnofauna	*Rock Lake Shale*	*Timberhill Siltstone*	*Lawrence Shale*
Asteriacites	X		X
Bergaueria	X		
Chevronichnus	X		
Chondrites	X	X	X
Cochlichnus	X		
Conostichus	X		
Curvolithus	X	?	
Cruziana	X		X
Didymaulichnus	X	X	X
Isopodichnus	X	X	X
Lingulichnus	X	X	
Pelecypodichnus	X	X	X
Microspherichnus	X		
Planolites	X	X	X
Rusophycus	X	?	
Scalarituba	X		X
Tigillites	X		
?Taenidium	X		
?Trichichnus	X	X	
Tomaculum	X		

EXPLANATION OF PLATE 1

FIG. A.— *Curvolithus*, top of bed in convex relief; Rock Lake Shale locality number 4.
B. — *Isopodichnus problematicus*, sole of bed in convex relief; Rock Lake Shale locality number 3.
C. — *Chondrites* in center of figure, single *Planolites* in upper left and lower right; top of bed in convex relief; Timberhill Siltstone locality number 1.
D. — *Microspherichnus linearis*, small stuffed burrows similar to those of Bromley and Asgaard (1979, Fig. 6C), top of bed in convex relief; Rock Lake Shale locality number 4.
E. — *Lingulichnus verticalis*, sole of bed in convex relief; Timberhill Siltstone locality number 2.
F. — *Cochlichnus* and *Pelecypodichnus*, top of bed in convex relief; Rock Lake Shale locality number 4.
All bar scales represent 10 mm. Figures A,B,D, and F are from Hakes (1976).

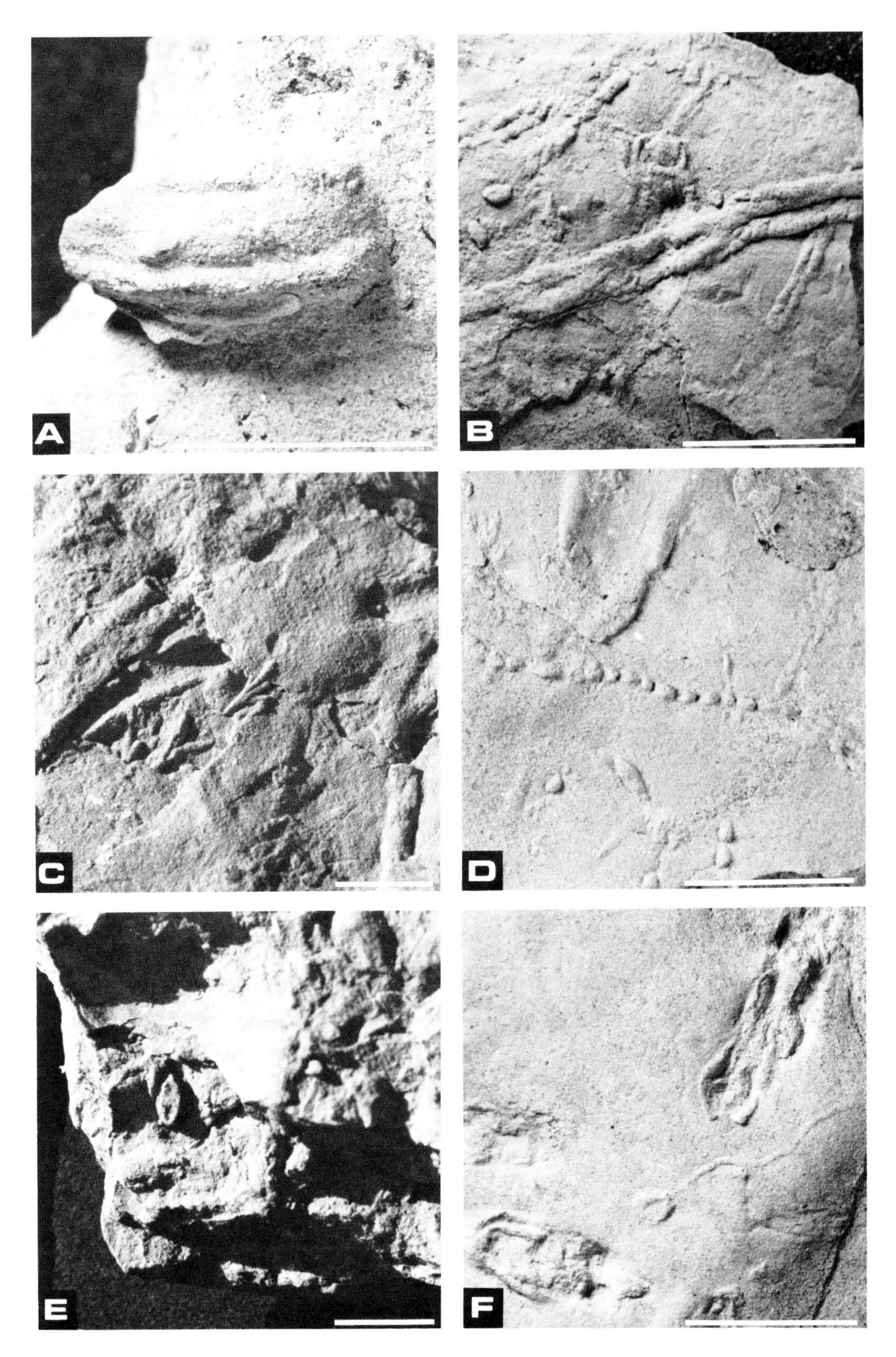
A
B
C
D
E
F

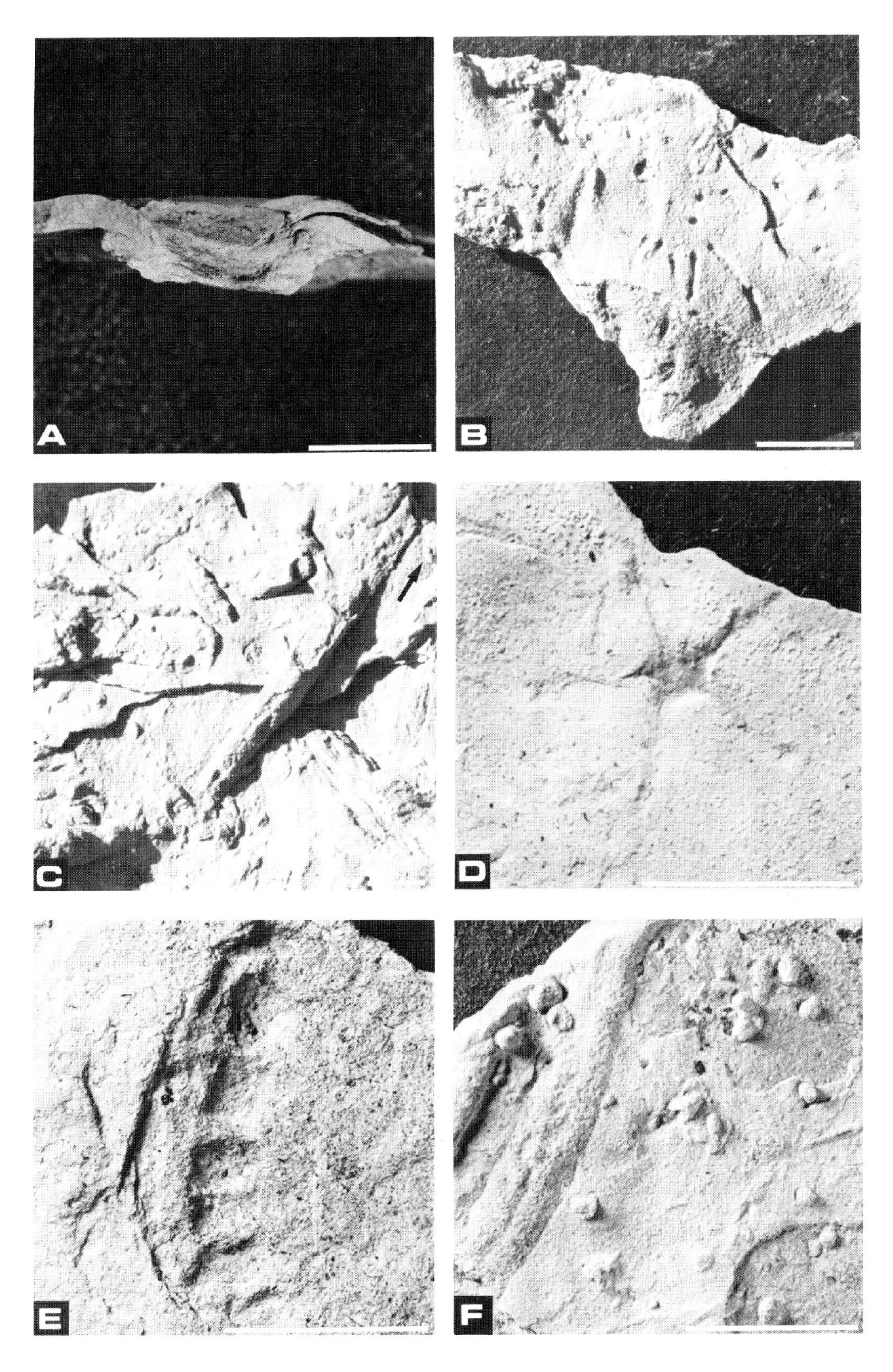
A
B
C
D
E
F

EXPLANATION OF PLATE 2

FIG. A.— *Bergaueria*, side view; Rock Lake Shale locality number 4.
B. — Small unfilled, random burrows, top of bed in concave relief; Timberhill Siltsone locality number 1.
C. — *Didymaulichnus*, arrow points to small *Pelecypodichnus*, sole of bed in convex relief, Timberhill Siltstone locality number 1.
D. — *Asteriacites lumbricalis*, top of bed in concave relief, Rock Lake Shale locality number 4.
E. — *Scalarituba*, top of bed in concave relief, Rock Lake Shale locality number 1.
F. — *Didymaulichnus* plus small stuffed or filled burrows (see Banks, 1970, Pl. 1a, b), sole of bed in convex relief; Rock Lake Shale locality number 4.
All bar scales represent 10 mm. Figures A,D, and E are from Hakes (1976).

The trace fossils listed in Table 1 belong to all of Seilacher's (1964) ethologic groups, except grazing trails (pascichnia) which are more characteristic of flysch deposits.

Timberhill Siltstone.—The Timberhill Siltstone possesses a much less diverse trace fossil assemblage than the Rock Lake Shale, but there are certain similar elements (Table 1). Most of the trace fossils present are relatively abundant and are small. *Chondrites* branches are less than 1 mm wide. The *Lingulichnus* burrow is 2 to 3 mm in transverse section. As in the Rock Lake Shale, *Isopodichnus* (1.5 to 2 mm wide), ?*Curvolithos* (4 mm wide), and *Didymaulichnus* (5 to 6 mm wide) are present. Forms of *Pelecypodichnus* are 1.5 by 4 mm in size. Again, grazing trails are absent.

Lawrence Shale.— Trace fossils in the Lawrence Shale were described by Hakes (1977) (Table 1). Marine indicators are *Asteriacites quinquefolis* (15 mm in diameter), *Chondrites* (branch width 1 to 2 mm), and *Scalarituba*. *Isopodichnus* is a nonmarine indicator. ?*Aulichnites* is 6 to 8 mm wide, and ?*Curvolithos* is approximately 5 mm wide. Dimensions of the trace fossils are similar (to those of the Timberhill Siltstone and the Rock Lake Shale) assemblages. Diversity does not approach that of the Rock Lake Shale, but there are numerous individuals throughout the unit. Grazing trials are again absent.

Relationship to body fossils.—Body fossils of marine organisms have been found within the Rock Lake Shale, Timberhill Siltstone, and Lawrence Shale, but they are not plentiful. Their general absence within these units therefore leads to the conclusion that the units were nonmarine in origin (Moore, 1966; Heckel, 1972). The subsequent discovery and description of marine trace fossils such as *Chondrites* and *Asteriacites* demonstrated that this was not always the case (Hakes, 1976, 1977).

Most commonly, a few molluscs, inarticulate brachiopods, or crinoid columnals are located stratigraphically near an under- or overlying limestone, as in the Lawrence Shale (Moore, 1929). Many of these body fossils are poorly preserved and deteriorate rapidly in outcrop (Heckel, 1972; Hakes, 1976). Heckel thought this was why the body fossil faunas had not been extensively studied. Heckel (1975) considered the Timberhill Siltstone to be poorly fossiliferous but reported a local occurrence of pinnid and nuculid bivalves, gastropods, and the brachiopod *Linoproductus* from one locality in southeastern Kansas.

The distribution of conodonts was studied by Heckel and Baesemann (1975, Fig. 4) in Missourian strata of eastern Kansas. The distribution of conodont elements within the Rock Lake Shale is of particular interest as they are absent in the trace fossil-bearing beds (Fig. 3). Their absence was considered by Heckel and Baesemann to indicate a nonmarine phase within the unit. Approximately 100 km south in Anderson County, the Rock Lake Shale contains a well-described brackish water, lagoonal assemblage of land plants, air-breathing arthropods, shallow marine (euryhaline) invertebrates, and amphibians. This is the *Garnettius* assemblage (Moore *et al.*, 1936; Moore, 1966).

Other Small-Size Ichnofauna

Trace fossils were originally described, no doubt, because they were easily found. In other words, they were obvious sedimentary features and on the whole were relatively large. It is not surprising that assemblages of small (generally less than 5 mm) traces have been overlooked and that, because of this, their paleoenvironmental usefulness has not been generally recognized. In their discussion of *Cruziana* and *Rusophycus* in the lower Paleozoic of Spain, Crimes *et al.* (1977) related trace width directly to bed thickness. They noted that small trilobites would preferentially colonize low energy environments characterized by thin bedding, and larger trilobites would colonize high energy environments with resultant thick bedding. Legg (this publication) observed a similar direct relationship in the width of the ichnogenera *Arenicolites* and *Diplocraterion* to bed thickness in the Middle Cambrian of northern Spain. If these relationships are applied to assemblages of small-sized trace fossils as discussed here, it would be possible to anticipate some control related to sedimentary dynamics.

Freshwater assemblages.—Freshwater invertebrate traces belong to the *Scoyenia* ichnofacies and occur in a variety of environments such as ephemeral ponds and lakes, rivers, river beds, and "permanent" lakes. These lebensspuren are still poorly understood, but data are beginning to accumulate from modern environments (Chamberlain, 1975).

The *Scoyenia* ichnofacies is not easily defined by invertebrate trace fossils alone. A summary of the characteristic traces and types of associated sediments was given by Frey and Seilacher (1980, table 4). Constituents of the ichnofacies are "insect and other arthropod traces; certain forms of *Isopodichnus* and *Planolites*; scattered snail and clam crawling traces and shallow burrows. Local diversity, abundance and complexity generally less than in marine or nearshore environments." The *Scoyenia* ichnofacies is readily recognized in cases where vertebrate tracks and trails are found in red bed sequences. If the origin of the sediments can be demonstrated to be nonmarine by nonichnological means, such as through body fossils, geochemistry, and/or regional setting, then the traces can be considered with some confidence to belong to the *Scoyenia* ichnofacies. It can be difficult to prove an environment was nonmarine by study of the invertebrate traces alone.

Trace fossil assemblages from periglacial lakes in the late Paleozoic of Natal (Savage, 1971) and Pleistocene of Britain (Gibbard and Stuart, 1974) appear dominated by arthropod resting impressions and locomotion trails. The largest of these is about 10 mm wide, and the smaller ones are only a few millimeters wide. Numerous small traces were figured by Stanley and Fagerstrom (1974) from the Miocene of central North America. This assemblage came from a fluvial setting and included horizontal meniscus-filled feeding burrows 3 to 4 mm wide plus branching and curved burrows 1 to 2 mm wide. Associated with these were various other vartical, curved, and spiral forms. Daley (1968) described a series of trace fossils from the Oligocene of Britain which too was characterized by its small traces, some of which are similar to those of Stanley and Fagerstrom (1974). Included in this assemblage are meniscate burrows 2 to 3 mm wide, plus branching and curved burrows approximately 1 mm wide.

Probably the most diversified assemblage of nonmarine invertebrate traces reported is from the Triassic of Greenland (Bromley and Asgaard, 1979). Preservation was excellent due to the fine-grained nature of the sediments. The major trace fossils were: *Isopodichnus problematicus* (= *C. problematica*), *Cylindricum* (placed in *Skolithos*), *Fuersichnus*, *Scoyenia* (a meniscus-filled burrow), *Pelecypodichnus*, *?Margaritichnus*, and *Arenicolites*. All were of similar size to the above-mentioned freshwater traces.

Freshwater-brackish assemblages.—Vossmerbäumer (1970) described a suite of small trace fossils from the brackish-water lower Liassic of Sweden. Lithologies were shale to fine-grained sandstone with flaser bedding, coals, and root horizons. The diversity of this assemblage is similar to that of the Rock Lake Shale, although ichnogeneric names were not generally used. As in the Rock Lake Shale, the Liassic trace fossils were in the millimeter and centimeter size range, commonly with horizontal orientation. The assemblage represented all of Seilacher's (1964) ethologic groups except grazing trails. Forms of *Pelecypodichnus* displayed rheotactic orientation and vertical repetition. Small locomotion trails approximately 1 mm wide were similar to *Aulichnites* or *Didymaulichnus*. Others were trilobate. *Phycodes* and *Teichichnus* were 1 to 1.5 mm wide. Small sac-like resting impressions similar to *?Margaritichnus* as figured by Bromley and Asgaard (1979, fig. 5c) also were reported. With respect to the presence of body fossils, Vossmerbäumer contended that the depositional setting was essentially freshwater with brackish water fluctuations recorded by the presence of a marine bivalve fauna able to tolerate lower salinity conditions. He also noted that the appearance of *Phycodes*, known only from marine environments, would signify the onset of at least weakly marine conditions.

Deep-water marine trace fossils.—Seilacher (1974, p. 241) described a "phylogenetic size decrease" with time in flysch trace fossils within the ichnogenera *Nereites*, *Oldhamia*, and *Paleodictyon*. As a general rule, Silurian traces are noticeably larger than those found in Cretaceous and Tertiary flysch deposits, even though large forms are found within these younger strata. Seilacher considered that the phenomenon was an evolutionary response to a limited food supply which was typical of the deep sea where organisms developed "compartmentalized systems" to acquire nutrients. Although it is not suggested that the evolutionary trends in flysch trace fossils are related to the reduced size of individuals found in marginal marine environments, the size reduction caused by environmental stress is worth noting.

Precambrian trace fossils.—Many trace fossils from Precambrian sediments are small in size. *Cochlichnus*-like trails from the Vendian of Eastern Europe figured by Fedonkin (1977, pl. 2d) are slightly larger than the Kansas Rock Lake Shale specimens, which are less than 1 mm wide (Pl. 1F; Hakes, 1976, pl. 5, fig. 5). Small string-like arrangements of balls identified as *Neonereites uniserialis* by Fedonkin (1977, pl. 1a) are the same size as strings of balls pictured by Hakes (1976, pl. 10, fig. 1a, b) as *Microspherichnus* (0.5 to 0.6 mm in diameter) (Pl. 1D). "Thread-like trails" from the Nama Group of South-West Africa (Germs, 1972, pl. 1, fig. 5, pl. 2, fig. 1) resemble the loop burrow of the Rock Lake Shale (Hakes, 1976, pl. 11, fig. 2a). Bioturbate textures with small, straight burrows and circular burrows were described by Banks (1970, pl. 1a, b) from the Late Precambrian of Finnmark, Norway. Recognizable ichnogenera such as *Rusophycus*, *Cruziana*, *Phycodes*, and *Diplocraterion* only appear in the Early Cambrian strata. The bilobed *Rugoinfractus* Palij (1974) occurs in the Riphaen of the Ukraine and was considered by Durham (1978) to be similar to *Didymaulichnus* (Pl. 2C,F). *Rugoinfractus* is up to 12 mm wide, and is wider than the Rock Lake Shale specimens, which are between 4 to 7 mm wide (Pl. 2F), but is similar in size to those in the Lawrence Shale (Hakes, 1977) and the Timberhill Siltstone (Pl. 2C).

Alpert (1977, Fig. 2) documented cosmopolitan changes in ichnofauna across the Precambrian-Cambrian boundary, as have Seilacher (1956), Rhoads and Morse (1971), Stanley (1976), and Brazier (1979). Brazier (1979, fig. 7) compared the widths of Precambrian (Riphaen and Vendian) trace fossils with those recorded for Lower Cambrian (Tommotian) forms plus body fossils. Riphean and Vendian age burrows produced by deposit-feeding worms ranged from 1 to 10 mm in width. Younger Tommotian burrow widths were nearly as small but could be up to nearly twice as wide. Similarly, epifaunal grazing or locomotion trails of Vendian worms were 1 to 25 mm and Vendian molluscs 5 to 30 mm. The width of Tommotian worm and mollusc traces were as great as 35 mm. A general pattern developed that Precambrian traces were smaller than those found in the overlying earliest Cambrian sediments.

Trace fossils from the Late Precambrian Pound Quartzite, South Australia, are not as small as those mentioned above (Glaessner, 1969), and therefore no axiom is proposed for the size of all Precambrian traces. It is sufficient to note that many tiny traces are preserved in these strata, and it would seem possible to assume that their size limitations were a response to less than optimum environmental conditions.

Reduced oxygen.—Rhoads and Morse (1971) and Byers (1977, plus references cited therein) observed that as the amount of dissolved oxygen within a water mass decreases, faunas will become improverished and reduced in body size. Byers studied the Upper Devonian Middlesex Shale from New York State in North America and compared bioturbation (abundance and size of burrows) with this model as developed from faunal distribution in modern stagnant basins. Three zones were recognized. The shallowest, the aerobic zone, is characterized by shelly faunas and bioturbated substrates. The next deepest, the dysaerobic zone, encompasses a zone of decreasing dissolved oxygen where the benthic community becomes correspondingly less diverse, less abundant, smaller in body size, less heavily calcified, and is dominated by infauna with a comparative increase in worm-like organisms. Benthic organisms are totally absent from the deepest, the anaerobic zone, where there is a complete lack of bioturbation. Burrows from the Middlesex Shale, which were interpreted to have formed within the dysaerobic zone, were shown to be less than 1 to 2 mm wide (Byers, 1977, fig. 5). However, this model for faunal distributions in the modern environment has been questioned (Tunnicliffe, 1981).

DISTRIBUTION AND INTERPRETATION OF TRACE FOSSILS

Distribution

Trace fossils from the Rock Shale, Lawrence Shale, and Timberhill Siltstones are listed in Table 1. Each unit contains a moderately diverse assemblage. *Planolites, Isopodichnus problematicus, Chondrites, Didymaulichnus lyelli*, and *Pelecypodichnus* are common to the Timberhill Siltstone and Rock Lake Shale. The ichnogenera *Asteriacites, Cruziana, Curvolithus*, and *Scalarituba* are common to the Lawrence Shale and Rock Lake Shale. The paleoenvironmental significance of the Rock Lake Shale and Lawrence Shale trace fossils was discussed by Hakes (1976, 1977).

Paleoenvironmental Significance

Seilacher (1963) stated that the interpretation of salinity cannot be done solely by the study of trace fossils because the activities which control trace morphology are related to sedimentary facies and not to salinity. In cases such as the starfish resting impression, *Asteriacites quinquefolis*, where a producer can be somewhat accurately deduced, the trace fossil will suffice; for instance, there are no known freshwater starfish. Usually such distinctions are not so clearly evident, and the exact producer of the trace is rarely known with confidence. If body fossils, not necessarily those of the trace fossil producers, are preserved with the trace fossils, they also must be studied so that an integrated picture evolves. Seilacher reasoned that trace fossils (even insignificant forms) could then be logged within stratigraphic sequences, and from this, their distribution could prove important. As an example, he noted the paleoenvironmental significance of *Planolites opthalmoides, Cochlichnus*, and "*Gyrochorte*" *carbonaria* in the Late Carboniferous cyclothems of the Ruhr Valley. These trace fossils occur stratigraphically between *Lingula*-bearing beds and beds containing freshwater bivalves.

Trace fossils from the Rock Lake Shale, Timberhill Siltstone, and Lawrence Shale formed in a brackish water (marginal marine) environment. Normal marine salinities were reduced by freshwater streams carrying in silts and fine-grained sand in which the trace fossils are now preserved. The trace fossil assemblages are characterized by individuals that are mainly indicative of marine conditions such as *Asteriacites* and *Chondrites* plus those with nonmarine (*Isopodichnus problematicus*) or marine to brackish affinities (*Lingulichnus*) (see Pl. 1E). In addition, most of the trace fossils are small in size (in the 1 to 10 mm range), and the individuals are smaller than the same ichnogenera found elsewhere (Table 2). This size reduction is considered to have been caused by a dwarfism of organisms as a result of the reduction in marine salinities.

There are five trace fossils common to these three units. They are *Chondrites, Didymaulichnus lyelli*,

Isopodichnus problematicus, *Pelecypodichnus*, and *Planolites*. Two of them, *Chondrites* and *Isopodichnus*, were formerly considered to be facies specific.

Chondrites is normally an indicator of fully marine conditions. It is, however, found throughout the complete bathymetric range from shallow marine to abyssal (Simpson, 1957; Frey and Chowns, 1972). *Chondrites* characteristically appears as a well-developed pattern of branching burrows with numerous bifurcations. Small-size burrow systems are not diagnostic of changing environmental parameters, since more than one size burrow system can occur in the same rock slab (Simpson, 1957, pl. XXI). However, the degree to which the system of branching is developed may be significant. In the Rock Lake Shale, *Chondrites* branches have the characteristic acute angle of the *Chondrites* network, but the number of bifurcations rarely exceeds two (Hakes, 1976, pl. 4, fig. 2b). This is partially explained because of the thin bedded (approximately 5 mm thick) lenticular sediments where intricate burrow patterns do not have sufficient space to develop along discontinuous surfaces. Poor patterns of branching are similarly seen in the Lawrence Shale (Hakes, 1977, pl. 2d) and the Timberhill Siltstone (Pl. 1C) in sediments with a lenticular nature, as in those of the Rock Lake Shale. In all three cases burrow diameter is within the millimeter size range. Because poorly developed branching is common to all three units, it is possible that the burrowing behavior of normal marine organisms was disrupted by a decrease in salinity thereby producing poorly developed *Chondrites*. In comparison, the partial *Chondrites* burrows of the Rock Shale are replaced stratigraphically upward in the lower part overlying the South Bend Limestone (Fig. 1) by more robust forms, which have numerous sets of bifurcations. The South Bend Limestone was deposited under marine conditions (Hakes, 1976).

There are two marine indicators common to at least two of the three Kansas units. They are *Asteriacites* (Pl. 2D; Seilacher, 1963) and *Scalarituba* (Pl. 2E; Conkin and Conkin, 1968). *Cruziana* and *Rusophycus* are not considered to have been formed by trilobites in these units as they do not possess paired (bifid) scratch marks. Therefore, no salinity ranges are inferred from their presence.

The ichnogenus *Isopodichnus* is a representative of the nonmarine *Scoyenia* ichnofacies (Seilacher, 1967, 1978; Trewin 1976; Frey and Seilacher, 1980). The stratigraphic association of *Isopodichnus* with many other fossils in the Kansas section that occur in marine environments sheds some doubt upon its restriction to the freshwater realm (Pl. 1B; Hakes, 1976, 1977). Bromley and Asgaard (1979) have since considered *Isopodichnus* to have been produced under both freshwater and marine conditions. They have accomplished this by placing various forms of *Isopodichnus* within the "trilobite" ichnogenera *Rusophycus* and *Cruziana* as *R. eutendorfensis* (= coffee bean *Isopodichnus*) and *C. problematica* (= band form *Isopodichnus*). Their nomenclatural proposals are discussed in greater detail elsewhere in this paper.

The trace fossil *Isopodichnus* is commonly considered to be a brackish to freshwater indicator (Häntzschel, 1975; Trewin, 1976; Frey and Seilacher, 1980). This is especially true for post-Paleozoic sediments where *Isopodichnus* was no longer produced contemporaneously with the marine trilobite trail *Cruziana* (Seilacher, 1978). Bromley and Asgaard (1979) considered *I. problematicus* (= *C. problematica*) to have been produced also in marine environments, based on three lower Paleozoic examples. Small forms of *Cruziana* could be grouped with *Isopodichnus* in Paleozoic sediments because the criterion for the taxonomic division of the two ichnogenera appears to be size (Seilacher, 1970). The small forms of *Isopodichnus problematicus* described in this paper are less than 3 mm wide, and *Cruziana* trails are rarely that narrow. Häntzschel (1975) listed the range of widths for *Cruziana* as between 5 and 80 mm. As an example, Crimes et al. (1977, fig. 8) measured the width of 261 specimens of *Cruziana* and *Rusophycus* from the Cambrian and Ordovician of Spain. Only one specimen of *Cruziana semiplicata* was less than 10 mm wide, and that specimen was greater than 5 mm in width. Häntzschel (1975) stated that *Isopodichnus* could be up to 6 mm wide, which does establish an overlap between the ichnogenera. In the Kansas material, *Isopodichnus* is between 1.2 and 2.8 mm wide and, because of this small size, the problem of misidentification is not considered to be serious.

Isopodichnus has been placed in brackish water settings (Linck, 1942; Glaessner, 1957). A brackish water origin fits very well for the Rock Lake Shale, Timberhill Siltstone, and Lawrence Shale, as in all three units marine conditions existed with freshwater influences. Perhaps the freshwater facies restrictions on the ichnogenus can be broadened at least for the "band-shaped" forms. If freshwater conditions did exist during these periods of shale deposition in the Late Pennsylvanian, they must have been very short-lived.

The brackish-marine interpretation for these three units relies in part on the appearance of *Isopodichnus*. Other trace fossils found there are not facies restricted. *Pelecypodichnus* and *Planolites* are known from both freshwater and marine sediments (Seilacher, 1955; Hakes, 1977; Bromley and Asgaard, 1979). *Didymaulichnus lyelli* is commonly marine, but because of its supposed molluscan origin, it may yet be found in nonmarine environments.

An additional factor supporting a brackish-marine interpretation is the small size of the individual traces which make up the entire assemblage. The range in size of many of these traces is shown in Table 2, and this range is compared with those of the same trace fossils from other geologic settings. In most cases, the Kansas material is the small-size end member or is well within the lower half of the size range. Some exceptions to this are *Planolites*, which is a well-documented facies crossing form, *Isopodichnus problematicus*, which is partially defined by its size and is not an indicator of full marine conditions, and *Tigillites*, which is larger than usual. As a general rule, most of these trace fossils are smaller than would normally be anticipated. The marine affinites of *Asteriacites, Bergaueria, Chondrites, Conostichus, Curvolithus, Didymaulichnus lyelli*, and *Scalarituba* are discussed elsewhere in this paper. Their uniquely small size could be related to the inability of their producers to reach full size as a response to a lowering of marine salinities within the basin. Examples are known from the modern environment.

Two well-documented modern examples of invertebrate dwarfism were described by Pearse and Gunter (1957) and Segerstråle (1957). In shallow marine environments that are influenced by freshwater run off, it is typical to observe a decrease in the number of stenohaline organisms. Juveniles of fully marine organisms are known to venture into these areas. As they mature, they lose their ability to tolerate reduced salinities and migrate back to fully marine conditions (Pearse and Gunter, 1957). Any traces produced there would be necessarily biased toward the small size ranges. In the Baltic Sea, which has areas of relatively stable salinity, brackish water conditions, the larvae of the starfish, *Asterias rubens*, and *Ophiura albida* are washed into brackish water areas where they survive but remain small and do not reach sexual maturity. There is a general reduction in body size of fully marine organisms in these lowered salinity areas (Segerstråle, 1957). The small size of the trace fossils from the units discussed in this paper is considered to be related to similar salinity reductions within the marginal marine clastics of the Late Pennsylvanian seas in Kansas.

The overall stratigraphic sequence within the Late Pennsylvanian strata has always been thought to represent fluctuating environmental conditions, as might be expected with the regular lithologic alternation of limestone and clastic units (Moore, 1966). A weakness in the original concepts associated with these cycles was that so much of the clastic section, composed of shales, was considered nonmarine even to the extent of being subaerially exposed with the development of soil horizons. Subsequent detailed paleontologic studies, specifically the work done on conodont distributions (Heckel and Baesemann, 1975), have increased the size of the marine faunal lists for the shales. The picture of a significantly more marine-dominated paleoenvironment developed. The general lack of body fossils in the trace fossil-bearing beds was considered to have been caused by substantially decreased salinity plus increased turbidity (Heckel, 1972, 1977). This fits very observations of Pearse and Gunter (1957).

Hudson (1980) mentioned the relationship of trace fossils to body fossils in the brackish water facies of the Middle Jurassic, Great Estuarine Series of Britain. As with body fossils, the trace fossils increase in diversity with higher salinities. The most common burrow was *Planolites*, which has no facies significance, and the same applies for *Pelecypodichnus*, which was also present. It is difficult to assess the significance of variations in diversity in the Kansas material. On the whole, diversity is highest in the Rock Lake Shale decreasing in turn to the Lawrence Shale and Timberhill Siltstone (Table 1). Based upon these relative abundances, it appears that the Rock Lake Shale was the most marine, followed by the Lawrence Shale, then by the Timberhill Siltstone.

OTHER SMALL-SIZE ICHNOFAUNAS—INTERPRETATION AND SIGNIFICANCE

Other small trace fossil assemblages discussed here are from freshwater, brackish to freshwater, flysch, Precambrian, and reduced oxygen environments. Of the five, the brackish to freshwater ichnofauna from the lower Liassic of Sweden is the closest in comparison, except that there is a general lack of marine trace fossils (Vossmerbäumer, 1970). Many of the Tertiary freshwater traces are very similar in size to those from Kansas (Daley, 1968; Stanley and Fagerstrom, 1974). The same is true for those from the Triassic of Greenland (Bromley and Asgaard, 1979), but there, too, marine traces were absent. A freshwater origin was well established by the body fossils found associated with those trace fossils. Size reduction in flysch trace fossils is considered significant in that it appears to be an adaptation to environmental stress (Seilacher, 1974). Small Precambrian traces are of importance for a similar reason as with reduced oxygen environments.

MARGINAL MARINE TRACE FOSSILS—SUMMARY OF CHARACTERISTICS

A unique trace fossil assemblage exists within the Upper Pennsylvanian of Kansas that is thought to be typical of a brackish-marine or marginal marine environment. In this environment, nearshore marine conditions are influenced by freshwater influx lowering marine salinities. The characteristics of the marginal trace fossil assemblage are listed below:

Small traces.—Trace fossil size within the Upper Pennsylvanian shale units is commonly in the millimeter size range (Table 2). Many of these trace fossils are indicative of marine environments and occur as much larger forms. It is thought that the small size is a response to a reduction in normal marine salinity due to freshwater influx.

Excellent preservation.—As with size, the degree of preservation is directly dependent upon lithology. Detailed structure cannot be preserved in sediments where

TABLE 2.—COMPARISON IN RANGE OF WIDTHS OR DIAMETERS FOR TRACE FOSSILS COMMONLY FOUND IN BRACKISH WATER ENVIRONMENTS IN THE UPPER PENNSYLVANIAN OF KANSAS WITH OTHER OCCURRENCES IN THE GEOLOGIC RECORD.

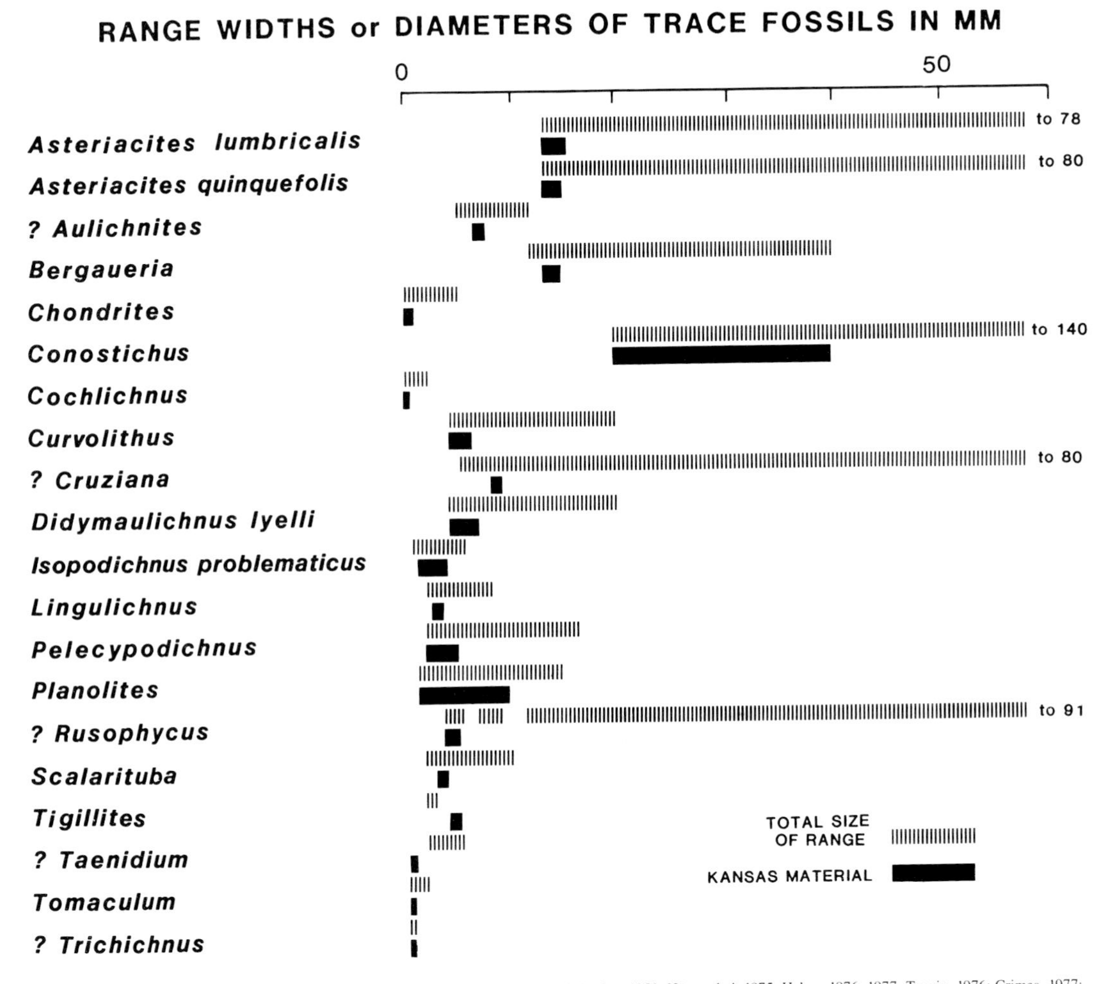

In the cases of *Lingulichnus* and *Pelecypodichnus*, width equals smaller dimension. (Seilacher, 1953; Häntzschel, 1975; Hakes, 1976, 1977; Trewin, 1976; Crimes, 1977; Crimes et al., 1977; Fedonkin, 1977; Palij et al., 1979; Pollard, 1981.)

grain size is similar or larger than the sculpture on the trace itself. Likewise, the preservational potential for small traces in fine-grained sediments is anticipated to be high. The Kansas traces are commonly preserved in siltstone or fine-grained sandstone.

High to moderate diversity.—These assemblages are characterized by their diversity (Table 1). As a result, it can prove difficult to assess what morphologic characteristics are taxonomically, not to mention paleoenvironmentally, significant. Many assemblages may be listed as "Miscellaneous Forms" until they are found in other areas and become more thoroughly understood.

Moderate abundance.—Once these assemblages are identified, the members are rarely represented by only one or two individuals. Abundance is not always related directly to the original body fossil population because one animal can be responsible for more than one trace, and increased rates of sedimentation will tend to dilute the ichnofauna that is eventually preserved. However, increased rates of sedimentation can result in a better quality of preservation, which is a distinct advantage for taxonomic identification (Middlemiss, 1962). The Upper Pennsylvanian traces from Kansas are preserved in thin, very fine-grained lentils which represent episodic sedimentation and which must have been relatively rapidly deposited, since they were not destroyed by bioturbation.

No preferential preservation.—There is no special mode of preservation for this assemblage, as its members are found on tops, soles, and within beds. Forms such as *Pelecypodichnus* or *Asteriacites* are preserved on both tops and bottoms in positive and negative relief, respectively. Vertical or oblique worm burrows or the linguloid burrow, *Lingulichnus*, which are commonly longer than the individual bed is thick, may be preserved as segments with a series of lentils.

All ethological groups except grazing trails.—Grazing trails are developed as a strategy to mine the sediment intensively for nutrients (Seilacher, 1967). These traces are normally found in deep water (abyssal) sediments where thin films of nutrients are locally developed in the sediments or at the sediment water interface (Röder, 1971; Seilacher, 1974). The absence of grazing trails here is, therefore, not surprising.

CONCLUSIONS

Trace fossil assemblages are known from three Late Pennsylvanian units (Rock Lake Shale, Timberhill Siltstone, and Lawrence Shale) which are typical of shallow water, marginal marine (brackish-marine) environments. The characteristics of the individuals within these assemblages are: (1) small size, (2) excellent preservation, (3) high to moderate diversity, (4) moderate abundance, (5) no preferential preservation, and (6) representation of all ethological groups except grazing trails. The trace fossils are preserved on siltstone and fine-grained sandstone lentils within predominantly shale units. It is thought that these trace fossil assemblages have been overlooked because of their size and may be found in similar shallow water environments.

ACKNOWLEDGMENTS

I should like to thank C. Teichert (University of Rochester) and T.P. Crimes (University of Liverpool) for early discussions about the ideas in this paper, and M.D. Brondos (Phillips Petroleum) for showing me the Timberhill Siltstone localities. This manuscript was critically reviewed by R.G. Bromley (Geologisk Centralinstitut, Copenhagen) and T.P. Crimes. The paper is published with the permission of Phillips Petroleum Company.

ADDENDUM

The following appeared since submission of the manuscript and deserves comment. An assemblage of Precambrian trace fossils was described in the Kuibus subgroup of southwest Africa (Crimes and Germs, 1982). Many of these trace fossils were unusually small and typically smaller than the sizes commonly quoted for them. These forms were considered to have been due, in part, to dwarfism of the producers in response to unfavorable conditions.

REFERENCES

ALPERT, S., 1973, *Bergaueria* Prantl (Cambrian and Ordovician), a probable actinian trace fossil: Jour. Paleontology, v. 47, p. 919–924.

______, 1977, Trace fossils and the basal Cambrian boundary, *in* Crimes, T.P., and Harper, J.C. (eds.), Trace Fossils 2: Geol. Journal Spec. Issue 9, Liverpool, Seel House Press, p. 1–8.

BANDEL, K., 1967, Isopod and limulid marks and trails in Tonganoxie Sandstone (Upper Pennsylvanian) of Kansas: Univ. Kansas Paleont. Contrib., Paper 19, p. 1–10.

BANKS, N.L., 1970, Trace fossils from the late Precambrian and Lower Cambrian of Finnmark, Norway, *in* Crimes, T.P., and Harper, J.C. (eds.), Trace Fossils: Geol. Journal Spec. Issue No. 3, Liverpool, Seel House Press, p. 19–34.

BRAZIER, M.D., 1979, The Cambrian radiation event, *in* House, M.R. (ed.), The Origin of Major Invertebrate Groups: Systematics Assoc. Spec. Vol. No. 12, London, Academic Press, p. 103–159.

BROMLEY, R., AND ASGAARD, U., 1979, Triassic freshwater ichnocoenoses from Carlsberg Fjord, East Greenland: Palaeogeogr., Palaeoclimatol., Palaeoecol., v. 28, p. 39–80.

BYERS, C.W., 1977, Biofacies patterns in euxinic basins, *in* Cook, H.E., and Enos, Paul (eds.), Deep-water Carbonate Environments: Soc. Econ. Paleontologists Mineralogists Spec. Publ. 25, p. 5–17.

CHAMBERLAIN, C.K., 1971, Morphology and ethology of trace fossils from the Ouachita Mountains, southeastern Oklahoma: Jour. Paleontology, v. 45, p. 212–246.

______, 1975, Recent lebensspuren in non-marine aquatic environments, *in* Frey, R.W. (ed.), The Study of Trace Fossils: New York, Springer-Verlag, p. 431–458.

CONKIN, J.E., AND CONKIN, B.M., 1968, *Scalarituba missouriensis* and its stratigraphic distribution: Univ. Kansas Paleont. Contrib., Paper 31, p. 1–7.

CRIMES, T.P., 1977, Trace fossils of an Eocene deep-sea sand fan, northern Spain, *in* Crimes, T.P., and Harper, J.C. (eds.), Trace Fossils 2: Geol. Journal Spec. Issue 9, Liverpool, Seel House Press, p. 71–90.

______, AND GERMS, G.J.B., 1982, Trace fossils from the Nama Group (Precambrian-Cambrian) of southwest Africa (Namibia): Jour. Paleontology, v. 56, p. 890–907.

______, LEGG, I., MARCOS, A., AND ARBOLEYA, M., 1977, ?Late Precambrian--low Lower Cambrian trace fossils from Spain, *in* Crimes, T.P., and Harper, J.C (eds.), Trace Fossils 2: Geol. Journal Spec. Issue 9., Liverpool, Seel House Press, p. 91–138.

CUBITT, J.M., 1979, The geochemistry, mineralogy and petrology of upper Paleozoic shales of Kansas: State Geol. Surv. Kansas, Bull, 217, p. 1–117.

DALEY, B., 1968, Sedimentary structures from a nonmarine horizon in the Bembridge Marls (Oligocene) of the Isle of Wight, Hampshire, England: Jour. Sed. Petrology, v. 38 p. 114–127.

DURHAM, J.W., 1978, The probable metazoan biota of the Precambrian as indicated by the subsequent record: Ann. Review Earth Planet Sciences, v. 6, p. 21–42.

Fagerstrom, J.A., and Burchett, R.R., 1972, Upper Pennsylvanian shoreline deposits from Iowa and Nebraska: Their recognition, variation, and significance: Geol. Soc. America Bull., v. 83, p. 367–388.
Fedonkin, M.A., 1977, Precambrian-Cambrian ichnocoenoses of the east European platform, *in* Crimes, T.P. and Harper, J.C. (eds.), Trace Fossils 2: Geol. Journal Spec. Issue 9, Liverpool, Seel House Press, p. 183–194.
Frey, R.W., and Chowns, T.M., 1972, Trace fossils from the Ringgold road cut (Ordovician and Silurian), Georgia: Georgia Geol. Survey, Guidebook 11, p. 25–55.
______, and Seilacher, A., 1980, Uniformity in marine invertebrate ichnology: Lethaia, v. 13 p. 183–207.
Germs, G.J.B., 1972, Trace fossils from the Nama Group, South-West Africa: Jour. Paleontology, v. 46, p. 864-870.
Gibbard, P.L., and Stuart, A.J., 1974, Trace fossils from proglacial lake sediments: Boreas, v. 3, p. 69–74.
Glaessner, M.F., 1957, Palaeozoic arthropod trails from Australia: Paläont. Zeitschr., v. 31, p. 103–109.
______, 1969, Trace fossils from the Precambrian and basal Cambrian: Lethaia, v. 2, p. 369–393.
Häntzschel, W., 1975, Trace fossils and problematica, *in* Teichert, Curt. (ed.), Treatise on Invertebrate Paleontology Part W, Supplement 1: Lawrence, Kansas, Geol. Soc. America and Univ. Kansas, 269 p.
Hakes, W.G., 1976, Trace fossils, and depositional environment of four clastic units, Upper Pennsylvanian megacyclothems, northeast Kansas: Univ. Kansas Paleont. Contrib., Article 63, 46 p.
______, 1977, Trace fossils in Late Pennsylvanian cyclothems, Kansas, *in* Crimes, T.P., and Harper, J.C. (eds.), Trace Fossils 2: Geol. Journal Spec. Issue 9, Liverpool, Seel House Press, p. 209–226.
Heckel, P.H., 1972, Recognition of ancient shallow marine environments, *in* Rigby, J.K., and Hamblin, W.K. (eds.), Recognition of Ancient Sedimentary Environments: Soc. Econ. Paleontologists Mineralogists Spec. Publ. 16, p. 226–286.
______, 1975, Stratigraphic and depositional framework of the Stanton Formation is southeastern Kansas: Kansas Geol. Survey, Bull. 210, 45p.
______, 1977, Origin of phosphatic black shale facies in Pennsylvanian cyclothems of Mid-Continent North America: Am. Assoc. Petroleum Geologists Bull., v. 61, p. 1045–1068.
______, and Baesemann, J.F., 1975, Environmental interpretation of conodont distribution in Upper Pennsylvanian (Missourian) megacyclothems in eastern Kansas: Am. Assoc. Petroleum Geologists Bull., v. 59, p. 486–509.
Hudson, J.D., 1980, Aspects of brackish-water facies and faunas from the Jurassic of north-west Scotland: Proc. Geol. Assn., v. 91, p. 99–105.
Linck, O., 1942, Die Spur *Isopodichnus*: Senckenbergiana, v. 25, p. 232–255.
Middlemiss, F.A., 1962, Vermiform burrows and rate of sedimentation in the Lower Greensand: Geol. Mag., v. 99, p. 33–40.
Moore, R.C., 1929, Environment of Pennsylvanian life in North America: Am. Assoc. Petroleum Geologists Bull., v. 13, p. 459–487.
______, 1936, Stratigraphic classification of the Pennsylvanian rock of Kansas: Kansas Geol. Survey Bull. 22, 256 p.
______, 1966, Paleoecologic aspects of Kansas Pennsylvanian and Permian cyclothems, *in* Merriam, D.F. (ed.), Symposium on Cyclic Sedimentation: Kansas Geol. Survey Bull. 169 (1964), v. 1, p. 287–380.
______, Elias, M.K., and Newell, N.D., 1936, A "Permian" flora from the Pennsylvanian rocks of Kansas: Jour. Geology, v. 44, p. 1–31.
Palij, V.M., 1974, On finding of the trace fossil in the Riphaen deposits of the Ovruch Ridge: Rep. Acad. Sci. Ukraine SSR, Ser. B, Geology, Geophysics, Chemistry, Biology, Jahrg., v. 36, p. 34–37 (in Ukranian).
______, Posti, E., and Fedonkin, M.A., 1979, Soft-bodied Metazoa and trace fossils of Vendian and Lower Cambrian, *in* Keller, B.M., and Rozanov, A. Yu. (eds.), Upper Precambrian and Cambrian Paleontology of East-European Platform: Akad. Science U.S.S.R., p. 49–82, 156–176, 204–210 (in Russian).
Pearse, A.S., and Gunter, G., 1957, Salinity, *in* Hedgpeth, J.W. (ed.), Treatise on Marine Ecology and Paleoecology: Ecology, Geol. Soc. America Mem. 67, v. 1, p. 129–158.
Pollard, J.E., 1981, A comparison between the Triassic trace fossils of Cheshire and south Germany: Palaeontology, v. 24, p. 555–588.
Röder, H., 1971, Gangsysteme von *Paraonis fulgens* Levinsen 1883 (Polychaeta) in ökologischer, ethologischer und aktuopaläontologischer Sicht: Senckenbergiana Maritima, v. 3, p. 3–51.
Rhoads, D.C., and Morse, J.W., 1971, Evolutionary and ecologic significance of oxygen-deficient marine basins: Lethaia, v. 4, p. 413–428.
Savage, N.M., 1971, A varvite ichnocoenosis from the Dwyka Series of Natal: Lethaia, v. 4, p. 217–233.
Schindewolf, O.H., 1928, Studien aus dem Marburger Buntsandstein. III-VII: Senckenbergiana, v. 10, p. 16–54.
Segerstrale, S.G., 1957, Baltic Sea, *in* Hedgpeth, J.W. (ed.), Treatise on Marine Ecology and Paleoecology: Geol. Soc. America Mem. 67, v. 1, p. 751–800.
Seilacher, A., 1953, Studien zur Palichnologie, II. Die Fossilen Ruhespuren (Cubichnia): Neues Jahrb. Geologie, Palaontolgie, Abhandl., v. 98, p. 87–124.
______, 1955, Spuren und Fazies im Unterkambrium, *in* Schindewolf, O.H., and Seilacher, A., Beiträge zur Kenntnis des Kambriums in der Salt Range (Pakistan): Acad. Wiss. Lit. Mainz. math.-nat. Kl. Abhandl., no. 10, 1955, p. 177–143.
______, 1956, Der Beginn des Kambriums als biologische Wende: Neues Jahrb. Geologie, Palaontolgie, Abhandl., v. 103, p. 155–180.
______, 1960, Lebensspuren als Leitfossilien: Geol. Rundschau, v. 49, p. 41–50.
______, 1963, Lebensspuren und Salinitätsfazies: Fortschr. Geol. Rheinld. u. Westfal., v. 10, p. 81–94.
______, 1964, Biogenic sedimentary structures, *in* Imbrie, J., and Newell, N.D. (eds.), Approaches to Paleoecology: New York, John Wiley & Sons, Inc., p. 296–316.

______, 1967, Bathymetry of trace fossils: Marine Geology, v. 5, p. 413–428.
______, 1970, *Cruziana* stratigraphy of "nonfossiliferous" Paleozoic sandstones, *in* Crimes, T.P., and Harper, J.C. (eds.), Trace Fossils: Geol. Journal Spec. Issue 3, Liverpool, Seel House Press, p. 447–476.
______, 1974, Flysch trace fossils: Evolution of behavioral diversity in the deep sea: Neues Jahrb. Geologie, Palaontologie, Monatsh. 1974, p. 233–245.
______, 1978, Use of trace fossil assemblages for recognizing depositional environments, *in* Basan, P.B. (ed.), Trace Fossil Concepts: Soc. Econ. Paleontologists Mineralogists Short Course 5, p. 185–201.
SIMPSON, S., 1957, On the trace fossil *Chondrites*: Geol. Soc. London, Quart. Jour., v. 112, p. 475–499.
STANLEY, K.O., AND FAGERSTROM, J.A., 1974, Miocene invertebrate trace fossils from a braided river environment, western Nebraska, U.S.A.: Palaeogeogr., Palaeoclimatol. Palaeocol., v. 15, p. 63–82.
STANLEY, S.M., 1976, Fossil data and the Precambrian-Cambrian evolutionary transition: Am. Jour. Science, v. 276, p. 56–76.
TOOMEY, D.F., 1969, The biota of the Pennsylvanian (Virgilian) Leavenworth Limestone, Midcontinent Region. Part 1. Stratigraphy, paleogeography, and sediment facies relationships: Jour. Paleontology, v. 43, p. 1001–1018.
TREWIN, N.T. 1976, *Isopodichnus* in a trace fossil assemblage from the Old Red Sandstone: Lethaia, v. 9, p. 29–37.
TROELL, A.R., 1969, Depositional facies of Toronto Limestone Member (Oread Limestone, Pennsylvanian), subsurface marker unit in Kansas: Kansas Geol. Surv. Bull. 197, 29 p.
TUNNICLIFFE, V., 1981, High species and abundance of the epibenthic community in an oxygen-deficient basin: Science, v. 294, p. 354–356.
VOSSMERBÄUMER, H., 1970, Untersuchungen zur Bildungsgeschichte des Unteren Lias in Schonen (Schweden): Geologica et Palaeontologica, v. 4, p. 167–193.
WANLESS, H.R., BAROFFIO, J.R., GAMBLE, J.C., HORNE, J.C., ORLOPP, D.R., ROCHA-CAMPOS, A., SOUTER, J.E., TRESCOTT, P.C., VAIL, R.S., AND WRIGHT, C.R., 1970, Late Paleozoic deltas in the central and eastern United States, *in* MORGAN, J.P. (ed.), Deltaic Sedimentation-Modern and Ancient: Soc. Econ. Paleontologists Mineralogists Spec. Publ. 15, p. 215–245.
ZELLER, D.E., ed., 1968, The stratigraphic succession in Kansas: Kansas Geol. Survey Bull. 189, 81 p.

EARLY LIASSIC TRACE FOSSIL ASSEMBLAGES FROM THE HOLY CROSS MOUNTAINS, POLAND: THEIR DISTRIBUTION IN CONTINENTAL AND MARGINAL MARINE ENVIRONMENTS

GRZEGORZ PIEŃKOWSKI
Geological Institute, Laboratory of Sedimentology, Department of Petrography, Mineralogy, and Geochemistry, 02–519 Warszawa, Rakowiecka 4, Poland

ABSTRACT

In Early Liassic continental deposits from the northern slope of the Holy Cross Mountains, Poland, trace fossils are very rare and the assemblage is not diverse. Only simple feeding structures (fodinichia) and crawling traces (repichnia) are present, and these are mostly associated with levee or distal crevasse deposits.

In brackish-marine deposits, which form the major part of the Holy Cross Lower Liassic sequence, trace fossils are much more numerous and diverse. However, compared to other Jurassic ichnocoenoses from fully marine environments, the Holy Cross Early Liassic brackish water forms are generally smaller and less diverse, and spreiten burrows are rare.

Based on detailed descriptions of 26 bore holes and numerous outcrops, the usefulness of trace fossils in differentiating shore-prograding from barrier-lagoon and deltaic sequences is confirmed. With bore hole materials, trace fossils are of primary importance in making this differentiation.

INTRODUCTION

Liassic deposits of the northern slope of the Holy Cross Mountains represent the easternmost margin of the Liassic basin in Europe. These deposits constitute an extension of the Liassic sedimentary cover along the Danish-Polish Trough (Fig. 1). The Holy Cross Lower Liassic in Poland consists for the most part of clastic, terrigenous deposits, sometimes containing lignites and siderites. Because of the lack of good index fossils, only lithostratigraphic subdivision exists for this sequence (Fig. 2 and Karaszewski, 1962; Dadlez, 1978; and Pieńkowski, 1981).

In the course of detailed sedimentological studies, 26 bore holes, with an average length of about 100 m, and numerous outcrops were examined (Fig. 1). Sedimentological analysis of the cyclicity of sedimentation, the sedimentary structures, paleocurrent directions (in the outcrops), distribution of fossil fauna and flora, clay mineralogy and boron content of the sediments all were evaluated in order to interpret the

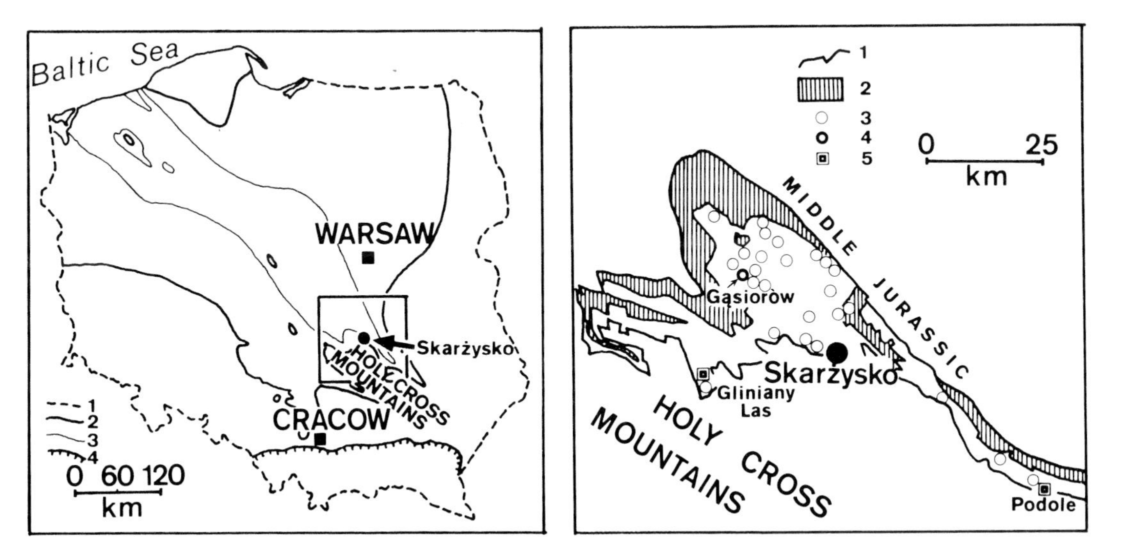

Fig. 1.—*Left*, Location of the Holy Cross Mountains in Poland. 1 = national boundary; 2 = boundary of Liassic deposits; 3 = boundary for Liassic deposits with thickness greater than 400 m; 4 = Carpathian overlap area (after Dadlez, 1978). *Right*, Map of inset area, left. 1 = boundary of Liassic deposits at the surface or below Quaternary deposits; 2 = Middle and Upper Liassic outcrop area (not investigated); 3 = boreholes studied; 4 = Gąsiorów borehole; 5 = outcrops studied.

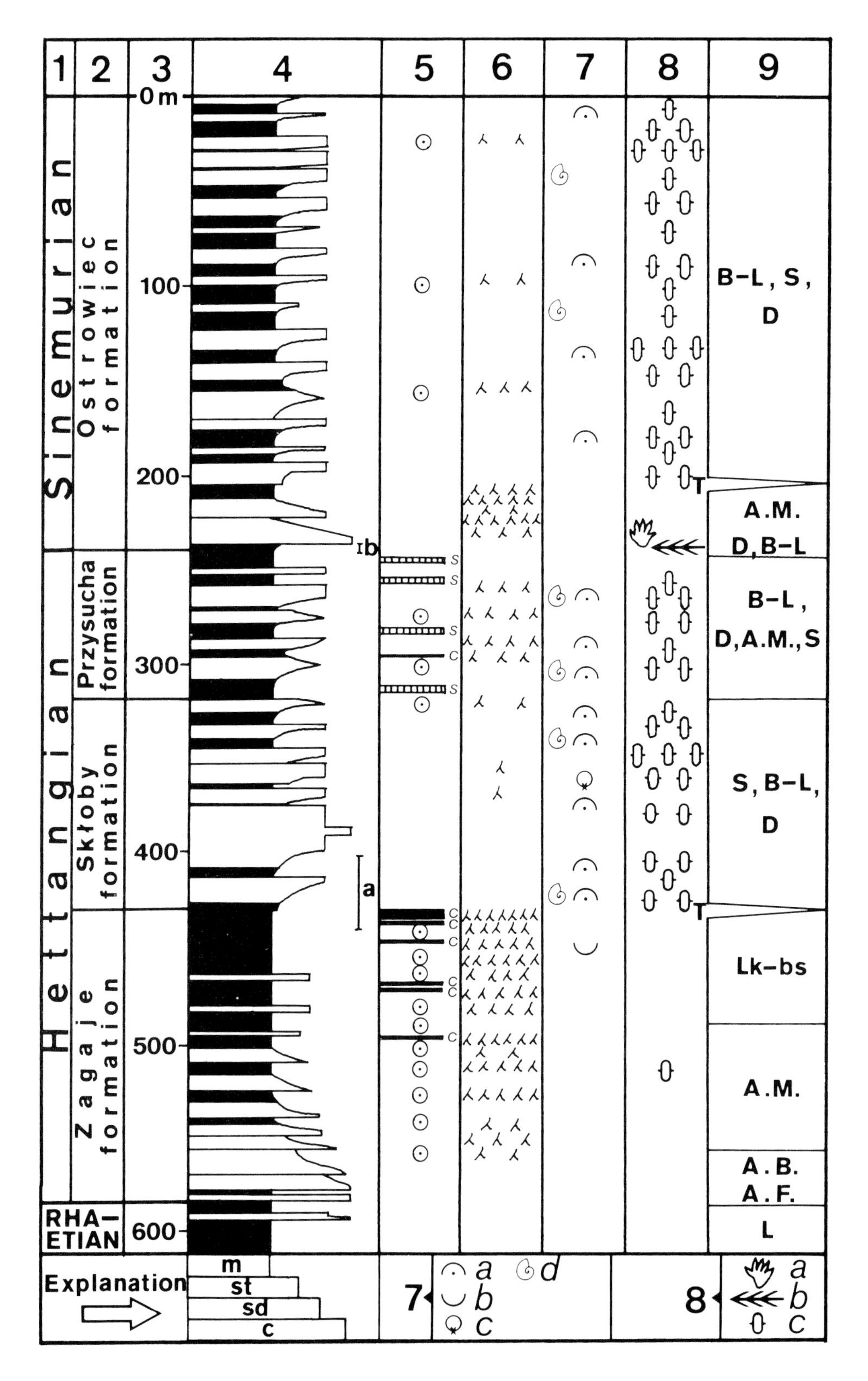

1
2
3
4
5
6
7
8
9
0 m
100
200
300
400
500
600
Sinemurian
Hettangian
RHAETIAN
Ostrowiec formation
Przysucha formation
Skłoby formation
Zagaje formation
B–L, S, D
A.M. D, B–L
B–L, D, A.M., S
S, B–L, D
Lk–bs
A.M.
A.B. A.F.
L
T
a
b
s
c
Explanation
m
st
sd
c
7 a b c d
8 a b c

Fig. 2.—Generalized profile of the Holy Cross Lower Liassic section. Column 1 = Liassic stages; 2 = stratigraphic units; 3 = thickness; 4 = simplified lithologic profile: m—mudstones, st—siltstones, sd—sandstones, c—conglomerates; a = section shown in Figure 3; b = section shown in Figure 6; 5 = presence of coal and siderite, for further explanation see Figure 3; 6 = presence of rootlets; 7 = faunal components: a—brackish-water bivalves, b—fresh-water bivalves, c—branchiopods, d— foraminifera; 8 = occurrence of trace fossils: a— reptile footprints, b—marks of fishes, c— other trace fossils; 9 = interpretation of paleoenvironments: A.F—alluvial fans, A.B— braided rivers, A.M—meandering rivers, L.K— lakes, bs—backswamps, D—deltas, B-L— barriers and lagoons, S—nearshore, T—transgressions.

evolution of this Early Liassic basin and its paleoenvironments (Figs. 2, 3). The occurrence of trace fossils throughout the sequence was carefully noted.

STRATIGRAPHIC SETTING

The lowermost stratigraphic unit—the Zagaje Formation (early Hettangian)—contains exclusively continental deposits that demonstrate a gradual lowering of depositional energy through time. The result is a vertical sequence of alluvial fan and braided river deposits, meandering river deposits, and lake and swamp deposits (Fig. 2). Toward the center of the sedimentary basin, where the thickness of the formation is greatest, braided river deposits may be absent, whereas lacustrine deposits are quite thick.

Strata at the base of the next formation, the Skłoby Formation, represent deposition by a widespread and rapid brackish-marine transgression (Figs. 2, 3). Lowered salinity at the margin of the transgressing sea is evidenced by a sparce bivalve fauna containing thin-shelled Cardiniidae, Mytilidae, and Pholadomyidae (Troedsson, 1951) and some agglutinated foraminifera (Jurkiewicz, 1967; Karaszewski and Kopik, 1970). Fossils of stenohaline organisms are entirely absent.

Shore-prograding, coarsening upward cycles prevail in the Skłoby Formation. These cycles represent sediments deposited under conditions of shallow water and high energy prograding on fine-grained sediments deposited under lower energy conditions in deeper, offshore waters (Fig. 3). Barrier-lagoon and deltaic cycles also are prominent here, but fining upward, fluvial cycle facies are rare and only occur at the very margins of the basin. These fluvial facies of the Skłoby Formation are closely connected with the deltaic cycles (Fig. 4).

The next unit, the ore-bearing Przysucha Formation, represents deposition under regressive conditions. The regression led to the creation of closed or semi-enclosed shallow embayments and lagoons. Iron-enriched water was delivered to these basins from adjacent, extensive lowlands covered with swamps. Under these reducing conditions, siderites and sideroplesites were deposited. Episodes of filling with coarser materials sometimes interrupted the sideritic mudstone sedimentation.

The Ostrowiec Formation (Sinemurian) is the uppermost formation in the Holy Cross Lower Liassic sequence. It begins with continental deposits containing numerous fining upward fluvial cycles. Beds above the fluvial cycles represent a second transgression in the Lower Liassic sedimentary sequence and are similar to those of the underlying Skłoby Formation (Fig. 2).

SYSTEMATIC ICHNOLOGY

The systematic distribution of the trace fossils described herein is presented in Figure 5. The trace fossil ichnogenera are grouped according to their ethologic affinities.

Dwelling Structures (Domichnia)

ARENICOLITES Salter, 1857

Arenicolites sp.

Pl. 1I

Description.—Vertically oriented, U- or J-shaped burrows. If U-shaped, a convex accumulation of sediment can occur around the posterior shaft.

Remarks.—These forms occur together with *Skolithos* and *Monocraterion*, but they are rare.

MONOCRATERION Torell, 1870

Monocraterion sp.

Pl. 1C

Description.—Straight or slightly curved, conical burrows. In vertical section, simple or composed of concentrical cones. Diameter up to 12 mm, length up to 20 cm.

Remarks.—The author frequently found *Monocraterion* in association with *Skolithos*. This confirms the view of Hallam and Swett (1966) that the difference between the burrows of these two ichnogenera is not due to different tracemaker organisms but to different rates of sedimentation. *Skolithos* is formed under conditions of slow sedimentation, and *Monocraterion* occurs with conditions of relatively rapid sedimentation. It should be noted here that the ichnogenus *Calycraterion* described by Karaszewski (1971) may represent the upper, conical parts of *Monocraterion* burrows.

SKOLITHOS Haldeman, 1840

Skolithos sp.

Pl. 1D,F,G

Description.—Vertical, simple tubes with constant diameter, filled with structureless sediment. Diameter 2–8 mm, length up to 20 cm.

Remarks.—Most common dwelling structures (domichnia) preserved in the Holy Cross Lower Liassic sequence. According to Seilacher (1967) and Crimes (1970), *Skolithos* is diagnostic of very shallow, high energy coastal environments. This interpretation is confirmed for the strata described in the present paper

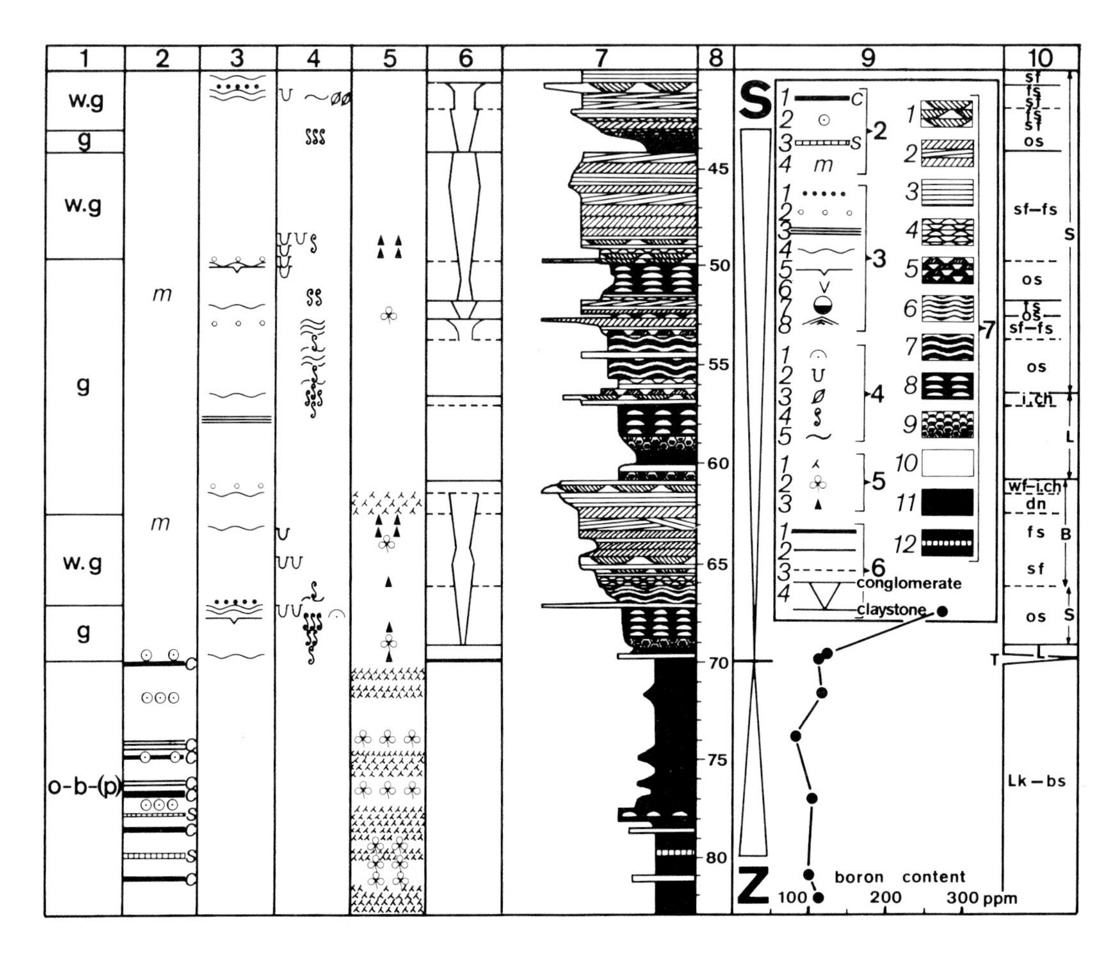

Fig. 3.—Section of the Gąsiorów borehole showing the contact between the continental Zagaje Formation and the brackish-marine Skłoby Formation. Column 1 = color: o-b-(p) = alternating colors, o = olive (dominant), b = black, (p) = pink (sporadic), g = gray, w.g. = white-gray. Column 2 = occurrence of selected lithologic types, explanation in inset frame: 1—coals and lignites, 2—siderite bands, 3—siderite concretions, 4—muscovite-rich layers. Column 3 = sedimentary structures: 1— interbeds of clay and mud, 2—interbeds of quartz grains larger than 2 mm, 3—varve-like laminations, 4—erosional surfaces, 5—small-scale erosional channels, 6—dessication cracks, 7—load casts, 8—oscillation ripples. Column 4 = fauna and trace fossils: 1—brackish-water bivalves, 2—dwelling structures, 3—resting traces, 4—feeding traces, 5—crawling traces. Column 5 = plant remains: 1— rootlets, 2—well-preserved flora, 3—plant debris. Column 6 = sedimentary cycles: 1—boundaries of formations, 2—boundaries of first-order sedimentary cycles, 3—boundaries of second-order sedimentary cycles, 4—changes in the diameter of mean grain size. Column 7 = lithologic and sedimentologic profile; width of the profile is proportional to the coarsest grain size, successively: claystone, mudstone, siltstone, fine-grained sandstone, medium-grained sandstone, coarse-grained sandstone, conglomerate; claystones and mudstones are black. Explanation of sedimentary structures in the inset frame: 1—trough cross bedding, 2—tabular bedding, 3—horizontal lamination, 4—horizontal wavy lamination, 5—ripple-drift cross lamination, 6—flaser bedding, 7—wavy bedding, 8—lenticular bedding, 9—disturbed bedding with some mud content, 10—structureless sediment, 11—horizontal laminations of claystones and mudstones, 12—main siderite bands. Column 8 = thickness in meters. Column 9 = boron content in ppm, Z— Zagaje Formation, S—Skłoby Formation. Column 10 = interpretation of sedimentary environments (see Fig. 2); more detailed explanation: os—offshore, sf—shoreface, dn—aeolian dunes, wf—washover fans, i.ch—inlet channels, T—transgression.

Fig. 4.—Stratigraphic section at the Podole outcrop (see Fig. 1) exhibits well-developed, coarsening-upwards deltaic sequences, sometimes joined with fining-upward, fluvial cycles (cycle "c"). Explanation: Column 1 = lithologic and sedimentologic profile, m—mudstone, st—siltstone, sd— sandstone, for other symbols see Figure 3. Column 2 = cycles and their boundaries. Column 3 = plant remains and coal seams. Column 4 = additional sedimentary structures (see Fig. 3). Column 5 = occurrence of *Palaeophycus* (densely packed tubes) and *Pelecypodichnus czarnockii* (see Plate 2B,C).

since *Skolithos* occurs only in the brackish-marine deposits, mainly in sandstones of coastal origin. Sometimes *Skolithos* was found to occur with desiccation cracks or with erosional structures (Pl. 1G).

Escape Structures (Fugichnia of Simpson, 1975)
Pl. 1B; Pl. 2A

Description.—Vertical burrows with strong, downward oriented deflections of adjacent laminations.

Remarks.—These structures are not assigned to a formal ichnotaxon. The majority of the Liassic escape structures (e.g. Pl. 1A) are believed to have been formed by bivalves in a manner similar to that described by Reineck (1958). In some instances, a connection between the escape structure and *Pelecypodichnus* is visible, and fossilized bivalves have been found at the ends of some of the escape structures. Other escape structures (Pl. 2A) are associated with *Chondrites* and appear to be related to episodes of rapid sedimentation. These structures seem analogous to the escape structures formed by modern polychaetes as described by Reineck (1958).

Resting Traces (Cubichnia)
PELECYPODICHNUS Seilacher, 1953
Pelecypodichnus amygdaloides Seilacher, 1953
Pl. 1A,E,H

Description.—Almond-shaped, elongate traces preserved in convex hyporelief; very rarely preserved in concave epirelief. In vertical section (Pl. 1H), they are U-shaped. Average dimensions; length about 2 cm, width about 0.8 cm.

Remarks.—Several workers have interpreted these almond-shaped traces as bivalve resting traces (Seilacher, 1953; Osgood, 1970; Hallam, 1970; Hakes, 1977). The regular, almond-like shape is often deformed. On the wider ends of the traces, additional bumps can be found which may reflect the motion of the bivalve's foot. The author also has observed bilateral, coffee bean-shaped forms (Pl. 1E).

Pelecypodichnus czarnockii (Karaszewski, 1974)
Pl. 2B

Description.—Hypichnial, asymmetric, strongly convex, drop-shaped trace fossils. One end is strongly convex, whereas the opposite end merges with the sole of the sandstone bed. The form is larger than *Pelecypodichnus amygdaloides*. Length averages 5 cm.

Remarks.—This larger form differs from *P. amygdaloides* by its larger dimensions and strongly asymmetric profile resembling a drop. This form was originally assigned to the ichnogenus *Umbonichnus* by Karaszewski (1974). However, it is the author's opinion that the form clearly falls within the definition of the ichnogenus *Pelecypodichnus*, and the ichnospecies is thus reassigned to this ichnogenus. The "smaller form" of *Umbonichnus czarnockii* described by Karaszewski (1974) should be regarded as synonymous with *P. amygdaloides*.

Crawling Traces (Repichnia)
COCHLICHNUS Hitchcock, 1858
Cochlichnus sp.
(not figured)

Description.—Horizontal, smooth, sinusoid trails.

IMBRICHNUS Hallam, 1970
Imbrichnus sp.
Pl. 2F

Description.—Trail with imbricated structures. Cylindrical bumps which form the trail sometimes may be slightly separated.

Remarks.—Hallam (1970) interpreted *Imbrichnus* as

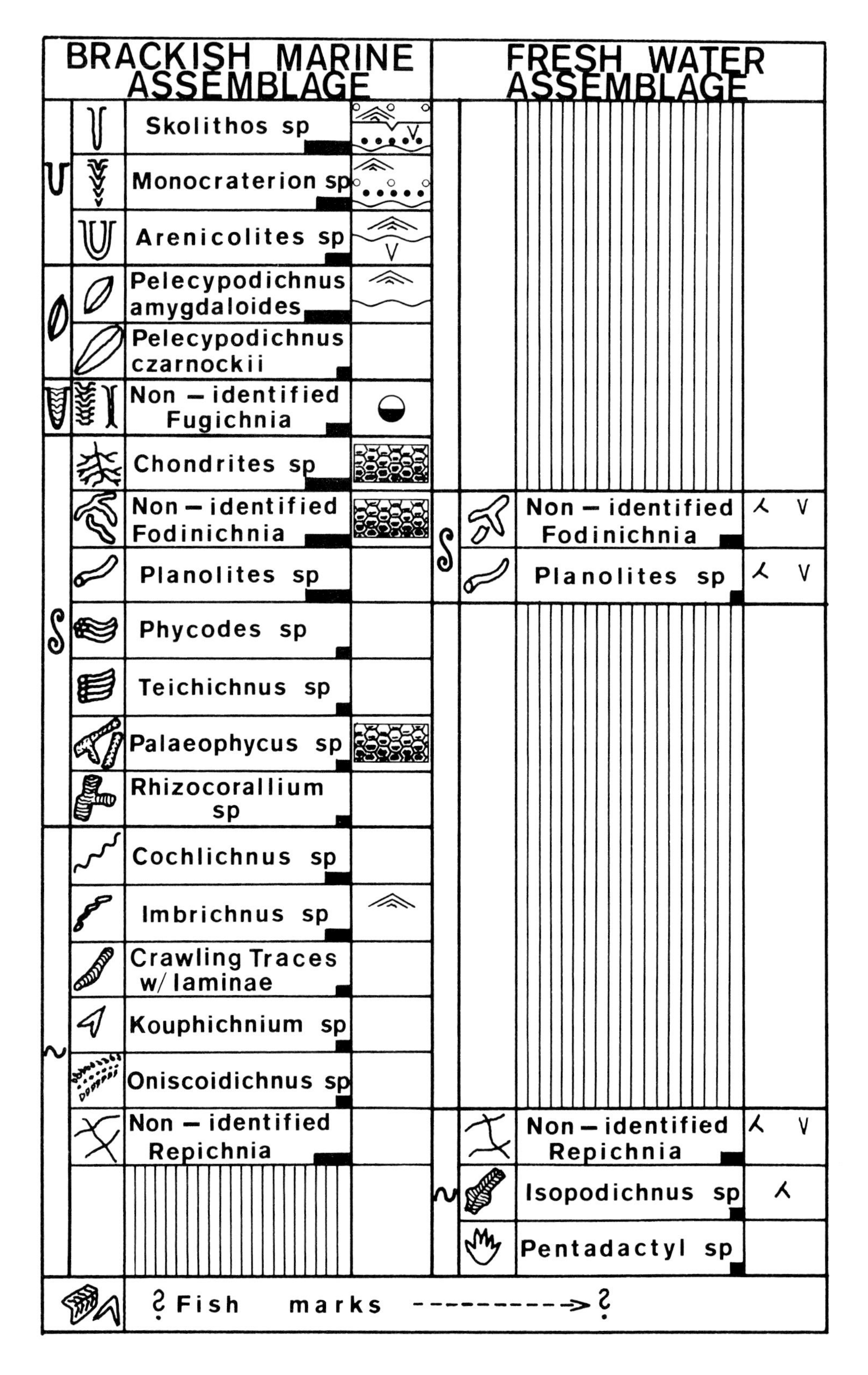
BRACKISH MARINE ASSEMBLAGE
FRESH WATER ASSEMBLAGE
Skolithos sp
Monocraterion sp
Arenicolites sp
Pelecypodichnus amygdaloides
Pelecypodichnus czarnockii
Non – identified Fugichnia
Chondrites sp
Non – identified Fodinichnia
Planolites sp
Phycodes sp
Teichichnus sp
Palaeophycus sp
Rhizocorallium sp
Cochlichnus sp
Imbrichnus sp
Crawling Traces w/ laminae
Kouphichnium sp
Oniscoidichnus sp
Non – identified Repichnia
Non – identified Fodinichnia
Planolites sp
Non – identified Repichnia
Isopodichnus sp
Pentadactyl sp
? Fish marks ----------> ?

Fig. 5.—List of Lower Liassic trace fossils from the northern slope of the Holy Cross Mountains. See Figure 7 for explanation of symbols of associated sedimentary structures. Beneath the trace fossil names, an index of their relative abundance in given. Length of the black bar is proportional to abundance of a given taxon: shortest bar denotes sporadic forms, double length = rare forms, three times longer = abundant forms, four times longer = very abundant forms.

a bivalve locomotion trail. Joint occurrence of *Imbrichnus* and *Pelecypodichnus* (Pl. 2F) would seem to confirm such an origin.

ISOPODICHNUS Bornemann, 1889
Isopodichnus sp.
Pl. 2H

Description.—Bilobate trail with transverse ornamentation.

Remarks.—*Isopodichnus* is known from non-marine deposits of Greenland (Bromley and Asgaard, 1972), Australia (Glaessner, 1957), and Scotland (Trewin, 1976). This type of trail is believed to have been made by an arthropod (Glaessner, 1957; Trewin, 1976) or by notostracans (Bromley and Asgaard, 1972).

KOUPHICHNIUM Nopcsa, 1923
Kouphichnium sp.
Pl. 2E

Remarks.—In this instance only a V-shaped fragment of the trail has been found. The trackway is believed to have been formed by a horseshoe crab or a similar arthropod (Malz, 1964; Winicierz, 1973).

ONISCOIDICHNUS Brady, 1949
Oniscoidichnus sp.
(not figured)

Description.—This trackway consists of two parallel rows of densely arranged, small wrinkles about 1 cm in diameter. Along the axis of the trackway, there is an additional row of wrinkles. The trackway is gently curved.

Remarks.—Brady (1949) compared *Oniscoidichnus* to those trackways formed by the modern isopod *Oniscus*.

Crawling Traces with Concentric Laminae
(not figured)

Description.—Straight or slightly curved, horizontal tubes with irregular, almost spreite-like structure.

Reptile Footprints—Pentadactyls
(not figured)

Remarks.—Karaszewski (1975a, 1975b) described two forms of reptile footprints from the lowermost part of the Ostrowiec Formation (see Figs. 2, 6).

Feeding Structures (Fodinichnia)
CHONDRITES Sternberg, 1883
Chondrites sp.
Pl. 2A

Description.—Systems of branching, dendritic tunnels, sometimes crossing each other.

Remarks.—Phobotaxis in *Chondrites* systems (Simpson, 1957) was questioned by Osgood (1970). In the Polish Liassic material, the author found many *Chondrites* systems where individual tunnels overlapped and crossed each other (Pl. 2A). These forms resemble those illustrated by Osgood (1970, Pl. 81, fig. 4) and designated *Chondrites* Type "C". The more regular forms (Pl. 2D) resemble Osgood's *Chondrites* Type "A" (1970, p. 355, fig. 10).

PALAEOPHYCUS Hall, 1847
Palaeophycus sp.
Pl. 2C

Description.—Cylindrical or subcylindrical, horizontally oriented burrows, stuffed with sediment. The burrows are straight or slightly curved, sometimes branching and covered with dense, irregular wrinkles. Along the axis of the burrow, additional parallel tubes can occur. The sediment fill is coarser than the surrounding matrix.

Remarks.—*Palaeophycus* has been interpreted as a crawling trace by Osgood (1970), but the bifurcations and coarser sediment filling of the burrows indicates that these trace fossils are more appropriately designated as feeding structures. These feeding structures from the Polish Liassic seem to be associated with horizons rich in organic matter. Such horizons may have served as a food source attractive to the tracemaker organisms. *Palaeophycus* has been recorded from just one location, namely the prodelta sediments of the Podole outcrop (Fig. 4).

PHYCODES Richter, 1850
Phycodes sp.
(not figured)

Description.—Bundles of closely packed tubes, filled with coarse material.

PLANOLITES Nicholson, 1873
Planolites sp.
Pl. 2J

Description.—Cylindrical burrows, variably oriented to the bedding plane and filled with structureless sediment. Diameter ranges from 3 to 5 mm.

Remarks.—*Planolites* has been interpreted as a facies-independent burrow by Crimes (1970). In the Polish Liassic sequence, *Planolites* occurs in both continental and brackish marine environments.

RHIZOCORALLIUM Zenker, 1836
Rhizocorallium sp.
(not figured)

Description.—Horizontally oriented, spreiten burrows; the burrows may ramify.

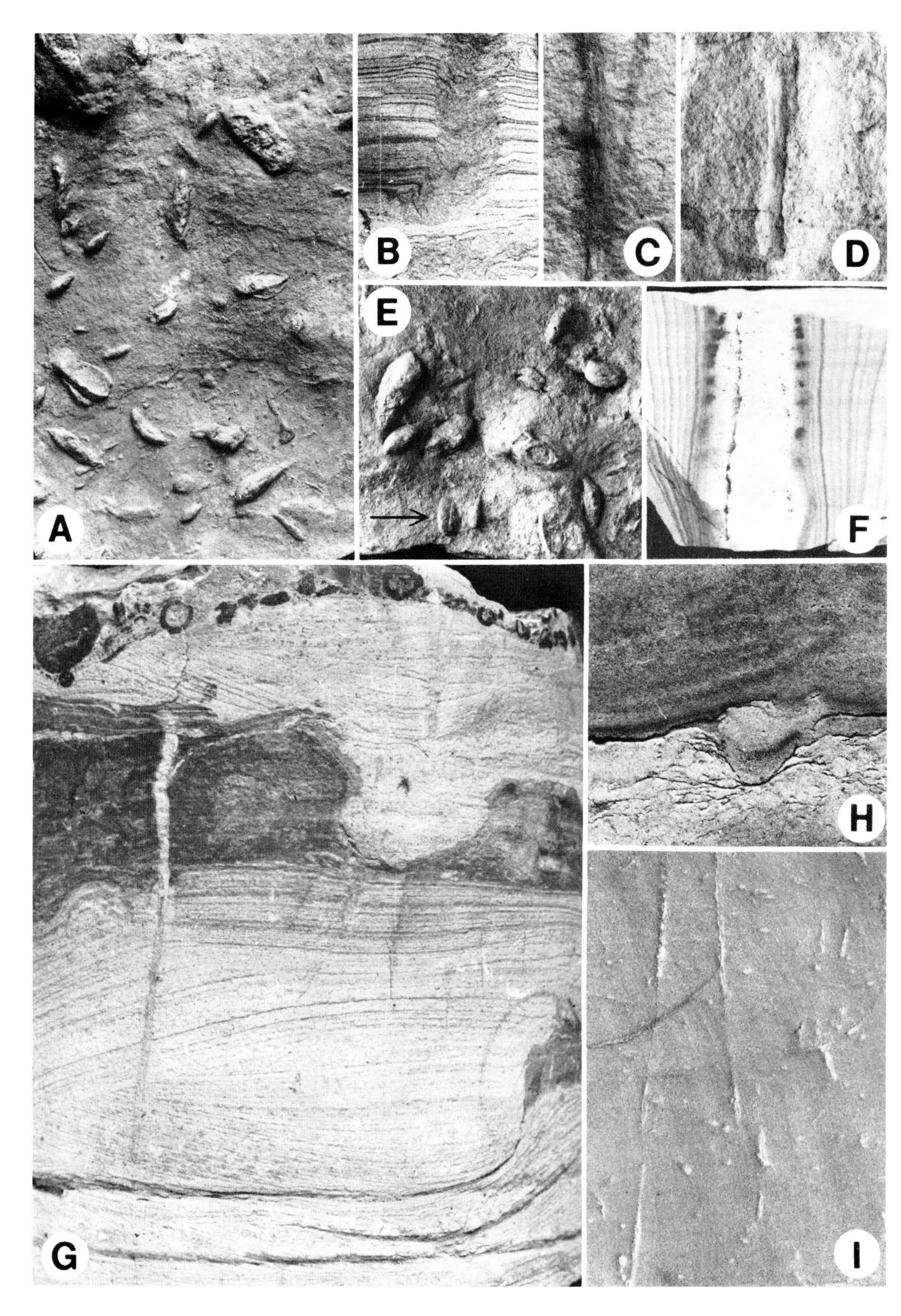
A
B
C
D
E
F
G
H
I

EXPLANATION OF PLATE 1

FIG. A.— *Pelecypodichnus amygdaloides*, Ruszkowice outcrop, Skłoby Formation. Sole, X0.5.
B.— Escape structure of a bivalve (Fugichnia), downward-oriented deflections of lamination are visible, Pogroszyn-1 borehole, Skłoby Formation. Vertical section, natural size.
C.— *Monocraterion* sp., upper, conical part of the burrow, Szwarszowice outcrop, Skłoby Formation. Vertical section, natural size.
D.— *Skolithos* sp., Szwarszowice outcrop, Skłoby Formation, Vertical section, natural size.
E.— *Pelecypodichnus amygdaloides*, Ruszkowice outcrop, Skłoby Formation. Some morphological varieties of this trace fossil can be observed, including bisymmetrical, coffee bean-shaped forms (arrow). Sole, X0.5.
F.— *Skolithos* sp., Mirzec outcrop, Skłoby Formation. The burrow is surrounded by Liesegang's concentric rings. Vertical section, natural size.
G.— Surface of a fragment of the core from Mroczków borehole, Przysucha Formation. Two erosional surfaces with small-scale channels, as well as *Skolithos* can be observed. Fragments of mineralized Skolithos tubes are included in the intraformational conglomerate. Natural size.
H.— Cross-section of *Pelecypodichnus* situated on an erosional surface. Gąsiorów borehole, Skłoby Formation. Natural size.
I.— *Arenicolites* sp., Mroczków borehole, Ostrowiec Formation. Vertical section, X0.5.

TEICHICHNUS Seilacher, 1955
Teichichnus sp.
(not figured)

Description.—Retrusive, vertical system of horizontal or slightly inclined tubes, with each tube touching the overlying and underlying tubes.

Marks made by fishes
Pl. 2G, I

Remarks.—In the lowermost part of the Ostrowiec Formation at Gliniany Las (Figs. 1, 6), above the sandstone layer with *Monocraterion* and below the sandstone-mudstone complex with reptile footprints (Karaszewski, 1975a, 1975b), there are some peculiar sole marks associated with sandstone interbeds. Two morphological types are present: herringbone-like marks and horseshoe-like marks. Gentle, fan-like structures occur around these marks (Pl. 2G). The sole of the sandstone bed shows a profusion of crawling traces. In vertical-longitudinal cross section, the herringbone-shaped and horseshoe-shaped marks have, in most cases, asymmetric profiles; the wider end of the mark merges gently with the sole of the sandstone bed, while the opposite end plunges into the bed. Herringbone-shaped marks show some morphological variety in planar view. There are regular marks with distinct axes and irregular marks without axes (Pl. 2G, a and c). Intermediate types also are present.

The marks in question permit speculation as to their origin. The herringbone-like marks probably were made by fish brushing against the bottom. It should be noted that some modern freshwater fish make shallow gouges on the bottom of ponds while searching for food (Dzułyński and Kinle, 1956). The pectoral fins or the gill cover of the fish could produce the herringbone-like marks during the back and forth motion of the fish digging into the bottom sediment. The horseshoe-shaped marks also may be ascribed to the gill cover, as it is similar in shape (Pl. 2I). Dzułyński and Kline (1956) have described markings from the Carpathian Flysch which have a probable fish-maker origin. These traces show a similar asymmetrical profile but lack ornamentation. Transverse ornamentation has been noted by Shäfer (1972) on some fish burrows. The gentle, fan-shaped structures also found here could result from water motion caused by fish hovering just above the bottom. The discovery by Maslankiewicz (1965) of the fish fossils *Pholidophorus angustus* (Agassiz) and *Semionotus* cf. *S. bergeri* (Agassiz) in the Przysucha Formation tends to support this explanation.

DISTRIBUTION OF THE TRACE FOSSILS ACCORDING TO DEPOSITIONAL ENVIRONMENTS

The depositional environments of the Holy Cross Lower Liassic sequence were reconstructed using patterns of cyclicity of sedimentation, analysis of physical sedimentary structures, and mineralogy and boron content, and the occurrence of trace fossils, rootlets, and coal and siderite (Fig. 3). Analysis of the trace fossil assemblages and their distribution proved highly useful in interpretation of the paleoenvironments. Distribution of the trace fossils in the different sedimentary environments is protrayed in Figures 5, 6, and 7.

Continental Deposits

In the continental deposits that occur in the lowermost part of the Zagaje Formation, as well as in the lowermost part of the Ostrowiec Formation and in some parts of the ore-bearing Przysucha Formation, trace fossils are sporadic, and the assemblage is not very diverse (Fig. 5). The assemblage consists of only some feeding structures (*Planolites* and unidentified feeding burrows) and a few crawling traces (*Isopodichnus* and unidentified locomotion trails), as well as reptile footprints. No clear pattern of relationships between sedimentary cycles, depositional environments, and trace fossil occurrence can be developed

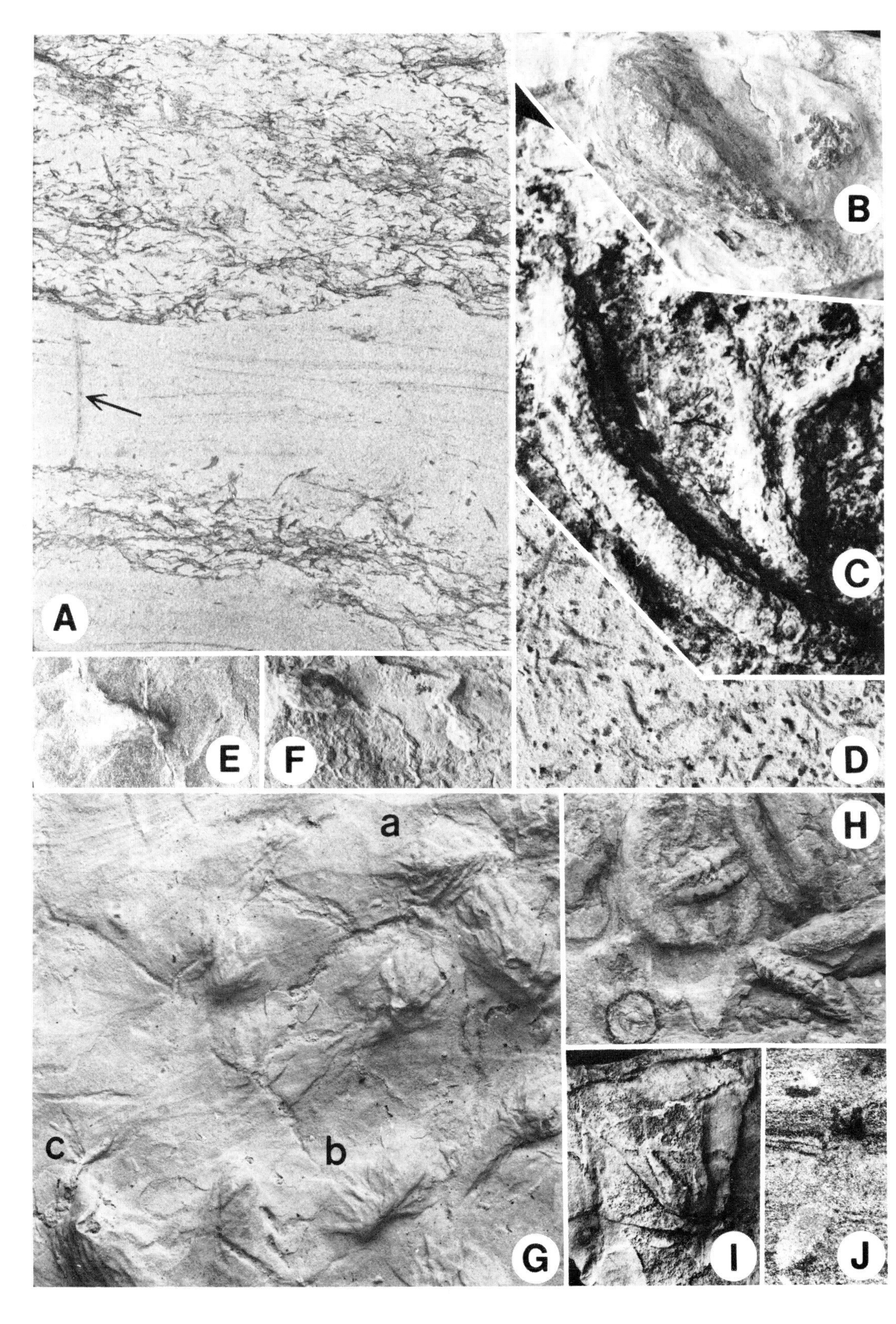
A
B
C
D
E
F
G
H
I
J
a
b
c

EXPLANATION OF PLATE 2

FIG. A.— Cross-section of core from the Pogroszyn 1 borehole, Skłoby Formation. Two horizons with abundant *Chondrites* are visible. Arrow points to vertical escape structure (Fugichnia) in cross-bedded sandstone. Vertical section, natural size.
B.— *Pelecypodichnus czarnockii*, Podole outcrop, Skłoby Formation. Sole, natural size.
C.— *Palaeophycus* sp., Podole outcrop, Skłoby Formation. Horizontal section, natural size.
D.— *Chondrites* sp., Zawada borehole, Skłoby Formation. Horizontal section, natural size.
E.— *Kouphichnium* sp., Pogroszyn-1 borehole, Skłoby Formation. Vertical section, natural size.
F.— *Imbrichnus* sp. associated with *Pelecypodichnus* sp., Szwarszowice outcrop, Skłoby Formation. Sole, X0.5.
G.— Marks made by fishes. *a*, Herringbone-like type; *b*, intermediate type; and *c*, irregular type. Between these marks, crawling traces are visible, as well as fan-like structures (upper left corner). Gliniany Las outcrop, lower part of the Ostrowiec Formation. Sole, X0.5.
H.— *Isopodichnus* sp.; in the lower left corner a cylindric section of a plant root is visible. Gleboka Droga 1 borehole, Zagaje Formation. Sole, natural size.
I.— Fish mark, horseshoe-like type. Gliniany Las outcrop, lower part of the Ostrowiec Formation. Sole, X0.5.
J.— *Planolites* sp. Huta borehole, Zagaje Formation. Vertical section, natural size.

owing to the general paucity of the ichnofauna. Levee deposits and the distal parts of crevasse splays appear to be the more favorable environments for the formation and preservation of these uncommon forms. Almost no trace fossils occurred in both coarse-grained channel deposits and very fine-grained fluvial plain (and/or lacustrine) deposits.

Brackish Water Deposits

With the beginning of marine transgression, represented by the base of the Skłoby Formation (Figs. 2, 3), trace fossils become much more numerous and well defined. All of the ethological categories, except for grazing traces are represented. Here, trace fossils are useful in differentiating particular sedimentary environments or cycles (Fig. 7). In the Skłoby Formation, Przysucha Formation, and the main part of the Ostrowiec Formation, three types of coarsening upward cycles dominate; namely shore-prograding, barrier-lagoon, and deltaic cycles. The total thickness of these cycles, derived from bore hole data, is 1,496 meters. Each cycle has been divided into two parts: the fine-grained lower part and the coarse-grained upper part. In all, 1,880 occurrences of trace fossils have been recorded; the data are summarized in Figure 3. Abundances for individual ethological categories were determined with relation to the total thicknesses of particular parts of the cycles; these are recorded in Figure 7.

Some significant patterns of trace fossil distribution can be observed. In the fine-grained parts of all three types of cycles, feeding structures dominate. Indeed, in the lagoonal deposits, that domination is absolute. A scarcity of oxygen in waters of the lagoonal environment as indicated by the precipitation of siderite on the lagoonal bottom (Wyrwicki, 1966) was lethal to suspension feeders, which would have constructed primarily dwelling burrows. Resting traces also are absent. In this setting, such traces would have been produced by bivalves, which require highly oxygenated waters (Rhoads, 1975). Thus only feeding structures formed by the more tolerant deposit feeders are present in the lagoonal facies.

In the deltaic, fine-grained deposits, the occurrence of feeding structures again is relatively high. The scarcity of dwelling structures and resting traces in these deposits is probably caused by the higher turbidity of waters near the river mouths. High turbidity is an inhibiting factor for suspension feeders (Rhoads and Young, 1970). However, in the fine-grained parts of the shore-prograding cycles, dwelling structures and resting traces become more abundant.

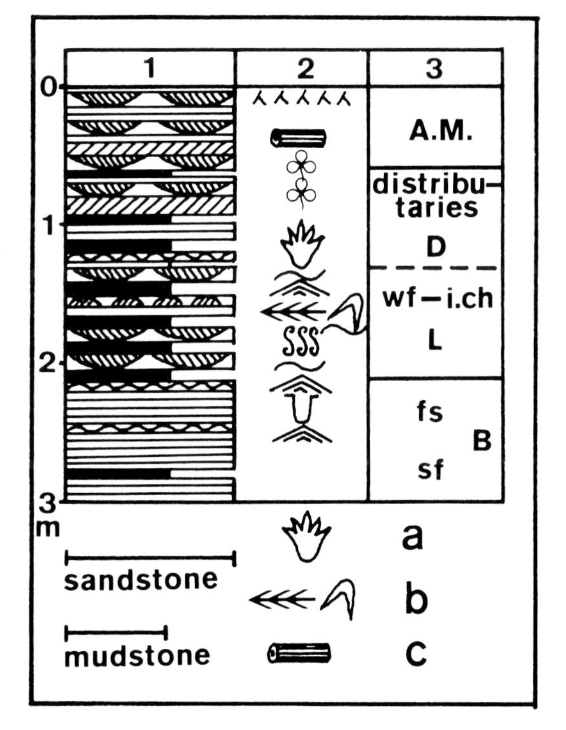

Fig. 6.—Profile of the Gliniany Las outcrop showing a regressive sequence with different trace fossils. Column 1 = lithologic and sedimentologic profile. Column 2 = trace fossils, sedimentary structures, and plant remains: a—reptile footprints, b—marks of fishes, c—large pieces of wood. For explanation of other symbols, see Figures 2 and 3.

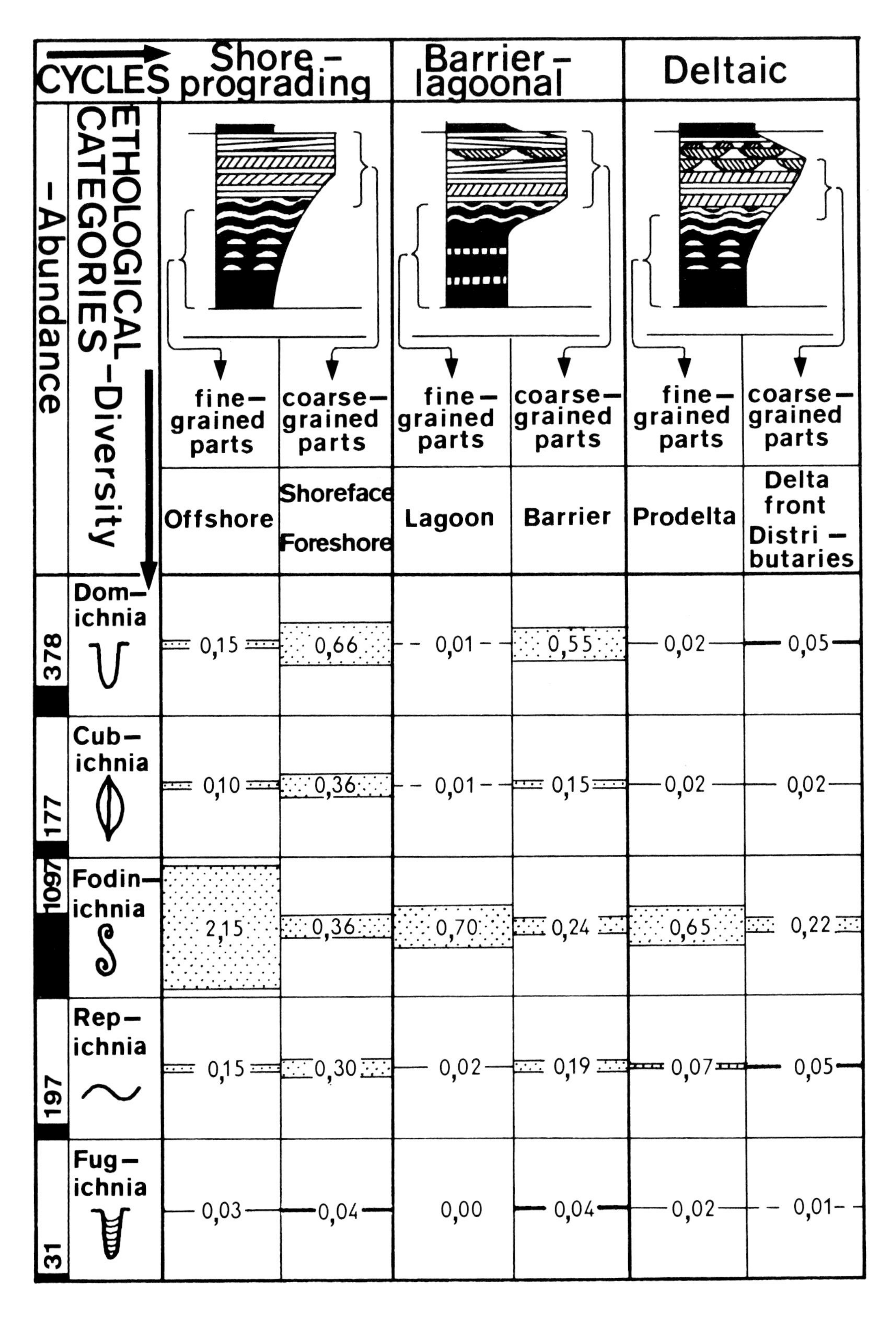
CYCLES
Shore - prograding
Barrier - lagoonal
Deltaic
ETHOLOGICAL CATEGORIES -Diversity
-Abundance
fine-grained parts
coarse-grained parts
Offshore
Shoreface
Foreshore
Lagoon
Barrier
Prodelta
Delta front
Distri-butaries
Dom-ichnia
378
0,15
0,66
0,01
0,55
0,02
0,05
Cub-ichnia
177
0,10
0,36
0,01
0,15
0,02
0,02
Fodin-ichnia
1097
2,15
0,36
0,70
0,24
0,65
0,22
Rep-ichnia
197
0,15
0,30
0,02
0,19
0,07
0,05
Fug-ichnia
31
0,03
0,04
0,00
0,04
0,02
0,01

Fig. 7.—Distribution of the Holy Cross Early Liassic trace fossils in different brackish water sedimentary environments. The frequency of occurrence of individual ethological categories has been determined in relation to the total thickness of each part of the given cycle, so that each number represents the number of occurrences for one meter. For further explanation of symbols, see Figure 3.

In the coarse-grained parts of these cycles, the similarity between the shore-prograding and barrier-lagoon sequences is characteristic. Dwelling structures and resting traces prevail, although in the uppermost parts of these cycles (consisting of higher energy foreshore, washover fan, and inlet channel deposits) the number of trace fossils rapidly decreases. Apparently these higher energy depositional environments were not favorable for the formation and preservation of trace fossils.

In the coarse-grained parts of the deltaic cycles (delta-front and distributary deposits), dwelling structures and resting traces are extremely scarce, again probably owing to higher turbidity of the water. Delta-plain deposits here are designated as continental, since their ichnologic characteristics are analogous to those of the continental trace fossil assemblages described earlier.

The use of trace fossils in differentiating shore-prograding, barrier-lagoon, and deltaic sequences can be summarized as follows. Beds representing shore-prograding conditions differ from the barrier-lagoonal deposits by the trace fossil assemblages characteristic of the lower, fine-grained parts of the cycles; namely, domination of the barrier-lagoonal sediments by feeding structures and a great diversity of trace fossil types in the shore-prograding sediments, including feeding and dwelling structures and resting traces. In differentiating deltaic sequences from the shore-prograding and barrier-lagoonal sequences, the coarse-grained parts of the cycles are important. Here, scarcity of dwelling structures and resting traces is diagnostic for the deltaic sequences.

Crawling traces are most abundant in sediments representing shore-prograding conditions and in coarse-grained parts of the barrier-lagoonal cycles. Thus their distribution pattern is similar to that for the dwelling structures and resting traces. The occurrence of escape structures was too spotty for use in such interpretations.

RELATIONSHIP BETWEEN TRACE FOSSILS, SMALL-SCALE ENERGY CHANGES, AND RATES OF SEDIMENTATION

In the Polish Liassic, dwelling structures often occur in beds with oscillation ripples and erosional structures (Fig. 4). Sometimes mineralized pipes of *Skolithos* were removed and redeposited with mud clasts to form an intraformational conglomerate (Pl. 1G). Such associations are indicative of higher energy, nearshore conditions.

Feeding structures, resting traces, and escape structures often are closely associated with small-scale changes in the rate of sedimentation. During stages of slow sedimentation, dense networks of feeding structures were developed (Pl. 2A). During periods of more rapid sedimentation, vertical escape structures, very similar to those described by Reineck (1958) for polychaetes, were developed.

In many cases, the joint occurrence of *Pelecypodichnus amygdaloides* and escape structures was observed. These escape structures (Pl. 1B) bear a resemblance to *Monocraterion* burrows, but they are more irregular. The author thinks they are bivalve escape structures and interprets their occurrence as corresponding with periods of rapid sedimentation, such as would occur in storm events. By contrast, stable and relatively slow rates of sedimentation enabled the development of many different types of trace fossils, which in turn obliterated most primary sedimentary structures.

CONCLUSIONS

Early Liassic trace fossil assemblages from the Holy Cross Mountains region in Poland show that in continental deposits trace fossils are scarce and the assemblage is not diverse; feeding structures (*Planolites*) and crawling traces (*Isopodichnus*) are the most common forms. These trace fossils occur rarely in deposits of mixed lithology (alternating mud and sand layers) that were deposited mainly in levees or in the distal parts of crevasse splays.

The trace fossil assemblage from low energy, brackish-marine basin deposits is much more diverse, with individual trace fossils more numerous; although in comparison to some other full marine ichnocoenoses, the Holy Cross assemblage is impoverished. The majority of specimens recorded are reduced in size, resting traces are not diverse, and spreiten burrows are very rare. The occurrence of dwelling structures and resting traces in the fine-grained parts of the coarsening upward cycles permits recognition of beds representing shore-prograding conditions. The absence of dwelling structures and resting traces in the coarse-grained parts of the cycles is helpful in defining the deltaic sequences. These criteria are particularly useful when examining bore hole materials.

In this setting, *Pelecypodichnus* and *Chondrites* have been found only in the brackish water deposits, and their occurrence here is of diagnostic importance. The brackish water trace fossil assemblage from the Holy Cross Lower Liassic resembles assemblages described by Vossmerbäumer (1970) from Early Jurassic (Hettangian) beds of Sweden and by Hallam (1970) from Middle Jurasic strata near Dorset, England.

ACKNOWLEDGMENTS

This paper is part of a doctoral dissertation submitted to the Department of Geology, Warsaw University. The full assistance of my major advisor, Professor Piotr Roniewicz is gratefully acknowledged. I also thank Zbigniew Kozydra, Geological Institute of Warsaw, for facilitating access to the bore hole materials, and Anna Pasieczna, my colleague at the Geological Institute, for determination of the clay mineralogy and boron content of the sediments. Wladyslaw Karaszewski aided with fruitful discussions of questions in the beginning stages of my work, and Professor Adolf Seilacher provided helpful discussion on matters related to trace fossil taxonomy.

Krzysztof Jaworowski and H. Allen Curran reviewed the manuscript and provided helpful criticism. Curran, Julie Sandeen, and Carolyn Estes of Smith College aided with revisions of the manuscript. Krystyna Ilska provided technical assistance with drafting of the figures.

REFERENCES

BROMLEY, R.G. and ASGAARD, U., 1972, Notes on Greenland trace fossils I. Fresh-water *Cruziana* from the Upper Triassic of Jameson Land, East Greenland: Grønlands Geol. Unders., v. 49, p. 7–13.

CRIMES, T.P., 1970, The significance of trace fossils in sedimentology, stratigraphy and palaeoecology with examples from Lower Palaeozoic strata, *in* Crimes T.P. and Harper J.C. (eds.), Trace Fossils, Geological Journal Special Issue 3, Liverpool, Seel House Press, p. 106–126.

DADLEZ, R., 1978, Stan litostratygrafii epikontynentalnej dolnej jury w Polsce i propozycje jej usystematyzowania: Kwartalnik Geologiczny, v. 22, p. 773–790.

DZUŁYŃSKI, S. and KINLE J., 1956, Problematic hieroglyphs of probable organic origin from the Beloveza beds (Western Carpathians): Ann. de la Soc. Géol. de Pol., v. XXVI, p. 265–272.

GLAESSNER, M.F., 1957, Paleozoic arthropod trails from Australia: Palaeontology, v. 31, p. 103–109.

HAKES, W.G., 1977, Trace fossils in Late Pennsylvanian cyclothems, Kansas, *in* Crimes T.P. and Harper J.C. (eds.), Trace Fossils 2: Geological Journal Special Issue 9, Liverpool, Seel House Press, p. 209–226.

HALLAM, A., 1970, *Gyrochorte* and other trace fossils in the Forest Marble (Bathonian) of Dorset, England, *in* Crimes T.P. and Harper J.C. (eds.), Trace Fossils: Geological Journal Special Issue 3, Liverpool, Seel House Press, p. 189–200.

______, and SWETT, K., 1966, Trace fossils from the Lower Cambrian Pipe Rock of the north-west Highlands: Scottish Journal of Geology, v. 2, p. 101–106.

JURKIEWICZ, I., 1967, The Lias of the western part of the Mesozoic Zone surrounding the Świętokrzyskie (Holy Cross) Mountains and its correlation with the Lias of the Cracow-Wieluń Range: Biuletyn Instytutu Geologicznego, v. 200, p. 5–132.

KARASZEWSKI, W., 1962, The stratigraphy of the Lias in the northern Mesosoic Zone surrounding the Świętokrzyskie (Holy Cross) Mountains, central Poland: Insytut Geologiczny, Prace, v. 30, p. 400–416.

______, 1971, Some fossil traces from the Lower Liassic of the Holy Cross Mountains, Central Poland: Bull. Acad. Polon. Sci., Sér. Sci. de la Terre, v. 19, p. 101–105.

______, 1974, A new trace fossil from the Lower Jurassic of the Holy Cross Mountains: Bull. Acad. Polon. Sci., Sér. Sci. de la Terre, v. 22, p. 157–160.

______, 1975a, O unikalnym odcisku łapy gada jurajskiego w dolnym liasie świętokrzyskim: Przeglad Geologiczny, v. 7, p. 358–360.

______, 1975b, Footprints of *Pentadactyl Dinosaurus* in the Lower Jurassic of Poland: Bull. Acad. Polon. Sci., Sér. Sci. de la Terre, v. 23, p. 135–136.

______, and KOPIK, J., 1970, Lower Jurassic, *in* Stratigraphy of Mesosoic of the Holy Cross Mountains: Instytut Geologiczny, Prace, v. 56., p. 65–98.

Malz, H., 1964, *Kouphichnium walchi*, die Geschichte einer Fährte und Tieres: Natur und Museum, v. 94, p. 81–104.

MAŚLANKIEWICZ, Z., 1965, *Semionotus* cf. *bergeri* Agassiz from the Lias of the Holy Cross Mountains, Poland: Acta Palaeontologica Pol., v. 10, p. 57–68.

OSGOOD, R.G., JR., 1970, Trace fossils of the Cincinnati area: Palaeontographica Americana, v. VI, no. 41, p. 1–444.

PIEŃKOWSKI, G., 1981, Sedymentologia dolnego liasu północnego obrzezaenia Gór Świętokrzyskich (Sedimentology of the Lower Liassic of the Northern Slope of the Holy Cross Mountains) [unpub. Ph.D. thesis]: Warsaw, Library of the Geology Department, Warsaw University, 146 p.

REINECK, H.E., 1958, Wühlbau Gefüge in Abhängigkeit von Sediment-Umlagerungen: Seckenbergiana Lethaea, v. 39, p. 1–24.

RHOADS, D.C., 1975, The palaeoecological and environmental significance of trace fossils, *in* Frey R.W. (ed.), The Study of Trace Fossils, New York, Springer-Verlag, p. 147–160.

______, and YOUNG, D.K., 1970, The influence of deposit-feeding benthos on bottom sediment stability and community trophic structure: Jour. Marine Res., v. 28, p. 150–178.

SHÄFER, W., 1972, Ecology and palaeoecology of marine environments, Craig, E., ed., Edinburgh and Chicago, Oliver and Boyd and University of Chicago Press, 568 p.

SEILACHER, A., 1953, Studien zur Palichnologie, II. Die Fossilen Ruhespuren (Cubichnia): Neues Jb. Geol. Palänt. Abh., v. 98, p. 87–124.

______, 1967, Bathymetry of trace fossils: Marine Geology, v. 5, p. 413–428.

SIMPSON, S., 1957, On the trace-fossil *Chondrites*: Quart. Jour. Geol. Soc. London, v. 112, p. 475–499.

______, 1975, Classification of trace fossils, *in* Frey R.W. (ed.), The Study of Trace Fossils, New York, Springer-Verlag, p. 39–54.

TREWIN, N.H., 1976, *Isopodichnus* in a trace fossil assemblage from the Old Red Sandstone: Lethaia, v. 9, p. 29–37.

TROEDSSON, G., 1951, On the Höganös Series of Sweden: Lunds Universitets Arsskrift., v. 47, p. 1–269.

VOSSMERBÄUMER, H., 1970, Untersuchungen zur Bildungsgeschichte des Unteren Lias in Schonen (Schweden): Geologica et Palaeontologica, v. 4, p. 167–193.

WINCIERZ, J., 1973, Küstensedimente und Ichnofauna aus dem oberen Hettangium von Mackendorf (Niedersachsen): Neues Jb. Geol. Pälant. Abh., v. 144, p. 104–141.

WYRWICKI, R., 1966, Osady zelaziste liasu świętokrzyskiego: Biuletyn Instytutu Geologicznego, v. 195, p. 71–158.

TRACE FOSSIL ASSEMBLAGES IN MIDDLE TRIASSIC MARGINAL MARINE DEPOSITS, EASTERN BORDER OF THE MASSIF CENTRAL, FRANCE

GEORGES R. DEMATHIEU
Institut des Sciences de la Terre, Université de Dijon, 21100 Dijon, France

ABSTRACT

An integrated study of trace fossil assemblages from Middle Triassic deposits of the eastern border of the French Massif Central, including analysis of primary sedimentary structures, plant imprints, invertebrate lebensspuren, and vertebrate trackways, has enabled reconstruction of the depositional environments. In this study, statistical methods are used to support morphological observations, permitting better definition of the ichnotaxa formed by reptilian vertebrates. In addition, by applying other statistical methods and the concept of "Euclidian distance," it is possible to estimate population sizes for the trackmakers. Using these data, a reconstruction of the biological relationships between the different trackmakers represented by tracks in the trackway areas and between the trackmakers and their environment is proposed.

Lithology, primary sedimentary structures, and lebensspuren indicate that the trackway areas of this Triassic borderland were large sandy shores at the edge of a sea or lagoon. The succession of footprint-bearing beds in the lower sandstones, or "Grès inférieurs du Lyonnais," suggests cycles of deposition due to repeated marine transgressions.

INTRODUCTION

The first vertebrate footprints (two slabs) discovered in the Triassic of the eastern border of the French Massif Central were described by Lortet (1892). However, studies of these deposits were neglected, and they were only briefly mentioned in general works describing the area (Roman, 1926, 1950; Ricour, 1962).

The discoveries in this area of new material containing abundant footprints began in 1960. These trace fossils were subsequently studied and described by Demathieu (1970). At the same time, lithologic, paleogeographic, and stratigraphic investigations of these rocks were conducted by Courel (1973).

The purpose of this paper is to complete and summarize the previous studies concerning the paleoenvironments in which the footprint-bearing beds were formed (Demathieu, 1970; Courel and Demathieu, 1973; Demathieu and Haubold, 1978; Courel et al., 1979). This study reconstructs the interrelationships between the paleoenvironments, the populations of trace-maker animals, and their surroundings. The setting described herein is specific to the outcrops of the study area. However, with further study, it may be possible to extend the findings to other similar Triassic sequences, for example, the German Buntsandstein.

THE GEOLOGIC SETTING

Stratigraphy

Rocks of Triassic age crop out in a narrow zone about 300 km in length along the eastern border of the Massif Central of France (Fig. 1). The outcrop belt is located in the Saône-Rhône Valley and extends from Dijon (Côte d'Or) to Largentière (Ardèche). The Triassic sequence is not thick (average 60 to 100 m), but it does thicken to about 200 m in the south. Following the work of Courel (1973), the Triassic sequence can be subdivided into five units (Fig. 2). These units are described briefly below, basal unit first.

Unit 1—"Grès inférieurs du Lyonnais"—Beds of quartz sandstone, very coarse-grained at the bottom and becoming more fine-grained (< 0.1 mm) toward the top; total thickness of 7 to 30 m, massively bedded, with individual beds 0.8 to 2 m thick. The upper part of the unit consists of much thinner but discrete beds 0.01 to 0.20 m thick. These beds exhibit distinct graded bedding. Trace fossils are well-preserved on bedding planes and are most numerous in the upper part of the unit. The trace fossil assemblage includes traces formed both by invertebrates and vertebrates, with the latter being the most common. Plant imprints also occur, but they are relatively rare.

Fig. 1.—Location of Middle Triassic deposits along the eastern border of the Massif Central, France.

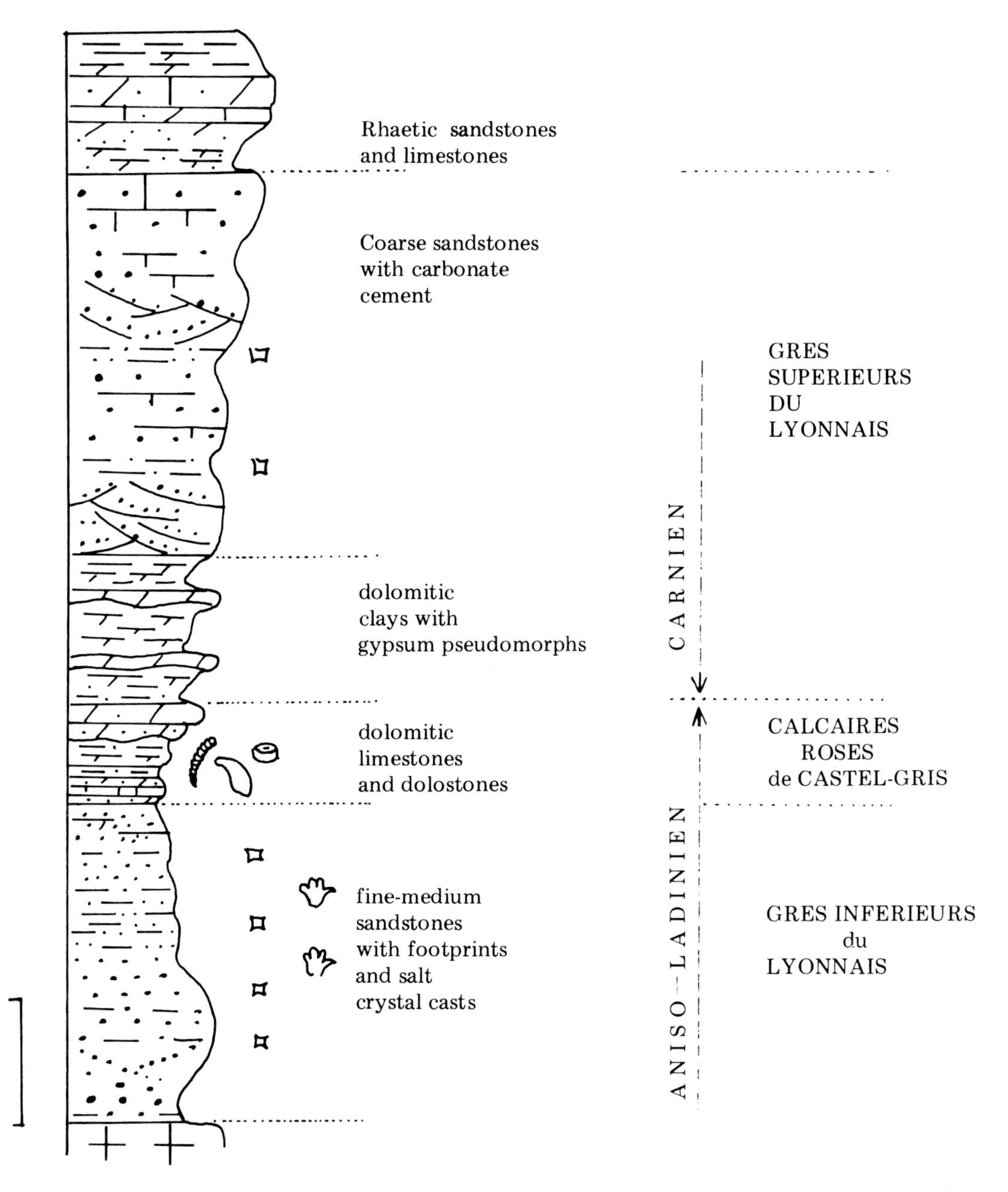

Fig. 2.—Generalized stratigraphic column for the Triassic rocks of the Massif Central region, France. Bar scale = 10 m. (after L. Courel)

Unit 2—"Calcaires roses de Castel gris"—Beds of pink, dolomitic limestone; total thickness of 10 to 12 m. This unit contains some body fossils, including bryozoans, linguloid brachiopods, mollusks, and fish teeth.

Unit 3—Thin-bedded, unfossiliferous, dolomitic clay shales—total thickness of 8 to 10 m. Gypsum pseudomorphs are abundant.

Unit 4—"Grès supérieurs du Lyonnais"—Very coarse-grained, dolomitic, quartz sandstone; total thickness of 25 to 30 m; non-fossiliferous.

Unit 5—Very thin bed of brown, fine-grained quartz sandstone; rarely exposed; total thickness of 15 to 20 cm near Lyon. This thin but lithologically distinct bed contains fish teeth and small fish bones (Roman, 1926).

Units 1 and 2 are of Middle Triassic (Late Anisian to Ladinian age; Courel et al., 1968; Demathieu, 1970; Courel, 1973). Units 3, 4, and 5 are of Late Triassic age (Unit 5 is Rhaetian; Courel and Demathieu, 1973). Trace fossils occur only in Unit 1, the formation "Grès inférieurs du Lyonnais", and it is this unit that will be considered exclusively herein.

Primary Sedimentary Structures

The primary sedimentary structures of Unit 1 have been described in detail by Courel et al. (1979). Such structures found in this unit are principally symmetric and asymmetric ripple marks, mud cracks, and salt crystal casts. Mud-crack casts and salt crystal casts can occur as convex epireliefs on the upper surfaces of slabs but the former normally are present as concave epireliefs. Courel et al. (1979) showed that the ripple marks formed under low energy conditions with a flow velocity less than 0.5 m/sec. Scour and tool marks also occur, but they are rare and their occurrence is localized.

Fossils

No body fossils have been recovered from Unit 1. However, the imprints of plants and a variety of trace fossils produced both by invertebrates and vertebrates are preserved on bedding plane surfaces in this unit. The plant fossils consist of molds of small fronds, without preserved organic matter. These molds are numerous locally but are, for the most part, poorly preserved. Most have been assigned by Grauvogel-Stamm (1977) to the genus *Voltzia* (Pl. 1I).

The trace fossils produced by invertebrates occur at the same stratigraphic levels as the vertebrate trace fossils. One often finds the trace fossils of both invertebrates and vertebrates on the same slab, but the former are not nearly as numerous as the latter. Ichnogenera present and attributed to the activity of invertebrates include *Cochlichnus*, *Isopodichnus*, *Planolites*, and *Skolithos* (Pl. 1A-H).

ANALYSIS OF VERTEBRATE FOOTPRINTS

Much information can be derived from the study of vertebrate footprints, contributing to our knowledge in the areas of stratigraphy (Courel et al., 1968; Demathieu, 1977b), paleoecology (Demathieu, 1970, 1975), and paleontology (Demathieu, 1970; Demathieu and Haubold, 1974, 1978). From the paleontologic standpoint, the footprints give some indication of the appendicular skeleton of the trackmakers and other aspects of the paleobiology of the vertebrates. It is usually difficult to match exactly an appendicular skeleton with a particular footprint because (1) the ichnospecies are more numerous than the known skeletons of pes and manus; (2) there are variations in size and divarication of digits of different skeletons so that no skeleton corresponds directly with the footprints; and (3) many of the reconstructed skeletons are incomplete or hypothetical.

Table 1 presents numerical data for the definitions of the main ichnogenera of Unit 1 of the Triassic borderland, and these ichnogenera are figured in Plate 2. The names used here are those of ichnogenera and are not the names of the animals which made the footprints (cf. Sarjeant and Kennedy, 1973; Basan, 1979).

Although it is not possible to correlate directly a given ichnospecies with its tracemaker species, the ichnologic research often permits one to assign the tracemakers to their higher taxons, such as family, suborder, or order. The ichnogenera are herein interpreted to represent the following reptilian groups:

Pseudosuchians: *Synaptichnium* (formed by *Ticinosuchus*, *Euparkeria*, or Aetosaurus)
Pseudosuchians: *Sphingopus* (formed by *Hesperosuchus* or *Ornithosuchus*)
Protosuchians or precrocodilians(?): *Brachychirotherium*
Coelurosaurians: *Anchisauripus* and *Coelurosaurichnus*, related to *Coelophysis* or *Halticosaurus*

The ichnogenera *Chirotherium* and *Isochirotherium* may represent a group of archosaurians (thecodonts) that were precursors to the dinosaurs (Demathieu and Haubold, 1978), but they do not represent pseudosuchians (sensu Krebs, 1976). The ichnogenera *Rhynchosauroides* and *Rotodactylus* probably were probably produced by small reptiles such as the lepidosaurians (the scales of inferior faces of the digits are rectangular and transverse), but this is not certain. It is possible that these ichnogenera could represent, pro parte, archosaurians.

The study of vertebrate footprints provides two useful sets of information (Peabody, 1948; Demathieu, 1970; Haubold, 1970). This information pertains to (1) gross morphology of the animal, including shape and arrangement of digits and the relationship between digits, claws, skin, and pads; and (2) size of the animal, including overall length and breadth, lengths of digits, and divarication between digits.

Descriptions and measurements of the morphological characters of the footprint trace fossils are relatively easy to make. However, it is necessary to determine the significance of the measurements because footprints are of variable size, even in a trackway. For this reason, the author has used statistical methods (Demathieu, 1970) to verify the homogeneity of footprint populations, including determination of the mean, standard deviation, variability (= standard deviation X 100/mean), confidence interval for the mean at the 5 percent level, and the Cramer homogeneity test (Van der Waerden, 1967, p. 238). The ratios between length characters also are compared, in order to reduce the influence of allometric growth, which is important for comparisons between two distinct species or between two species that cannot easily be differentiated (Courel and Demathieu, 1976).

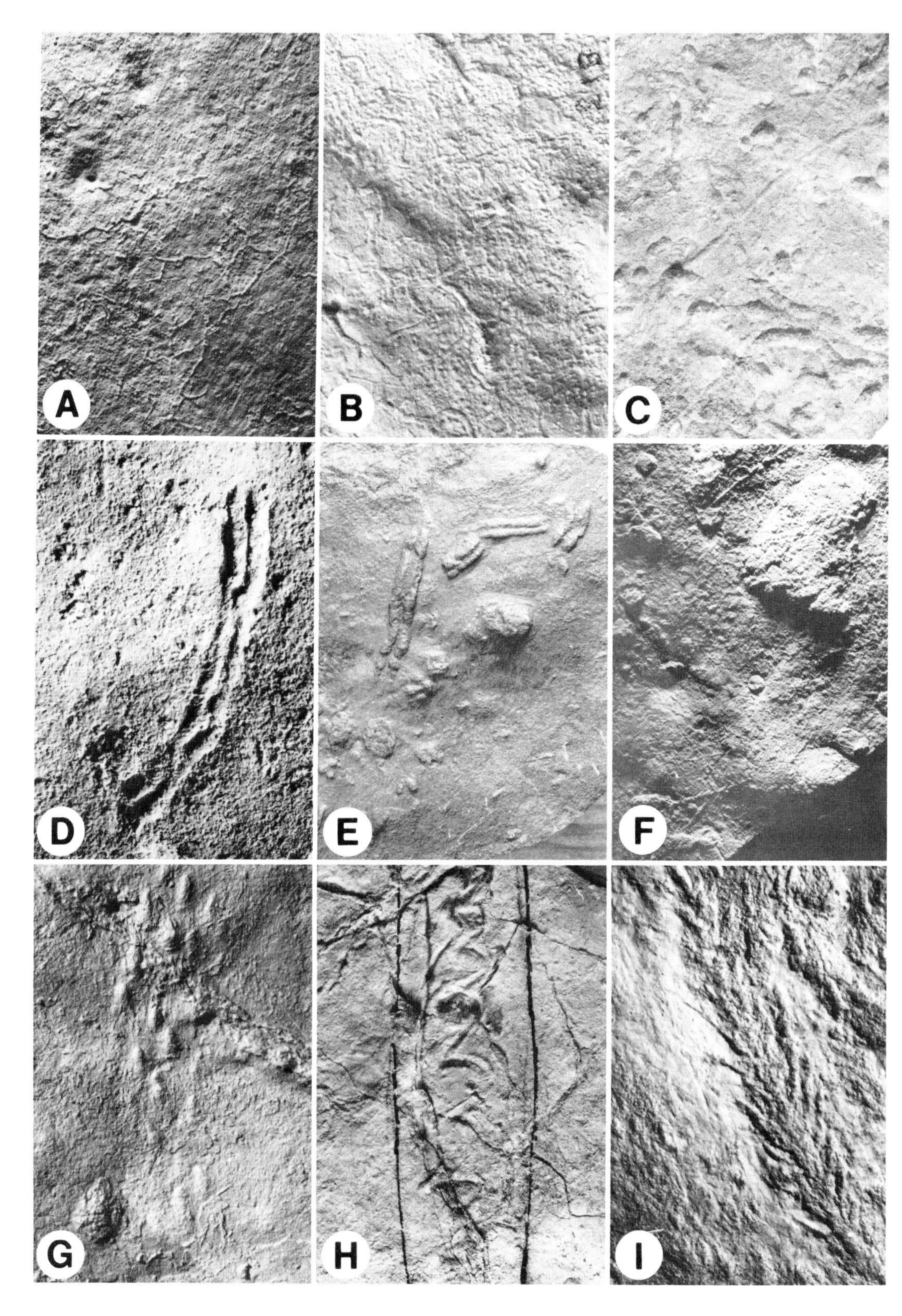
A
B
C
D
E
F
G
H
I

Footprint ichnospecies are defined not only on the basis of morphology, but also statistically. One might suppose that some differences in measurement would occur if measurements were made by different workers, possibly leading to significant error. However, in using measurements made by different colleagues, the author has found that the error introduced is not significant. Such measurements can be useful to all workers if the procedures of measurement are clearly defined.

Tables 2, 3, 4, and 5 show results concerning the ichnospecies *Brachychirotherium gallicum* Willruth, a particularly abundant form in the Triassic of Lodève (France). The relatively high variabilities reflect the influence of allometric growth of the trackmakers which seems to be important here (Table 3). But this parameter decreases when the ratios are considered (Table 4). The correlations between length characters of the footprints show that the true correlations which would be observed for the feet of living animals are well preserved by the footprints.

The parameters of the frequency distributions give some important information:

1) Variability is a coefficient that indicates whether or not the distribution has a wide spread. It depends on (a) the physical conditions and nature of the sediment, (b) the patterns of placement and retraction of the feet during locomotion, and (c) the size differences for the trackmakers.

In a trackway, the causes of the variability are points a and b above, and this coefficient is less than 10 percent. In the case of animals of different sizes, forming the same ichnospecies, variability can be over 20

TABLE 1.—SHORT DIAGNOSES OF THE MAIN FOOTPRINT ICHNOGENERA FROM THE GERMAN BUNTSANDSTEIN AND THE MIDDLE TRIASSIC OF FRANCE

Ichnogenera	Short Diagnoses
Rhynchosauroïdes Maidwell 1911	IV > III > II. Pentadactyl heteropod footprints; pes often tridactyl and digitigrad. Manus smaller than the pes and semi-plantigrad or plantigrad or digitigrad. Pes pace angulation 110°.
Rotodactylus Peabody 1948	III ÷ IV > II > I. Pes pentadactyl, digitigrad, I and V represented by their claws; the latter lies far behind the set of the first 4. Manus similar and smaller than the pes and behind it. Pes pace angulation 120–150°.
Synaptichnium Nopcsa 1923	III ÷ IV > II > I. Pes semi-plantigrad, pentadactyl; toe V straight behind and lateral to the set of the first 4; cross axis: 55–70°. Manus digitigrad, pentadactyl, smaller (⅓ to ½) than the pes. Pes pace angulation 100–150°.
Chirotherium Kaup 1835	III > IV ≥ II > I. Pes pentadactyl, semi-plantigrad; toe V short and curved toward the outside, behind the set of the first 4; cross axis 80–90°. Manus digitigrad, smaller than the pes (= ⅓). Pes pace angulation 165–180°.
Brachychirotherium Beurlen 1950	III > II ≥ IV > I. Pes pentadactyl, semi-plantigrad; reduced toe V; cross axis 60–85°. Manus digitigrad, smaller than the pes (⅓ to ½). Pes pace angulation 95–170°.
Isochirotherium Haubold 1970	III ÷ II > I ≥ IV. Pes pentadactyl, plantigrad to digitigrad; reduced toe V; cross axis: 70–90°. Manus digitigrad, smaller than the pes (⅕). Pes pace angulation: 160–180°.
Sphingopus Demathieu 1966	III > II > IV > I. Pes pentadactyl digitigrad; reduced toes I and V; cross axis 90°. Manus digitigrad (when present), smaller than the pes: ⅓–¼. Pes pace angulation: 150–170°.
Batrachopus Hitchcock 1845	III > IV ≥ II > I. Pes tetradactyl, digitigrad; cross axis; 75–90°. Manus tetradactyl, digitigrad, smaller than the pes. Pes pace angulation 140–170°.
Coelurosaurichnus Huene 1941	III > IV > II. Bipedal, scarcely quadrupedal trackways, pes tridactyl; toes relatively thick. Manus tetradactyl. Pes pace angulation: 170–180°.
Anchisauripus Lull 1904	III > IV > II. Bipedal trackways; pes tridactyl, or tetradactyl when hallux impresses. Little divarication of the digits. Pace angulation 170–180°.
Grallator Hitchcock 1858	III > IV > II. Bipedal trackways; pes tridactyl; III clearly longer than II and IV; digits subparallel. Pace angulation 170–180°.

EXPLANATION OF PLATE 1

Fig. A.— Trails formed by worm-like animals; convex hyporeliefs, X0.33.
B.— *Cochlichnus* sp; convex hyporeliefs, X0.5.
C.— *Planolites* sp.; convex hyporeliefs, X0.33.
D.— Bilobate trace fossils with a resting trace at the upper end; convex hyporelief, X0.6.
E.— *Isopodichnus* sp.; convex hyporeliefs, X0.9.
F.— *Skolithos* sp.; convex hyporeliefs, X0.6.
G.— Arthropod tracks (?); convex hyporeliefs, X0.9.
H.— Problematica; concave epireliefs, X0.6.
I.— Plant imprints of *Voltzia* sp.; convex hyporeliefs, X0.8.

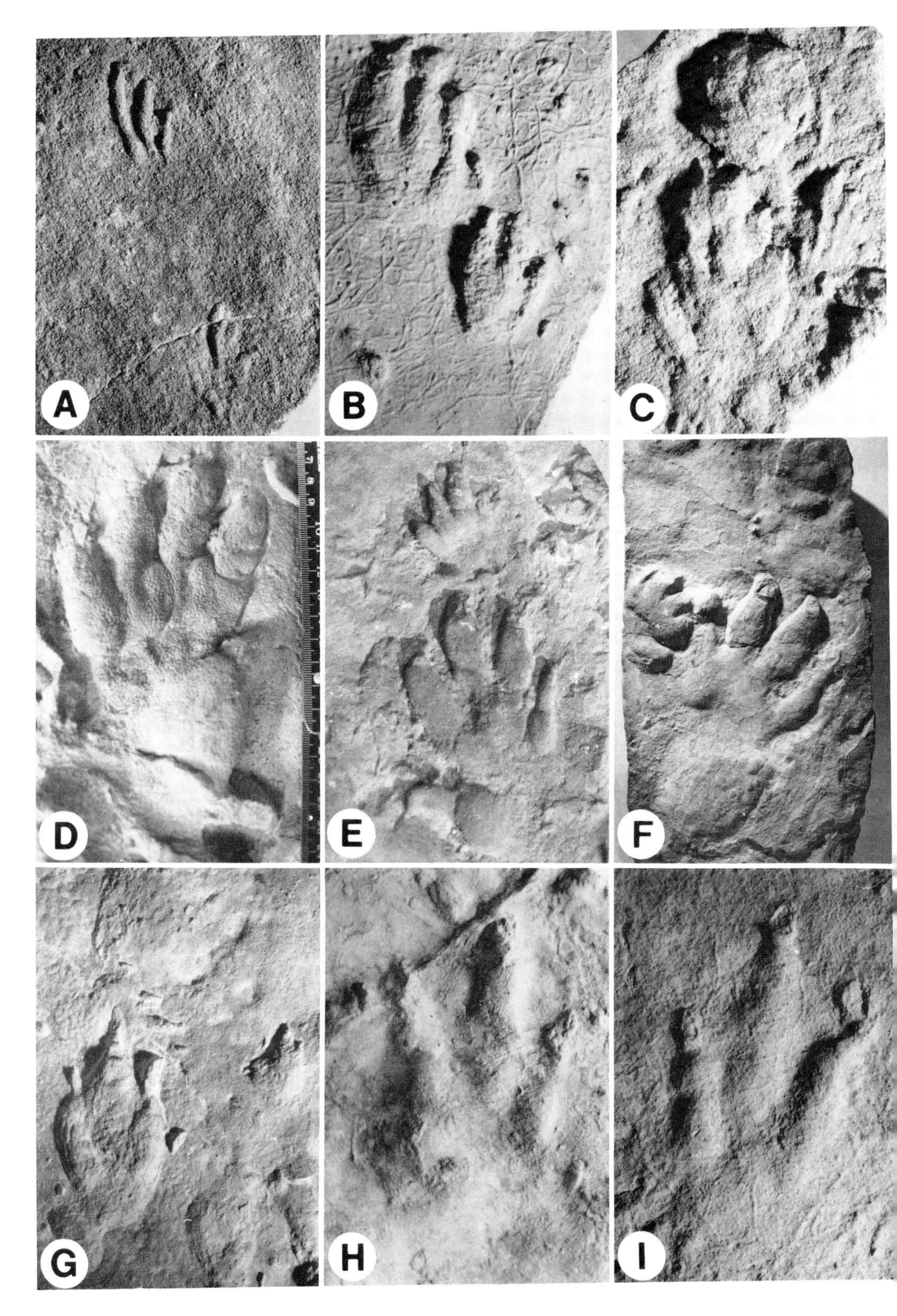
A
B
C
D
E
F
G
H
I

TABLE 2.—NUMERICAL DATA FROM THE MEASUREMENT OF 34 FOOTPRINTS OF BRACHYCHIROTHERIUM GALLICUM

Localities		Length of Digits (mm)					Footprint (mm)		Set of the first 4 digits (mm)		Divarications (degrees)			
		I	II	III	IV	V	Length L	Width l	Length M	Width m	II-IV t	I-IV t'	I-V f	Cross axis q
Lodève	1	56	77	81	57	/	/	/	89	83	31	39	/	78
(Museum)	2	59	78	88	72	112	185	105	98	84	26	36	60	65
	3	48	61	65	52	66	129	90	76	76	33	52	60	66
	4	65	79	82	69	93	165	103	93	102	35	45	60	70
	5	56	78	84	69	115	185	102	107	92	20	48	55	65
	6	59	71	79	66	94	175	107	97	100	36	46	50	55
	7	43	58	69	58	70	145	105	77	77	25	30	60	75
Lodève	8	55	75	77	57	94	170	98	90	90	35	45	60	70
(Grammont)	9	60	70	81	66	85	160	93	100	93	26	36	48	65
	10	56	73	75	64	115	182	94	95	85	29	39	34	67
Lyon	11	55	62	71	62	92	155	101	81	79	30	48	43	70
(Museum)	12	65	87	102	82	111	223	130	115	101	17	30	36	70
	13	66	77	88	71	92	180	109	108	100	27	50	50	70
	14	64	84	90	70	/	170	/	106	106	32	48	/	60
	15	60	68	70	60	112	150	95	85	84	35	52	/	68
	16	62	74	92	74	92	195	120	114	111	28	35	42	69
Lyon	17	58	82	92	70	89	180	106	95	99	30	52	40	65
(Univ.)	18	53	78	86	50	81	190	99	99	93	25	55	35	60
	19	79	96	96	85	112	190	130	118	119	30	46	70	70
	20	51	72	73	56	82	155	112	114	93	25	44	70	80
	21	66	82	98	81	86	190	114	118	109	30	50	50	70
	22	76	99	106	86	92	180	109	108	102	30	54	60	75
Genève	23	68	88	94	79	94	180	120	110	102	32	46	55	70
(Museum)	24	74	95	104	84	120	195	130	114	111	38	50	55	70
	25	64	84	86	74	102	170	105	94	97	32	48	55	65
Mines de	26	43	67	76	65	86	152	104	82	79	20	28	55	70
Paris	27	72	100	107	82	110	195	125	115	110	30	45	60	65
Paris	28	59	75	91	73	78	156	100	90	84	17	30	53	68
(Museum)	29	83	96	102	85	121	200	140	114	117	28	36	60	69
	30	57	69	74	66	76	145	97	94	85	30	50	64	62
	31	51	69	76	65	83	157	101	85	92	29	36	53	68
	32	84	95	102	87	113	200	138	117	112	20	28	42	64
	33	70	83	89	70	94	170	117	94	93	27	38	43	66
	34	77	92	97	87	100	190	120	105	103	30	48	57	69

←

EXPLANATION OF PLATE 2

Fig. A.— *Rhynchosauroides petri* Demathieu, 1966, right set of impressions; convex hyporeliefs, X0.75.

B.— *Rotodactylus rati* Demathieu, 1971, left pes and manus; convex hyporeliefs. Note the claw of the 5th toe behind the set of 4 digits and the lattice of worm trails, X0.6.

C.— *Synaptichnium priscum* Demathieu, 1970, right set of impressions; convex hyporeliefs, X0.5.

D.— *Brachychirotherium gallicum* Willruth, 1917, left pes. Note scales under the toe pads; convex hyporeliefs, X0.4.

E.— *Chirotherium barthii* Kaup, 1835, right set of impressions; convex hyporeliefs, X0.2.

F.— *Isochirotherium coureli* Demathieu, 1970, left set of impressions with an isolated manus beside the pes; convex hyporeliefs, X0.2.

G.— *Sphingopus ferox* Demathieu, 1966, right set of impressions. The manus is smaller and at the right of the pes; convex hyporeliefs, X0.6.

H.— *Coelosaurichnus largentierensis* Courel and Demathieu, 1976, right pes; convex hyporeliefs, X0.35.

I.— *Anchisauripus bibractensis* Demathieu, 1971, left pes; convex hyporelief, X0.7.

TABLE 3.—PARAMETERS OF FREQUENCY DISTRIBUTION FROM THE MEASUREMENT OF FOOTPRINT CHARACTERS OF BRACHYCHIROTHERIUM GALLICUM

Characters	N	Mean (mm)	Standard deviation (mm)	Coefficient of variability (%)	Confidence interval for the mean at the level 5% (mm)		Cramer's test		
							Asymmetry	Test variable	Probability (%)
I	34	62.2	10.385	16.7	58.5	65.8	0.323	0.80	50
II	34	79.2	11.206	14.1	75.3	83.2	0.206	0.56	58
III	34	86.6	11.712	13.5	82.4	90.7	0.054	0.14	89
IV	34	70.4	10.606	15.1	66.7	74.1	−0.010	0.02	98
V	32	95.7	14.809	15.5	90.3	101.1	0.026	0.06	95
L	33	174.7	20.140	11.5	167.5	181.9	−0.071	0.17	86
l	32	110.0	13.434	12.2	105	115	0.718	1.73	9
M	34	99.9	12.662	12.7	95.5	104.4	−0.184	0.46	65
m	34	96.0	11.872	12.4	91	101	0.091	0.23	82
t	34	28.5	5.195	18.2	26.6	30.3	−0.599	1.49	14
t'	34	43.0	8.058	18.7	40.2	45.9	−0.512	1.27	20
f	31	52.7	9.595	18.1	49.2	56.3	−0.344	0.82	41
q	34	67.9	4.194	7.2	66.4	69.4	0.008	0.02	98

percent when the sample contains both young and mature tracemakers, but the norm is between 10 and 20 percent, as seen in the example presented in Table 3. Note that in Table 3 the characters do not have equal variability. The variability for the length characters originates mainly from the part played by the digits when the pes touches the substrate. Here, digit III has the lowest variability, which shows the prominent role of this digit during locomotion.

The highest values of variability are found with the divarications of the digits, although these characters are not influenced by allometric growth. All of the chirotherid footprint ichnospecies have high variability values, and many have values greater than those shown here (Demathieu, 1970; Courel and Demathieu, 1976). The cross axis angle (Peabody, 1948), formed by the axis of digit III and the digit-metapodial line, is a very important character for comparisons with skeletons (Demathieu and Haubold, 1978). Its variability is always lower than 10 percent.

2) Confidence interval for the mean at the 5 percent probability level shows the accuracy with which the mean is known.

3) Cramer's test, easier to calculate than the Chi square test, indicates whether or not the distribution is homogeneous. The probability is the value that exceeds the test variable calculated from the asymmetry (Fisher, 1958). A high probability suggests good homogeneity. The homogeneity is doubtful between 10 percent and 5 percent, and lower than 5 percent there is

TABLE 4.—RATIOS OF LENGTHS OF FOOTPRINT CHARACTERS OF BRACHYCHIROTHERIUM GALLICUM

Characters	N	Mean (mm)	Standard deviation (mm)	Coefficient of variability (%)	Confidence interval for the mean at the level 5% (mm)		Correlations	
							Coefficient	Minimum coefficient (5%)
III/I	34	1.405	0.1360025	9.7	1.35	1.46	0.825	0.330
III/II	34	1.10	0.0605	5.5	1.07	1.12	0.925	0.330
III/IV	34	1.24	0.1083	8.8	1.19	1.28	0.878	0.330
II/IV	34	1.13	0.1146	10.1	1.09	1.18	0.832	0.330
IV/I	34	1.15	0.1110	9.7	1.10	1.18	0.860	0.330
II/I	34	1.28	0.145	8.1	1.24	1.32	0.885	0.330
V/III	32	1.11	0.1687	15.2	1.05	1.18	0.520	0.339
L/1	32	1.60	0.1425	8.9	1.54	1.65	0.737	0.339
M/m	34	1.04	0.0689	6.6	1.01	1.07	0.863	0.330
L/M	32	1.76	0.1300	7.4	1.70	1.81	0.800	0.339
l/m	32	1.149	0.0903	7.8	1.11	1.19	0.813	0.339

no homogeneity. For an ichnospecies, one must take into consideration the whole set of tests. Just one or two tests may not be significant and cannot permit one to conclude that an ichnospecies is invalid, because one or two distributions can be biased by unknown external factors. In the cited example, only one character shows a doubtful homogeneity.

4) In comparing ichnospecies from different layers, it is not advisable to compare the true measurements because of possible variations due to allometric growth. In such cases, comparison of the ratios between length characters is more appropriate (Courel and Demathieu, 1976). Such comparisons can be an important aid in the definition of an ichnogenus or ichnospecies. Table 6 shows a comparison of this sort between *Brachychirotherium gallicum* and *Brachychirotherium circaparvum*.

Table 4 presents the principal ratios. The variability of these measurements is lower than that for the lengths. The "Student's" test (Fisher, 1958) is used for the comparison of means (Table 6). The cross axis has been used as a main anatomical character. Among the 12 tested ratios, 5 show significant differences; this is sufficient evidence to conclude that two true ichnospecies are present, thus confirming the conclusion based on morphological appearance.

In Table 5, comparison between the mean length of the toes of *Brachychirotherium gallicum* shows the order of length of the toes used in preparing diagnoses for the main footprint ichnogenera from strata of Early and Middle Triassic age of Europe and North America. The diagnoses presented in Table 1 are the result of both morphologic studies and statistical studies of the type described herein.

PALEOECOLOGIC ANALYSIS

Estimates of Population Numbers for the Trackmakers

Reconstruction of the depositional environments for the Middle Triassic lower sandstones (Unit 1) also leads to a paleoecologic study. For this, estimates of the numbers of vertebrate trackmakers is useful to enable recognition of the various feeding types (trophic groups). However, the problem is difficult. One can say that one trackway has one tracemaker, but it is incorrect to say *a priori* that *n* trackways of the same ichnospecies have *n* tracemakers.

Several methods can be used to determine the correct relationships between trackways and tracemakers (Demathieu, 1970). One method is based on observation of the trackways, relief of the imprints, and measurements of the footprints in the trackways. The depth of the track can indicate if multiple trackmakers walked together or not. Because the substrate surface of the trackway areas becomes increasingly dry after waters recede, it is possible to recognize if two trackways were made by one animal or by two animals of the same size. This method has been applied in the study of the Largentière's layer (Courel and Demathieu, 1976; Demathieu, 1977a).

A new approach to this research is presented here. This method is founded on the notion of "reduced Euclidian distance". The method is the most useful when footprints lie on separate slabs, which is the case for *Brachychirotherium gallicum* specimens on the footprints level from Fozières, near Lodève. Only the length characters of the digits and footprints are considered because one cannot mix length and divarication measurements.

The reduced Euclidean distance between two footprints is given by the formula:

$$d = \left[\sum_{k=1}^{n} \frac{1}{S_k^2} \left(X_{ik} - X_{jk} \right)^2 \right]^{1/2}$$

TABLE 5.—COMPARISON BETWEEN THE LENGTHS OF THE FIRST FOUR DIGITS OF BRACHYCHIROTHERIUM GALLICUM

	II-III	II-IV	I-IV
Standard deviation	11.639	11.079	10.711
Difference of means (D)	7.3	8.8	8.0
Theoretical difference (d)	5.7	5.5	5.3
Conclusion	d<D	d<D	d<D
Result	II<III	IV<II	I<IV
be..................	.. I<IV<II<III		

TABLE 6.—COMPARISON OF MEANS OF THE RATIOS OF LENGTH CHARACTERS FOR BRACHYCHIROTHERIUM GALLICUM AND B. CIRCAPARVUM FOOTPRINTS

RATIOS	III/I	III/II	III/IV	II/IV	IV/I	II/I	V/III	L/l	M/m	L/M	l/m	q
Standard deviation	0.1785	0.0558	0.1103	0.1167	0.1528	0.1554	0.1805	0.1390	0.0775	0.1366	0.0872	5.1535
Difference of means (D)	0.26	0.03	0.00	0.03	0.22	0.20	0.33	0.03	0.04	0.05	0.05	5.3
Calculated theoretical difference (d)	0.12	0.04	0.07	0.08	0.10	0.11	0.12	0.10	0.05	0.09	0.06	3.5
Conclusion	d<D	d>D	d>D	d>D	d<D	d<D	d<D	d>D	d>D	d>D	d>D	d<D

Table 2 gives measurements for the *B. gallicum* footprints. In this table, i and j are lines (i.e., footprints) and k represents columns (i.e., characters). Here, X_{ik} is the measurement of character X from the line i and column k, and S is the standard deviation (given in Table 3) of the character from column k. From the data of Table 2, all the distances can be calculated between the footprints on a two by two basis. With 32 useable footprint units, the number of footprint units and the number of distance units that can be calculated equals 32 × 31/2 = 496 distance units.

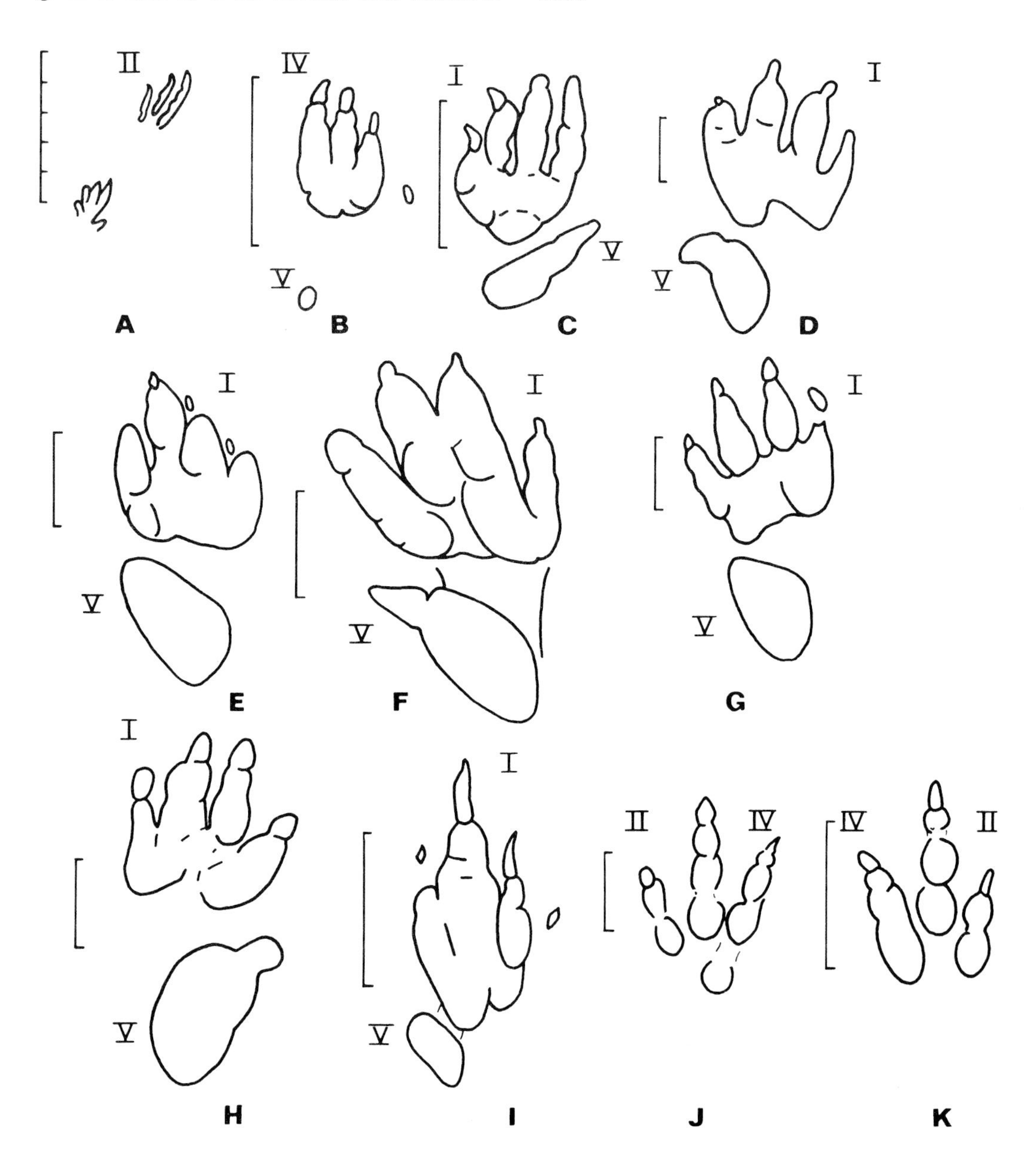

Fig. 3.—Vertebrate footprint ichnospecies of the Triassic deposits from the Massif Central, France: *A*, *Rhynchosauroides petri*, two impressions of right pes; *B*, *Rotodactylus rati*, left pes; *C*, *Synaptichnium priscum*, right pes; *D*, *Chirotherium barthii*, left pes; *E*, *Brachychirotherium circaparvum*, left pes; *F*, *Brachychirotherium gallicum*, left pes; *G*, *Isochirotherium coureli*, left pes; *H*, *Isochirotherium felenci*, right pes; *I*, *Sphingopus ferox*, left pes; *J*, *Coelosaurichnus largentierensis*, right pes; *K*, *Anchisauripus bibractensis*, left pes.

However, one must be cautious because two footprints of a trackway can differ greatly when a character is poorly preserved on a footprint. Preservational problems can cause error in the distance calculations.

The succession of investigation and analysis is as follows: (1) search for the smallest footprint; and (2) calculate n—1 distances from this reference footprint. If the distribution of the distances is not homogeneous and it appears that they are grouped in small sets, then the number of these sets can give an estimate of the number of tracemaker animals. In this example, it was determined that approximately 8 animals made the 32 measured footprints.

The various methods of footprint analysis are not mutually exclusive and can be applied jointly. They give only approximate numbers, and there is no means for absolute verification. The analyses show that the number of the trackmakers was approximately proportional to the number of footprints. This is owing to the fact that the areas of the footprint-bearing surfaces studied are generally small compared to the large and flat areas where the reptiles wandered. In addition, the probability that an animal walked twice over the same ground is low. The fact that the number of footprints is approximately proportional to the number of tracemakers is important because it allows one to use percentages of ichnospecies or genera from the same layer or same level for comparisons with other layers or levels. It is not a random phenomenon that the main layers of Chasselas (Saône-et-Loire), Chasselay (Rhône), and Largentière (Ardèche; see Fig. 1) contain approximately equal percentages (Demathieu, 1977b). This strongly suggests that the reptile faunas of these localities were similar.

Using both morphological observations and statistical data derived from the footprints, it is possible to restore the relationships between the trackmakers and their ecological position within the community. In doing so, it is assumed that all species of a given ichnogenus will have a similar place in the trophic web structure of their respective communities; however, in this study, only ichnogenera are considered.

Based primarily on the relative census estimates of the trackmakers derived from study of the footprints, it is suggested that the respective ichnogenera represent trackmakers with the following characteristics:

The footprints *Rhynchosauroides* Maidwell, 1911 and *Rotodactylus* Peabody, 1948 were formed by herbivores of small size (0.5 to 1.5 m length) and were very numerous, often forming herds (Fig. 3 A,B; Pl. 2 A,B).

Footprints of *Isochirotherium* Haubold, 1970 were formed by large, bulky herbivores (3 to 5 m length), often forming herds. These were the most numerous animals present (Fig. 3 G,H; Pl. 2 F).

Footprints of *Chirotherium* Kaup, 1835 were formed by large (3 to 5 m length) predators. These animals were very rare in Middle Triassic time and are represented by solitary trackways around those of smaller reptiles (Fig. 3 D; Pl. 2 E).

The footprints *Anchisauripus* Lull, 1904 (Fig. 3 K; Pl. 2 I); *Coelurosaurichnus* Huene, 1941 (Fig. 3 J; Pl.

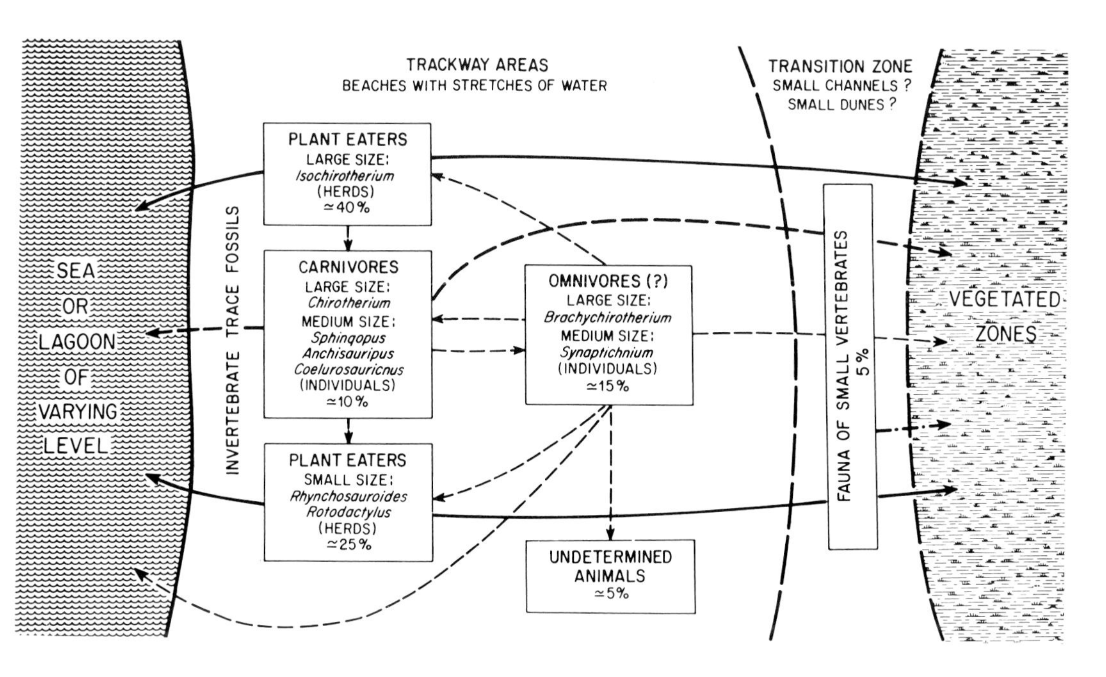

Fig. 4.—Generalized reconstruction of relationships between the trackmakers and their environment.

2 H); and *Sphingopus* Demathieu, 1966 (Fig. 3 I; Pl. 2 G) were predators of small to medium size (1 to 3 m length). These trackways often are found in association with trackways of large footprints.

Footprints of *Brachychirotherium* Beurlen, 1950 possibly represent an omnivorous animal. The animal produced erratic, solitary trackways and was of medium to large size (2 to 5 m length; Fig. 3 E,F; Pl. 2 D).

The footprints *Synaptichnium* Nopsca, 1923 (Fig. 3 C; Pl. 2 C) and *Batrachopus* Hitchcock, 1845 were formed by omnivores or carnivores. The interpretation is not clear because these footprints are very rare and are usually only found with the small *Rotodactylus* or *Rhynchosauroides* footprints, in trackways formed by herds.

Figure 4 presents a summary of the reconstructed Middle Triassic paleoecologic relationships. It is based on data from the main footprint layers of Largentière, Lyon, and Macon. A single reconstruction model is possible because the numerical proportions of the footprints of the tracemakers were quite similar for all of the layers studied (Demathieu, 1977b). The percentages given in Figure 4 are the arithmetic means of the calculated percentages of footprint types from each layer. In a study of this type, it is necessary to make estimates. The population numbers presented here are gross approximations only, but they do aid in creating a more complete and revealing view of Triassic life in the environment along the margin of a sea or lagoon. The study shows that these environments were populated by a well-balanced reptilian fauna and that the margin areas of these seas or lagoons were not desert regions.

PALEODEPOSITIONAL ENVIRONMENT

By combining information from the trackways with information from the invertebrate trace fossils, plant imprints, and physical sedimentary structures, it is possible to reconstruct the paleodepositional environment of the trackway areas. Aquatic invertebrate lebensspuren (*Cochlichnus, Isopodichnus, Planolites, Skolithos*) are present in some parts of the trackway areas. These trace fossils, along with ripple marks, give evidence of shallow waters at the border of the trackway areas. Salt crystal casts indicate that these waters probably were marine, and that a sea or lagoon bounded the trackway areas on one side. The molds of small fronds of *Voltzia* that probably were transported by wind or by water in small channels indicate that there were vegetated zones on dry land areas adjacent to the trackway areas, on the side opposite the sea or lagoon.

Microvertebrate footprints (Demathieu, 1977a), which often occur on the same slabs with archosaur footprints, confirm the presence of very small animals. Because of their reduced size, these animals could not travel long distances and thus would not move far from their resting places (small channels, small dunes), which were probably at or near the boundary between the trackway areas and dry land. These large trackway areas were not entirely tidal flats, because the effect of the tides would have destroyed all traces as soon as they were made. Spring or storm tides sometimes could have overflowed these areas, but this cannot be verified.

Studies of the orientation of the trackways for each level in the Middle Triassic sequence has revealed that the majority of the reptiles followed a preferential direction (Demathieu, 1970; Courel and Demathieu, 1976; Gand, 1978). This, combined with the fact that the trackways are generally rectilinear, and that no resting traces have ever been discovered, suggests that these areas were probably pathways between land and water.

The depositional environment of these areas changed with time. Superposed footprint levels (Fig. 5) can occur in the same layer. Two successive levels are separated by a deposit now represented by the thickness of a slab (1 to 20 cm). If one examines the bedding of a slab on its undersurface (Fig. 5), one can see a thin bed of clay or silt overlain by grains of increasing size up to the middle of the slab. At this point, grain diameter rarely exceeds 0.5 mm. Above this, grain size decreases, and at the top of the bed it is similar to that at the base.

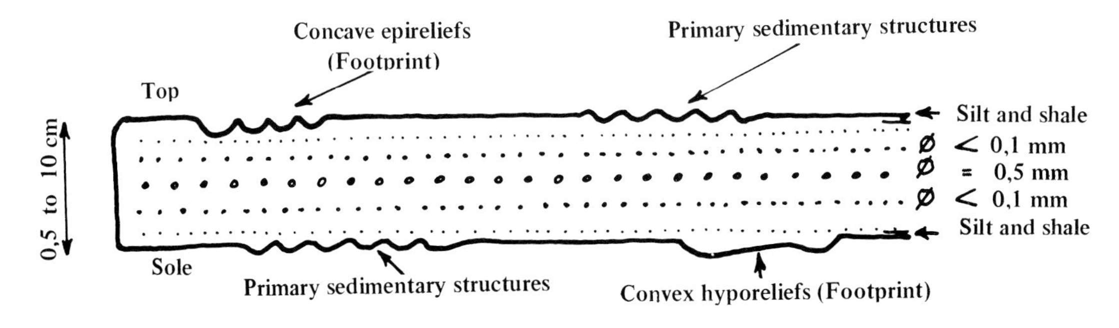

Fig. 5.—Transverse section through a slab of trace fossil-bearing quartz sandstone. The different trace fossils of the two surfaces suggest that the lower layer was consolidated when the concave epireliefs of the upper layer were formed.

Each slab represents a sequence of sedimentation. The deposits could not be the result of tidal deposition because tidal deposits normally are thinner. River flooding could not have produced these deposits either because of the presence of graded bedding and the geographically widespread regularity of the sequence. Rather, a transgression is indicated. The vertical succession of grain sizes (Fig. 5) shows that, at first, the energy of the transgressive waters was very low, then energy increased to a moderate level, and, finally, energy decreased to a very low level, as at the beginning of the cycle.

All surfaces of the slabs do not bear footprints or marks. Some slabs are flat and without any trace fossils whereas others exhibit ripple marks only. However, each of the footprint-bearing levels, whatever the density of trace fossils may be (up to 25 footprints of great size per square meter for the level of Chasselay, Rhône), shows ichnospecies which belong to the reference ichnofauna (Y level of Chasselay). These considerations point out the permanence of this ecosystem and its geographical extent through the course of Middle Triassic time.

CONCLUSIONS

The Middle Triassic trackway areas described in this study were fine-grained sandy areas situated between water (a sea or lagoon) and dry land. The landward limit of tidal flat areas is marked by the terminal occurrence of invertebrate trace fossils, whereas the seaward boundary of vegetated dry land is indicated by the final occurrence of microvertebrate footprints. Sedimentation between successive footprint-bearing levels resulted from minor transgressions followed by regressions. These events witness the beginning of a major Mesozoic transgression on the eroded Hercynian Massif Central.

Although it is not usually possible to match a given ichnospecies or ichnogenus with its proper tracemaker species or genus, ichnologic research often permits the assignment of the tracemaker to its proper higher taxonomic level. Furthermore, morphologic and statistical analyses of the reptilian footprints can be used to derive population estimates for the various tracemakers. These numbers then permit one to develop a trophic structure for the reptilian paleocommunity.

ACKNOWLEDGMENTS

The author's thanks are extended to H. Allen Curran, Smith College, for his kind and careful contributions to the writing and review of this paper; to Louis Courel, University of Dijon, for his contribution to the stratigraphic section; and to Liliane Gallet-Blanchard, University of Paris IX, and an anonymous reviewer for their helpful criticism of an earlier version of the manuscript. Dillon Scott and Carolyn Estes, Smith College, assisted with the editing of the manuscript.

REFERENCES

Basan, P. B., 1979, Trace fossil nomenclature: The developing picture: Palaeogeography, Palaeoclimatology, Palaeoecology, v. 28, p. 143–167.

Courel, L., 1973, Modalités de la transgression mésozoique: Trias et Rhétien de la bordure Nord et Est du Massif Central français: Soc. Géol. France, Mem. 118, 152 p.

______, and Demathieu, G., 1973, Données récentes sur le Trias du Mont d'Or lyonnais dans les domaines de la Stratigraphie et de l'Ichnologie: Géobios, v. 6, p. 5–25.

______, and ______,. 1976, Une ichnofaune reptilienne remarquable dans les grès triasiques de Largentière (Ardèche, France): Paleontographica, Abt. A, Bd. 151, p. 195–216.

______, ______, and Buffard, R., 1968, Empreintes de pas de Vertébrés et stratigraphie du Trias: Bull. Soc. Géol. France, v. X, p. 275–281.

______, ______, and Gall, J-C., 1979, Figures sédimentaires et traces d'origine biologique du Trias moyen de la brodure orientale du Massif Central, signification sedimentologique et paléoécologique: Géobios, v. 12, p. 379–397.

Demathieu, G., 1970, Les empreintes de pas de Vertébrés du Trias de la bordure N-E. du Massif Central: Cahiers de Paléontol., CNRS, Paris, 211 p.

______, 1975, Reconstitutions paleoecologiques à partir de données ichnologiques, possibilités et difficultés: Bull. Soc. Géol. France, v. XVII, p. 896–898.

______, 1977a, Des microvertébrés dans le Trias moyen du Lyonnais et du Maconnais, révélés par leurs empreintes; signification paléoécologique: Géobios, v. 10, p. 351–367.

______, 1977b, La palichnologie des Vertébrés. Developpement recent et rôle dans la stratigraphie du Trias: Bull. BRGM (2), IV, v. 3, p. 269–278.

______, and Haubold, H., 1974, Evolution und Lebensgemeinschaft terrestrischer Tetrapoden nach ihren Fährten in der Trias: Freiberger Forschungshefte, C 298, p. 51–72.

______, and ______, 1978, Du problème de l'origine des Dinosauriens d'après les données de l'Ichnologie du Trias: Géobios, v. 11, p. 409–412.

Fisher, R.A., 1958, Statistical Methods for Research Workers: Edinburgh-London, Oliver and Boyd, 356 p.

Gand, G., 1978, Sur le matériel ichnologique récolté dans le Muschelkalk de Culles-les-Roches (S-et-L): Bull. Soc. Hist. Nat. Creusot, v. XXXV (2), p. 21–44.

Grauvogel-stamm, L., 1977, Découvertes d'empreintes de conifères dans le Trias de la bordure Nord-Est du Massif Central français: Sci. Géol. Bull., v. 30, p. 75–78.

HAUBOLD, H., 1970, Ichnia amphibiorum et reptiliorum fossilium: Handb. Paläoherpetologie, Teil 18, Stuttgart-Portland, 124 p.
KREBS, B., 1976, Pseudosuchia: Handb. der Paläoherpetologie, Teil 13: Stuttgart, p. 40–98.
LORTET, L., 1892, Les Reptiles fossiles du bassin du Rhône: Arch. Mus. Hist. Nat. Lyon, v. 5, p. 1–139.
PEABODY, F.E., 1948, Reptile and amphibian trackways from the Lower Triassic Moenkopi formation of Arizona and Utah: Univ. Calif. Publ. Geol. Sci., v. 27, no. 8, p. 295–468.
ROMAN, F., 1926, Géologie lyonnaise: Paris, P.U.F., 82 p.
______, 1950, Le Bas Vivarais: Paris, Hermann, 150 p.
RICOUR, J., 1962, Contribution à la révision du Trias français: Paris, Imprimerie Nationale, 471 p.
SARJEANT, W.A.S. AND KENNEDY, J., 1973, Proposal of a code for the nomenclature of trace fossils: Jour. Canadien Sci. de la Terre, v. 10, p. 460–475.
VAN DER WAERDEN, B. L., 1967, Statistique mathematique: Paris, Dunod, 371 p.

BIOGENIC STRUCTURES AND DEPOSITIONAL ENVIRONMENTS OF A LOWER PENNSYLVANIAN COAL-BEARING SEQUENCE, NORTHERN CUMBERLAND PLATEAU, TENNESSEE, U.S.A.

MOLLY FRITZ MILLER AND LARRY W. KNOX
Geology Department, Vanderbilt University, Nashville, Tennessee 37235; and Department of Earth Sciences, Tennessee Technological University, Cookeville, Tennessee 38505

ABSTRACT

Sedimentary rocks of the Fentress Formation and lower Rockcastle Conglomerate include sandstones, shales, siltstones, and coals which are divisible into 6 facies. Facies characteristics, including rock type and physical and biogenic sedimentary structures, as well as vertical and lateral relationships between the facies, are consistent with interpretation of the sequence as deposited in back-barrier lagoons and tidal channels; tidal flats; and tidal inlet, channel, and delta environments. Superposition of burrowed sandstones, whose modern analogues are formed primarily in estuarine or marine environments, over coals in the middle of the sequence, and the occurrence of thinly bedded sandstones of tidal flat origin near the top, reflect the importance of marine influence during deposition of the sequence. Of the 20 ichnogenera found, 15 occur elsewhere only in rocks deposited in marine environments. Presence of characteristically marine trace fossils and absence of exclusively non-marine trace fossils give strong evidence of marine domination. This is corroborated by the lack of features characteristic of alluvial (channel or overbank) deposits or of sediments deposited on delta plains in strongly constructional, fluvially dominated delta systems. The Fentress Formation-Rockcastle Conglomerate sequence could have been deposited within a wave- or tide-dominated deltaic system or within a destructional portion of a broadly constructional delta, although an overall deltaic setting is not recognizable at the scale of this study.

Many trace fossils found in the sequence are restricted to rocks deposited in particular environments. Three assemblages are recognized: (1) a *Palaeobullia-?Thalassinoides* assemblage, occurring in tidal flat sandstones, (2) a *Conostichus-Rosselia* assemblage restricted to highly burrowed back-barrier sandstones, and (3) a *?Lennea* assemblage including forms most abundant in sandstones interpreted as deposited on a lower tidal flat, perhaps adjacent or transitional to a tidal delta or channel.

Narrow facies ranges and marine affinities of trace fossils in these otherwise unfossiliferous rocks give further evidence that trace fossils are important aids in paleoenvironmental reconstruction.

INTRODUCTION

The upper Crab Orchard Mountains Group in north central Tennessee comprises a Lower Pennsylvanian sequence of coals, shales, sandstones, and conglomerates containing a moderately diverse and abundant ichnofauna. Extensive studies of Carboniferous rocks of much of the Appalachian Plateau have been completed. However, there has been no detailed investigation of rocks exposed on the northern Cumberland Plateau of Tennessee, nor have biogenic structures been systematically described and extensively used in the reconstruction of the Carboniferous depositional setting, except in one localized area (Basan et al., 1979). This paper presents an integrated paleoenvironmental interpretation for a portion of the Lower Pennsylvanian sequence in northern Tennessee (Figs. 1, 2) using lithologic, sedimentologic, and biologic data, and discusses the distribution of the common biogenic structures and their role in environmental reconstruction.

STRATIGRAPHY

Regional Setting

Basic paleodepositional model.—Tennessee Carboniferous stratigraphy is similar to that of other areas in the Appalachian Basin in broadly consisting of Mississippian carbonates overlain by Pennsylvanian clastics. Although the sharp lithologic boundary was originally interpreted as representing a major regional unconformity, more recent work has shown interfingering relationships between the carbonates and clastics and has documented the presence of Mississippian fossils in sandstones previously considered to be Pennsylvanian (Milici, 1974). Re-evaluation of this information as well as consideration of newly acquired lithologic and sedimentologic data led Ferm and coworkers to develop a new model of barrier systems for deposition of Carboniferous rocks in the Appalachian Basin.

According to the barrier model (Ferm et al., 1972; Hobday, 1974; Milici, 1974; Ferm, 1974; Milici et al., 1979), deposition in the southern Appalachian Basin was essentially continuous during the Carboniferous. Sharp lithologic boundaries are interpreted as a result of localized environmental changes. Orthoquartzitic sandstones overlying carbonates and marine shales and underlying coal-bearing sequences consisting of graywackes, siltstone, shales, and coals are interpreted as barrier island, and tidal delta and channel deposits. The underlying carbonates were offshore carbonate mounds surrounded by clays. Landward of the barrier complex, graywackes, siltstones, shales, and coals were deposited in back-barrier, lower delta plain, upper

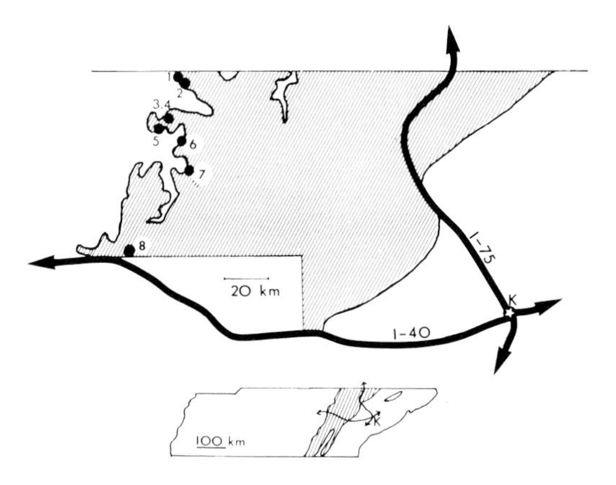

Fig. 1.—Location of outcrops of Cumberland Plateau of Tennessee. Striped area is zone of outcrop of Pennsylvanian rocks. K = Knoxville; numbered lines represent interstate highways.

delta plain, and alluvial plain environments. Superposition of carbonates by clastics indicates large-scale progradation; however, interfingering of facies shows that smaller-scale movements of the strand line were not uncommon. Strike of the shoreline was generally northeast-southwest from Alabama to Pennsylvania (Hobday and Horne, 1977), although Horne (1979a) suggested the existence of a large northwest-southeast trending embayment in Tennessee.

Tennessee Carboniferous stratigraphy.—The Carboniferous stratigraphic nomenclature of Tennessee is shown in Figure 2. The sequence investigated in this study on the northern Cumberland Plateau includes the lower-most Rockcastle Conglomerate and the immediately subadjacent sandstones, shales, siltstones, and coals of the upper Fentress Formation. These rocks are included in the Crab Orchard Mountains Group which, at its type section approximately 100 km south-southeast of the study area, is 200 m thick. Wilson et al. (1956) reported that rocks of the Group are 100 m thick in the study area and thicken to a maximum of 300 m 85–100 km to the southeast. Sandstones within the Crab Orchard Mountains Group correspondingly thin to the west and northwest; the Rockcastle Conglomerate thins from 100 m to 30 m from east to west. In the northern Cumberland Plateau the Sewanee Conglomerate is not well developed and the term Fentress Formation is used for all the rocks between the Mississippian units and the Rockcastle Conglomerate. Roughly correlative rocks to the north are included in the Lee Formation (Ferm et al., 1971).

Tennessee stratigraphy: Relation to littoral model.—Ferm et al. (1972) and Milici (1974) have related rocks of the southern Cumberland Plateau in Tennessee to the littoral (barrier) model, and Carlson (1979) and Milici et al. (1979) have extended the interpretation into the northern plateau area. Ferm et al. (1972) and Milici (1974) interpret the sequence as broadly progradational

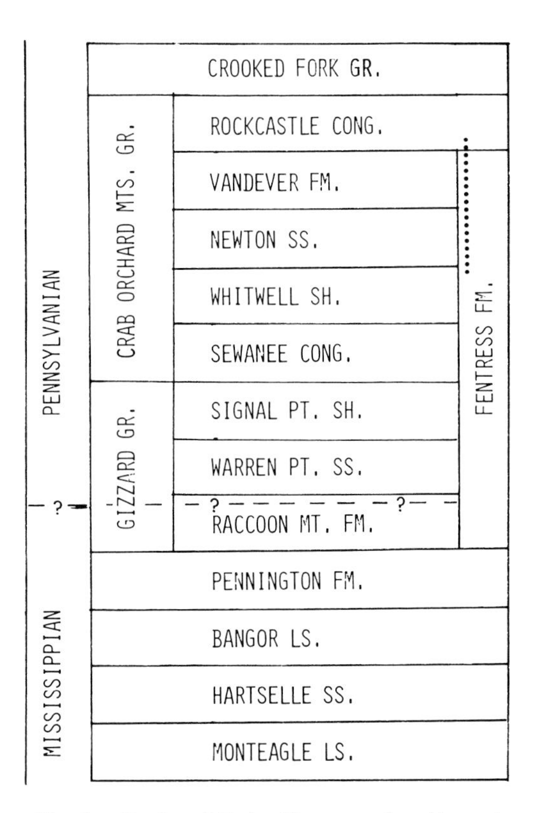

Fig. 2.—Portion of Carboniferous stratigraphic section, Tennessee. Vertical dotted line indicates section studied. From Milici, 1974.

from north to south with the Mississippian limestones deposited as offshore carbonate mounds and the Pennington Formation rocks as reflecting deposition in shoreface environments. The Gizzard Group represents tidal flat, washover, and coastal marsh sediments, whereas the thick Sewanee sandstones were deposited as barrier sands (Milici, 1974, p. 127). Deposition in back-barrier environments, including lagoon and marsh, is reflected in the Whitwell Shale, Newton Sandstone, and Vandever Formation. The stratigraphic location of the transition from predominantly marine to predominantly fluvial deposition is not known; it may occur in the overlying Crooked Fork Group.

The Sewanee Conglomerate thins to the north and is missing in northern Tennessee where its lateral equivalents are the back-barrier deposits of the Fentress Formation (Milici et al., 1979). Carlson interpreted the origin of similar but younger orthoquartzitic sandstones in the Wartburg Basin approximately 90 km to the east. These sand bodies are linear, trending northwest-southeast, and of barrier origin, with back-barrier deposits to the northeast; associated coal beds are pod-shaped and generally parallel to shoreline (Carlson, 1979, p. 422). Milici et al. (1979, p. G24) pointed out that shale beds within the Rockcastle Conglomerate increase in number and thickness to the east. They interpret the Rockcastle Conglomerate as barrier deposits in the west, where orthoquartzitic sandstones predominate, and as back-barrier deposits further to the east.

FACIES AND ENVIRONMENTS OF DEPOSITION

General

Six facies are distinguishable within the upper Fentress Formation and lowermost Rockcastle Conglomerate of the northwestern Cumberland Plateau. The vertical sequence of facies varies from outcrop to outcrop, and not all facies are represented at all locations. A generalized section showing approximate average thicknesses of facies and a common vertical sequence is given in Figure 3. Locations of outcrops and the facies exposed at each are listed in the Appendix.

Interpretation of the depositional environments represented by these rocks is hampered by limited exposure in scattered outcrops. Vertical and lateral relationships between rock units can be interpreted with confidence only on the scale of a single outcrop. Detailed models of Pennsylvanian deposition elsewhere on the Appalachian Plateau have been developed from excellent exposures allowing large-scale three dimensional reconstruction of facies relationships (e.g. Baganz et al., 1975; Horne, 1979b), but interpretation(s) at such a large scale is not possible on the northwestern Cumberland Plateau. Superposition of burrowed sandstone containing marine trace fossils overlying coals indicates a coastal setting, but whether this coast was a destructional part of a deltaic system or part of a barrier system is uncertain. Barrier terminology is used in Figure 3 because there is no evidence for deposition within a river-dominated delta; recognizable distributary mouth bar, interdistributary bay, and crevasse splay deposits are lacking and there is little indication of unidirectional currents. However, the Fentress Formation-Rockcastle Conglomerate sequence could have been deposited within a wave- or tide-dominated deltaic system; at the scale of this study such deposits could closely resemble barrier deposits. The general picture of the Carboniferous shorelines in

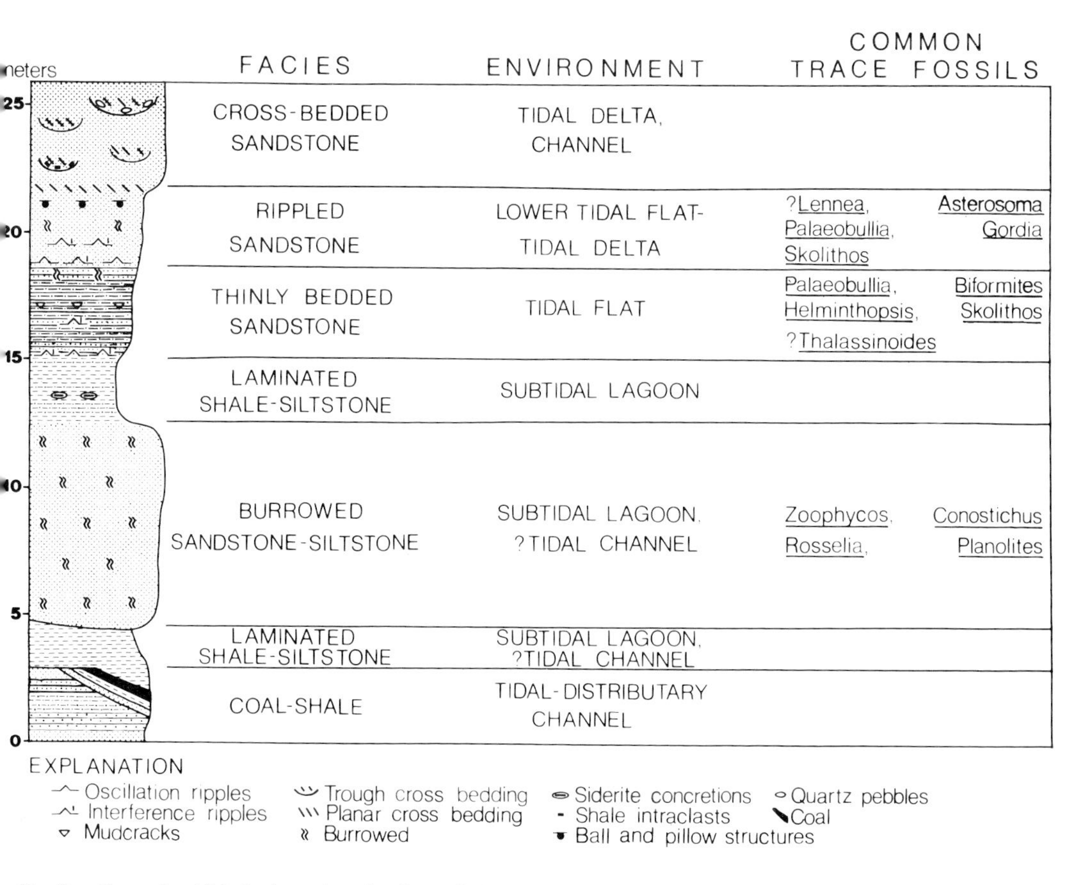

Fig. 3.—Generalized lithologic section of rocks studied, showing facies, common trace fossils and inferred environments of deposition.

the Appalachian region is that of a complex mosaic of intergrading deltaic and barrier environments (Hobday, 1974; Ferm, 1974; Hobday and Horne, 1977). The equivocal nature of the Fentress Formation and Rockcastle Conglomerate in terms of their depositional setting is expectable in light of the gradational character between facies associated with destructional deltaic systems and barrier complexes.

Coal—Underclay Facies

Description.—Because the best exposed and most complete sections are in strip mines, the basal facies generally is coal. Where the coals are well exposed (Localities 1, 2, 6) they occur as part of channel-fill sequences. At Locality 6 the channel is approximately 20 m wide and 5 m deep, and at section 2 the depth of the channel is 2 to 3 m. Maximum thickness of the coal, 0.7 m to 1 m, occurs in the middle of the channel. The relationship of the coal to the rest of the channel-fill sequence is obscure. Only at Locality 6 is the base of the channel exposed. Here, a 0.7 m thick sandstone bed floors the channel and is overlain by a 0.8 m thick layer of light-colored mudstone with abundant wood fragments; this mudstone underlies the coal and presumably is the underclay. At Locality 6, as well as at other localities where the coal is exposed, coal is overlain by black shale, which forms the top of the channel-fill sequence. At Locality 1 the shale interfingers with the coal near the margin of the channel. At several localities (1, 6, 8) the coals were originally mined by hand in tunnels that presumably followed the axes of the channels. The channel axes have a strong northerly component, as determined by alignment of apparent axes with original tunnel openings as reported by local residents.

Interpretation.—Most modern coastal peat deposits accumulated in marsh or swamp lands between distributaries on delta plains (Frazier and Osanik, 1969) or on peat islands behind barriers in estuarine settings (Staub and Cohen, 1979), rather than in channels, but organic-rich deposits are accumulating in channels in several modern depositional systems. Abandoned lower delta plain distributary channels on the St. Bernard delta (Mississippi Delta) are filling with organic muck and clay (Saxena, 1979). Horne et al. (1978) generalized that distributaries on lower delta plains tend to be straight rather than meandering and are often abandoned due to loss of gradient and resulting distributary switching. After abandonment they are more likely to be filled with fine-grained sediment, including organic material, than are meandering channels which gradually are filled with point bar deposits. Organic material could also accumulate in more marine-dominated settings, some of which grade into lower delta plain environments and would yield very similar deposits. Saxena (1979, p. 335) reported that the Chandeleur Islands, formed by reworking of distributary-mouth sands after abandonment of the St. Bernard delta, are backed on their landward side by shallow sounds in which clay and organic muck are accumulating. Continued infilling under such conditions allows plants to become established, leading to increased deposition of organic material. In the wave-dominated Senegal Delta, where strong littoral currents and wave activity rework river-born sand and deposit it as a barrier which blocks outlet to the sea, the river flows parallel to the coastline before eventually breaching the barrier (Coleman and Wright, 1975). Continued progradation has led to the development of a series of alternating abandoned channels and beach ridges which parallel the shore. The channels are being filled with organic detritus (Coleman and Wright, 1975).

Abandoned tidal channels could also be loci for coal deposition. Many tidal creeks in the Niger Delta have been captured, leaving abandoned channels which may be filled with fine-grained (including organic-rich) material (Allen, 1970, p. 145); the Fentress coals may have been deposited in similar Pennsylvanian settings.

Some coals in the Appalachian Plateau have been interpreted as deposited in back-barrier environments (Milici, 1974; Horne et al., 1974; Mathew, 1977). These coals characteristically are thin and discontinuous (Horne et al., 1978), as are those of the Fentress Formation. Some are considered to have accumulated on peat islands (Mathew, 1977). Others are interpreted as abandoned tidal channel-fill deposits (Horne et al., 1974), although a detailed model for their origin is not available.

Shale—Siltstone Facies

Description.—Dark, laminated to fissile shales and siltstones are abundant, constituting up to 50 percent of the section, although abundance varies laterally within the outcrops. Traces are rare and include only small (less than 0.5 cm in diameter) vertical, oblique, and horizontal burrows. Plant fragments are abundant, but no root structures were observed. Disc-shaped siderite concretions up to 10 cm in diameter are common at Locality 1. One contained a well-preserved specimen of a bivalve, probably the brackish to fresh water species *Anthroconauta phillipsii* (Rogers, 1965).

Layers of fissile shale up to 1.5 m thick overlie the coals at Localities 1, 2, 6, 8. At Localities 1, 2, and 6 the shales clearly are part of the channel fill; the contact between them and the overlying bioturbated sandstone facies is abrupt and, at Locality 1, erosional. Shales and siltstones up to 7 or 8 m in thickness occur higher in the section, above the burrowed sandstone facies, at Localities 1, 2, and 6 (Fig. 4a). A 0.75 m bed of siltstone is present between the laminated sandstone and rippled sandstone beds at Locality 3a. Where the shale-siltstone facies occurs high in the section, the contacts between it and the overlying thinly-bedded sandstone facies or rippled sandstone facies is gradational. At Locality 1, the shale-siltstone facies abruptly overlies burrowed sandstones (Fig. 4a).

Interpretation.—Paucity of structures in this facies makes environmental interpretation difficult, although

Fig. 4.—*A*, Burrowed Sandstone between siltstone and shale at Locality 1. Person in lower right for scale. *B*, Burrowed sandstone at Location 2, with specimen of *Skolithos*. Lens cap for scale.

clearly it was deposited under quiet water conditions. The presence of a well-preserved, articulated brackish or fresh water bivalve suggests less than normal salinity. In Middle Pennsylvanian rocks farther east in Tennessee, disc-shaped siderite nodules are restricted to rocks deposited under marine to brackish conditions (Milici et al., 1979, p. G31). Rarity of biogenic structures may be due in part to poor exposure rather than solely to lack of trace-producing animals.

In modern coastal settings, fine-grained sediments are deposited in a variety of environments. Lagoonal sediments accumulating landward of barriers consist of clays and fine silts which commonly are burrowed but which may also be laminated (Elliott, 1978). Extensively burrowed gray silty clays make up lagoonal deposits landward of barriers along the Virginia coast (Morton and Donaldson, 1973). Fine-grained sediments deposited in tidal channels far from the inlet in Mugu Lagoon, California, are also bioturbated (Warme, 1971), as are fine-grained point bar deposits in tidal creeks along the South Carolina coast (Barwis, 1979). Clays and silts are accumulating in shallow bays behind barriers (Chandeleur Islands) in destructive portions of the Mississippi Delta (Saxena, 1979).

Fine-grained sediments are also deposited in delta plain marshes and interdistributary bays associated with constructional deltas, such as the active portions of the Mississippi Delta (Morgan, 1970). Oomkens (1974, p. 201) distinguished between silts deposited in marine to brackish water environments of the Niger Delta and those deposited in fresh to brackish water environments on the basis of the presence/absence of a marine fauna, the boron content of the sediments, and on the relative proportions of mangrove and grass pollen. However, he noted that the distinction is often difficult to make, a conclusion not surprising in light of the gradational transition from back- barrier to interdistributary environments in modern delta systems.

Silt and clay are deposited in intertidal flats associated both with barrier and deltaic systems. Grain size of sediment tends to decrease with increasing height above sea level (Evans, 1965; Reineck and Singh, 1973) on intertidal flats. Extensive tidal flats develop in areas of high tidal range, and tide-dominated deltaic systems such as those of the Colorado, Ord, and Klang Rivers (Meckel, 1975; Coleman and Wright, 1975) have large intertidal areas where fine-grained sediment is being deposited. The presence or absence of a flora and fauna as well as of desiccation cracks and evaporites depends upon factors such as height above mean low water, climate, and rate of sedimentation (Coleman and Wright, 1975). Burrowed fine-grained sediments also are found on the margins of salt marsh estuaries in Georgia (Mayou and Howard, 1975) and laminated muddy and silty sands are accumulating along the axes of these channels.

Shales of the upper Fentress Formation do not have desiccation cracks, evaporite deposits, root impressions, or any other indication of subaerial deposition. However, because the rocks are too fractured to identify mudcracks or rooting even if they were present, and because Holocene intertidal areas populated by marsh grasses (not present in the Paleozoic) may have been barren in the Pennsylvanian, an intertidal/supratidal origin for these shales and siltstones cannot be completely ruled out. Some of the shales clearly were parts of distributary or tidal channel-fill sequences with coals. The relatively great lateral extent of others of these shales and siltstones (over 2 km^2 at Locality 1) strongly suggests that they were deposited in back-barrier or in distributary lagoons or bays, such as those which occur along the Atlantic coast of the U.S. (Morton and Donaldson, 1973), or were associated with wave- or tide-dominated portions of the Mississippi (Saxena, 1979) or Niger deltas (Oomkens, 1974).

A back-barrier origin has been suggested for other Pennsylvanian sequences elsewhere on the Appalachian Plateau consisting of fine-grained clastic rocks interbedded with thin coals (Horne et al., 1978) and has also been suggested for the siltstones and shales of the Whitwell Shale and Vandever Formations of the Crab Orchard Mountains Group in southern Tennessee (Milici, 1974).

Burrowed Sandstone—Siltstone Facies

Description.—Rocks of this facies are fine- and very fine-grained sandstones and coarse siltstones which have been extensively burrowed. The sandstones have abundant fine-grained material in the matrix. Medium to thick bedding is preserved (Fig. 4a), but smaller-scale bedding has been obliterated by burrowing (Fig. 4b). Trace fossils present include *Rosselia* (A = abundant), *Conostichus* (C = common), *Olivellites* (R = rare), *Skolithos* (C), *Planolites* (R-C), and *Zoophycos*, which is abundant in a layer 0.5 m thick near the top of the facies at Locality 1 and in a bed of similar thickness and stratigraphic position at Locality 7. Discrete traces are less abundant than the nearly ubiquitous burrowed texture. Fossilized plant material is abundant and ranges from nearly ubiquitous finely comminuted debris to fragments of *Calamites* over a foot long (Localities 2, 8) to coalified logs (Locality 8).

Where exposed (Localities 1,2), the contact between the bioturbated sandstone facies and the underlying shale is abrupt (Fig. 3) and burrows in the sandstone do not extend downward into the shale. Good exposures at several stripped areas along the summit of Golman Mt. (Locality 1) show that the bioturbated sandstone is a channel-fill deposit. The channel axis trends north to northeast and the channel is approximately 300 m wide, with a maximum thickness of 8 to 10 m of channel fill (bioturbated sandstone). At Locality 1 the horizontal beds are 0.2 - 0.5 m thick and there is no well-defined upward decrease or increase either in bed thickness or in grain size. At Localities 2 and 8 there is no indication that the burrowed sandstones and siltstones are part of a channel-fill sequence; maximum thickness of the facies at these localities is 20 to 22m. Toward the south (Localities 7,8), siltstone predominates, and at Locality 7, a 0.5 m thick sandstone bed containing abundant *Zoophycos* forms a resistant layer at the top of the siltstone. The upper contact with the overlying shale (Localities 1, 2, 7) is either gradational (2) or abrupt (1, 7); at Locality 8 it grades into the overlying cross-bedded sandstone.

Interpretation.—Highly burrowed rocks reflect deposition in areas where wave and current stratification was overshadowed by biological reworking (Howard, 1972). Such quiet water conditions could have existed either offshore, or in a protected shoreline environment, such as a back-barrier lagoon. Burrowing in some modern lagoonal sediments is intense (Warme, 1971; Reineck and Singh, 1973; Ronan, 1975). In Mugu Lagoon, California, the rate of bioturbation by the deposit-feeding thalassinid shrimp *Callianassa californiensis* greatly exceeds the rate of sedimentation, resulting in a sand-mud sequence whose texture and structures are controlled primarily by biologic reworking (Warme, 1971; Miller, 1982a).

The extensively burrowed texture and channel-fill geometry of some of the burrowed sandstones and siltstones of the Upper Fentress Formation put constraints on the environments in which they may have been deposited. First, the overlying or interstitial water must have been brackish, or, more likely, of near normal marine salinity. Studies of bioturbation in modern estuaries along the Georgia coast have shown that the abundance and diversity of biogenic structures increases seaward, especially in estuaries with fresh water influx (Howard, 1975); sediments deposited in low salinity environments (e.g., inner estuarine) generally are not extensively bioturbated (Dörjes and Howard, 1975). Paleoecological evidence corroborating this interpretation is given by the fact that most of the trace fossils occurring in the bioturbated sandstone facies of the Fentress Formation are found in marine rocks of diverse ages from many geographically widespread areas.

Deposition within channels occurs in a number of coastal environments. Alluvial channels on delta plains may be filled with point bar deposits, but the resulting channel fill is usually characterized by fining upward sequences and a paucity of bioturbation (Barwis, 1979). Distributary mouth bars are typically lens-shaped and become finer-grained, less massively bedded, and more burrowed from the center to the margins (Gould, 1970; Horne, 1979b). These characteristics differ from those of the burrowed sandstone facies. However, the burrowed sandstones could have originated as reworked distributary mouth bar deposits in an abandoned distributary or tidal channel. Much of the tidal channel fill in the Niger Delta originates as distributary mouth bar sand which is reworked and redeposited in tidal channels, the range of width to depth ratio of which includes that of the Fentress Formation channel (Oomkens, 1974). Redistribution of distributary-carried sediment by tidal currents is also a major factor in determining sand body geometry and distribution in strongly tide-dominated destructional delta systems such as the Mekong, Irawaddy, and Ord deltas (Fisher et al., 1969; Coleman and Wright, 1975).

Modern tidal channel and tidal creek deposits differ from the burrowed sandstone and siltstone facies in some important respects. Whereas most tidal channel and tidal creek point bar deposits show a general upward decrease in grain size and in bedding thickness (Oomkens, 1974; Meckel, 1975; Barwis, 1979), the grain size of the burrowed sandstone facies is quite constant at a given locality. Cross-bedding is common in active fill tidal channel deposits (Oomkens, 1974; Coleman and Wright, 1975; Barwis, 1979), but it is not apparent in the burrowed sandstones and siltstones of the Fentress Formation, where it may have been obliterated by biologic reworking. Sediments accumulating in tidal and estuarine channels within the Colorado Delta are horizontally bedded or laminated, although burrowing is restricted to thin clay interbeds (Meckel, 1975).

The extensive burrowing in the burrowed sandstone facies suggests near normal marine salinity and an area protected enough to allow biologic reworking to dominate over wave and current stratification. Such conditions are met in modern tidal creeks (Barwis, 1979), salt marsh estuaries (Howard, 1975), and lagoons. In the Niger Delta, some tidal channels merge subaqueously into lagoons (Oomkens, 1974). A lagoon-tidal channel origin for the burrowed sandstone and siltstone facies of the Fentress Formation could account for the textural differences within the facies, with the sands deposited in channels and the silts in lagoons. The sandstone bed bioturbated by *Zoophycos* overlying the burrowed siltstones at Locality 7 may have been a washover deposit, which are common in back-barrier environments, particularly along microtidal (< 2 m tidal range) coasts (Hayes, 1975). The *Zoophycos* producer may have been able to colonize areas soon after environmental perturbation; Miller and Johnson (1981) suggested that the producer of the similar trace fossil *Spirophyton* was able to quickly exploit ephemeral patches with favorable environmental conditions within a Devonian estuarine/deltaic setting.

The alternative interpretation, that the burrowed sandstone was deposited in an offshore or offshore-shoreface transition environment is less attractive, even though burrowed sands are found in these modern environments (Howard, 1972). The Fentress Formation burrowed sandstones do not have the parallel to burrowed sequences commonly found in modern offshore-shoreface transition deposits, nor are they capped by the trough cross-bedded sands typical of lower shoreface deposits (Howard, 1972; Reineck and Singh, 1973).

Fig. 5.—*A*, Very thinly bedded sandstone facies overlain by cross-bedded sandstone facies at Locality 3b. *B*, Close-up of sandstone with thin shale interbeds at Locality 3b.

Very Thinly Bedded Sandstone Facies

Description.—This facies includes horizontally bedded very fine-grained sandstones with shale interbeds (Fig. 5a). Bedding thicknesses of the sandstones range from 0.3 cm to 10 cm; most beds are between 1 cm and 2.5 cm thick. Beds grade upward into shales, the maximum thickness of which is a few millimeters (Fig. 5b). Shale thickness varies inversely with sandstone thickness. Long wavelength (10 to 15 cm), low amplitude (0.2 to 0.3 cm) interference ripples are abundant, and desiccation cracks are present but rare. Very small pieces of plant debris are abundant, especially toward the tops of graded beds. Although a few beds are highly bioturbated, most have been only slightly disrupted by biological activity (Fig. 5b). Traces on bedding planes are abundant and well preserved, but vertical burrows are less common. At Localities 3a, 3b, and 4 the rock has been quarried for building stone, and quarried slabs provide abundant bedding plane exposure. At Locality 3a *Palaeobullia* is the most abundant trace fossil. *Biformites*, a small form of *Skolithos* and *Planolites*, are common, and *Asteriacites*, *Kouphichnium*, and *Helminthopsis* are rare. *Olivellites* occurs in profusion in one thin bed. A few hundred meters away, at Locality 3b, *Thalassinoides* is the most abundant trace, although *Palaeobullia* is also common. At Locality 4 (200 to 300 m from Locality 3b) and at Localities 1, 2, 5, 6, 8 the shale interbeds are reduced or absent, interference ripples are more common and of larger amplitude, and there is less fine-grained matrix. Traces include rare horseshoe tracks and resting traces (*Kouphichnium* and *Limulicubichnus*), ?*Lennea* (C), *Gordia* (R), *Skolithos* (R-C), *Caprionichnus* (R), *Planolites* (R), and *Helminthopsis* (R), as well as *Palaeobullia* (C).

The very thinly bedded sandstone facies is up to 3 to 4 m thick at Localities 3a and b (Fig. 5a); elsewhere it is no thicker than 1.5 m. At Locality 3b, it grades upward into rocks of the rippled sandstone and cross-bedded sandstone facies. At Locality 3a, it grades laterally and vertically into sandstones of the rippled sandstone facies, which grades into a .75 m thick unity of dark laminated siltstone. The lower contact is not exposed at any of the localities.

Interpretation.—Several characteristics of rocks of this facies indicate deposition on a tidal flat. Micrograded beds and mud drapes record deposition during decreased energy conditions associated with high tide, and interference ripples reflect diverse directions of water movement during various stages of the tidal cycle (Klein, 1970; Reineck and Singh, 1973). The paucity of bioturbation suggests deposition on upper portions of a tidal flat where burrowing is less intense than lower on the flat where the density of infaunal animals is high. The presence of desiccation cracks supports this interpretation. Cleaner sandstones with thinner shale partings and higher amplitude ripples such as those exposed at Locality 4 probably represent deposition lower on the tidal flat, where bed shear (and inferentially, current velocity) was higher.

Alternatively, the laminated sandstone facies may represent distributary-mouth bar deposits or splay deposits into a tidal lagoon. Although distributary-mouth bar sandstones show a general coarsening upward (Donaldson et al., 1970), individual beds, especially those located toward the margin, are commonly graded (Fisher et al., 1969). The bedding in the very thinly bedded sandstone facies closely resembles that of interdistributary bay siltstones and distributary-mouth bar sandstones illustrated by Horne (1979b, p. 654, Figs. 14, 20). However, lack of exposure prohibits lateral tracing of the very thinly bedded sandstone facies of the Fentress Formation, and there is insufficient evidence of the proximal, more highly cross-bedded and channeled portion of a distributary mouth bar (Fisher et al., 1969; Elliott, 1978) to warrant this interpretation. A series of splays into a tidal lagoon could have characteristics very similar to tidal flat sandstones, especially if they accumulated above the mean low water level.

Rippled Sandstone—Siltstone Facies

Description.—Rocks of this facies are very thinly to thinly bedded, fine- to very fine-grained sandstones and coarse siltstones. Thin sections reveal more silica cement and less finely disseminated matrix than that found in sandstones of the very thinly bedded sandstone facies, and in hand sample they are also lighter in color. Shale partings and interbeds are less abundant than in the very thinly bedded sandstone facies. Plant debris is common, but there is no evidence of rooting. At the small stone quarry (Locality 5) where rocks of this facies are best exposed, many bedding surfaces are rippled (Fig. 6). Oscillation ripples are most abundant. Some are straight and sharp-crested (amplitude 5–6 cm; wavelength 9–10 cm), whereas others are flat-topped (amplitude 2–3 cm, wavelength 6–9 cm) and merge laterally into interference ripples. At Locality 5, the trend of ripple crests is NE to EW (15 measurements). Current ripples are found only at Locality 1 where interference ripples are much more abundant. Flaser bedding occurs at Locality 7. Bedding thicknesses range from 0.5 cm to 15 cm, although most beds are 1 to 4 cm thick; at Locality 7 the beds are thicker (up to 40 cm). Ball and pillow structures over a meter in longest dimension occur in one bed at both Locality 1 and Locality 5. Some layers at Localities 5 and 7 have been intensely bioturbated, but these constitute only about 10 percent of the section. Distinctive trace fossils include *Palaeobullia* (A, Localities 5, 7), ?*Lennea* (A-5), *Gordia* (R-C, 5), *Planolites* (C, 5, 7), *Skolithos* (R-C, 5, 7), *Conostichus* (R-C, 7), *Bergaueria* (C-7), *Asterosoma* (A-7), *Rhizocorallium* (R-7), *Rosselia* (R-3b), and *Scalarituba* (R-7).

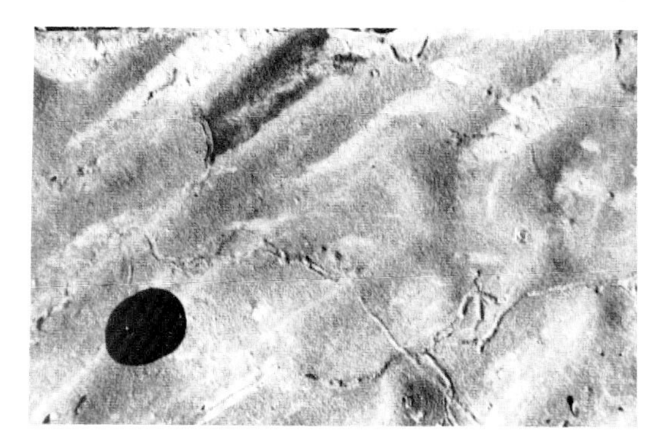

Fig. 6.—Rippled sandstone bed at Locality 5. Note *Palaeobullia* crossing ripple crests. Lens cap for scale.

Maximum thickness of this facies is 20 m at Locality 7, although some of this interval is covered. It is best exposed at Locality 5, where, as at Locality 3b, rippled sandstones grade upward into sandstones of the cross-bedded sandstone facies.

Interpretation.—The thinly bedded sandstones and siltstones overlying the thinly bedded sandstone facies were deposited under higher energy conditions than those of the thinly bedded sandstone facies. Flat topped and interference ripples are common on intertidal sand flats (Klein, 1970; Knight and Dalrymple, 1975). Flaser bedding occurs in lower intertidal sands along the North Sea (Reineck, 1975) and in intertidal sediments elsewhere (Evans, 1975; Larsonneur, 1975). The presence of loaded beds suggests periods of rapid deposition of fine sands (Blatt et al., 1980), and soft sediment deformational features are common in modern tidal creek deposits (Evans, 1975; Barwis, 1979). The abundance of ripples, and, in comparison with the very thinly bedded sandstones, reduction in number of shale interbeds and in amount of fine-grained matrix seen in thin section are indicative of deposition lower on a tidal flat, possibly grading into the margins of a flood-tidal delta or tidal channel point bar.

Cross-Bedded Sandstone Facies

Description.—Cross-bedded, fine- to medium-grained conglomeratic sandstones make up the Rockcastle Conglomerate and cap the section studied. The sandstones are quartz-rich, with very little fine-grained matrix or finely comminuted plant debris. At Localities 3a and 3b quartz pebbles are abundant on toesets, and at Locality 8 shale clasts up to 8 cm across occur in the lower sandstones. Thin (15 to 25 cm thick) discontinuous beds of highly bioturbated, finer-grained sandstone as well as thin shale stringers occur sporadically through the lower part of this facies at Locality 8.

Bedding is complex in rocks of the cross-bedded sandstone facies, but is characteristically trough cross-bedding. However, some tabular cross-bed sets occur at Locality 3b, and low angle planar cross-bedding is present, although rare, at Localities 1, 5, and 8. Troughs generally are 1.5 m to 2 m wide and from 0.25 to 0.5 m deep. The most common cross-bed dips are to the southwest and southeast, although only 20 measurements are available. Channels 5 to 7 m wide and up to several meters deep are present at Localities 3a, 3b, 5, 7, and 8; axial trends are in a north-northeast to south-southwest direction. The channels generally are filled with trough cross-bedded sands. Beds of finer-grained sandstone at Locality 8 are micro-cross-laminated; load features are also present.

Traces are rare in the cross-bedded sandstone facies. Those that occur are either in rocks transitional to the rippled sandstone facies (*Rosselia* ar Locality 3b) or exposed on the soles of sandstone beds overlying layers of finer-grained sandstone or siltstone (horizontal traces at Locality 8); a few vertical to oblique burrows cut through cross-bedding at Locality 8.

The cross-bedded sandstone facies overlies rocks of the burrowed sandstone-siltstone facies, the shale-siltstone facies, or the rippled sandstone facies. Where it overlies rippled sandstones (Localities 1, 2, 3b, 5) the contact is gradational over 2m or so. Where cross-bedded sandstones overlie finer-grained rocks the contact commonly is abrupt and in some places clearly erosional (Localities 2, 3a, 8), although in some areas at Locality 8 the basal beds of the lowest trough cross-bed sets are more burrowed and composed of finer-grained sandstones than the upper beds. The nature of the contact varies laterally at Localities 2 and 8.

Interpretation.—The cross-bedded sandstone facies has many characteristics of tidal inlet-channel-delta deposits, including basal conglomeratic lag deposits, and abundant complex trough cross-bedding (Kumar and Sanders, 1974; Barwis and Hayes, 1979). The trough cross-beds result from megaripples which could form on flood or ebb tidal deltas, or in tidal inlets or channels. Large-scale channels with basal conglomerates, which occur at Locality 7 and higher in the Rockcastle at a Locality 20 km to the northeast, represent deeper parts of tidal channels, whereas the cross-bedded sands above the rippled sandstone facies at other localities (such as 1, 2, 3, 5, 8) may be flood-tidal delta deposits. Differences in the nature of the contact between the cross-bedded sandstone facies and the underlying rocks and variations within the cross-bedded sandstone facies are consistent with deposition in a tidal inlet, channel, or delta because although the same tidal processes control deposition in these environments, there is a gradual change in depositional conditions from inlet into lagoon and sluggish tidal channels (Reinson, 1979). Extent, morphology, and thickness of sands deposited in flood and ebb tidal deltas, tidal channels, and tidal inlets depend on a variety of factors, of which tidal range might be the most important. Mesotidal estuarine systems (tidal range 2–4 m) have abundant and deep inlets, well-developed meandering tidal channel systems, extensive tidal deltas, and few washover deposits, whereas microtidal estuaries (tidal range less than 2 m) are characterized by fewer and shallower inlets, less well developed tidal channels and ebb-tidal deltas, and extensive washover deposits (Hayes, 1975; Barwis and Hayes, 1979). Lack of sandstones of probable washover origin and the abundance of quartz-rich sandstones with structures similar to those in modern tidal channels and deltas indicate a mesotidal system. Available cross-bed data, although limited, suggest paleocurrents approximately normal to the NW-SE trending embayed shoreline proposed for the Pennsylvanian of north-central Tennessee (Horne, 1979a), which is also consistent with a mesotidal interpretation (Horne, 1979a).

Alternatively, the cross-bedded sandstone may have been deposited along a macrotidal coast (tidal range greater than 4 m). Macrotidal estuaries are characterized by funnel-shaped mouths, linear sand ridges in the center of the estuary, and extensive mudflats along the margin (Hayes, 1975). More abundant tidal flat deposits than are present in the Fentress Formation-Rockcastle Conglomerate might be present if deposition occurred along a macrotidal estuary. In the Ord River Delta of western Australia, bedforms on the long narrow sand bodies are of the same scale as those in cross-bedded sandstone facies (Wright et al., 1975). However, too little is known about: 1) the internal structure of these ridges and those in similar settings such as the Klang and Colorado Deltas (Coleman and Wright, 1975; Meckel, 1975), 2) the bedforms present in the intervening channel deposits in these modern environments, or 3) the sequence and orientation of bedforms, and the vertical and lateral relationships of subfacies within the cross-bedded sandstone facies of the Rockcastle Conglomerate to allow meaningful comparisons to be made. A detailed study of the sedimentary structures and facies relationships within the

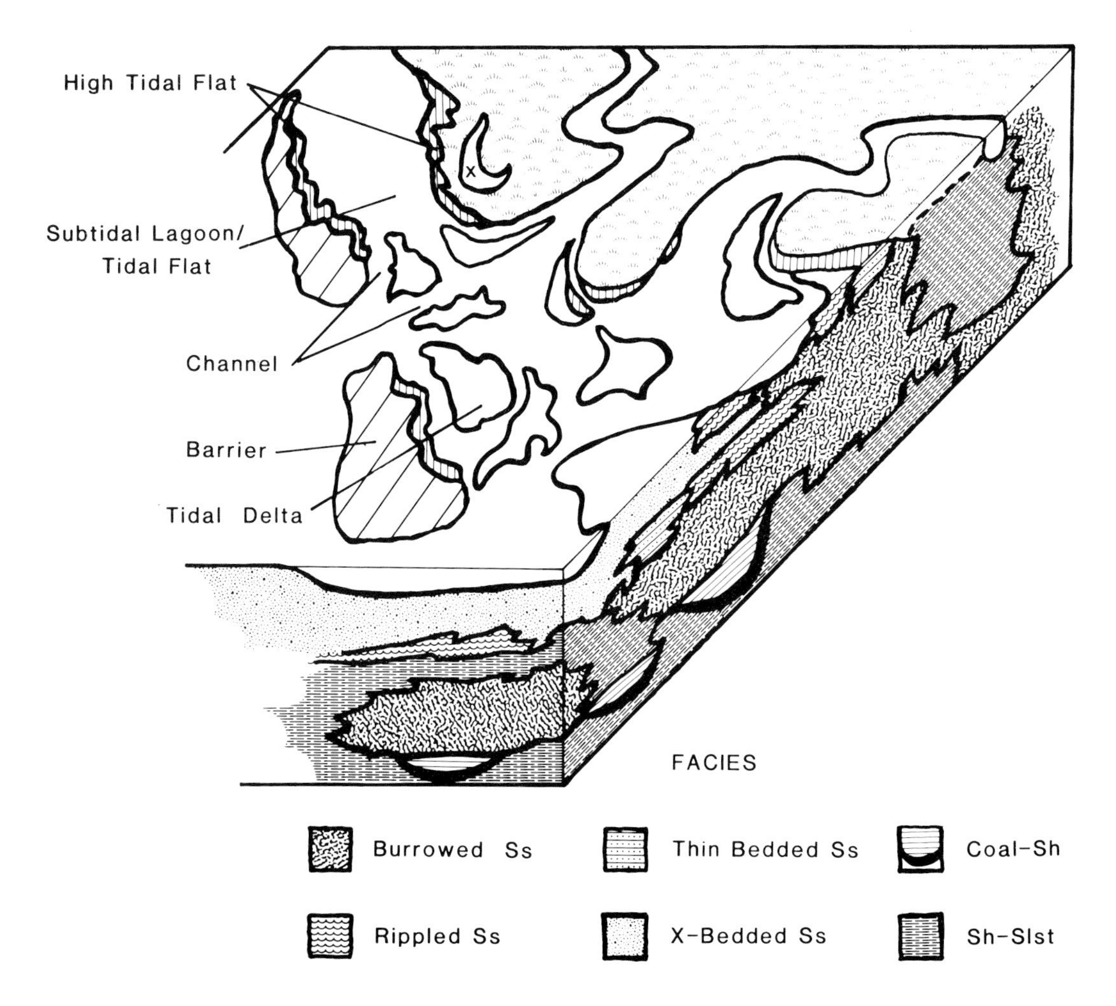

Fig. 7.—Possible depositional setting and facies distributions for upper Fentress Fm.—lower Rockcastle Conglomerate. Inferred depositional environments for facies are given in description for Figure 3.

entire Rockcastle Conglomerate has been undertaken by S.R. Jackson, and her results should indicate whether deposition occurred along a microtidal, mesotidal, or macrotidal coastline.

Summary of Facies and Depositional Environments

A schematic diagram of possible depositional environments for the upper Fentress Formation-lower Rockcastle Conglomerate is given in Figure 7. In this model the coals and some of the shales and siltstones were deposited in back-barrier abandoned tidal or distributary creeks. Shales and siltstones which are not obviously part of channel-fill sequences, as well as burrowed sandstones and shales, would have been deposited in subaqueous back-barrier lagoons which merged into tidal channels and creeks. Very thinly bedded sandstones with thin shale interbeds were probably deposited on high tidal flats, perhaps in areas protected from currents, whereas rippled sandstones accumulated on lower tidal flats. Cross-bedded sandstones of the lower Rockcastle Conglomerate probably were deposited in tidal delta, channel, and inlet complexes.

The generalized depositional setting illustrated in Figure 7 is essentially that of a mesotidal barrier setting (Barwis and Hayes, 1979) and is consistent with the available lithologic, sedimentologic and paleontologic data. The shifting of environments shown in Figure 7 could result in the generalized vertical sequence of facies (Fig. 3). If organic material was accumulating at point X (Fig. 7), for instance, migration of the tidal inlet would result in deposition of back-barrier muds, silts, and burrowed sands. With continued migration, lower tidal flat, tidal delta, and eventually tidal channel and inlet sands would be deposited at this location. Because the base of the very thinly bedded sandstone facies (high tidal flat) is nowhere exposed, its relationships with the lagoonal facies are not clear. However, the lateral variation with the facies at the exposures where it is best developed suggest that the protected high tide flats were not extensive and that the exposure at Localities 3 and 4 represent an isolated remnant. This model also accounts for the variations in contacts between facies. Back-barrier lagoonal sediments could either grade upward into tidal delta sands, as represented at Locality 8, where the bottom beds of some (tidal delta) cross-bed sets are interbedded with siltstones and burrowed, or they could be eroded by higher shear stresses generated by higher currents within the tidal channels. Evidence for a mesotidal setting includes the abundance of interference and flat-topped ripple marks, and the abundance of marine trace fossils, indicating near-normal marine salinity and strong tidal influence. This interpretation is also consistent with Horne's (1979a) reconstruction of the Pennsylvanian shoreline showing a major (mesotidal) embayment in Tennessee. A microtidal setting is less likely because oscillation ripples, although common, especially at Locality 5, are not as abundant as interference ripples and because there is little evidence of extensive washover deposits, which characterize modern microtidal barrier systems (Barwis and Hayes, 1979).

A problem with this model is that it proposes a barrier system for a sequence which lacks a barrier facies (foreshore and backshore dune deposits; Reinson, 1979). This does not invalidate a barrier interpretation, however, because tidal inlet migration would lead to the selective preservation of subaqueous tidal inlet and channel facies; emergent beach and dune sands are more likely to be eroded and reworked into subaqueous deposits (Hoyt and Henry, 1967). Ancient barrier systems with few or no barrier deposits have been recognized. Barwis and Makurath (1978) interpreted the Upper Silurian Keyser Limestone in Virginia as a tidal inlet sequence, based on: (1) resemblance of the sequence to modern tidal inlet fill, (2) abrupt lithologic break between the underlying shallow marine rocks and the tidal inlet deposits, and (3) the corresponding faunal break from the diverse marine fauna to a depauperate (lagoonal) fauna. No facies directly attributable to a barrier itself was recognized. Similarly, Ferm et al (1972) interpreted Lower Pennsylvanian sands on the southern Cumberland Plateau as deposits of a barrier system, even though sandstones of the foreshore facies are rare.

Alternatively, the sandstones, shales, and coals of the Fentress Formation could have been deposited along a macrotidal coastline. Shoals and sandbars separated by sandy channels and extensive tidal flats are characteristic of tidally dominated coasts, which typically are in narrow embayments; examples include the Colorado, Ord, Irawaddy, Klang, and Mekong deltas (Fisher et al., 1969; Meckel, 1975; Coleman and Wright, 1975). The fact that there is abundant evidence of tidal influence within the Fentress Formation-Rockcastle Conglomerate sequence, and that there is no barrier facies represented, support this model. However, the model shown in Figure 7 is favored on the grounds that there is little evidence of the extensive tidal flats developed along macrotidal coasts, that the embayment would have been wider than most areas with very high tidal ranges, and that it has been suggested that sufficient organic material to produce coal only accumulates if barriers are present to reduce marine influence (Ferm, 1974). Results of current study of the Rockcastle Conglomerate (by S.R. Jackson) should clarify the depositional environment(s) of the unit as well as those of the subjacent sequence.

Based on the abundance of physical and biogenic structures similar to those found in marine-dominated coastal environments, we conclude that the upper Fentress Formation-lower Rockcastle Conglomerate sequence was deposited under predominantly marine conditions. During this study no features characteristic of river-dominated deltaic sedimentation, such as well-defined distributary mouth bar deposits, fining upward channel-overbank deposits, and crevasse-splay depos-

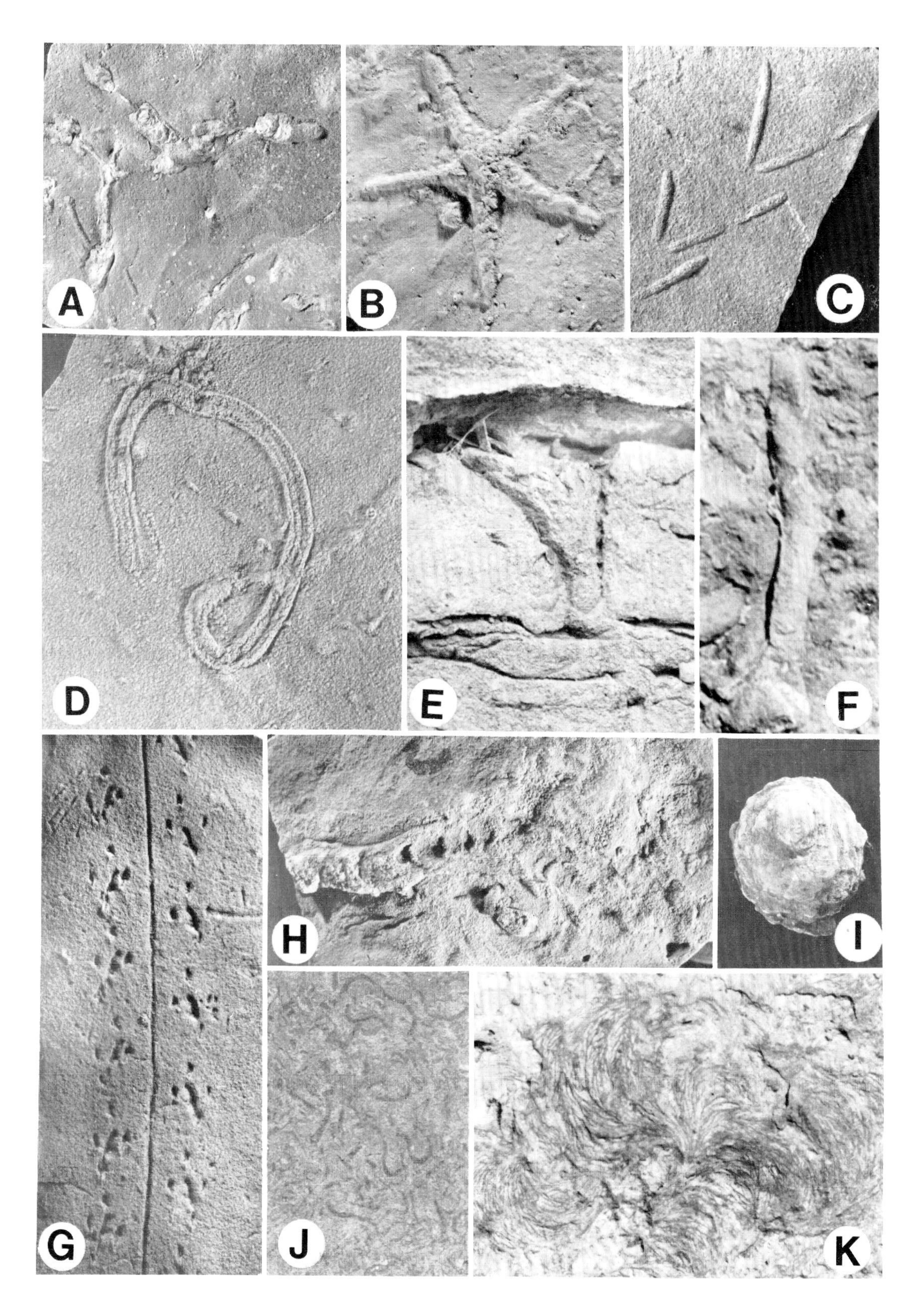
A
B
C
D
E
F
G
H
I
J
K

ts, were recognized. In addition, no evidence of progradation was identified. This does not imply that the sequence was not deposited within a deltaic setting, but rather that a deltaic system is not recognizable at the scale of this study. Many modern deltas, even highly river-dominated deltas such as the Mississippi (Saxena, 1979), have areas where marine processes dominate, and in many these marine-dominated areas are extensive (Fisher et al., 1969; Coleman and Wright, 1975). Because the change from destructive to constructive environments is gradational in modern deltas, the resulting facies changes would also be gradational and difficult to recognize. Study of the sandstones of the Rockcastle Conglomerate and of the sandstones and shales of the overlying Crooked Fork Group should give further evidence of the nature of the coastal setting of which the upper Fentress Formation-lower Rockcastle Conglomerate sequence reflects a small part.

TRACE FOSSILS

All specimens illustrated in Plates 1 and 2 are stored in the Vanderbilt University Department of Geology Repository unless otherwise indicated. Facies distributions of trace fossils from the Fentress Formation and Rockcastle Conglomerate are given in Figure 8. In Figures 9 and 10 compilations of paleoenvironmental distributions for these ichnogenera reported in the literature are compared with the inferred paleoenvironmental distributions during deposition of the Fentress-Rockcastle sequence. Toponomic terms describing position of the traces relative to bedding are defined by Simpson (1975, Fig. 3.1).

Ichnogenus ASTERIACITES von Schlotheim, 1820
Ichnospecies *Asteriacites quinquefolis* Quenstedt, 1876
Pl. 1B

Description.—Star-like traces 9 cm to 14 cm in diameter consisting of 5 arms projecting outward from a central area approximately 2 cm in diameter. Surface texture is coarsely knobby. Preserved in concave epirelief and convex hyporelief.

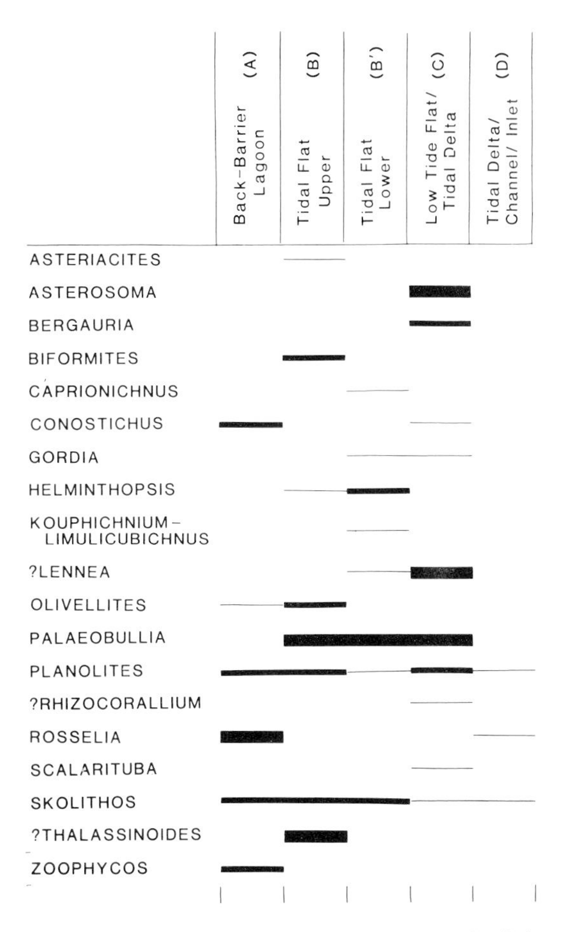

Fig. 8.—Paleoenvironmental distribution of trace fossils in the upper Fentress Fm.—lower Rockcastle Conglomerate. Width of line indicates abundance: thin = rare, thick = abundant, intermediate = common. Letters correspond to facies: A = burrowed sandstone facies; B = very thinly bedded sandstone facies, particularly that with prominent shale interbeds; B' = very thinly bedded sandstone facies, but with less well-developed shale interbeds and more abundant ripple marks than that included in B; C = rippled sandstone facies; D = cross-bedded sandstone facies.

EXPLANATION OF PLATE 1

FIG. A.— *Thalassinoides* on upper bedding surface. From very thinly bedded sandstone facies. Locality 3b, × 2.
B.— Latex mold of *Asteriacites* from very thinly bedded sandstone facies, Locality ca, × 0.5.
C.— *Caprionichnus* preserved on underside of bed (convex hyporelief). From very thinly bedded sandstone facies, Locality 4, × 0.7.
D.— *Palaeobullia*: preservation, facies, and locality as for 1C. Note change in morphology along length of trail, × 0.5.
E.— *Rosselia* in highly bioturbated sandstone, Locality 1, × 1.
F.— *Planolites* in highly bioturbated sandstone, Locality 2, × 0.5.
G.— Latex mold of *Kouphichnium*. Original is preserved in convex hyporelief. From very thinly bedded sandstone facies at Locality 4. U.S. National Museum #314846, × 0.5.
H.— *Scalarituba* in sandstone of the rippled sandstone and siltstone facies, Locality 7, × 0.8.
I.— *Conostichus* from highly bioturbated sandstone facies, Locality 8, × 0.5.
J.— *Helminthopsis* in sandstone from very thinly bedded sandstone facies at Locality 1. Orientation unknown, × 1.
K.— *Zoophycos* on slab of sandstone in float at Locality 7, probably from rippled sandstone facies, × 0.25.

Discussion and interpretation.—Two specimens of *Asteriacites* were found in the Fentress Formation at Locality 3a. One of the arms of the larger specimen (14 cm diameter) is forked, indicating lateral movement of one of the producer's arms.

Seilacher (1953) described and interpreted two varieties of star-like traces. *Asteriacites lumbricalis* is small and characterized by transverse striations on the arms; Seilacher suggested that it is produced by ophiuroids or slender asteroids. The larger form, *A. quinquefolis*, which includes the Fentress Formation specimens, has less regular ornamentation and is interpreted as an asteroid resting trace.

Environmental distribution.—In the Fentress Formation *Asteriacites* occurs in very thinly bedded sandstone facies interpreted as tidal flat deposits (Fig. 8). Hakes (1977) attributed the small size of specimens from the Lawrence Shale (Pennsylvanian, Kansas) to restricted marine conditions, based on the occurrence of small, sexually immature asterozoans in brackish water of the Baltic Sea. The large size of the Fentress Formation specimens may suggest normal marine salinity, especially since *Asteriacites* most commonly has been reported from shallow marine deposits of presumed normal marine salinity. However, some modern ophiuroids and asteroids inhabit waters with salinities

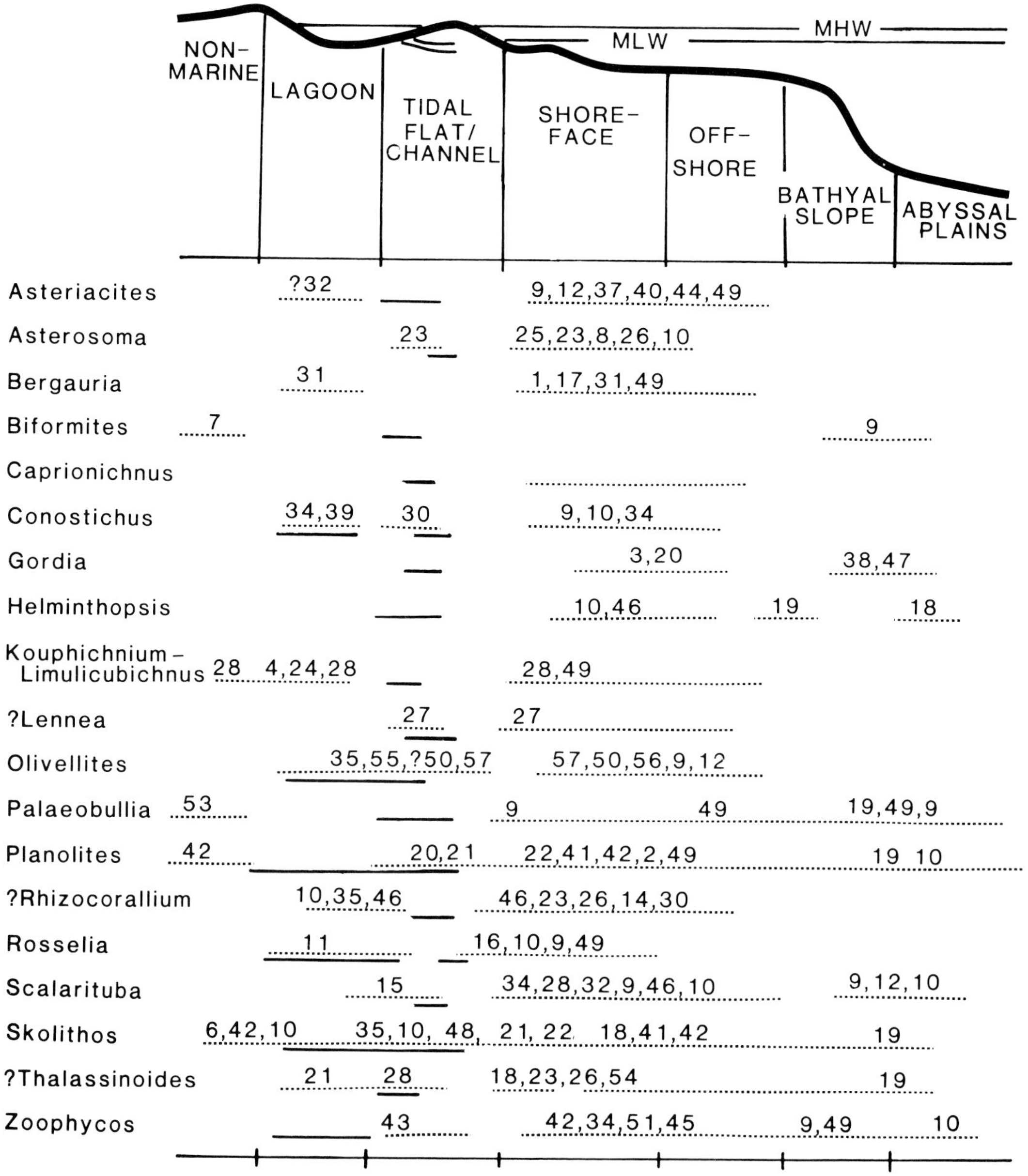

NON-MARINE
LAGOON
TIDAL FLAT/ CHANNEL
SHORE-FACE
OFF-SHORE
BATHYAL SLOPE
ABYSSAL PLAINS
MLW
MHW

Asteriacites ?32 9,12,37,40,44,49
Asterosoma 23 25,23,8,26,10
Bergauria 31 1,17,31,49
Biformites 7 9
Caprionichnus
Conostichus 34,39 30 9,10,34
Gordia 3,20 38,47
Helminthopsis 10,46 19 18
Kouphichnium–Limulicubichnus 28 4,24,28 28,49
?Lennea 27 27
Olivellites 35,55,?50,57 57,50,56,9,12
Palaeobullia 53 9 49 19,49,9
Planolites 42 20,21 22,41,42,2,49 19 10
?Rhizocorallium 10,35,46 46,23,26,14,30
Rosselia 11 16,10,9,49
Scalarituba 15 34,28,32,9,46,10 9,12,10
Skolithos 6,42,10 35,10, 48, 21, 22 18,41,42 19
?Thalassinoides 21 28 18,23,26,54 19
Zoophycos 43 42,34,51,45 9,49 10

as low as 8 percent (Meyer, 1980), and the same may have been true of Pennsylvanian forms.

Ichnogenus ASTEROSOMA Von Otto, 1854
Pl. 2A

Description.—Circular structure 8 to 10 cm in diameter consisting of a series of burrows radiating outward from a central "burrow" 1.5 cm in diameter. Preserved as convex hyporelief on sandstone beds; maximum relief is 3 cm.

Discussion and interpretation.—A notable feature of specimens of *Asterosoma* from the Fentress Formation is the abrupt termination at the apex of the structure, suggesting that a burrow may have extended into the sediment. A specimen of the type species illustrated by Häntzschel (1975, Fig. 25, 1a) shows the same sharp termination. In the Fentress Formation, vertical burrows with laminated, flared upper portions are commonly exposed in vertical sections. Because the presence of a radial structure on the bedding plane cannot be documented, these speciments are included in the ichnogenus *Rosselia* (described below) rather than *Asterosoma*. Although Seilacher (1960, in Häntzschel, 1975) regarded *Rosselia* as a junior synonym of *Asterosoma,* we follow Häntzschel (1975) in maintaining it as a separate ichnogenus.

A variety of forms, including radiating, cylindrical, rod-shaped, zoned, brached, and funnel-shaped burrows have been assigned to the ichnogenus *Asterosoma* (Frey and Howard, 1971a; Farrow, 1966; Chamberlain, 1978). *Asterosoma* from the Fentress Formation most closely resembles the type species *A. radiciforme,* (the radiating form), especially as illustrated by Chamberlain (1978, Fig. 79) from the Pennsylvanian of Ohio. *Asterosoma* is interpreted as a burrow with radiating feeding trails (Chamberlain, 1971a, Tex-Fig. 8); Mesozoic forms have been interpreted as produced by decapod crusaceans (Häntzschel, 1975).

Environmental distribution.—In the Fentress Formation, *Asterosoma* occurs abundantly within discrete horizons of the rippled sandstone-siltstone facies which probably was deposited in a lower tidal flat, perhaps transitional to a tidal delta (Fig. 8). Elsewhere, other forms of *Asterosoma* occur in tidal flat, lagoonal, nearshore, shoreface, and off-shore deposits (Fig. 9). Chamberlain (1978, Fig. 131), however, indicated that the radial form, similar to the Tennessee specimens, may be restricted to a zone just below wave base.

Ichnogenera BERGAUERIA Prantl, 1946,
and CONOSTICHUS Lesquereux, 1876
Pl. 2J, K; Pl. 1I

Description.—Conical or subcylindrical vertically oriented burrows 4 to 6 cm in diameter and 3 to 6 cm high. Exterior marked by transverse wrinkles. Entire

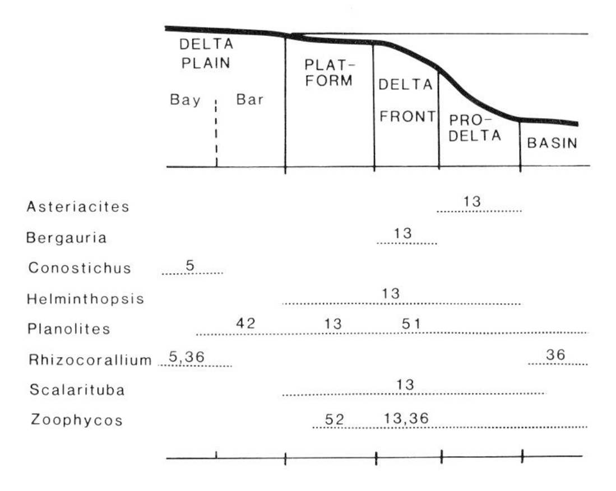

Fig. 10.—Distribution of trace fossils occurring in the Fentress-Rockcastle sequence and in rocks of deltaic origin described in the literature. These distributions were not plotted on Figure 9 because the authors used delta terminology, which cannot be related readily to the terminology used in Figure 9. This does not imply that there is not substantial overlap in environmental conditions and trace producer distributions in the two shoreline regimes. Numbers refer to references given in description for Figure 9.

Fig. 9.—Onshore-offshore distribution of trace fossils occurring in Lower Pennsylvanian rocks of Tennessee. Solid lines indicate ranges within the upper Fentress Fm. -lower Rockcastle Conglomerate, dashed lines indicate ranges of occurrences in literature. Numbers refer to references below and location of numbers reflects approximate position, if indicated. Only a few references are given for Mesozoic and Cenozoic occurrences of *Thalassinoides* and *Rhizocorralium*, and for well-known facies-crossing forms (e.g., *Planolites*). Occurrences of traces considered herein to be junior synonyms are plotted in the distributions of the ichnogenera in which they are included (e.g., the distribution of *Psammichnites* of Chamberlain, 1971b is included with that of *Olivellites*.) The distribution shown for *Palaeobullia* is for the entire Scolicia group. Format modified from Chamberlain, 1978. 1. Alpert, 1973; 2. Alpert, 1975; 3. Bandel, 1967a; 4. Bandel, 1967b; 5. Basan et al., 1979; 6. Boyer, 1979; 7. Bromley and Asgaard, 1979; 8. Campbell, 1971; 9. Chamberlain, 1971b; 10. Chamberlain, 1978; 11. Chamberlain, 1980; 12. Chamberlain and Clark, 1983; 13. Chaplin, 1980; 14. Chisholm, 1968; 15. Conkin and Conkin, 1968; 16. Cotter, 1973; 17. Crimes, 1970; 18. Crimes, 1975; 19. Crimes, 1977; 20. Crimes et al., 1977; 21. Curran and Frey, 1977; 22. Curran and Martino, 1980; 23. Farrow, 1966; 24. Fischer, 1978; 25. Frey and Howard, 1970a; 26. Fürsich, 1975; 27. Goldring and Langenstrassen, 1979; 28. Goldring and Seilacher, 1971; 29. Grasso and Wolff, 1977; 30. Gutschick and Rodriguez, 1977; 31. Hakes, 1976; 32. Hakes, 1977; 33. Harland, 1978; 34. Henbest, 1960; 35. Horne, 1979a; 36. Hubert et al., 1972; 37. Jones, 1935; 38. Ksiazkiewicz, 1970; 39. Lesquereux, 1876; 40. Lewarne, 1964; 41. Miller, 1977; 42. Miller, 1979; 43. Miller and Johnson, 1981; 44. Osgood, 1970; 45. Osgood and Szmuc, 1972; 46. Rodriguez and Gutschick, 1970; 47. Roniewicz and Pienkowski, 1977; 48. Scott and Basan, 1980; 49. Seilacher, 1964; 50. Sheldon, 1968; 51. Siemers, 1975; 52. Thayer, 1974; 53. Turner, 1978; 54. Warme and Olson, 1971; 55. Whaley and Hester, 1979; 56. Williamson and Williamson, 1968; 57. Yochelson and Schindel, 1978.

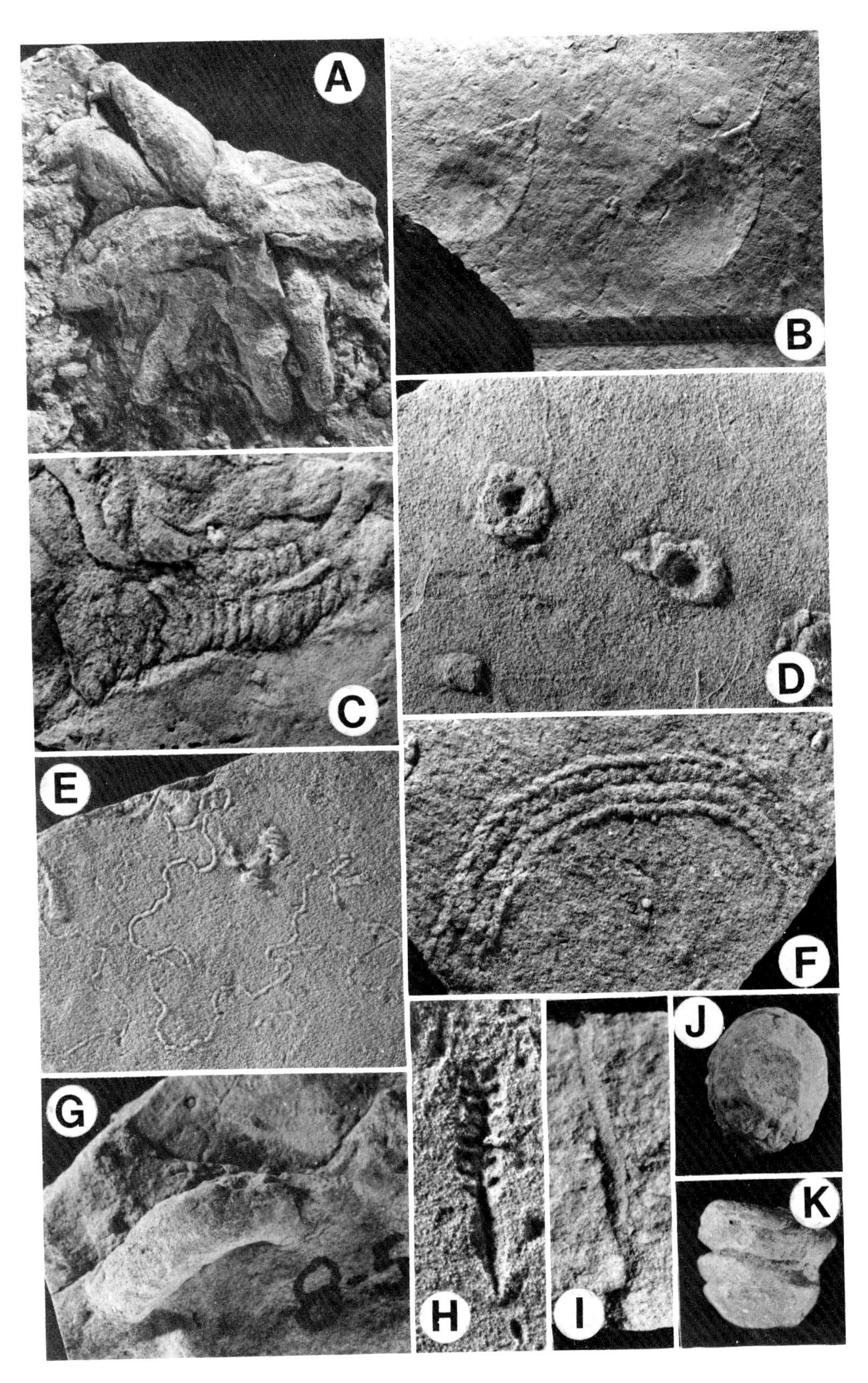
A
B
C
D
E
F
G
H
I
J
K

burrows commonly are weathered out; some also are preserved in concave epirelief or convex hyporelief.

Bergaueria.—Subcylindrical hemispherical burrows 2.5 cm to 4 cm in diameter and 2 to 3 cm high. Exterior marked by few transverse constrictions. Radial ridges may be present on bottom. (Pl. 2J, K).

Conostichus.—Cone-shaped burrow up to 4.5 cm in diameter and 5 cm high, with coarse transverse wrinkles and laminated internal structure. Lower terminus is a narrow bulb. (pl. 11).

Discussion and interpretation.—The major difference between these two trace fossils is in overall shape. *Conostichus* is more conical than *Bergaueria,* which tends to be shaped like a gum drop (Alpert, 1973; Crimes et al., 1977). *Conostichus* is also larger and lacks both the basal radial ornamentation and ther terminal depressiona found in some specimens of *Bergaueria* (Alpert, 1973). One specimen of *Conostichus* found in place at Locality 7 has a dome-shaped cone; all other have slightly depressed upper surfaces.

Bergaueria and *Conostichus* are both generally considered to be sea anemone burrows (Arai and McGugan, 1969; Chamberlain, 1971a; Alpert, 1973), although Bransom (1962) interpreted *Bergaueria* as the burrow of a scyphomedusan. Chamberlain (1972a, Text-Fig. 4) interpreted the terminal bulb of *Conostichus* as the reflection of the expanded physa of the anemone acting as a holdfast. Alpert (1973) suggested that the lamination in *Conostichus* reflects upward movement of the anemone through the sediment, which is supported by reports of vertically stacked specimens (Pfefferkorn, 1971). The *Bergaueia* producer apparently was not capable of escape if covered by sediment.

Conostichus is predominantly a Pennsylvanian form (Chamberlain, 1971a; Pfefferkorn, 1971). *Bergaueria* occurs in rocks of Cambrian through Cretaceous age (Chamberlain, 1978, p. 146), but it is a particularly common Cambrian trace fossil (Arai and McGugan, 1968; Alpert, 1973; Boyd, 1974; Crimes et al., 1977).

Environmental distribution.—In the Fentress Formation *Bergaueria* is rare to common at only one locality (7) in the rippled sandstone facies, and is thought to have been deposited on a lower tidal flat, perhaps transitional to a tidal delta (Fig. 8). *Conostichus* is common in the bioturbated sandstone facies of the Fentress Formation and also occurs, although less abundantly, in tidal flat-delta transition sandstones with *Bergaueria* (Fig. 8). Elsewhere, *Bergaueria* and *Conostichus* occur in rocks deposited in shallow marine environments and deltaic environments (Figs. 9, 10).

Ichnogenus Biformites Linck, 1949
Pl. 2H

Description.—Elongate horizontal trace consisting of segments of two types: (1) smooth burrows 1 to 2 mm in diameter, and 10 to 18 mm long; and (2) elongate series of 2 to 7 chevrons that are 2 to 4 mm in width and 3 to 9 mm long. Trace fossil is composed of two or more segments. Single pairs of segments are straight; those in linear series are straight to curved, or, rarely, circular. Preserved in convex hyporelief.

Discussion and interpretation.—In specimens from the Fentress Formation the V's of the chevrons point away from the smooth segment; this differs from the type specimens illustrated in Häntzschel (1975, Fig. 3a, b) in which the limits of the transverse ridges are perpendicular to the main structure. Cut slabs of specimens revealed no disruption of the casting medium (sandstone). This indicates that the burrows either were not produced by an animal burrowing intrastratally through the overlying sandstone, or that the original overlying sediment was eroded and the present casting sandstone subsequently deposited.

Chamberlain (1971a) described *Biformites* from Pennsylvanian rocks of Oklahoma, but his specimens are larger (7 mm wide) than the Fentress specimens and lack their sharply defined transverse ridges. Our specimens more closely resemble the specimen illus-

←

Explanation of Plate 2

Fig. A.— *Asterosoma* preserved on underside of sandstone bed from rippled sandstone facies at locality 7, × 0.5.
B.— *Limulicubichnus* on upper bedding surface. From very thinly bedded sandstone facies, Locality 4. Scale is 30 cm long. *Kouphichnium* (Pl. 1G) is on under surface of same bed. U.S. National Museum #314833 (left), #314845 (right), × 0.5.
C.— *Olivellites* on upper bedding surface. From bioturbated sandstone facies, Locality 2, × 1.
D.— ?*Lennea* on upper bedding surface. Burrows are vertical, with horizontal components. From very thinly bedded sandstone facies, Locality 4, × 1.5.
E.— *Gordia*. Preservation, facies, and locality as for Plate 2D; × 1.
F.— *Palaeobullia*. Preservation, facies and locality as for Plate 2D, E, × 1.
G.— ?*Rhizocorallium* on underside of sandstone bed, Locality 7. Spreite can be seen in vertical section along cut surface in left of photograph, × 0.8.
H.— Latex mold of *Biformites* preserved on underside of sandstone bed, very thinly bedded sandstone facies, Locality 3a, × 2.
I.— Vertical section of *Skolithos* in slab of sandstone containing *Lumulicubichnus* and *Kouphichnium* (Pl. 1G, 2B). U.S. National Museum #314847, × 2.
J.— Bottom view of *Bergaueria* from rippled sandstone facies, Locality 7, × 1.
K.— Side view of specimen of Figure J, × 1.

trated by Boyer (1979, Fig. 1) from the Jurassic of New Jersey, although the chevrons on the Tennessee specimens are better defined and the smooth portion is shorter. Morphological differences between the Fentress Formation trace fossils and other described forms are great and may be behaviorally significant.

The producer and function of *Biformites* are not well understood. Seilacher (1955; *in* Häntzschel, 1975) interpreted it as a dwelling burrow. However, Chamberlain (1971a) interpreted it as the feeding burrow of a worm-like animal, and Boyer (1979) suggested that it is both a resting and a crawling trace, possibly of an arthropod. The two distinct divisions imply that the producer's body had at least two well defined regions, but whether it was a soft-bodied animal, such as a sipunculid or an arthropod, cannot be determined.

Environmental distribution.—In the Fentress Formation *Biformites* is restricted to rocks of the very thinly bedded sandstone facies, deposited on a tidal flat (Fig. 8). Elsewhere, it occurs in both non-marine and deep water marine deposits (Fig. 9). However, environmental distributions of this trace should be used carefully because of the great morphological variation of forms included in the ichnogenus.

Ichnogenus CAPRIONICHNUS Harland, 1978
Ichnospecies *Caprionichnus steinvikensis* Harland, 1978
Pl. 1C

Description.—Series of chevron-shaped burrows, each 2 mm wide.

Discussion and interpretation.—Harland (1978) established the ichnogenus *Caprionichnus* and ichnospecies *C. steinvikensis* for chevron-shaped burrows occurring in Ordovician rocks of Norway. Other than having more distinctly nested chevrons, these closely resemble the Tennessee forms. Harland (1978) interpreted *Caprionichnus* as representing probings of a deposit-feeding animal moving in a horizontal burrow at or above the surface with the chevron markings. The behavior represented is similar to that reflected by some forms of *Chondrites* (Harland, 1978, p. 99). In the Tennessee specimens the burrows do not actually meet to form a "V". This is consistent with Harland's interpretation because the producer probably would initiate lateral tunneling before reaching the level of the sand-shale interface, along which the chevrons were formed.

Environmental distribution.—*Caprionichnus* is rare in the Fentress Formation. It was found only in the very thinly bedded sandstone facies at Locality 4, which is interpreted as having been deposited on a tidal flat, although somewhat lower on the flat than the very thinly bedded sandstones at Localities 3a and b (Fig. 8). The Ordovician mudstone sequence from which the type species was found was deposited under offshore marine conditions (Harland, 1978).

Ichnogenus GORDIA Emmons, 1844
Pl. 2E

Description.—Long, irregularly curving and bending burrows 1 mm wide preserved in convex hyporelief on bedding planes.

Discussion and interpretation.—The most notable characteristics of specimens from the Fentress Formation are their uniform width and high length-to-width ratio. They are included in the ichnogenus *Gordia* because they lack the regular sinuous curves of *Cochlichnus*, the loose meanders of *Helminthopsis* (Häntzschel, 1975, p. W52, W70) and the regular meanders of *Cosmorhaphe* (Ksiazkiewicz, 1970, p. 292). However, the Tennessee *Gordia* closely resembles *Helminthopsis tenuis* as illustrated by Ksiazkiewicz (1970, Fig. 3b). Crimes et al. (1977) interpret *Gordia* as the feeding burrow of a worm-like animal. Many modern polychaetes are long and narrow and may move along sand-mud interfaces, but it has not been demonstrated that they pack their burrows with fecal matter (Ronan, 1977; Ronan et al., 1981). The nemertean *Paranemertes peregrina* produces branching but otherwise similar traces on surficial intertidal sediments in Bodega Harbor (Ronan et al., 1981).

Environmental distribution.—In the Fentress Formation *Gordia* occurs in the very thinly bedded and rippled sandstone facies in association with *Kouphichnium, Limulicubichnus, ?Lennea, Skolithos* and *Palaeobullia* (Fig. 8). Elsewhere *Gordia* has been recorded from both shallow- and deep-water marine deposits (Fig. 9).

Ichnogenus HELMINTHOPSIS Heer, 1877
Pl. 1J

Description.—Irregular traces of low relief, 1 to 2 mm in diameter. Preserved on top and bottom bedding plane surfaces.

Discussion and interpretation.—Specimens of *Helminthopsis* from the Fentress Formation appear as dark ribbons of fine-grained material on sandstone bedding planes (Pl. 1J). They occur in the upper portions of micrograded sandstones and siltstones. Abundance of discontinuous burrows on bedding planes indicates that the producers moved obliquely up and down within the sediment. This is supported by the fact that very small, dark flattened burrows can be seen in cut vertical sections of very fine-grained sandstones.

These traces closely resemble *Helminthoida* illustrated by Chamberlain (1978, Fig. 118) from the Oligocene of Oregon and by Chaplin (1980, Plate 6, Fig. 4) from the Mississippian of Kentucky. However, the specimens from Tennessee are placed in the ichnogenus *Helminthopsis* because they lack the regular, tight meandering pattern of *Helminthoida*. The distinction between regular meanders (*Helminthoida*) and irregular meanders (*Helminthopsis*) seems useful, and perhaps the Oligocene and Mississippian traces also should be designated as *Helminthopsis*. The Tennessee specimens differ from *Helminthopsis* as illustrated by Ksiazkiewicz (1970) and Häntzschel (1975) in being much smaller. Chamberlain (1971a, p. 227) noted, however, that *Helminthopsis* from the Ouachita Mountains varies from 0.5 mm to 40 mm in diameter.

Chamberlain (1971a, p. 216) interpreted *Helminthopsis* as produced by a vagile deposit feeder and filled with mud, presumably fecal in origin. Polychaetes, the most extensively studied modern "worms," have been observed to deposit fecal matter on the sediment-water interface outside of their burrows rather than packing it within the burrow (Ronan, 1977). In light of this, flattened *Helminthopsis* from the Fentress Formation may be interpreted as the collapsed walls of a burrow produced by a worm-like animal.

Environmental distribution.—In the Fentress Formation, *Helminthopsis* occurs exclusively in the very thinly bedded sandstone facies interpreted as tidal flat in origin (Fig. 8). It is common in flysch deposits, although the very similar forms described by Chamberlain (1978) and Chaplin (1980) both occur in shallow-water sediments. (Figs. 9, 10).

Ichnogenus Kouphichnium Nopsca, 1923
Pl. 1G

Description.—Trackway 4 cm wide consisting of a median furrow separating clusters of 3 to 4 oval pits surrounding a bifid impression.

Discussion and interpretation.—One specimen over a meter long was found on the underside (convex hyporelief) of a sandstone bed at Locality 4; the description above refers to the trace fossil as it would appear on an upper bedding surface.

In a restudy of trails interpreted as amphibian in origin, which included comparison with modern limulid traces, Caster (1938) documented the diversity of trace morphologies which could be produced by horseshoe crabs. Modern limulid walking appendages consist of 4 pairs of legs with bifid terminations and a fifth posterior "pusher" pair with four narrow processes forming a stellate terminus (Barnes, 1974, p. 379). Ancient limulids had similar walking appendages (Størmer et al., 1955). The genal spines and long spike-like posterior telson may also make traces. An "ideal" track may be composed of anteriorly oriented chevrons. Each side of the "V" is composed of 4 oval pits (walking leg impression), a radiate impression (pusher mark) and a medial groove (telson mark). Comparison of specimens figured by Caster (1938, 1944), Bandel (1967b), and Goldring and Seilacher (1971) illustrates the variety of traces limulids can produce.

In studying well-preserved specimens in thinly laminated sediments, Goldring and Seilacher (1971) found that the traces commonly are preserved as undertracks, produced on subsurface interfaces by penetration of surficial sediment by the animals' appendages. Most or all of the component parts are preserved in surface tracks, but not clearly. Undertracks are clearer, but the shallowly produced components are lost at depth. Goldring and Seilacher reexamined a number of specimens from Upper Paleozoic and Triassic rocks and determined the order of component fallout with increasing depth; genal grooves are lost first, followed by telson groove, pusher impressions, and finally front leg impressions.

In specimens from the Fentress Formation, a well-defined pusher impression is not preserved. The bifid marks may have been made by the pushers; however, the front legs also may produce bifid impressions (Goldring and Seilacher, 1971, Fig. 3). Remarkably, the telson groove is continuous and well preserved; this is unusual because it commonly is preserved only on near-surface traces which are generally not as clear as the Tennessee specimen. Caster (1938, Fig. 5) noted that the telson of modern limulids may be used as a pushing organ, which would result in a deeper groove. Perhaps the Tennessee limulid was using its telson as an aid in locomotion.

Environmental distribution.—Only one specimen of *Kouphichnium* was found in the Fentress Formation. It occurred in the very thinly bedded sandstone facies, which is interpreted as a tidal flat deposit, although at this locality (4) the presence of ripple marks and diminished thickness of shale interbeds suggests these were deposited lower on the flat than beds in the adjacent quarries.

Størmer et al. (1955) suggested that the environment in which horseshoe crabs lived has changed between the Devonian and Recent. Devonian forms were marine (Caster, 1938), whereas those from the Carboniferous and Permian were fresh water (Størmer et al., 1955; see also King, 1965; Goldring and Seilacher, 1971). Some occurrences of Triassic limulids are marine, but many occur in brackish to fresh water deposits (Fig. 8). Goldring and Seilacher (1971) restudied limulid tracks and suggested that the apparent paleoenvironmental distribution reflects preservational bias as well as (or instead of) original environmental distribution of the trace producers. Preservation of the tracks would be most likely in areas affected by small-scale turbidity currents, such as non-marine lakes, and unlikely in most marginal marine environments. The fact that limulid traces are occasionally preserved in marine deposits (e.g., Jurassic Solenhofen Limestone, Germany, Goldring and Seilacher, 1971; Pennsylvanian Tonganozie Sandstone, Kansas, Bandel, 1967b) suggests that limulids have lived in marine and marginal marine environments, as they do today, for much of their geological record. The occurrence in the Crab Orchard Mts. Group of *Kouphichnium* and *Limulicubichnus* with trace fossils found elsewhere in marine deposits (*Palaeobullia, ?Lennea, Gordia*) gives further support to Goldring's and Seilacher's (1971) hypothesis.

Ichnogenus ?LENNEA Kräusel and Weyland, 1932
Pl. 2D

Description.—Thick-walled shaft 5 mm in diameter with horizontal and oblique branches. Burrow fill is finer-grained and lighter in color than the surrounding rock.

Discussion and interpretation.—Häntzschel (1975) described and illustrated type specimens which are larger than the Fentress Formation specimens and in which the branches are narrower than the vertical shaft, consistently intersect the shaft at right angles, and bifurcate and branch downward away from the central structure. Because of these differences, the Tennessee specimens are only tentatively included in this ichnogenus; however, the overall similarity indicates that the behavioral patterns represented are similar. Goldring and Langenstrassen (1979) sketched drawings of two types of *?Lennea*, one of which has nonbifurcating branches and superficially resembles the Fentress forms.

The burrow filling in the speciments of *?Lennea* is lighter in color and much finer grained than the surrounding rock. It is not obviously pelleted and gives no indication of having been biologically reworked. Unfortunately, no burrows were found in place, and it is not known whether clay layers consistently overlie the *?Lennea*-bearing beds. Presence of clay layers would support an interpretation of the structures as passively-filled dwelling burrows.

Environmental distribution.—*?Lennea* is common in rocks of the rippled sandstone facies and also occurs in the cleanest sandstones of the very thinly bedded facies. These rocks are interpreted as having been deposited in tidal flat to lower tidal flat-tidal delta transition zone (Fig. 8). *?Lennea* also has been reported from Devonian sandstones in Germany and intertidal sand-shoal deposits (Fig. 9).

Ichnogenus LIMULICUBICHNUS Miller, 1982
Ichnospecies *Limulicubichnus serratus* Miller, 1982
Pl. 2B

Description.—Tear-drop shaped depression on upper bedding surface 9 to 11 cm wide and 15 to 17 cm long. Margins at posterior (narrow) end serrate; terminus is narrow and elongate. Type specimens (Pl. 2B) are in the U.S. National Museum.

Discussion and interpretation.—The ichnogenus *Limulicubichnus* was erected by Miller (1982b) for shallow depressions with features clearly indicative of production by limulids. Limulid resting traces are not common. Hardy (1970) described such traces from the Carboniferous of England and named them *Kouphichnium rossendalensis*. His specimens differ from those from the Fentress Formation in their lunate outline and small size. They more closely resemble the small limulid burrows described and named *K. cordiformis* by Fischer (1978) from the Harding Sandstone (Ordovician) of Colorado; Miller (1982b) included both of these ichnospecies in *Limulicubichnus*.

The modern horseshoe crab *Limulus polyphemus* is abundant along the Gulf and Atlantic coasts of North America. Adults live subtidally along sandy shores, extending to depths of 30 m or more, but they are most abundant from 5 to 6 m (Rudloe, 1979). They migrate onshore to spawn on the supratidal areas of beaches. After hatching, the juveniles move to intertidal and then to subtidal areas, where they mature (Rudloe, 1979). Young horseshoe crabs are nocturnal (Eldredge, 1970, p. 12) and spend their days in shallow burrows.

The mechanics of burrowing by juvenile *Limulus polyphemus* under laboratory conditions was studied by Eldredge (1970). He found that the prosoma (front portion) is bent downward as the walking legs push forward. In flexed position the frontal margin is sharp and, when pushed into the sediment, creates a large pile of sand which is spread over the prosoma. Sediment under the animal is removed by the walking legs. The complete burial is effected by downward arching of the telson into the sediment, followed by forceful pushing of the opisthoma into the sediment, creating a depression and a cloud of sediment which settles as a blanket over previously uncovered portions of the animal (Eldredge, 1970, p. 6).

Features of *Limulicubichnus serratus* are consistent with this mode of burrowing. Slightly raised posterior and lateral margins represent accumulations of sediment swept out by the walking legs. The shallowly sloping anterior portion of the burrow probably was formed as sand slumped in under the prosomal flexure as a result of oversteepening during burrowing.

Environmental distribution.—In the Fentress Formation, two specimens of *Limulicubichnus serratus* were found in the very thinly bedded sandstone facies at Locality 4 (Fig. 8). In other areas *Limulicubichnus* occurs in nonmarine and estuarine deposits of Ordovician and Carboniferous age (Hardy, 1970; Fischer, 1978; see Fig. 9). The behavior of subtidal limulids has not been well studied. If shallow burrowing, whether for protection, for avoidance of desiccation, or for spawning, is more common in the intertidal zone than in the subtidal, the ratio of number of specimens of *Limulicubichnus* to number of specimens of *Kouphichnium* may be a sensitive paleoenvironmental indicator.

Ichnogenus OLIVELLITES Fenton and Fenton, 1937
Ichnospecies *Olivellites plummeri*
Fenton and Fenton, 1937
Pl. 2C

Synonymy.—*Psammichnites* Chamberlain, 1971a; *Cruziana* Whaley and Hester, 1979; *Scolicia* Horne, 1979b.

Description.—Sinuous horizontal trace 1.1 to 1.5 mm wide with thin (1 to 2 mm in diameter) longitudinal median ridge on upper surface. Fine transverse markings bisect structure longitudinally but do not cut medial ridge. Preserved in concave epirelief.

Discussion and interpretation.—Häntzschel (1975, p. W106) included *Olivellites* in the "*Scolicia* group", a group of trace fossils which are similar to *Scolicia* but are not true synonyms. Retention of *Olivellites* as a separate ichnogenus is warranted because *Olivellites* is morphologically distinct and because specimens from widely spaced geographical areas closely resemble one another. Specimens listed in the synonymy and all other reported occurrences are from Carboniferous rocks, indicating that *Olivellites* may be a stratigraphically restricted trace fossil. Fenton and Fenton (1937) interpreted *Olivellites* as the feeding burrow of the gastropod *Olivella* produced just beneath the sediment surface, but Yochelson and Schindel (1978, p. 795) concluded that *Olivellites* was not made by a gastropod. As evidence they cited MacGinitie and MacGinitie's (1949) observation that surface depressions rather than ridges are left by shallow burrowing by *Olivella*, pointed out that modern snails do not pass sediment under the foot, as would be required to produce the concave upward transverse markings, and noted that there are no known Pennsylvanian siphonate snails of a size which could produce a trace with the dimensions of *Olivellites*. The median ridge of the specimens from the Fentress Formation is composed of cleaner sand than the remainder of the burrow or the surrounding rock. This sediment may have been biologically reworked and may be indicative of deposit-feeding activity by an organism whose identity remains unkown.

Environmental distribution.—*Olivellites* occurs in the burrowed sandstone facies but is not abundant (Fig. 8). This facies probably was deposited in a protected back-barrier lagoon. It is also abundant in a thin zone within the very thinly bedded sandstone facies at Locality 3a, which probably was deposited on the upper portion of a tidal flat. In occurrences outside of Tennessee *Olivellites* is found in shallow marine deposits or in coal measures (Fig. 9).

Ichnogenus PALAEOBULLIA Götzinger and Becker, 1932
Pl. 1D, 2F

Description.—Horizontal traces 5 mm to 10 mm wide of variable morphology preserved in concave epirelief and convex hyporelief. Morphological end members preserved on undersides of beds include: (1) single smooth, slightly V-shaped ridges; (2) trails with 2 to 4 lobes and evenly spaced, oblique transverse ridges; (3) four longitudinal rows of tiny knobs.

Discussion and interpretation.—Traces assigned to the ichnogenus *Palaeobullia* are abundant in the Fentress Formation. When preserved in concave epirelief they are nondescript furrows. When preserved on the undersides of beds, the morphology varies between the end members listed above. Differences in morphology are interpreted as resulting from variation in depth of production and in substrate consistency (Knox and Miller, this volume).

Häntzschel (1975, p. W106) subdivided *Scolicia*-like traces into two groups: (1) the *Palaeobullia* group, interpreted as surface trails of gastropods, and (2) the *Subphyllochorda* group, consisting of snail traces produced intrastratally. Häntzschel included an illustration from Götzinger and Becker (1934) showing morphotypes of *Palaeobullia*; this range is similar to that displayed by specimens from the Fentress Formation. Both groups include many genera, the relationships between which have yet to be determined.

Although traces in the *Scolicia* group generally are attributed to gastropod activity, the gastropod origin for at least one member (*Olivellites*) has been disputed (Yochelson and Schindel, 1978). *Pteridichnites*, a trace quite similar to *Palaeobullia*, is interpreted as produced by an arthropod (Häntzschel, 1975, p. W99), and Rodriguez and Gutschick (1970, Plate 7d) illustrated a conchostracan cyziciid trail which resembles some specimens of *Palaeobullia* from the Fentress Formation. However, several lines of evidence, including the facts that specimens of *Palaeobullia* display a range of variation comparable to that of modern snail traces, and that they are very similar to specimens of *Aulichnites*, a gastropod trace fossil (Bandel, 1967a), suggest that the Fentress Formation specimens were produced by gastropods (see Knox and Miller, this volume, for more complete discussion).

Environmental distribution.—In the Fentress Formation, *Palaeobullia* is abundant in sandstones of the very thinly bedded sandstone facies and of the rippled sandstone and siltstone facies (Fig. 8). Its distribution in rocks of different ages and from diverse geographic areas is difficult to assess, because of taxonomic confusion. Seilacher (1964) indicated that "*Scolicia*" is a facies-crossing trace fossil, and it is this distribution which is illustrated in Figure 9 for *Palaeobullia*.

Ichnogenus PLANOLITES Nicholson, 1873
Pl. 1F

Description.—Unbranched horizontal burrows 3 to 10 mm in diameter. Burrow fill is same as matrix. Burrows are unlined, but boundaries are well defined. Some entire burrows weather out, others are preserved in convex hyporelief.

Discussion and interpretation.—Nondescript horizontal burrows of various sizes from the Fentress Formation are included in this ichnogenus. Most commonly, *Planolites* is exposed on the undersurface of beds, but at Locality 1 many whole, isolated burrows can be found in float; these are nearly identical to those described by Chaplin (1980, Pl. 4–3) from the Mississippian of Kentucky. Larger (8 to 10 mm in diameter) specimens tend to be shorter (< 15 cm), which might suggest that their producers did not move for as long distances along sediment interfaces as did their smaller counterparts.

Planolites is interpreted as produced by the deposit-feeding activity of a worm-like animal (Häntzschel, 1975). However, in these specimens there is no evidence that the burrow filling is fecal material. Horizontal burrows filled by material that has not passed through an animal's gut commonly are included in the ichnogenus *Palaeophycus*, but Alpert (1975, p. 512) has suggested that this name be reserved for branching forms. Many modern polychaetes and other "worms" produce burrows with horizontal components, and none of those that have been extensively studied are known to fill their burrows with fecal material (Ronan, 1977). The Tennessee specimens could have resulted from passive filling with sediment of abandoned burrows.

Environmental distribution.—*Planolites* is widely distributed and moderately common in rocks of the Fentress Formation and Rockcastle Conglomerate (Fig. 8). It is a well-known facies-crossing trace (e.g., Seilacher, 1964), and only a few of its many occurrences are included in Figures 9 and 10.

Ichnogenus ?RHIZOCORALLIUM Zenker, 1836
Pl. 2G

Description.—U-shaped burrows 1.8 cm in diameter with horizontally oriented spreite between the U-tube. Burrow preserved in convex hyporelief.

Discussion and interpretation.—*Rhizocorallium* includes spreite-bearing U-tubes which are generally oblique to bedding; tubes are usually thick, the spreite is generally protrusive, and scratch marks, presumably made by crustaceans, are common (Häntzschel, 1975, p. W101). The one specimen from the Fentress Formation tentatively included in the ichnogenus consists of a U-shaped burrow with a diameter similar to that of *Rhizocorallium*. A vertical surface adjacent to the burrow shows a spreite. However, more specimens are needed for reconstruction of the entire burrow network.

Rhizocorallium is interpreted as a feeding and/or dwelling burrow (Häntzschel, 1975). Scratch marks on some specimens indicate a crustacean origin for these specimens. However, the occurrence of this trace in the Cambrian and other Paleozoic rocks much older than those in which crustaceans first appear suggests that other animals produced some Paleozoic *Rhizocorallium*.

Environmental distribution.—*?Rhizocorallium* occurs in the rippled sandstone facies (Fig. 8). In other areas it is found in rocks of diverse ages and deltaic settings (Figs. 9, 10).

Ichnogenus ROSSELIA Dahmer, 1937
Pl. 1E

Description.—Funnel-shaped oblique or vertical burrows preserved in full relief. Upper flared portion up to 4 cm in diameter and characterized by concentrically laminated fill. Vertical burrow may extend through upper funnel.

Discussion and interpretation.—Funnel morphology in the Tennessee specimens is variable. Many specimens are asymmetrical (Pl. 1E); in this regard it resembles the "*Asterosoma*-form helical funnel" described by Frey and Howard (1970a). Total height of the Fentress Formation specimens is not known. In a few specimens exposed on bedding planes, the vertical burrow can be seen to extend upward through the funnel. Chamberlain (1978, p. 152) also noted continuation of the burrows through the funnels and was able to discern that they become horizontally oriented on layers above and below the funnel structure.

Seilacher (*in* Häntzschel, 1975, p. W101) considered *Rosselia* to be a junior synonym of *Asterosoma*. Chamberlain (1971a, Text-fig. 8) interpreted *Rosselia* as a feeding burrow made by a worm-like animal and *Asterosoma* as representing fundamentally the same behavior in a horizontal plane. Alternatively, some specimens of *Rosselia* may be dwelling burrows (Chamberlain and Clark, 1973, p. 680). If so, *Rosselia* should be retained as a separate ichnogenus to include these specimens.

Environmental distribution.—In the Fentress Formation, *Rosselia* is abundant in the burrowed sandstone facies interpreted as perhaps deposited in a back-barrier lagoon or tidal channel. *Rosselia* is rare in the basal part of the cross-bedded sandstone facies (Rockcastle Conglomerate) at Locality 3b where it gradationally overlies the rippled sandstone facies. Here, deposition probably was in the distal portions of a tidal delta or in a tidal channel.

Ichnogenus SCALARITUBA Weller, 1899
Pl. 1H

Description.—Burrow, 10 cm wide, marked by curved transverse ridges spaced 3 to 5 mm apart.

Discussion and interpretation.—*Scalarituba*-like trace fossils are common in the Fentress Formation, but only a single well-preserved specimen was found. It closely resembles a specimen of *Scalarituba* illustrated by Conkin and Conkin (1968, Pl. 4, Fig. 7) and occurs on an upper bedding surface.

Scalarituba is a common Paleozoic trace. Its expression varies depending on substrate type and orientation (Conkin and Conkin, 1968; Chaplin, 1980). Chamberlain (1971a, p. 228) suggested that 2 species of *Neonereites* are synonymous with *Scalarituba*; he attributed morphological differences among the taxa to variations in preservation and orientation rather than to differences in fundamental behavior. Many tentatively identified specimens in the Fentress Formation are similar but lack the characteristic scalariform ridges. The single specimen resembles *?Thalassinoides* from the Fentress Formation (Pl. 1A) but differs in having coarser transverse markings and in lacking branches.

Chamberlain (1971a, Fig. 5) interpreted *Scalarituba* as the result of deposit feeding by an infaunal worm-like animal; Conkin and Conkin (1968, p. 5) also attributed it to a marine worm.

Environmental distribution.—*Scalarituba* occurs in the rippled sandstone facies at Locality 7, interpreted as having been deposited on a lower tidal flat (Fig. 8), perhaps transitional to a tidal delta. Conkin and Conkin (1968, p. 5) implied that *Scalarituba* may be restricted to tidal flat deposits, although Seilacher (1964) designated it as a facies-crossing trace fossil (Fig. 9). It also has been reported from deltaic deposits (Fig. 10).

Ichnogenus SKOLITHOS Haldemann, 1840
Pl. 21

Description.—Vertical to oblique burrows ranging from 1 to 8 mm in diameter.

Discussion and interpretation.—A variety of types of vertical burrows are here included in the ichnogenus *Skolithos*. Lumping of specimens in *Skolithos* is justified because it is not clear how much variation in burrow characteristics of size and nature of the lining is related to factors such as ontogenetic differences or substrate characteristics. However, three broad groups are recognizable: (1) very small (< 1 mm) unlined burrows expressed on upper bedding surfaces as small pits and on cut vertical sections as thin zones of disrupted sediments; (2) larger, well-defined but unlined burrows 2 to 3 mm in diameter (Pl. 2I); and (3) still larger (7 to 8 mm) vertical to oblique burrows which may be lined or partly lined with clay of carbonaceous material. Sediment filling the burrows is the same grain size as the surrounding sandstone.

Skolithos is not abundant in any of the facies and never occurs in closely packed aggregates as it does in Cambrian rocks from Sweden (Häntzschel, 1975, p. W107) and in other Paleozoic rocks. Many worm-like animals live in vertical burrows and tubes (Barnes, 1974, p. 486). These include both suspension feeders, such as phoronids, and deposit feeders, such as maldanid polychaete worms (Ronan et al., 1981).

Environmental occurrence.—The occurence of *Skolithos* in all facies of the Fentress Formation and the lower Rockcastle Conglomerate (Fig. 8) is similar to its facies-independent distribution elsewhere (Fig. 9).

Ichnogenus ?THALASSINOIDES Ehrenberg, 1944
Pl. 1A

Description.—Horizontal branching burrows 7 to 12 mm in diameter. Branches are generally Y-shaped. Burrow filling consists of concentric layers of sandstone and shale. Preserved in convex hyporelief and concave epirelief.

Discussion and interpretation.—?*Thalassinoides* in the Fentress Formation was produced by an animal burrowing along an interface and filling its burrow as it moved. The meniscate nature of the fill is best seen on upper bedding surfaces (Pl. 1A). Burrow fill is composed of sandstone from the lower and upper portions of the graded beds in which the traces occur. Although movement was generally horizontal, undulations along the interface resulted in variations in diameter of burrows exposed on bedding planes and the preservation of only short segments. Associated with ?*Thalassinoides* in the Fentress Formation are knobs on the undersides of beds 1.5 to 3 cm in diameter. The relationship between these and ?*Thalassinoides* is not clear.

Thalassinoides is a common and widespread ichnogenus in Mesozoic and Tertiary rocks (see Frey, 1975 for references). Typically, the branching networks form polygons with swellings common at points of bifurcations (Häntzschel, 1975, p. W116– W117). Because neither of these features is shown by specimens from the Fentress Formation, they are only tentatively included in this ichnogenus. *Thalassinoides* is interpreted as a feeding/dwelling burrow of one or more crustaceans, an interpretation corroborated by occurrences of thalassinid shrimp within burrows (Häntzschel, 1975).

Thalassinoides only rarely has been reported from Paleozoic rocks and meniscate fillings have not been described from Paleozoic specimens, although they occur in younger specimens (Kern and Warme, 1974; Farmer and Miller, 1981). Gutschick and Rodriguez (1977, p. 203, Pl. 2d) describe and illustrate *Thalassinoides* from Upper Devonian rocks of Montana which are similar to the specimens of this study but lack the meniscate fill. Chamberlain and Clark (1973) reported a *Thalassinoides*-like structure from the Pennsylvanian of Utah, and Pickerill and Roulston (1977) also described a form similar to *Thalassinoides* from Silurian rocks of the Gaspé Peninsula. Silicified branching networks of probable biogenic origin also occur in Ordovician rocks of Nevada (Miller, 1977). A major difficulty in including these forms in the ichnogenus *Thalassinoides* is that the presumed producers, thalassinids and possibly other crustaceans, are not known as fossils in rocks older than Jurassic age (Glaessner, 1969). Either these animals' traces were preserved better than their skeletons, or other animals produced similar burrows during the Paleozoic. The former is supported by the recent report of decapods in Devonian rocks by Schram et al. (1978) and their suggestion that knowledge of early history of decapods is still incomplete. In any case, morphological similarity between the Paleozoic and younger traces is great enough to warrant inclusion in the same ichnogenus.

Environmental distribution.—?*Thalassinoides* is abundant in the very thinly bedded sandstone facies, interpreted as tidal flat deposits (Fig. 8). It is abundant at Locality 3b, but much less common a few hundred meters away at Localities 3a and 4. In other areas outside of Tennessee it occurs in both shallow and deep water marine deposits (Fig. 9).

Ichnogenus Zoophycos Massalongo, 185
Pl. 1K

Description.—Fan-shaped structure on bedding plane consisting of a layer of thin, curved, concentric but non-intersecting lamellae.

Discussion and interpretation.—*Zoophycos*-bearing sandstones from the Fentress Formation at Localities 1 and 7 were so intensely reworked by the producer of *Zoophycos* that individual complete specimens are very rare. However, the trace fossils clearly fall within the morphological range displayed by shallow water specimens of *Zoophycos* from the Devonian of New York (Miller, 1978) and also resemble specimens from the Mississippian of Kentucky (Chaplin, 1980).

Although a number of different origins have been proposed for *Zoophycos* (Plumstead, 1967; Plicka, 1968; Bradley, 1973), it is generally regarded as a result of deposit feeding by a worm-like animal (Sarle, 1906; Bischoff, 1968; Simpson, 1970; Häntzschel, 1975).

Environmental distribution.—*Zoophycos* is found in sandstone beds about 0.5 m thick capping the burrowed sandstone facies at Localities 1 and 7. These rocks probably were deposited in a back-barrier lagoon or tidal channel. The sandstone bed in which *Zoophycos* occurs at Locality 7 overlies burrowed siltstones and perhaps originated as a washover deposit.

Zoophycos was originally considered indicative of deposition in bathyal environments (Seilacher, 1964; Chamberlain, 1971b). However, its occurrence in shallow water deposits subsequently has been well documented (Seilacher, 1967; Osgood and Szmuc, 1972; Thayer, 1974; Miller, 1978, 1979; see Figs. 9, 10). The significance of its occurrence in the Fentress Formation is that it gives further evidence that the *Zoophycos* producer(s) could live in extremely shallow-water, probably back-barrier environments. Previous records of its occurrence in shoreline deposits include brief allusions to its occurrence in siltstones alternating with coals (Carboniferous, England; Kennedy, 1975, p. 394) and in intertidal siltstones (Devonian, New York; Miller and Johnson, 1981).

PALEOENVIRONMENTAL SIGNIFICANCE OF THE FENTRESS FORMATION-LOWER ROCKCASTLE CONGLOMERATE ICHNOFAUNA

The significant feature of the Fentress Formation-lower Rockcastle Conglomerate ichnofauna is that it is composed primarily of marine trace fossils, although it occurs in a coal-bearing sequence of sandstones, shales and siltstones. As shown in Figures 9 and 10, only *Biformites, Kouphichnium, Planolites, Scolicia*, and *Skolithos* have been recorded from rocks deposited under non-marine conditions. The other 15 trace fossils are known only from marine deposits. The single non-marine occurrence of *Scolicia* is questionable, because the trace fossil in question was only tentatively identified as *Scolicia* (Turner, 1978). Most other occurrences of *Skolithos* are in rocks also deposited under marine conditions. Significantly, trace fossils often found elsewhere in non-marine deposits, including *Scoyenia, Isopodichnus, Pelecypodichnus, Cylindricum, Steinichnus, Fuersichnus* (Trewin, 1976; Bromley and Asgaard, 1979) and other unnamed traces (Daley, 1968; Stanley and Fagerstrom, 1974) are absent. Presence of a marine ichnofauna and absence of a non-marine ichnofauna indicate that the sequence was deposited under predominantly marine conditions. Trace fossil data are consistent with deposition of the lower Pennsylvanian clastic rocks in a barrier complex or in a tide- or wave-dominated deltaic system.

Many of the trace fossils in the Fentress Formation are facies-restricted (Fig. 8). Three broad trace assemblages are recognizable: (1) the *Palaeobullia*—*?Thalassinoides* assemblage; (2) the *Conostichus*—*Rosselia* assemblage; and (3) the *?Lennea* assemblage.

The *Palaeobullia*—*?Thalassinoides* assemblage consists of *Palaeobullia, ?Thalassinoides, Helminthopsis, Biformites, Asteriacites, Kouphichnium, Limulicubichnus, Caprionichnus, Planolites, Lennea*, and *Skolithos*. *Olivellites* is abundant in one layer at one locality. The *Palaeobullia*—*?Thalassinoides* assemblage is restricted to rocks of the very thinly bedded sandstone facies, interpreted as tidal flat deposits. Of particular interest is the occurrence of two specimens of *Asteriacites* produced by an actively burrowing asteroid. Because echinoderms generally are stenohaline (Beerbower, 1968), it is tempting to use this occurrence as indicative of normal marine salinity. However, even starfish occurrences must be interpreted with care, for some modern asteroids and ophiuroids are known to tolerate brackish water (Meyer, 1980).

An important characteristics of this assemblage is that the dominant ichnogenus changes from outcrop to outcrop. At Locality 3a *Palaeobullia* is the most abundant trace fossil and *Biformites* is common, whereas at Locality 3b, only a few hundred meters away, *?Thalassinoides* is the most abundant trace. *Helminthopsis* is most abundant in this facies at Locality 1, and *?Lennea, Kouphichnium*, and *Limulicubichnus* occur at Locality 4 in sandstones slightly cleaner than those at Localities 3a and 3b. These sandstones probably were deposited lower on the tidal flat. Variation in trace abundances within the same facies from outcrop to outcrop or even within a single outcrop (e.g., the restriction of *Olivellites* within the very thinly bedded sandstone facies to one thin bed at Locality 3a, where it is profuse) suggests that the trace producers were very sensitive to microenvironmental changes. These may have been either slight changes in the physical regime, such as substrate consistency, salinity, length of exposure (related to elevation), or biological interactions (such as variations in intensity of competition or predation). Faunal distributions on modern tidal flats have been shown to be related to height above mean low water and substrate consistency (Ricketts and Calvin, 1968; Frey and Howard, 1970b; Dörjes, 1977). They have also been found to be controlled by biologic interactions such as interference competition (Rhoads and Young, 1970; Ronan, 1975; Woodin, 1976) and predation (Wiltse, 1980).

It is also possible that the observed occurrences of these trace fossils reflects random or patchy distributions of the trace producer rather than responses to variations in microenvironmental conditions. Many modern invertebrates on tidal flats occur in clumps. In Mugu Lagoon the echinoid *Dendraster* is found in great densities in small areas (approximately 100 m^2) on the lower tidal flat, and in Barnstable Harbor, as well as in other marshes along the east coast, the snail *Nassarius obsoletus* swarms on the edges of tidal creeks in densities of hundreds per square meter. Patchy distributions are especially common in populations of immature animals, and random distributions are common for adults (Jackson, 1968; Levinton, 1977). However, given the low probability for trace preservation and limited outcrop, originally randomly distributed traces may appear finally as clumped trace fossils.

The *Conostichus*—*Rosselia* assemblage consists of these two ichnogenera as well as *Olivellites, Zoophycos, Planolites* and *Skolithos* and occurs in the burrowed sandstone-siltstone facies. *Planolites* and *Skolithos* are ubiquitous, but the abundance of *Rosselia, Conostichus, Zoophycos*, and *Olivellites* are variable. *Rosselia* is abundant, and *Conostichus*, throughout the facies at Locality 1, whereas *Zoophycos* is restricted to one layer, and *Olivellites* is absent. It is tempting to attribute the absence of *Olivellites* to reduced salinity at Locality 1, possibly caused by great distance from an inlet. *Olivellites* occurs in shallow marine deposits not associated with coals, and in Pennsylvanian rocks of Texas and other areas. This distribution differs from that of *Conostichus*, which is common in rocks associated with coals, suggesting that its producer may have been tolerant of brackish or fresh water. Temporarily increased salinity may have allowed the uppermost sands of this facies to be colonized by the *Zoophycos*-producer at Localities 1 and 7. However, relating variations in distribution and abundances of individual ichnogenera within this facies to salinity changes is speculative because there is no physical record of these changes within the rocks. The fact that physical sedimentary structures have been nearly obliterated by burrowing makes it difficult to detect gradients which may have existed in physical environmental parameters, and thus to relate them to ichnofaunal distributions. On a world-wide scale, the ichnogenera included in this assemblage from the Fentress Formation are most abundant in fine sandstones deposited in protected areas. The presence of *Zoophycos* in the Fentress Formation gives further evidence that a *Zoophycos* producer was capable of inhabiting very shallow water environments during the Paleozoic (Osgood and Szmuc, 1972; Grasso and Wolff, 1977; Miller, 1979; Miller and Johnson, 1981).

The ?*Lennea* assemblage includes ?*Lennea, Gordia, Asterosoma, Bergaueria, Conostichus, Scalarituba, Palaeobullia, ?Rhizocorallium, Planolites, Skolithos*, and perhaps *Zoophycos. Scalarituba, ?Rhizocorallium*, and ?*Thalassinoides* are rare. The assemblage occurs in rocks deposited on a lower tidal flat, or in an area transitional between a tidal flat and tidal delta or channel. ?*Lennea* dominates in the clean sandstones at Locality 5, which may have been closer to the tidal channel. *Asterosoma* occurs in profusion in one bed at Locality 7.

Very few biogenic structures occur in the cross-bedded sandstone facies of the Rockcastle Conglomerate. Only one specimen of *Rosselia* and rare specimens of the facies-independent *Skolithos* and *Planolites* were found. The absence of trace fossils, even in the sandstones sufficiently fine grained to allow preservation of traces, is related to the high bed-shear conditions which probably were inimical to most trace producers and, during sediment erosion and transportation, obliterated traces that were formed.

Variations in trace fossil distributions and abundances occur on two scales within the Fentress Formation. On a small scale, abundance of traces change within a single facies over a lateral distance of a few hundred meters. This may have been caused by one or more of several factors. First, the trace fossil distributions may reflect patchy or random distributions of the trace producers and be unrelated to small-scale environmental changes; such distributions are known to occur on modern tidal flats (Jackson, 1968; Levinton, 1977). Second, the observed distributions may result from interactions such as competition or predation. Predation may have affected where the trace producers could live, but its effects cannot be interpreted from trace fossils. Competition probably is only effective in structuring high density modern intertidal benthic communities (Wiltse, 1980); it is unlikely to have been important in controlling distributions of trace producers in the sparsely populated Pennsylvanian tidal flats, but it may have been a factor affecting communities inhabiting the back-barrier lagoons or tidal creeks. Thirdly, the trace fossil distributions may reflect subtle changes in environmental conditions. Unlike other marine environments, such as in the deep-sea, coastal environments are characteristically a complex mosaic of subenvironments with distinct substrate, exposure, and salinity (to name a few) characteristics. Distributions of modern intertidal invertebrates have been shown to correlate with environmental patchiness (Fenchel, 1977). Correlation between subtle environmental changes interpreted from rock type and sedimentary structures with variations in trace fossil abundances, especially in the very thinly bedded sandstone and ripple sandstone facies of the Fentress Formation suggest that physical environmental factors were the dominant controls on trace distributions.

On a larger scale, different facies within the Fentress Formation and lowermost Rockcastle Conglomerate are characterized by different trace fossil assemblages. Sharp lateral and vertical lithologic changes within the sequence indicate that changes in physical environmental conditions were abrupt. Ichnofacies are broadly correlative with lithofacies, indicating that the trace producers were sensitive to changes in the physical environmental regime. It is under these conditions where environmental boundaries were distinct, rather than where the environmental gradient was low (Miller, 1979), that trace fossils are most valuable as aids in environmental reconstruction.

SUMMARY

Using lithologic, sedimentologic, and trace fossil data, coal-bearing rocks on the northern Cumberland Plateau of Tennessee are interpreted as having been deposited in back-barrier, tidal flat, and tidal channel or delta subenvironments within a barrier or a marine-dominated deltaic system. Contribution of the trace fossils in interpreting the depositional environment is threefold. First, the majority of traces found in the otherwise nearly unfossiliferous Fentress Formation occur elsewhere only in marine rocks, giving strong evidence for a generally marine-dominated depositional regime. Secondly, because many of the traces are facies-restricted and because the vertical and lateral facies boundaries are sharply defined, the trace fossils serve as sensitive and reliable indicators of subenvironments within the depositional setting. Finally, because they yield important environmental information in spite of a lack of the good exposures necessary for the accumulation and interpretation of sedimentologic data, trace fossils are an especially useful tool for interpreting the paleodepositional environments of rocks of the northern Cumberland Plateau and in other areas where outcrops are small and scattered.

ACKNOWLEDGMENTS

John Wheeler and Oakley Hinds allowed access to their building stone quarries and generously donated slabs. We very much appreciate the help of Peggy Wrenne for preparing the photographs, Susan Jackson and Sandra Ward for drafting the figures, and Kelly Boyte for the typing the manuscript. J.H. Barwis, H.A. Curran, and R.W. Scott made many helpful comments on an earlier version of this manuscript, and we thank them for their thoughtful suggestions.

APPENDIX

All localities are in the state of Tennessee (see Fig. 1) and can be located on U.S. Geological Survey 7½' topographic quadrangle maps using the following directions.

Locality 1.—Moodyville Quadrangle. NE part of map, in strip mines on north and northwest sides of Golman Mt. All facies exposed.

Locality 2.—Pall Mall Quadrangle. NW part of map, in strip mine on southwest side of Goodman Ridge near head of Shellotte Branch (tributary of Wolf River). All facies exposed except rippled sandstone-siltstone facies.

Locality 3.—Riverton Quadrangle. NC-NE part of map, in small unmarked building stone quarry approximately 0.5 miles (0.8 km) east-southeast of Hinds Chapel. Jeep trail shown on map off of paved road leads to Locality 3a; 3b is on opposite side of small gully, a few hundred yards away. Very thinly bedded sandstone facies, rippled sandstone facies, shale-siltstone facies, and cross-bedded sandstone facies are exposed.

Locality 4.—Riverton Quadrangle. NC part of map, in small overgrown quarry in woods, about 30 feet west of jeep trail leading northwest from paved road, about 0.8 mi (1.2 km) east of Hinds Chapel. Jeep trail leads road just east of radio tower at sharp bend in road; outcrop is approximately 0.15 mi (0.25 km) northwest of paved road. Very thinly bedded sandstone facies exposed.

Locality 5.—Riverton Quadrangle. NC part of map, in building stone quarry 0.1 mile northwest of cemetery and 0.7 mi (1.1 km) southwest of Hinds Chapel. Rippled sandstone facies and cross-bedded sandstone facies exposed.

Locality 6.—Jamestown Quadrangle. SC part of map, in roadcut along paved road, 2.8 mi (4.5 km) west of junction of road with Hwy. 28 at Livingstone Center. All facies present, but generally poor exposure.

Locality 7.—Grimsley Quadrangle. NC part of map, along dirt road and jeep trail extending west from Hwy. 28, 1 mi (1.6 km) north of Jamestown Municipal Airport. Dirt road (with gate) leads to Layton Hood Coal Distributors; jeep trail descends escarpment formed by Rockcastle Conglomerate. Exposures are in abandoned quarry on hillside. Rippled sandstone facies and cross-bedded sandstone facies exposed.

Locality 8.—Obey City Quadrangle. EC part of map, in strip mine on south side of paved road, 0.6 mi (1 km) east of Cliff Springs. Mine extends south for more than 0.25 mi (0.4 km). All facies except rippled sandstone-siltstone facies exposed.

REFERENCES

ALLEN, J.R.L., 1970, Sediments of the modern Niger Delta: A summary and review, *in* Morgan, J.P. (ed.), Deltaic Sedimentation—Modern and Ancient: Soc. Econ. Paleontologists Mineralogists Spec., p. 138–151.

ALPERT, S.P., 1973, *Bergaueria* Prantl (Cambrian and Ordovician), a probable actinian trace fossil: Jour. Paleontology, v. 47, p. 919–924.

______, 1975, *Planolites* and *Skolithos* from the Upper Precambrian-Lower Cambrian, White-Inyo Mountains, California: Jour. Paleontology, v. 49, p. 508–521.

ARAI, M.N., AND MCGUGAN, A., 1969, A problematical Cambrian coelenterate (?): Jour. Paleontology, v. 42, p. 205–209.
BAGANZ, B.P., HORNE, J.C., AND FERM, J.C., 1975, Carboniferous and Recent Mississippi lower delta plains—A comparison: Gulf Coast Assoc. Geol. Soc., Trans., v. 37, p. 556–591.
BANDEL, K., 1967a, Trace fossils from two Upper Pennsylvanian sandstones in Kansas: Univ. Kansas Paleontol. Contr., Paper 18, 13 p.
______, 1967b, Isopod and limulid marks and trails in Tonganoxie Sandstone (Upper Pennsylvanian) of Kansas: Univ. Kansas Paleontol. Contr., Paper 19, 10 p.
BARNES, R.D., 1974, Invertebrate Zoology, 2nd ed.: Philadelphia, W.B. Saunders, Inc., 870 p.
BARWIS, J.H., 1979, Sedimentology of some South Carolina tidal-creek point bars, and a comparison with their fluvial counterparts, *in* Miall, A.D. (ed.), Fluvial Sedimentology: Can. Soc. Petroleum Geologists, Mem. 5, p. 129–160.
______, AND HAYES, M.O., 1979, Regional patterns of modern barrier island and tidal inlet deposits as applied to paleoenvironmental studies, *in* Ferm, J.C., and Horne, J.D. (eds.), Carboniferous Depositional Environments in the Appalachian Region: Columbia, S.C., Carolina Coal Group, p. 472–498.
______, AND MAKURATH, J.H., 1978, Recognition of ancient tidal inlet sequences: An example from the upper Silurian Keyser Limestone in Virginia: Sedimentology, v. 25, p. 61–82.
BASAN, P.B., MASON, C.E., AND CHAPLIN, J.R., 1979, Deltaic trace fossil associations in the lower tongue of Breathitt (Pennsylvanian) near Morehead, Kentucky: 9th Intl. Cong. of Carboniferous Stratigraphy and Geology, Abstracts of Papers, Urbana, Univ. of Illinois, p. 11–12.
BEERBOWER, J.R., 1968, Search for the Past, 2nd ed.: Englewood Cliffs, N.J., Prentice-Hall, Inc., 512 p.
BISCHOFF, B., 1968, *Zoophycos*, a polychaete annelid, Eocene of Greece: Jour. Paleontology, v. 42, p. 1439–1443.
BLATT, H., MIDDLETON, G., AND MURRAY, R., 1980, Origin of Sedimentary Rocks, 2nd ed.: Englewood Cliffs, N.J., Prentice-Hall, Inc., 782 p.
BOYD, D.W., 1974, Wyoming specimens of the trace fossil *Bergaueria*: Contr. to Geology, v. 13, p. 11–15.
BOYER, P.S., 1979, Trace fossils *Biformites* and *Fustiglyphus* from the Jurassic of New Jersey: Bull. New Jersey Acad. Science, v. 24, 73–77.
BRADLEY, J., 1973, *Zoophycos* and *Umbellula* (Pennatulacea): Their synthesis and identity: Palaeogeogr., Palaeoclimatol., Palaeoecol., v. 13, p. 103–128.
BRANSOM, C.C., 1962, *Conostichus*, a scyphomedusan index fossil: Oklahoma Geol. Notes, v. 22, p. 251–253.
BROMLEY, R., AND ASGAARD, U., 1979, Triassic freshwater ichnocoenoses from Carlsberg Fjord, East Greenland: Palaeogeogr., Palaeoclimatol., Palaeoecol., v. 28, p. 39–80.
CARLSON, G.D., 1979, Depositional modeling of Carboniferous rocks applied to coal exploration, northern Cumberland Plateau, Tennessee, *in* Ferm, J.C., and Horne, J.D. (eds.), Carboniferous Depositional Environments in the Appalachian Region: Columbia, S.C., Carolina Coal Group, p. 422–427.
CASTER, K.E., 1938, A restudy of the tracks of *Paramphibius*: Jour. Paleontology, v. 12, p. 2–60.
______, 1944, Limuloid trails from the Upper Triassic (Chinle) of the Petrified Forest National Monument, Arizona: Am. Jour. Science, v. 242, p. 78–84.
CHAMBERLAIN, C.K., 1971a, Morphology and ethology of trace fossils from the Ouachita Mountains, southeastern Oklahoma: Jour. Paleontology, v. 45, p. 212–246.
______, 1971b, Bathymetry and paleoecology of Ouachita Geosyncline of southeastern Oklahoma as determined from trace fossils: Am. Assoc. Petroleum Geologists Bull., v. 55, p. 34–50.
______, 1978, Recognition of trace fossils in cores, *in* Basan, P.B. (ed.), Trace Fossil Concepts: Soc. Econ. Paleontologists Mineralogists Short Course No. 5, Lecture Notes, p. 133–183.
______, 1980, The trace fossils of the Cretaceous Dakota hogback along Alameda Avenue, west of Denver, Colorado, *in* Basan, P.B. (ed.), Trace Fossils of Nearshore Depositional Environments of Cretaceous and Ordovician Rocks, Front Range, Colorado: Soc. Econ. Paleontologists Mineralogists Field Trip No. 1, Guidebook, p. 30–41.
______, AND CLARK, D.L., 1973, Trace fossils and conodonts as evidence for deep-water deposits in the Oquirrh Basin of central Utah: Jour. Paleontology, v. 47, p. 663–682.
CHAPLIN, J.R., 1980, Stratigraphy, trace fossil associations and depositional environments in the Borden Formation (Mississippian), northeastern Kentucky: Kentucky Geol. Soc., 1980 Ann. Fall Field Trip, Lexington, Kentucky Geol. Survey, 114 p.
CHISHOLM, J.I., 1968, Trace fossils from the Geological Survey boreholes in East Fife, 1963–64: Geol. Survey Great Britain Bull., no. 28, p. 103–119.
COLEMAN, J.M., AND WRIGHT, L.D., 1975, Modern river deltas: Variability of processes and sand bodies, *in* Broussard, M.L. (ed.), Deltas: Models for Exploration: Houston, Houston Geological Society, p. 99–149.
CONKIN, J.E., AND CONKIN, B.M., 1968, *Scalarituba missouriensis* and its stratigraphic distribution: Univ. Kansas Paleontol. Contr., Paper 31, 7 p.
COTTER, E., 1973, Large *Rosselia* in the Upper Cretaceous Ferron Sandstone, Utah: Jour. Paleontology, v. 47, p. 975–978.
CRIMES, T.P., 1970, The significance of trace fossils in sedimentology, stratigraphy, and paleoecology with examples from Lower Paleozoic strata, *in* Crimes, T.P., and Harper, J.C. (eds.), Trace Fossils: Geol. Jour., Spec. Issue 3, Liverpool, Seel House Press, p. 101–126.
______, 1975, The stratigraphical significance of trace fossils, *in* Frey, R.W. (ed.), The Study of Trace Fossils: New York, Springer-Verlag, p. 109–130.
______, 1977, Trace fossils of an Eocene deep-sea sand fan, northern Spain, *in* Crimes, T.P., and Harper, J.C. (eds.), Trace Fossils 2: Geol. Jour., Spec. Issue 9, Liverpool, Seel House Press, p. 71–90.
______, LEGG, F., MARCOS, A., AND ARBOLEYA, M., 1977, ?Late Precambrian-low Lower Cambrian trace fossils from Spain,

in Crimes, T.P. and Harper, J.C. (eds.), Trace Fossils 2: Geol. Jour., Spec. Issue 9, Liverpool, Seel House Press, p. 91–138.

Curran, H.A., and Frey, R.W., 1977, Pleistocene trace fossils from North Carolina (U.S.A.), and their Holocene analogues, *in* Crimes, T.P., and Harper, J.C. (eds.), Trace Fossils 2: Geol. Jour., Spec. Issue 9, Liverpool, Seel House Press, p. 139–162.

_____, and Martino, R.L., 1980, Trace fossil assemblages of Upper Cretaceous sand units, Delaware and New Jersey [abstr.]: Amer. Assoc. Petroleum Geologist Bull., v. 64, no. 5, p. 694–695.

Daley, B., 1968, Sedimentary structures from a non-marine horizon in the Bembridge Marls (Oligocene) of the Isle of Wight, Hampshire, England: Jour. Sed. Petrology, v. 38, p. 114–127.

Donaldson, A.C., Martin, P.H., and Kanes, W.H., 1970, Holocene Guadalupe Delta of Texas Gulf Coast, *in* Morgan, J.P. (ed.), Deltaic Sedimentation: Modern and Ancient: Soc. Econ. Paleontologists Mineralogists Spec. Pub. No. 15, p. 107–137.

Dörjes, J., 1977, Marine macrobenthic communities of the Sapelo Island, Georgia region, *in* Coull, B.C. (ed.), Ecology of Marine Benthos: Columbia, S.C., Univ. of South Carolina Press, p. 399–421.

_____, and Howard, J.D., 1975, Estuaries of the Georgia Coast, U.S.A.: Sedimentology and biology. IV. Fluvial-marine transition indicators in an estuarine environment, Ogeechee River-Ossabaw Sound: Senckenbergiana Marit., v. 7, p. 137–179.

Eldredge, N., 1970, Observations on burrowing behavior in *Limulus polyphemus* (Chelicerata, Merostomata), with implications on the functional anatomy of trilobites: Am. Mus. Novitates, No. 2436, 17 p.

Elliott, T., 1978, Clastic shorelines, *in* Reading, H.G. (ed.), Sedimentary Environments and Facies: New York, Elsevier, p. 143–177.

Evans, G., 1965, Intertidal flat sediments and their environments of deposition in the Wash: Quart. Jour. Geol. Soc. London, v. 121, p. 209–245.

_____, 1975, Intertidal flat deposits of the Wash, western margin of the North Sea, *in* Ginsburg, R.N. (ed.), Tidal Deposits: New York, Springer-Verlag, p. 13–20.

Farmer, J.D., and Miller, M.F., 1981, A deep-water trace fossil assemblage from the German Rancho Formation, Stump Beach, Salt Point State Park, *in* Frizzell, V. (ed.), Guidebook, Ann. Meeting, Soc. Econ. Paleontologists Mineralogists Field Trip, V. 3: Los Angeles, Pacific Coast Section, Soc. Econ. Paleontologists Mineralogist, p. 2–14.

Farrow, G.E., 1966, Bathymetric zonation of Jurassic trace fossils from the coast of Yorkshire, England: Palaeogeogr., Palaeoclimatol., Palaeoecol., v. 2, p. 103–151.

Fenchel, T., 1977, Competition, coexistence and character displacement in mud snails, *in* Coull, B.C. (ed.), Ecology of Marine Benthos: Columbia, S.C., Univ. of South Carolina Press, p. 229–243.

Fenton, G.L., and Fenton, M.A., 1937, *Olivellites*, a Pennsylvanian snail burrow: Am. Midland Naturalist, v. 18, p. 452–453.

Ferm, J.C., 1974, Carboniferous environmental models in eastern United States and their significance, *in* Griggs, G.B. (ed.), Carboniferous of the Southeastern United States: Geol. Soc. America Spec. Paper 148, p. 79–95.

_____, Horne, J.C., Swinchatt, J.P., and Whaley, P.W., 1971, Carboniferous depositional environments in northeastern Kentucky: Lexington, Kentucky Geol. Soc. Guidebook, Ann. Spring Conf., 30 p.

_____,Milici, R.C., and Eason, J.E., 1972, Carboniferous depositional environments in the Cumberland Plateau of southern Tennessee and northern Alabama: Tennessee Div. Geology, Rept. Invest. 33, 32 p.

Fischer, W.A., 1978, The habitat of the early vertebrates: Trace and body fossil evidence from the Harding Formation (Middle Ordovician), Colorado: Mountain Geologist, v. 15, p. 1–26.

Fisher, W.L., Brown, L.F., Jr., Scott, A.J., and McGowen, J.H., 1969, Delta systems in the exploration for oil and gas: Austin, Bur. of Economic Geol., Univ. Texas, 78 p.

Frazier, D.E., and Osanik, A., 1969, Recent peat deposits—Louisiana coastal plain, *in* Dapples, E.C., and Hopkins, M.E. (eds.), Environments of Coal Deposition: Geol. Soc. America Spec. Paper 114, p. 63–85.

Frey, R.W., 1975, The Study of Trace Fossils: New York, Springer-Verlag, 562 p.

_____, and Howard, J.D., 1970a, Comparison of Upper Cretaceous ichnofaunas from siliceous sandstones and chalk, Western Interior Region, U.S.A., *in* Crimes, T.P., and Harper, J.C. (eds.), Trace Fossils: Geol. Jour., Spec. Issue 3, Liverpool, Seel House Press, p. 141–166.

_____, and _____, 1970b, A profile of biogenic sedimentary structures in a Holocene barrier island-salt marsh complex, Georgia: Gulf Coast Assoc. Geol. Soc., Trans. v. 19, p. 427–444.

Glaessner, M.F., 1969, Decapoda, *in* Moore, R.C. (ed.), Treatise on Invertebrate Paleontology, Part R: Arthropoda 4: Lawrence, Kansas, Univ. Kansas Press and Geol. Soc. America, p. R400–R657.

Goldring, R., and Langenstrassen, F., 1979, Nearshore clastic facies, *in* House, M.R., Scrutton, C.T., and Bassett, M.G. (eds.), The Devonian System: Spec. Rept. in Paleontology 23, Paleontol. Assoc. London, p. 81–98.

_____, and Seilacher, A., 1971, Limulid undertracks and their sedimentological implications: Neues. Jahr. Geol. Paläont., v. 137, p. 422–442.

Gould, H.R., 1970, The Mississippi Delta complex, *in* Morgan, J.P. (ed.), Deltaic Sedimentation—Modern and Ancient: Soc. Econ. Paleontologists, Mineralogists Spec. Pub., p. 3–30.

Gotzinger, G., and Becker, H., 1932, Zur Geologischen Gliederung des Wienerwaldflysches (Neue Fossilfunde): Geol. Bundesanst. Wien, Jahrb, v. 82, p. 343–396.

Grasso, T.X., and Wolff, M.P., 1977, Paleoenvironments of the Marcellus and lower Skaneateles Formations of the Otsego County Region (Middle Devonian), *in* Wilson, P.H. (ed.), Guidebook to Field Excursions, 49th Ann. Mtg.: N.Y. State Geol. Assoc., p. A1–A50.

GUTSCHICK, R.C., AND RODRIGUEZ, J., 1977, Late Devonian-Early Mississippian trace fossils and environments along the Cordilleran Miogeocline, Western United States, *in* Crimes, T.P., and Harper, J.C. (eds.), Trace Fossils 2: Geol. Jour., Spec. Issue 9, Liverpool, Seel House Press, p. 195–208.

HAKES, W.G., 1976, Trace fossils and depositional environment of four clastic units, Upper Pennsylvanian megacyclothems, northeast Kansas: Univ. Kansas Paleontol. Contr., Art. 63, 46 p.

______, 1977, Trace fossils in Late Pennsylvanian cyclothems, Kansas, *in* Crimes, T.P., and Harper, J.C. (eds.), Trace Fossils 2: Geol. Jour., Spec. Issue 9, Liverpool, Seel House Press, p. 209–226.

HÄNTZSCHEL, W., 1975, Trace fossils and problematica, 2nd ed., *in* C. Teichert (ed.), Treatise on Invertebrate Paleontology, Pt. W., Miscellanea, Supplement: Boulder, Colo. and Lawrence, Kansas, Geol. Soc. America and Univ. Kansas Press, 229 p.

HARDY, P.G., 1970, New xiphosurid trails from the upper Carboniferous of northern England: Palaeontology, v. 13, p. 188–190.

HARLAND, T.L., 1978, *Caprionichnus*, a new trace fossil from the Caradocian of southern Norway: Geol. Jour., v. 13, p. 93–99.

HAYES, M.O., 1975, Morphology of sand accumulation in estuaries, an introduction to the symposium, *in* Cronin, L.E. (ed.), Estuarine Research, vol. 11: New York, Academic Press, Inc., p. 3–22.

HENBEST, L.G., 1960, Fossil spoor and their environmental significance in Morrow and Atoka series, Pennsylvanian, Washington County, Arkansas: U.S. Geological Survey Professional Paper 400-B, p. B383–B385.

HOBDAY, D.K., 1974, Beach and barrier-island facies in the Upper Carboniferous of Northern Alabama, *in* Griggs, G.B. (ed.), Carboniferous of the Southeastern United States: Geol. Soc. America Spec. Paper 148, p. 209–224.

______, AND HORNE, J.C., 1977, Tidally influenced barrier island and estuarine sedimentation in the Upper Carboniferous of southern West Virginia: Sed. Geology, v. 18, p. 97–122.

HORNE, J.C., 1979a, The effects of Carboniferous shoreline geometry on paleocurrent distribution, *in* Ferm, J.C., and Horne, J.C. (eds.), Carboniferous Depositional Environments in the Appalachian Region: Columbia, S.C., Carolina Coal Group, p. 509–514.

______, 1979b, Field Guide: The Hazard-Hyden area, *in* Ferm, J.C., and Horne, J.C. (eds.), Carboniferous Depositional Environments in the Appalachian Region: Columbia, S.C., Carolina Coal Group, p. 692–706.

______, FERM, J.C., CARUCCIO, F.T., AND BAGANZ, B.P., 1978, Depositional models in coal exploration and mine planning in Appalachian region: Am. Assoc. Petroleum Geologist Bull., v. 62, p. 62, p. 2379–2411.

______, ______, AND SWINCHATT, J.P., 1974, Depositional model for the Mississippian boundary in northeastern Kentucky, *in* Briggs, G. (ed.), Carboniferous of the Southeastern United States: Geol. Soc. America Spec. Paper 148, p. 97–114.

HOWARD, J.D., 1972, Trace fossils as criteria for recognizing shorelines in the stratigraphic record, *in* Rigby, J.K., and Hamblin, W.K. (eds.), Recognition of Ancient Sedimentary Environments: Soc. Econ. Paleontologists Mineralogists Spec. Pub. No. 16, p. 215–225.

______, 1975, Estuaries of the Georgia Coast, U.S.A.: Sedimentology and biology, IX. Conclusions: Senckenbergiana Marit., v. 7, p. 297–305.

HOYT, J.H., AND HENRY, V.J., 1967, Influence of inlet migration on barrier island sedimentation: Geol. Soc. America Bull., v. 78, p. 77–86.

HUBERT, J.P., BUTERA, J.G., AND RICE, R.F., 1972, Sedimentology of the Upper Cretaceous Cody-Parkman Delta, Southwestern Powder River Basin, Wyoming: Geol. Soc. America Bull., V. 83, p. 1649–1670.

JACKSON, J.B., 1968, Bivalves: Spatial and size frequency distributions of two intertidal species: Science, v. 161, p. 479–480.

JONES, D.L., 1935, Some asteriaform fossils from the Francis Formation of Oklahoma: Am. Midland Naturalist, v. 16, p. 427–428.

KENNEDY, W.J., 1975, Trace fossils in carbonate rocks, *in* Frey, R.W. (ed.), The Study of Trace Fossils: New York, Springer-Verlag, p. 377–398.

KERN, J.P., AND WARME, J.E., 1974, Trace fossils and bathymetry of the Upper Cretaceous Point Loma Formation, San Diego, California: Geol. Soc. America Bull., v. 85, p. 893–900.

KING, A.F., 1965, Xiphosurid trails from the Upper Carboniferous of Bude, North Cornwall: Proc. Geol. Soc., v. 1626, p. 162–165.

KLEIN, F. DE V., 1970, Depositional and dispersal dynamics of intertidal sand bars: Jour. Sed. Petrology, v. 40, p. 1095–1127.

KNIGHT, R.J., AND DALRYMPLE, R.W., 1975, Intertidal sediments from the south shore of Cobequid Bay, Bay of Fundy, Nova Scotia, Canada, *in* Ginsberg, R.N. (ed.), Tidal Deposits: New York, Springer-Verlag, p. 47–55.

KNOX, L.W., AND MILLER, M.F., in press, Environmental control of trace fossil morphology, *in* Curran, H.A., ed., Biogenic Structures: Their Use in Interpreting Depositional Environments: Soc. Econ. Paleontologists Mineralogists Spec. Pub. No. 34.

KSIAZKIEWICZ, M., 1970, Observations on the ichnofauna of the Polish Carpathians, *in* Crimes, T.P., and Harper, J.C. (eds.), Trace Fossils: Geol. Jour., Spec. Issue 3, Liverpool, Seel House Press, p. 283–322.

KUMAR, N., AND SANDERS, J.E., 1974, Inlet sequences: A vertical succession of sedimentary structures and textures created by the lateral migration of tidal inlets: Sedimentology, v. 21, p. 491–532.

LARSONNEUR, C., 1975, Tidal deposits, Mont Saint-Michel Bay, France, *in* Ginsburg, R.N. (ed.), Tidal Deposits: New York, Springer-Verlag, p. 21–30.

LESQUEREUX, L., 1876, Species of fossil marine plants from Carboniferous measures: Ind. State Geologist 7th Ann. Report, p. 134–145.

LEVINTON, J.S., 1977, Ecology of shallow water deposit-feeding communities Quisset Harbor, Massachusetts, *in* Coull, B.C. (ed.), Ecology of Marine Benthos: Columbia, S.C., Univ. of South Carolina Press, p. 191–227.

LEWARNE, G.C., 1964, Starfish traces from the Namurian of County Clare, Ireland: Palaeontology, v. 7, p. 508–513.

MacGinitie, G.E., and MacGinitie, N., 1949, Natural History of Marine Animals: New York, McGraw-Hill Book Co., Inc., 473 p.
Mathew, D., 1977, Geology of the Beckley Coal Seam in the Eccles #5 Mine near Eccles, West Virginia [Unpubl. Ph.D. dissert]: Univ. of South Carolina, 87 p.
Mayou, T.V., and Howard, J.D., 1975, Estuaries of the Georgia Coast, U.S.A.: sedimentology and biology. VI Animal-sediment relationships of a salt marsh estuary—Doboy Sound: Senckenbergiana Marit., v. 7, p. 205–236.
Meckel, L.D., 1975, Holocene sand bodies in the Colorado Delta, Salton Sea, Imperial County, California, *in* Broussard, M.L. (ed.), Deltas: Models for Exploration: Houston, Houston Geological Society, p. 239–265.
Meyer, D.L., 1980, Ecology and biogeography of living classes, *in* Broadhead, T., and Waters, J.A. (eds.), Echinoderms, Notes for a Short Course: Univ. Tennessee Dept. Geol. Sci., Studies in Geology 3, 235 p.
Milici, R.L., 1974, Stratigraphy and depositional environments of Upper Mississippian and Lower Pennsylvanian rocks in the southern Cumberland Plateau of Tennessee, *in* Griggs, G.B. (ed.), Carboniferous of the Southeastern United States: Geol. Soc. America Spec. Paper 148, p. 115–133.
_____, Briggs, G., Knox, L.M., Sitterly, P.D., Statler, A.T., 1979, The Mississippian and Pennsylvanian (Carboniferous) systems in the United States—Tennessee: U.S. Geological Survey Professional Paper 1110G, p. G1–G-38.
Miller, M.F., 1977, Middle and Upper Ordovician biogenic structures and paleoenvironments, southern Nevada: Jour. Sed. Petrology, v. 47, p. 1328–1338.
_____, 1978, Ethology and ecology of some Devonian shallow water *Zoophycos* and some possible implications for trace fossil evolution: Geol. Soc. America Ann. Mtg. Abs. with Programs, v. 10, p. 457.
_____, 1979, Paleoenvironmental distribution of trace fossils in the Catskill deltaic complex, New York State: Palaeogeogr., Palaeoclimatol., Palaeoecol., v. 28, p. 117–141.
_____, 1982a, Origin of massive appearing quartz arenites: mechanisms for maintenance of large populations of deposit feeders in quartz-rich sands: Geol. Soc. America Ann. Mtg. Abs. with Programs, v. 14, p. 217.
_____, 1982b, *Limulicubichnus*: A new ichnogenus of limulid resting traces: Jour. Paleontology, v. 56, p. 429–433.
_____, and Johnson, K.G., 1981, *Spirophyton* in alluvial-tidal facies of the Catskill deltaic complex: possible biological control of ichnofossil distribution: Jour. Paleontology, v. 55, p. 1016–1027.
Morgan, J.P., 1970, Depositional processes and products in the deltaic environment, *in* Morgan, J.P. (ed.), Deltaic Sedimentation—Modern and Ancient: Soc. Econ. Paleontologists Mineralogists Spec. Pub. 15, p. 31-47.
Morton, R.A., and Donaldson, A.C., 1973, Sediment distribution and evolution of tidal deltas along a tide-dominated shoreline, Wachapregue, Virginia: Sed. Geology, v. 10, p. 285–299.
Oomkens, E., 1974, Lithofacies relations in the Late Quaternary Niger delta complex: Sedimentology, v. 21, p. 195–222.
Osgood, R.G., 1970, Trace fossils of the Cincinnati area: Paleontogr. Americana, v. 6, p. 281–444.
_____, and Szmuc, E.J., 1972, The trace fossil *Zoophycos* as an indicator of water depth: Bull. Am. Paleontology, v. 62, no. 271, 22 p.
Pfefferkorn, H.W., 1971, Note on *Conostichus broadheadi*: Lesquereux (trace fossil: Pennsylvanian): Jour. Paleontology, v. 45, p. 888–892.
Pickerill, R.K., and Roulston, B.V., 1977, Enigmatic trace fossils from the Silurian Chaleurs Group of the southeastern Gaspé Peninsula, Quebec: Canadian Jour. Earth Sciences, v. 14, p. 1719–1736.
Plicka, M., 1968, *Zoophycos*, and a proposed classification of sabellid worms: Jour. Paleontology, v. 42, p. 836–849.
Plumstead, E.P., 1967, A general review of the Devonian fossil plants found in the Cape System of South Africa: Paleontologia Africana, v. 10, 78 p.
Reineck, H.E., 1975, German North Sea tidal flats, *in* Ginsburg, R.N. (ed.), Tidal Deposits: New York, Springer-Verlag, p. 5–12.
_____, and Singh, I.B., 1973, Depositional Sedimentary Environments: New York, Springer-Verlag, 439 p.
Reinson, G.E., 1979, Facies Models 6. Barrier Island Systems, *in* Walker, R.G. (ed.), Facies Models: Geoscience Canada, Reprint Series 1, p. 74–57.
Rhoads, D.C., and Young, D.K., 1970, The influence of deposit-feeding organisms on sediment stability and community trophic structure: Jour. Mar. Research, v. 28, p. 150–178.
Ricketts, E.F., and Calvin, J., 1968, Between Pacific Tides, 4th ed., revised by J.W. Hedgpeth: Stanford, Stanford Univ. Press, 614 p.
Rodriguez, J., and Gutschick, R.C., 1970, Late Devonian-early Mississippian ichnofossils from Western Montana and Northern Utah, *in* Crimes, T.P., and Harper, J.C. (eds.), Trace Fossils: Geol. Jour., Spec. Issue 3, Liverpool, Seel House Press, p. 407–438.
Rogers, M.J., 1975, A revision of the species of nonmarine bivalvia from the Upper Carboniferous of eastern North America: Jour. Paleontology, v. 39, p. 663–686.
Ronan, T.E., Jr., 1975, Structural and paleoecological aspects of a modern marine soft-sediment community [Unpubl. Ph.D. dissert]: Univ. of Calif., Davis, 220 p.
_____, 1977, Formation and paleoecologic recognition of structures caused by marine annelids: Paleobiology, v. 3, p. 389–403.
_____, Miller, M.F., and Farmer, J.D., 1981, Organism-sediment relationships on a modern tidal falt, Bodega Harbor, California, *in* Frizell, V. (ed.), Guidebook, Ann. Meeting, Soc. Econ. Paleontologists Mineralogists, p. 15–31.
Roniewicz, P., and Pienkowski, G., 1977, Trace fossils of the Podhale Flysch Basin, *in* Crimes, T.P., and Harper, J.C. (eds.), Trace Fossils 2: Geol. Jour., Spec. Issue 9, Liverpool, Seel House Press, p. 273–288.
Rudloe, A., 1979, *Limulus polyphemus*: A review of the ecologically significant literature, *in* Biomedical Applications of the

Horseshoe Crab (limulidae): New York, A.R. Liss, Inc., p. 27–35.
SARLE, G.J., 1906, Preliminary note on the nature of *Taonurus*: Rochester Acad. Proc., v. 4, p. 211–214.
SAXENA, R.S., 1979, Depositional environments on the lower delta plain of the Mississippi River, *in* Ferm, J.C., and Horne, J.C. (eds.), Carboniferous Depositional Environments in the Appalachian Region: Columbia, S.C., Carolina Coal Group, p. 344–367.
SCHRAMM, F.R., FELDMANN, R.M., AND COPELAND, M.J., 1978, The Late Devonian Palaeopalaemonidae and the earliest decapod crustaceans: Jour. Paleontology, v. 53, p. 1375–1387.
SEILACHER, A., 1953, Studien zur Palichnologie II. Die fossilen Ruhespuren (Cubichnia): Neues. Jahrb. Geologie, Paläontologie, Abhandl., v. 98. p. 87–124.
______, 1964, Biogenic sedimentary structures, *in* Imbrie, J., and Newell, N.D. (eds.), Approaches to Paleoecology: New York, J. Wiley and Sons, p. 296–316.
______, 1967, Bathymetry of trace fossils: Marine Geol., v. 5, p. 413–428.
SHELDON, R.W., 1968, Probable gastropod tracks from the Kinderscout Grit of Soyland Moor, Yorkshire: Geol. Mag., v. 105, p. 365–366.
SIEMERS, C.T., 1975, Paleoenvironmental analysis of the Upper Cretaceous Frontier Formation, northwestern Bighorn Basin, Wyoming: Wyo. Geol. Assoc. Guidebook, 27th Ann. Field Conf., p. 85–99.
SIMPSON, S., 1970, Notes on *Zoophycos* and *Spirophyton*, *in* Crimes, T.P., and Harper, J.C. (eds.), Trace Fossils: Geol. Jour., Spec. Issue 3, Liverpool, Seel House Press, p. 505–514.
______, 1975, Classification of trace fossils, *in* Frey, R.W. (ed.), The Study of Trace Fossils: New York, Springer-Verlag, p. 39–54.
STANLEY, K.O., AND FAGERSTROM, J.A., 1974, Miocene invertebrate trace fossils from a braided river environment, western Nebraska, U.S.A.: Palaeogeogr., Palaeoecol., Palaeoclimatol., v. 15, p. 63–82.
STAUB, J.R., AND COHEN, A.D., 1979, The Snuggedy Swamp of South Carolina: a back-barrier estuarine coal-forming environment: Jour. Sed. Petrology, v. 49, p. 133–144.
STØRMER, L., PETRUNKEVITCH, A., AND HEDGPETH, J.W., 1955, Chelicerata with sections on Pycnogonida and Palaeoisopus, *in* Moore, R.C. (ed.), Treatise on Invertebrate Paleontology, Part P: Arthropoda 2: Lawrence Kansas, Univ. Kansas Press and Geol. Soc. America, p. 4–41.
THAYER, C.W., 1974, Marine paleoecology in the Upper Devonian of New York: Lethaia, v. 7, p. 121–155.
TREWIN, N.H., 1976, *Isopodichnus* in a trace fossil assemblage from the Old Red Sandstone: Lethaia, v. 9, p. 29–37.
TURNER, B.R., 1978, Trace fossils from the Triassic fluviatile Molteno Formation of the Karoo (Gondwana) Supergroup, Lesotho: Jour. Paleontology, v. 52, p. 959–963.
WARME, J.E., 1971, Paleoecological aspects of a modern coastal lagoon: Univ. Calif. Publ. Geol. Sci. 87, 110 p.
______, AND OLSON, R.W., 1971, Stop 5: Lake Brownwood spillway, *in* Perkins, B.F. (ed.), Trace Fossils: A Field Guide to Selected Localities in Pennsylvanian, Permian, Cretaceous, and Tertiary Rocks of Texas and Related Papers: Soc. Econ. Paleontologists Mineralogists Field Trip, 1971, Baton Rouge, LSU School Geoscience, Misc. Publ. 71–1, p. 27–46.
WHALEY, P.W., AND HESTER, N.C., 1979, Road log— first day of field trip, *in* Whaley, P.W., Hester, N.C., Williamson, A.D., Beard, J.G., Pryor, W.A., Potter, P.E. (eds.), Depositional Environments of Pennsylvanian Rocks in Western Kentucky: Ann. Field Conf., Geol. Soc. Kentucky, October 11–13, 1979, p. 7–25.
WILSON, C.W., JR., JEWELL, J.W., AND LUTHER, E.T., 1956, Pennsylvanian geology of the Cumberland Plateau: Nashville, Tennessee Div. Geol. (unnumbered folio), 21 p.
WILTSE, W.I., 1980, Predation by juvenile *Polinices duplicatus* (Say) on *Gemma Gemma* (Totten): Jour. Exp. Mar. Biol. Ecol., v. 42, p. 187–199.
WOODIN, S.A., 1976, Adult-larval interactions in dense infaunal assemblages: Patterns of abundance: Jour. Mar. Research, v. 34, p. 25–41.
WRIGHT, L.D., COLEMAN, J.M., AND THOM, B.G., 1975, Sediment transport and deposition in a macrotidal river channel: Ord River, Western Australia, *in* Cronin, L.E. (ed.), Estuarine Research, vol. II: New York, Academic Research, p. 309–321.
YOCHELSON, E.L., AND SCHINDEL, D.E., 1978, A reexamination of the Pennsylvanian trace fossil *Olivellites*: Jour. Research, U.S. Geological Survey, v. 6, p. 789–796.

TRACE FOSSIL ASSEMBLAGES AND THEIR OCCURRENCE IN SILESIAN (MID-CARBONIFEROUS) DELTAIC SEDIMENTS OF THE CENTRAL PENNINE BASIN, ENGLAND

R.M.C. EAGAR[1], J.G. BAINES[2], J.D. COLLINSON[3], P.G. HARDY[4], S.A. OKOLO[5], AND J.E. POLLARD[1]
The Manchester Museum, The University, Manchester M13 9PL, England[1]; Union Oil Company of Great Britain, 32 Cadbury Road, Sunbury-on-Thames, Middlesex TW16 7LU, England[2]; Geologisk Institutt, Avd. A., Universitetet i Bergen, Allégaten 41, 5000 Bergen, Norway[3]; Department of Extra-Mural Studies, University of Bristol, 32 Tyndall's Park Road, Bristol BS8 1HR, England[4]; and Pan Ocean Corporation, P.O. Box 93, Lagos, Nigeria[5]

ABSTRACT

In the Silesian rocks of the Central Pennine Basin, three types of ancient delta sequence are recognized. Each contributed to the progressive filling of the Basin and to the gradual development of fluvial/paralic conditions in Westphalian time.

The turbidite-fronted delta of the lowest Namurian, of the Pendleian Stage of the Skipton area in the north of the Basin, shows three depth-related sedimentary associations which correspond with overlapping but distinct trace fossil assemblages. The Turbidite Association contains a *Rhizocorallium-Planolites-Bergaueria* assemblage on the base or top of thin-bedded turbidites; the Slope Association consists of *Lophoctenium* and *Curvolithus* in laminated sandstones and siltstones; and the Delta Top Association is characterized only by *Monocraterion-Skolithos* and *Pelecypodichnus* in parallel-bedded and cross-bedded sandstones. The Turbidite and Delta Slope Associations appear to belong to the *Zoophycos* ichnofacies of Seilacher (1967), and the Delta Top to the *Cruziana* and *Skolithos* ichnofacies. Deltaic deposition thus advanced into water a few hundreds of meters deep, probably of nearly fully marine salinity. Trace fossils of the deeper-water *Nereites* ichnofacies are lacking. Sedimentological factors such as energy level, substrate and food supply, rather than bathymetry alone, may have influenced the distribution of trace fossils.

In the south of the Basin, the later Lower Kinderscoutian delta is similar sedimentologically to that of the Pendleian, but is devoid of trace fossils, except for *Planolites* and *Pelecypodichnus* assemblages in the upper part of the delta slope and on the delta top. The absence of trace fossils with obvious marine affinities is consistent with the interpretation that, in intervals between marine inundations, basin water was less saline than it was in the Skipton area during Pendleian time.

During the Upper Kinderscoutian, Marsdenian, and Yeadonian stages, the Central Pennine Basin was filled, mainly from the north and east, by shallow-water sheet deltas, and by two shallow-water elongate deltas from the west. Trace fossils in the delta-plain sediments contain assemblages which can be assigned to the *Cruziana*, or rarely, *Zoophycos* ichnofacies, together with a variety of facies-crossing forms. They show progressive colonization of the delta-top paleoenvironments. They also suggest evolution of certain animal groups during this time. Thus bivalve escape shafts, attributed to cf. *Sanguinolites*, a marine genus in the Upper Kinderscoutian, show a steady increase in vertical extent, or height, throughout the period. By late Marsdenian time they were evidently formed by the non-marine genus *Carbonicola*. On independent evidence *Carbonicola* appears to have evolved from the bivalves which made the earliest escape shafts.

Lower Westphalian sediments indicate a gradual increase upwards in fluvial and swamp dominance of the extensive delta top. This is well substantiated by trace fossils, as far as they have been studied. They play a significant part in elucidating the general sedimentary environment; for instance, *Pelecypodichnus* escape shafts suggest seasonal flooding, as a result of monsoonal condition, in the west Lancashire coalfield. Xiphosurid traces (*Kouphichnium* and *Limulicubichnus*), which range from the Marsdenian upwards, provide insights into the more ephemeral aspects of sedimentation and paleoenvironments. Freshwater arthropod traces and vertebrate footprints (*Scoyenia* ichnofacies) are poorly known in the Westphalian of the Pennine area when compared with the roughly contemporary ichnofaunas of Nova Scotia, Canada, but such traces occur in the latest Silesian and early Permian rocks in other areas of Britain.

INTRODUCTION: BACKGROUND PALEOGEOGRAPHY AND SEDIMENTOLOGY

The Silesian rocks of northern England were deposited in an area of block and basin topography. This had developed in late Devonian time, but some tectonic movement was probably still taking place in the Carboniferous Period. The main thickness of deltaic sediments accumulated in what has been termed the "Central Province". This basinal area was bounded to the north by a fault-controlled margin against the Askrigg Block (Fig. 1), whilst to the south, a less well defined boundary lay along the northern edge of the Midlands Landmass ('M. L.', inset of Fig. 1). To the east of the area of present-day exposure, the pattern of basinal development is not well known, but it seems likely that several "gulfs" and "ridges" extended westwards from a basin margin. To the west, the basin continued into, or at least connected with basins in Ireland. Within the main basin, subsidiary topographic features such as the Derbyshire Massif and surround-

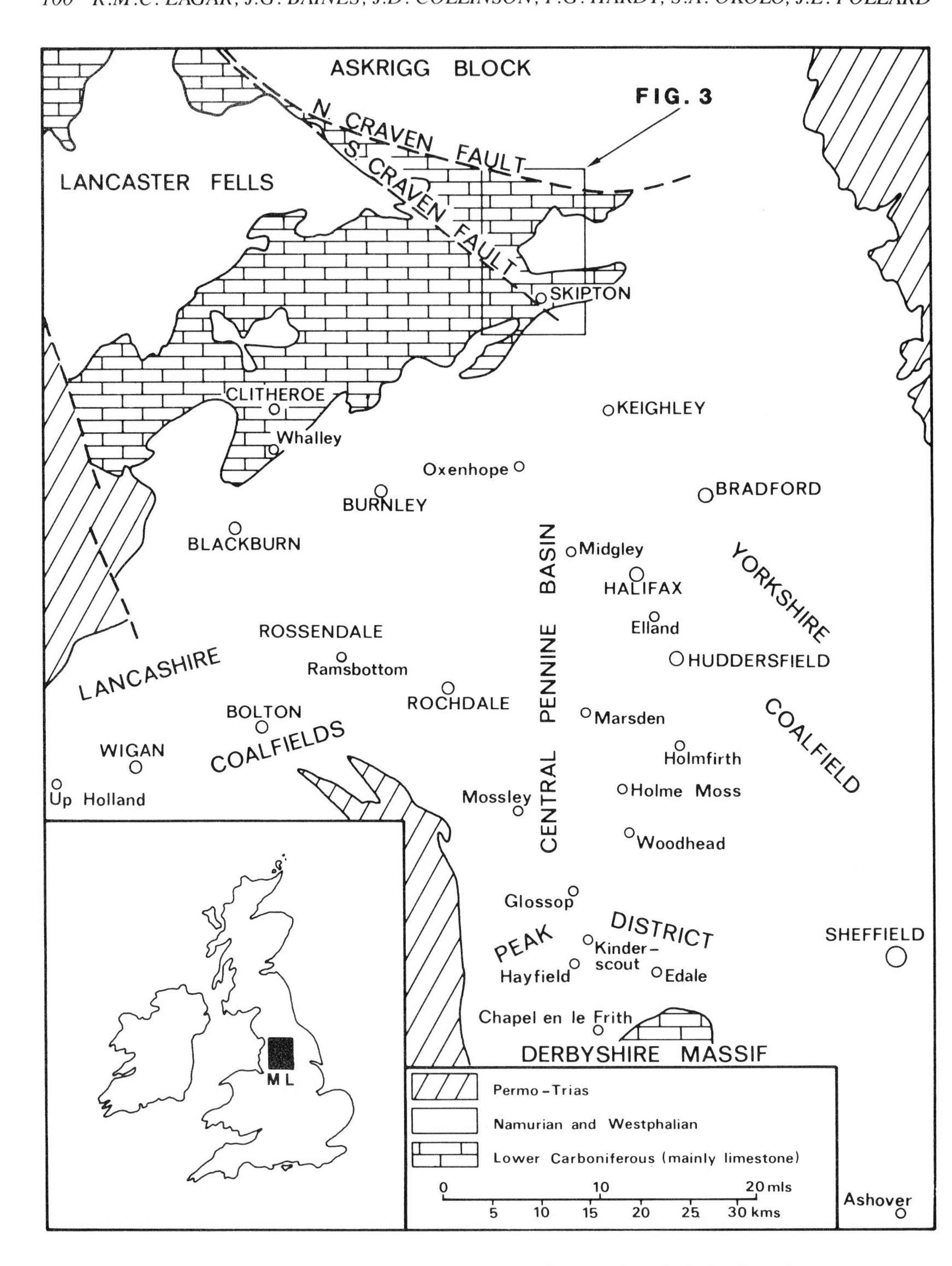

Fig. 1.—Sketch map of the Central Pennine Basin, England. Major faults are shown by broken lines. Circles denote centers of towns referred to in the sections of Fig. 2. The center line of the Pennine Basin coincides approximately with the north-south anticlinal axis of the Pennines.

ing "gulfs" had been defined clearly enough to allow marked facies differences to develop during the Lower Carboniferous. At the outset of deposition of clastic sediment, in the beginning of the Namurian (Fig. 2), it seems likely that the deeper parts of the basinal areas reached depths of several hundreds of meters, while higher topographic blocks lay under shallower water.

The theme of Namurian and early Westphalian sedimentation in the Central Province is the filling of this complex topography so that, by early Westphalian time, more or less uniform Coal Measure alluvial plain environments were established across the area (Calver, 1969; Kelling and Collinson, in press). This is very broadly indicated in Figure 2.

This paper is confined to the deposits of the main Central Pennine Basin which lie between the Askrigg Block and the Derbyshire Massif. The rocks are exposed in a north-south trending outcrop along the line of the Pennine Anticline and in some isolated areas to the west. In much of the basin fill, trace fossils provide the only organic evidence of the environmental conditions. Except from the middle of the Marsdenian Stage upward, where non-marine bivalve bands are occasionally found, body fossils throughout the Namurian are confined to widely spaced "marine bands", mostly units of fissile carbonaceous shale. Many of these contain a goniatite fauna which allows detailed correlation of those particular horizons (e.g., Bisat, 1924; Ramsbottom, 1969) (Fig. 2, left) and detailed reconstruction of paleogeographies. Even in sequences of basinal mudstones, fossils are confined to marine bands and these are separated by units of barren shale. This suggests that basin waters were subject to fluctuating salinities, being somewhat distantly connected with the open ocean. During periods of delta advance, water in the basin was probably of less than normal marine salinity. Trace fossil assemblages appear to be one of our main criteria for assessing these basinal salinity changes. The purpose of this paper is to describe the trace fossil assemblages recently discovered in these Silesian sediments and to assess their usefulness in elucidating both salinity changes and also other details of these ancient deltaic environments.

Silesian sediments in the Central Pennine Basin are entirely detrital, ranging in grain size from claystones to very coarse pebbly sandstones. Supply was from an active source area which lay to the north and northeast and which provided abundant feldspathic detritus (Sorby, 1859; Gilligan, 1920). Deposition of this material began in earliest Namurian time and, from then on, a series of deltas advanced to the south and west, eventually filling the Basin (Kelling and Collinson, in press). Periods of delta advance throughout the Namurian and early Westphalian are recorded by sequences tens to hundreds of meters thick which show a broadly upwards coarsening trend. Marine bands lie directly above these sequences and coincide with periods of delta abandonment and transgression, possibly associated with eustatic rises in sea level (Ramsbottom, 1979, 1981). The overall Silesian succession has been described as "cyclic" (e.g., Wright et al., 1927), but sedimentological analysis suggests that, at least in the Namurian, three types of sequences are present, each representing the advance of a different type of delta into the Basin. These sequences have been termed by Collinson (1976):

1) Deep-water, turbidite-fronted delta sequence
2) Shallow-water, sheet delta sequence
3) Shallow-water, elongate delta sequence

By Westphalian time, conditions had become more uniform over the area. The Coal Measure succession shows widespread but thinner sequences within which marine bands are much less common and seat-earths and coals are more abundant. Fossiliferous horizons associated with the abandonment of areas of coal swamp more commonly have a fauna of non-marine bivalves, a continuation of those appearing in the upper Namurian. However, widespread marine bands still commonly punctuate the sequence in the lowest part of the Westphalian (Calver, 1968a).

Namurian

The broad sedimentological features of the major types of Namurian sequence are outlined below:

1) *Deep-water deltaic sequences* are several hundreds of meters thick and comprise a lower turbidite unit, a middle, dominantly silty, upwards coarsening sequence, and an upper part dominated by mutually erosive channels filled with coarse sandstones (Walker, 1966a, b; Collinson, 1969, 1970; McCabe, 1975, 1978; Baines, 1977). The sequences, of which examples are provided in the Pendleian and Lower Kinderscoutian, record the advance of large deltas fed by major distributary channels. Their unusual feature is the turbidite unit, which is interpreted as having been deposited from a series of laterally coalescent turbidite lobes advancing in front of the main prograding delta slope. The lobes appear to have been fed by channels which linked back to the distributary channels and which thereby allowed coarse sediment to by-pass the delta slope and to be laid down in deeper water at its foot. If river-generated turbidity currents produced such a system (Collinson, 1970), the water in the basin might have been of reduced salinity.

2) *Shallow-water sheet delta sequences* are commonly tens of meters thick and consist of a lower unit which coarsens upward from mudstone and an upper sandstone unit made up of channel fills and unchannelized mouth-bar sandstones (Mayhew, 1967; Benfield, 1969; Okolo, 1982). The sequences are commonly capped by rootlet horizons and thin coal seams. The deltas which produced these sequences lacked turbidite precursors and their delta tops were extensive, complex areas made up of mouth-bar and channel

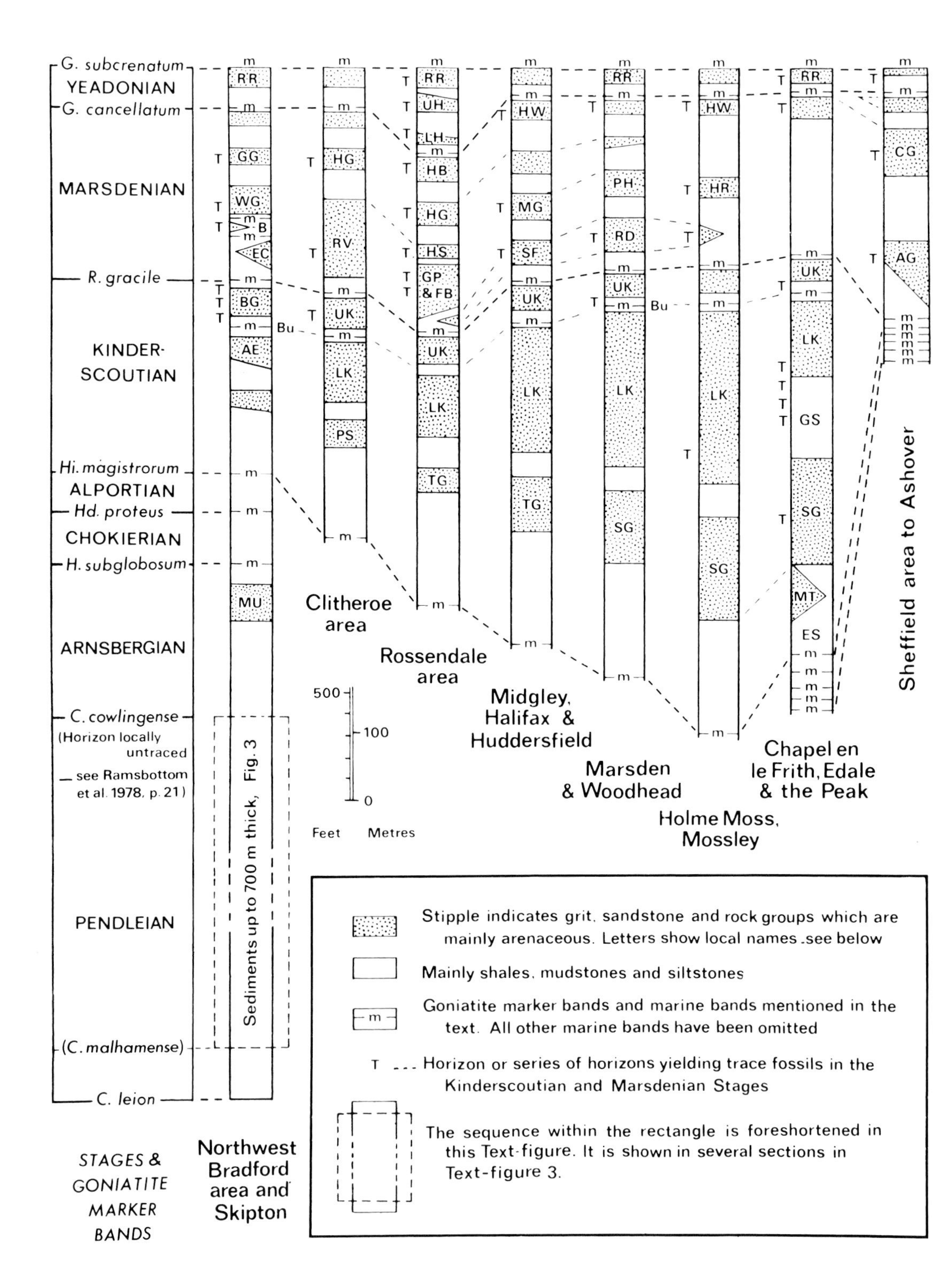

G. subcrenatum
YEADONIAN
G. cancellatum
MARSDENIAN
R. gracile
KINDER-SCOUTIAN
Hi. magistrorum
ALPORTIAN
Hd. proteus
CHOKIERIAN
H. subglobosum
ARNSBERGIAN
C. cowlingense
(Horizon locally untraced — see Ramsbottom et al. 1978, p. 21)
PENDLEIAN
(C. malhamense)
C. leion
STAGES & GONIATITE MARKER BANDS
Northwest Bradford area and Skipton
Sediments up to 700 m thick, Fig. 3
Clitheroe area
Rossendale area
Midgley, Halifax & Huddersfield
Marsden & Woodhead
Holme Moss, Mossley
Chapel en le Frith, Edale & the Peak
Sheffield area to Ashover
Feet
Metres
500
100
0
Stipple indicates grit, sandstone and rock groups which are mainly arenaceous. Letters show local names see below
Mainly shales, mudstones and siltstones
Goniatite marker bands and marine bands mentioned in the text. All other marine bands have been omitted
Horizon or series of horizons yielding trace fossils in the Kinderscoutian and Marsdenian Stages
The sequence within the rectangle is foreshortened in this Text-figure. It is shown in several sections in Text-figure 3.

sands. They probably compared with the Lafourche delta of the Holocene Mississippi (Kolb and van Lopik, 1966). Paleocurrents measured through inferred slope deposits are orientated in the direction of delta advance, suggesting down-slope underflow and, thereby, reduced basin water salinity.

3) *Shallow-water, elongate delta sequences* occur in two localized examples in the Haslingden Flags (Yeadonian) of Lancashire (Fig. 2). The sequences are upwards coarsening and several tens of meters thick. The fine-grained lower parts pass gradationally upwards into cross-laminated and cross-bedded sandstone units which are highly elongate and parallel to the paleocurrent direction. The sand bodies pinch out from a maximum thickness of around 30 m, giving them a lenticular cross-section. The sandstones have been interpreted as "bar finger" sands associated with delta distributary channels similar to those of the present-day birdsfoot delta of the Mississippi. Paleocurrents determined from cross-lamination throughout the sequence trend parallel to the direction of progradation and again suggest reduced basin salinity (Collinson and Banks, 1975).

At any one point in the basin, the lowest deltaic sequence is usually of turbidite-fronted, deep-water type, overlying basinal mudstones. It is commonly the case that all later sequences are of shallow-water type. The turbidite-fronted deltas filled the basin to emergent or near-emergent conditions and, thereafter, subsidence and compaction seldom produced water deep enough for the recurrence of this type of sequence. Shallow-water deltas dominate the upper part of the Namurian succession, occurring as a series of laterally extensive "cycles" whose distribution is uninfluenced by the topography which had controlled earlier Namurian sedimentation.

Westphalian

Westphalian sediments of the Central Province include some of the main coal deposits of the United Kingdom and their exploitation has provided a wealth of local stratigraphical information from which to draw a picture of the depositional setting. The elimination of early Namurian topography allowed rather uniform conditions of deposition to be established over the Central Province. Regional subsidence centered on Lancashire, diminishing to the basin margins and westward into Ireland (Calver, 1969; Eagar, 1975). The base of the Westphalian is not marked by any sudden change of depositional conditions as compared with the Namurian. Basal Westphalian A sediments occur for the most part as widespread and laterally uniform, upwards - coarsening sequences, several tens of meters thick with marine bands at their bases. The sequences are capped by widespread sheets of sandstone and, overall, they compare with the shallow-water sheet delta sequences of the late Namurian. In addition to the lateral uniformity of the thicker sequences, even minor intercalated sequences also show a remarkable lateral continuity (Eagar, 1952, Pls. I, II).

The lower Westphalian A succession shows an overall gradual transition towards a more typical pattern of Coal Measure sedimentation, which is best developed in the overlying Westphalian A-B strata. Sequences between coal seams and rootlet horizons are generally thinner, seldom more than 20 m thick and averaging about 10 m in the East Midlands Coalfield. Locally the coals may be up to several meters thick. Marine bands are much less abundant and faunal horizons commonly are of non-marine bivalves. The changing nature of the succession reflects a gradual withdrawal of marine influence. Sequences between coal seams are variable, both in thickness and in character, both laterally and between "cycles". They have a general tendency to coarsen upwards, but sediments between many seams are entirely of mudstone. Cutting through the succession of coal seams and detrital sediments are elongate channel sandstones, commonly up to 10 m thick and with widths of up to a few hundreds of meters.

The overall facies assemblage records the development of extensive swamps through which flowed large

Fig. 2.—Schematic sections of Namurian sediments in the Central Pennine Basin (Fig. 1), mainly after Ramsbottom (1966), with modifications from later work by Collinson and Banks (1975) and others.
In the left hand column abbreviations of goniatite marker bands are: G = *Gastrioceras*, R = *Reticuloceras*, Hi = *Hodsonites*, Hd = *Hudsonoceras*, H = *Homoceras*, C = *Cravenoceras*. Other goniatite bands are omitted (see Fig. 5).
In the sections, from top left to right: RR = Rough Rock, GG = Guiseley Grit, WG = Woodhouse Grit, B = "Bluestone", EC = East Carlton Grit, BG = Bramhope Grit, AE = Addingham Edge Grit, MU = Marchup Grit; HG = Hazel Greave Grit, RV = Revidge Grit, UK = Upper Kinderscout Grit, LK = Lower Kinderscout Grit, PS = Parsonage Sandstone; UH = Upper Haslingden Flags, LH = Lower Haslingden Flags, HB = Holcombe Brook Grit, HS = Helmshore Grit, GP = Gorpley Grit, FB = Fletcher Bank Grit, TG = Todmorden Grit; HW = Huddersfield White Rock; MG = Midgley Grit, SF = Scotland Flags; PH = Pule Hill Grit, RD = Readycon Dean Series; Bu = Butterley Marine Band; GS = Grindslow Shales, SG = Shale Grit, HR = Heyden Rock; MT = Mam Tor Sandstones; ES = Edale Shales; CG = Chatsworth Grit; AG = Ashover Grit.

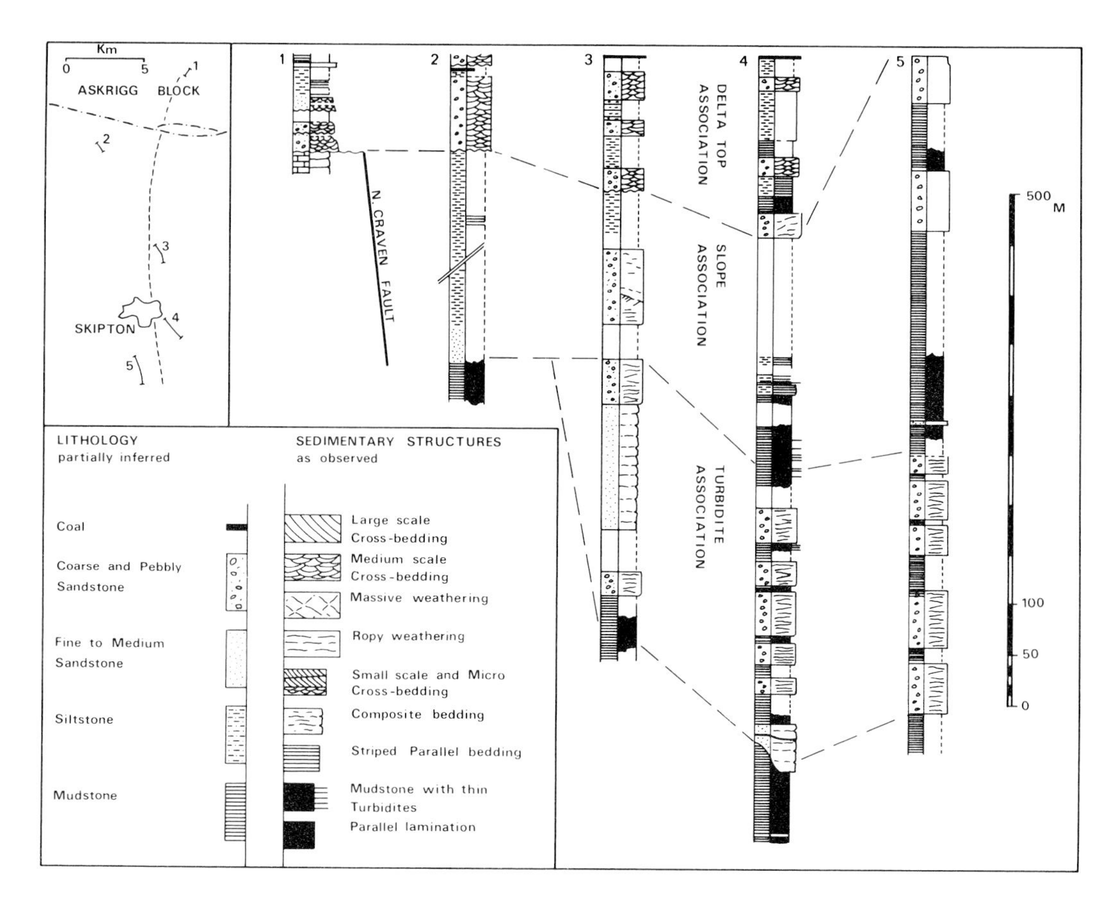

Fig. 3.—Sections of some Pendleian sediments of the Skipton area, simplified from Baines (1977).

distributary streams (Scott, 1980). Between channels, shifting and changing freshwater lakes and lagoons were, from time to time, filled by the advance of small deltas as river flow and sediment load were diverted. The episodes of lake filling gave rise to clastic sediment sequences and to the near-emergent conditions necessary for plant colonization, the plant communities being closely related to their sedimentary setting. When plant growth and productivity were able to keep pace with subsidence, thick peats accumulated to become preserved below an elevated water table. Eventual death of the swamp vegetation led rapidly to the re-establishment of lakes suitable for recolonization by non-marine bivalves.

DEEP-WATER, TURBIDITE-FRONTED DELTAIC SEQUENCES

The Pendleian Sequence around Skipton

The Pendleian sequence of the Skipton area (Figs. 3, 4) has yielded an extremely diverse ichnofauna which is distributed throughout the three major lithological units which compose it. The lowest unit, the Pendle Grits, consists of interbedded sandstones and finer sediments, the former being turbidites. It represents the first influx of clastic sediment into the basin following a period of deposition of mostly deep-water limestones and mudstones in Lower Carboniferous time. The middle unit, the Pendle Shales, is dominated by siltstones with a broadly coarsening upwards trend. Thin turbidite sandstones are present in the lower part and deep, sandstone-filled channels cut the sequence. This unit is interpreted as the deposit of the prograding delta slope advancing over the Pendle Grits with the channels being conduits for transfer of sand to deeper water, thereby largely by-passing the open slope. The upper unit, the Grassington Grits, is a delta-top association made up principally of laterally coalescing channel-fill deposits of fluviatile/delta distributary origin. Minor amounts of finer-grained sediment are preserved between the channels, including a thin coal seam at the top of the unit. The overall sequence, some 700 m thick in the Skipton area, represents the results of a complete deltaic progradation which took place in a few hundreds of thousands of years (Ramsbottom, 1979). Some of the lateral variability within it is shown in Figure 3.

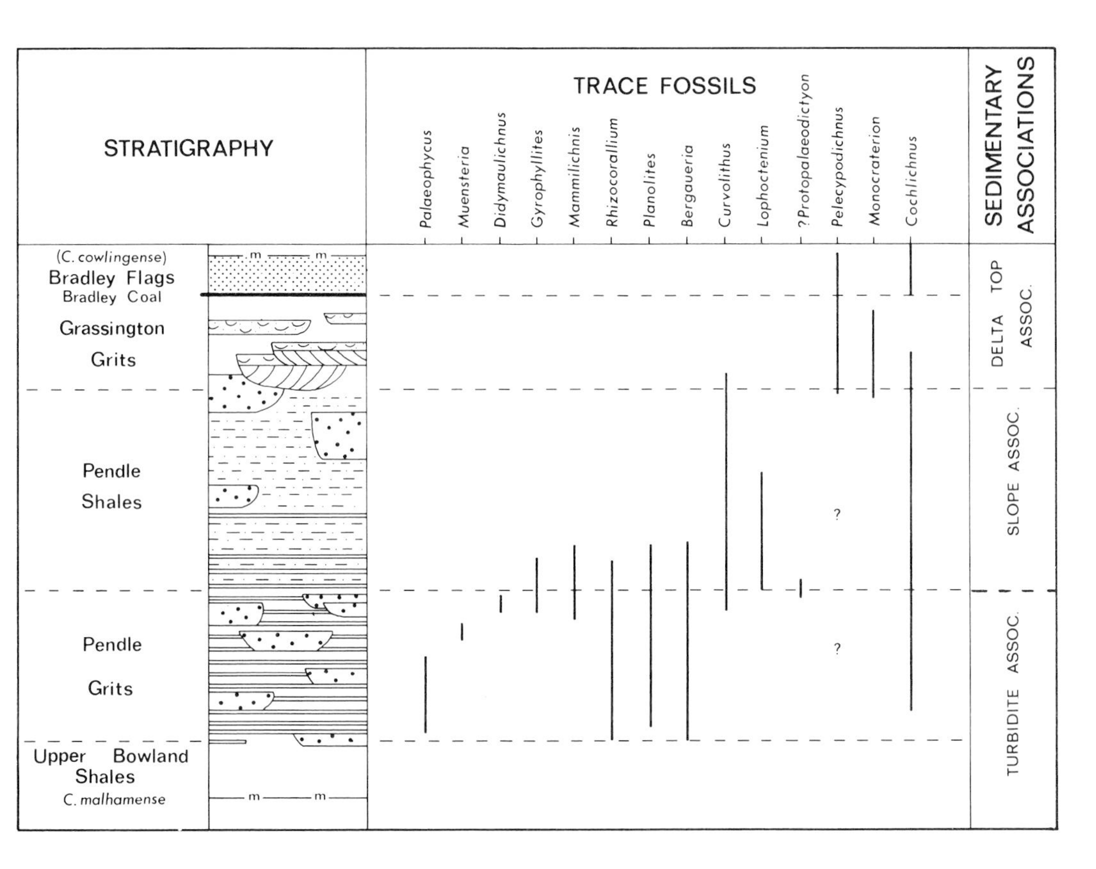

Fig. 4.—Vertical distribution of trace fossils through the Pendleian Skipton Moor Grits and the Bradley Flags, from Baines (1977). In the left hand column standard graphic log symbols are used (see Fig. 3).

Distribution of Trace Fossils in the Pendleian

The distribution of ichnogenera in terms of overall stratigraphical sequence is illustrated in Figure 4. It will be seen that there is wide variation in the observed ranges of these forms. The lower part of the Turbidite Association has relatively few ichnogenera but diversity increases in the upper part of the Association. This carries over into the lower part of the Slope Association, where diversity is greatest. This diversity decreases in the upper slope sediments, probably as a result of higher depositional rates compared with those of the lower slope. The Delta Top has the lowest diversity, characterized by *Pelecypodichnus* and *Monocraterion* (*Skolithos*) (Pl, 4A-C, E). *Curvolithus* and *Cochlichnus* (Pl, 2A, B; Pl. 3C) appear to be the forms with the greatest tendency to cross facies.

In the Turbidite Association, preservation of trace fossils is almost entirely restricted to thin, turbidite sandstone beds. Trace fossils are very rare in the interbedded mudstones. Preservation in hyporelief of the two forms *Planolites* and *Cochlichnus* (Pl. 2C; Pl. 3C) is limited by the thickness of the sandstone bed: beds thicker than about 10 cm lack these forms. *Bergaueria*, *Didymaulichnus* and *?Protopalaeodictyon* (Pl, 1A, B; Pl. 2D; Pl. 6D) may occur on the bases of thicker sandstone beds, but only *Bergaueria* is common in such cases. *?Protopalaeodictyon*, if it is a true graphoglyptid burrow, requires a degree of erosion associated with the turbidity current for its preservation, but some uncertainty attaches to the nature of its forms found in this sequence (see below).

The most abundant trace fossils found as epireliefs on the turbidite sandstones are two forms of *Rhizocorallium*. Both are interfacial burrows and their development was probably limited by the thickness of the overlying mudstone, *?R*. cf. *?R. jenense* (Pl. 5C; Pl. 6A, B) is found fairly low in the Turbidite Association while *R*. cf. *R. irregulare* (Pl. 5A, B) extends upwards into the base of the Slope Association. The great abundance and diversity of trace fossils within the lower part of the Slope Association is directly attributable to the abundance of thin turbidite sandstones at this level compared with their absence higher in the Association. *Gyrophyllites* and *Mammilichnis* (Pl. 1C, D) come in at the base of the Slope Association. This Association is characterized by the incoming of thicker units of

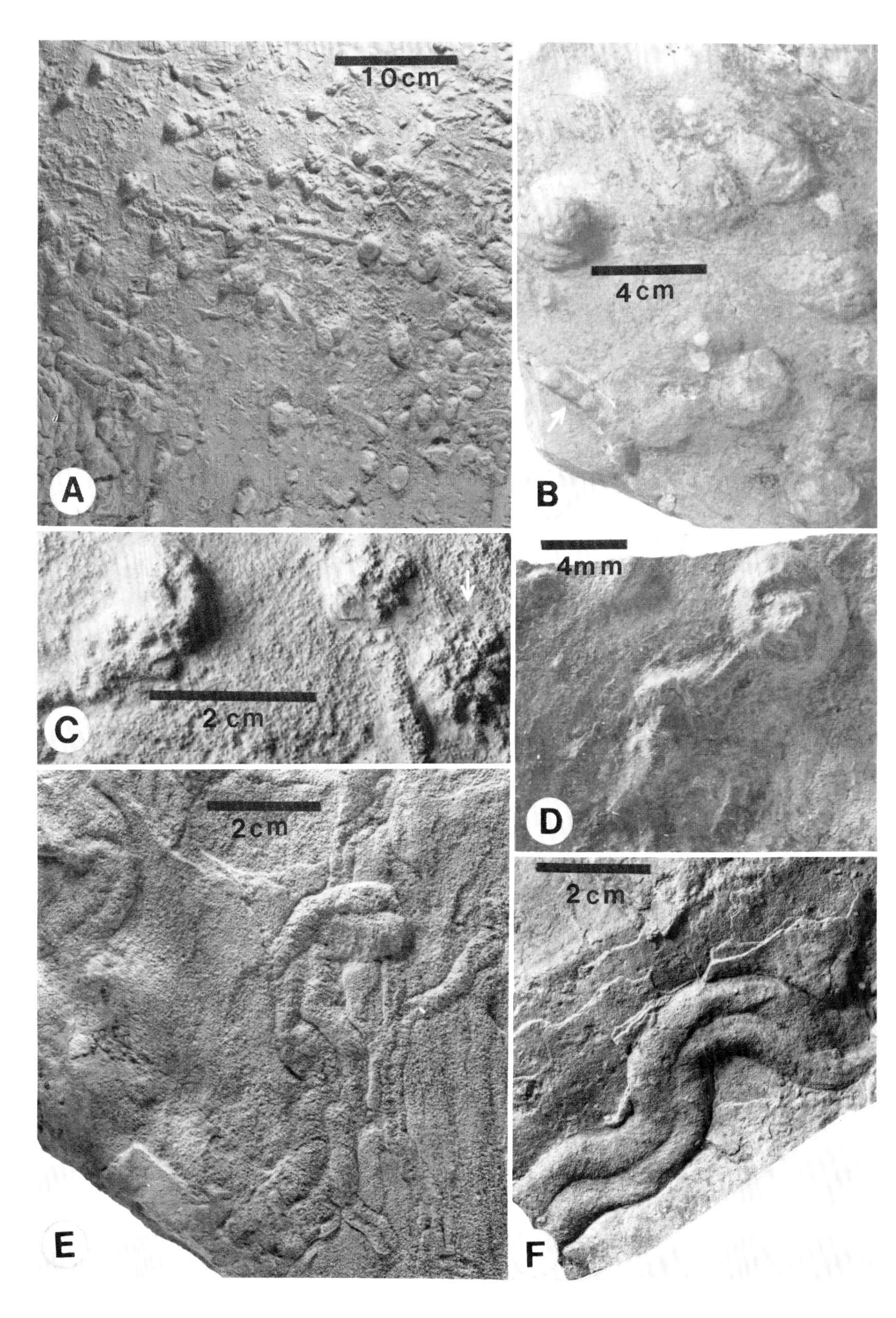
10cm
A
4cm
B
2 cm
C
4mm
D
2cm
E
2cm
F

parallel-bedded, striped siltstones which carry trace fossils preserved in full relief on splitting planes. *Lophoctenium* (Pl. 3A, B, D, E), *Cochlichnus*, and *Curvolithus* all occur in this way (Baines, 1977).

At the top of the Slope Association, where it passes into the Delta Top Association, small forms of *Pelecypodichnus* and *Monocraterion* are found in parallel-bedded siltstones and in ripple cross-laminated sandstones (Pl. 4A, C, E). They are commonly associated with *Cochlichnus* and *Curvolithus*. In the Delta Top Association, *Pelecypodichnus*, *Monocraterion*, *Cochlichnus* and rarely cf. *Aulichnites* (Pl. 1E, F) are the only forms recognized in the poorly exposed, finer grained lithologies between the major channel sandstones (Baines, 1977).

Interpretation of the Pendleian Assemblages

At first sight, the distribution of trace fossils through the Pendleian sequence suggests possible bathymetric control of these assemblages. Further consideration, however, indicates that the mode of deposition of the host sediments, rather than bathymetry may exert the more significant influence. Trace fossils of the Pendle Grit turbidites occur mainly on the bases and tops of thin-bedded sandstones. Bases have both predepositional forms (*Bergaueria, Gyrophyllites, Mammilichnis, ?Protopalaeodictyon*) and post-depositional interface feeding traces (*Planolites, Muensteria, Palaeophycus* (Pl. 6C, E). Upper interface mining traces characterize the tops of the sandstones (*Rhizocorallium*). A fourth assemblage of trace fossils characteristically occurs in thinly bedded, striped siltstones (*Curvolithus, Lophoctenium, Cochlichnus*), whereas ripple cross-laminated sandstones and associated siltstones carry *Monocraterion* and a large form of *Pelecypodichnus*. Some of these lithologies reflect relative depth in the context of this basin fill. However, the modes of deposition which give sand-mud interfaces, muddy food-rich substrates, more homogenous siltstones, or higher energy conditions at the sediment-water interface may have exerted a more significant control on the trace fossil distribution.

Many of the trace fossils are facies-crossing forms and only a few allow assignment to the bathymetric ichnofacies of Seilacher (1963) and Frey and Seilacher (1980). *Pelecypodichnus* and *Monocraterion*, both of which characterize the *Cruziana* and *Skolithos* ichnofacies of shallow-water origin, occur in sediments of the upper delta shope and delta top origin. *Lophoctenium* has been related to *Zoophycos* (Chamberlain, 1971b) and therefore suggests the presence of the *Zoophycos* ichnofacies, thought to form in water of intermediate depth. The association of the possible "deep-water forms" *?Protopalaeodictyon* and *Mammilichnis* (Crimes et al., 1981) with "shallow-water forms" *Bergaueria* and *Rhizocorallium* (Crimes, 1977) in the Turbidite Association does have a parallel with trace fossil assemblages described recently from deep-sea sand fans in Cretaceous-Eocene flysch sequences. Crimes (1977) and Crimes et al. (1981) explain such mixed assemblages in terms of control by topography and energy rather than by depth. They suggest that the mixed ichnofaunas were situated on depositional fan lobes swept by turbidity currents. Such a situation is comparable to that envisaged for the distal part of the Turbidite Association. However, the analogy must not be pressed too far, because in the Pendleian we are dealing with a small intracratonic basin with relatively modest water depth and true *Nereites* ichnofacies traces are absent.

Comparison with the bathymetric ichnofacies of Seilacher (1967) is based upon the concept of a marine

EXPLANATION OF PLATE 1

Figured specimens are housed in collections with the following abbreviations: University of Manchester, The Manchester Museum, M.Mus; Geology Department Special Collections, MGSF; University of Keele, Geology Department, Baines Collection, U.K.JGB; Okolo Collection, U.K.SAO; Collinson Collection, U.K.JDC; British Geological Survey, B.G.S.; Bolton Museum and Art Gallery, Bn.Mus. All figured specimens are from Baines (1977).

Fig. A,
B.— *Bergaueria*. *A*, Circular hypichnial protuberances on the tool-marked base of a thin turbidite sandstone. Pendleian, Turbidite Association. Catlow Gill, Carlton, near Skipton, Yorkshire. U.K.JGB.84. *B*, Enlarged view of positive hypichnia with granular bases (left), associated with transversely back-filled *Muensteria* burrow (arrow). Provenance as for Figure A. U.K.JGB.96.

C.— *Gyrophyllites*; positive hypichnia consisting of radially arranged bifid lobes around a central protuberance. Most complete specimen (right) shows eight distinct petal-like lobes. Pendleian, junction of Turbidite and Slope Association. Cawder Gill, Skipton Moor. U.K.JGB.210.

D.— *Mammilichnus*; positive circular hypichnial dome with central terminal projection on base of thin turbidite sandstone. Pendleian, Turbidite Association. ?Catlow Gill, Carleton, U.K.JGB.91.

E, F.— cf. *Aulichnites*. *E*, Positive epichnial sinuous bilobed trails in thinly laminated fine-grained sandstone. Two width classes are present. Some smaller specimens appear to show faint transverse sediment packing. Pendleian, Delta Top Association. Rams Gill, Embsay Moor, Yorkshire. U.K.JGB.135. *F*, Epichnial expression of a large sinuous trail with deep median groove and smooth flattened longitudinal lobes. Pendleian, Delta Top Association. Pike Stones, Hazlewood, Yorkshire. U.K.JGB.18.

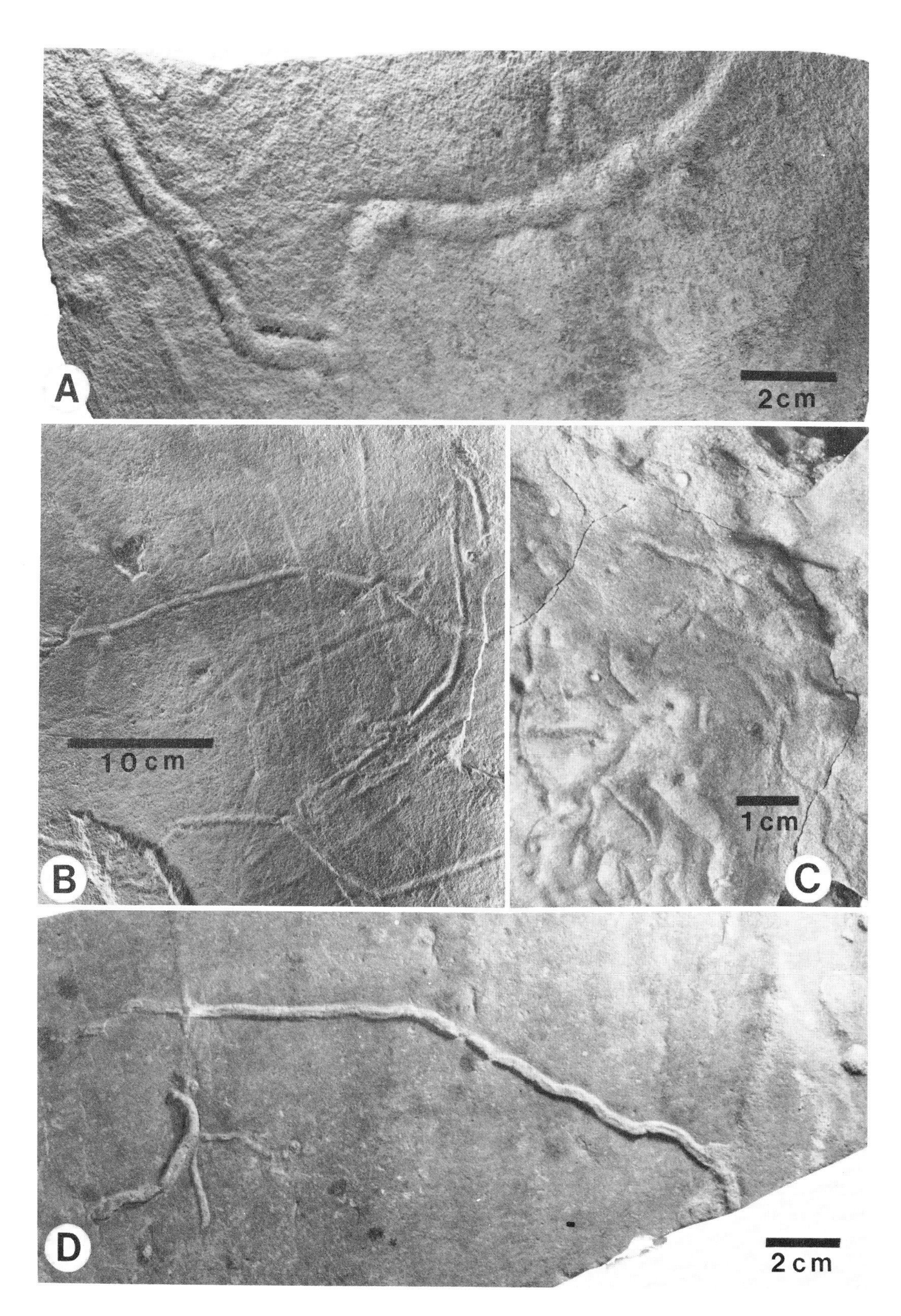
A
2cm
B
10cm
C
1cm
D
2cm

basin. The occurrence of goniatite-bearing marine bands in the Namurian separated by thick units of barren strata, comprising about 95 percent of the total of the Namurian succession, leads to possibilities of reduced salinity during deltaic progradation. The diversity of the trace fossil assemblage in the Pendleian sequence suggests that salinity was not much reduced, but models for comparison are lacking, and the evidence for paleosalinity changes suggested by trace fossils is ambiguous. *Bergaueria* and *Rhizocorallium* are considered marine, while *Cochlichnus* and *Pelecypodichnus* have been taken as indicators of reduced salinity in Carboniferous shelf sediments (Hakes, 1976; and this volume, p. 138, 142–3).

The Lower Kinderscoutian of North Derbyshire and South Yorkshire

The Lower Kinderscoutian of the southern part of the Central Pennine Basin is sedimentologically very similar to the Pendleian sequence described above (Reading, 1964; Walker, 1966a, b; Collinson, 1969; McCabe, 1978). It is illustrated by the succession below the Butterly Marine Band in Figure 2 (Chapel en le Frith, Edale, and the Peak). Distal turbidites of the Mam Tor Sandstones come in quite sharply above the Edale Shales with their discrete and widely spaced marine bands. The Mam Tor Sandstones apparently lack trace fossils. The overlying Shale Grit, which is thought to represent more proximal turbidite deposition, shows only very rare trace fossils and even rarer body fossils. In the middle part of the Shale Grit, marine bivalves have been found on two horizons in the Kinderscout region and *Planolites* occurs sparsely in thin bands close to these faunal horizons (Eagar, 1977). The fauna comprises *Sanguinolites* and cf. *Sanguinolites* Hind *non* M'Coy, which are regarded as reflecting marginally marine conditions. In a mudstone unit in the upper part of the Shale Grit of Blackden Brook, just northwest of Kinderscout, the bivalve *Promytilus* has been found on one horizon (Walker, 1964). Both these bivalve faunas, but particularly the latter, may indicate slightly reduced or "estuarine" salinities.

Trace fossils become abundant only in the upper part of the Grindslow Shales, the inferred product of progradation of the main delta slope. Silty sandstones and muddy siltstones containing very densely distributed burrows, referable to *Planolites* (Pl. 7F), extend over a stratigraphical interval of some 20 m. At a roughly similar stratigraphic level fairly common *Pelecypodichnus* also come in but these are restricted to units of ripple cross-laminated fine to medium-grained sandstone. The two trace fossils are mutually exclusive in terms of their host lithology, but can alternate in an interbedded sequence. Thus their occurrence may be sedimentologically controlled. The greater proportion of organic-rich mud, essential for *Planolites*, may have been inimical to bivalves, since it choked their gills (Eagar, 1948, p. 133).

In the highest part of the Grindslow Shales, interpreted as uppermost delta slope or subaqueous delta top, *Pelecypodichnus* occurs as virtually the only trace fossil. Burrows, interpreted as resting, that is, pedal anchoring places, occasionally indicate some upward movement of their former occupants (cf. Eagar, 1977, figs. 53, 57). Very short but unquestionable escape shafts, up to 1 cm in height, are present in some of the more coarsely grained sandstones of the Shales. At the base of the sequence, about the middle of the Grindslow Shales, burrows are small and all appear to have been resting places, within which the shell lay with its anterior end downward and long axis variably inclined to the substratum. Sizes and vertical extents of burrows tend to increase upward from the middle of the Grindslow Shales to the Lower Kinderscout Grit, a sequence which in general coarsens upward. But the largest and deepest burrows, (indicating merely larger shells. Pl. 7C, E) as well as the greatest height or vertical extent of the escape shafts, measured at right angles to the bedding planes, occur in a siltstone parting with fine-grained sandstone leaves within the lowest third of the Lower Kinderscout Grit (Fig. 2, Mossley). Thus, in addition to a correlation between burrow size and vertical extent of escape shafts with grain size, there is also a stratigraphical increase upwards in burrow size and height of escape shafts. Both are more commonly

EXPLANATION OF PLATE 2

All figured specimens from Baines (1977).

Fig. A,

B.— ?*Curvolithus*. *A*, Epichnial parting plane expression of low positive ribbon trail with marginal furrows. This is interpreted as the mold of the base of an overlying endichnial trilobed burrow of *Curvolithus* type (see Heinberg, 1970, fig. 3). Pendleian, Slope Association, Low Snaygill, Skipton. U.K.JGB.267. B, Crossing straight and curved ribbon trails in semi- relief on a parting plane within parallel-bedded fine-grained sandstone and siltstone. Pendleian, Slope Association, Howgill Beck, Beamsley, 10 km ENE of Skipton, Yorkshire. U.K.JGB. Negative No. 73.

C.— *Planolites*; short, curved epichnial grooves on linguoid ripples at the top of a thin turbidite sandstone. Some of the grooves appear to terminate in shallow pits or burrows; others approach *Cochlichnus* in form (lower right). Pendleian, Turbidite Association, Cawder Gill, Skipton Moor. U.K.JGB.94.

D.— *Didymaulichnus*; straight to slightly sinuous hypichnial trail with a narrow median groove. Base of a thin turbidite sandstone. Pendleian, Turbidite Association; provenance as for Figure C. U.K.JGB.152.

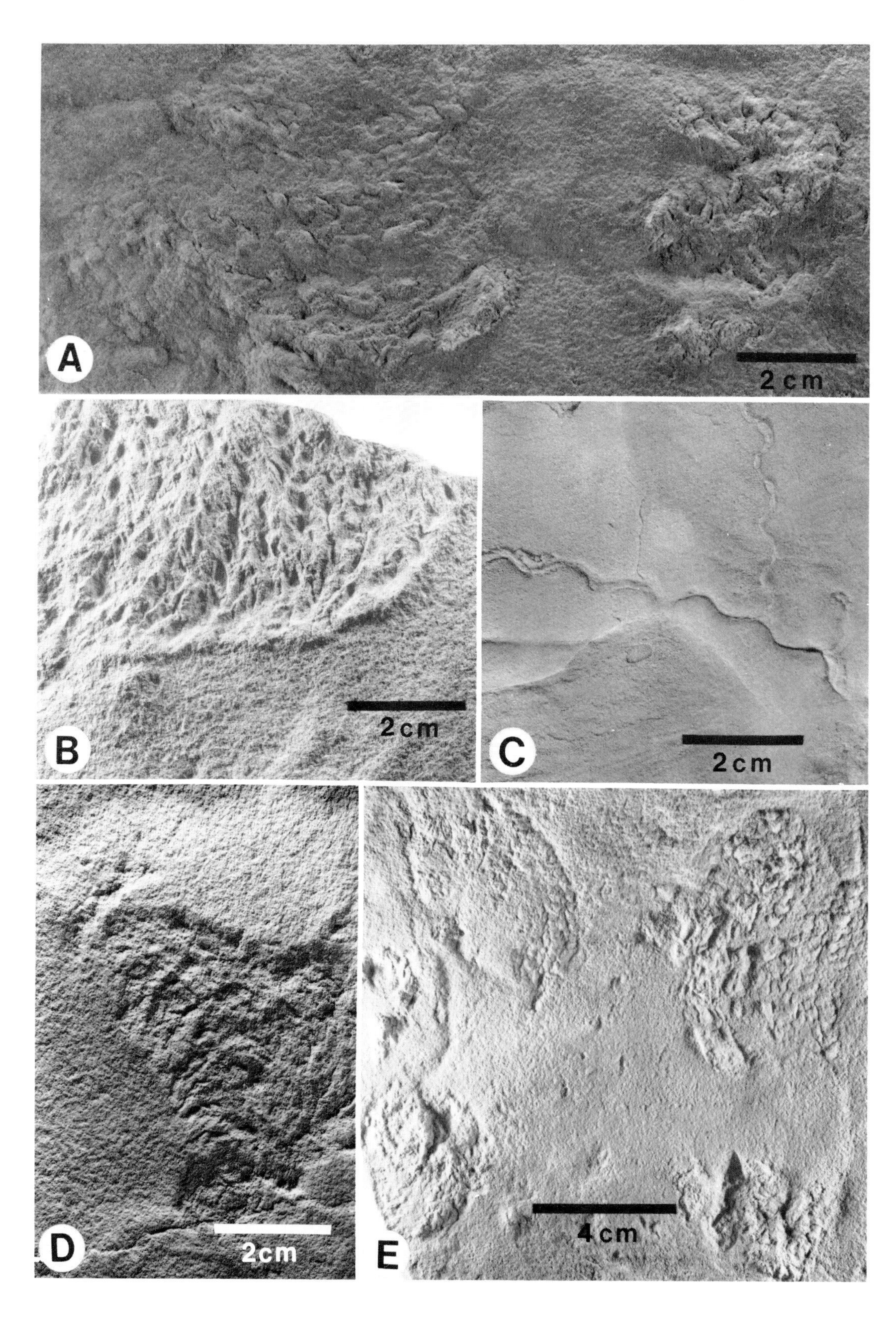
A
2 cm
B
2 cm
C
2 cm
D
2 cm
E
4 cm

oblique (Pl. 7E) than straight (Pl. 7D), and their long axes show some measure of preferred orientation on bedding planes, broadly compatible with regional paleocurrent trends (e.g. Fig. 7A, B).

At only two localities have trace fossils other than *Pelecypodichnus* and *Planolites* been found. On the northern side of Kinderscout, meandering trails referable to *Cochlichnus* occur on the tops of sharp-based sandstone beds in sediments between the main channel sandstones of the Kinderscout Grit. The sandstones are interpreted as crevasse splay deposits in an interdistributary setting.

Immediately north of Woodhead, at about the base of the Kinderscout Grit, one upper bedding plane in an interbedded sandstone and siltstone unit shows abundant stellate patterns of radiating, mud-filled depressions (Pl. 8D). The traces are referable to cf. *Asterichnus* (cf. Pl. 9C) and may indicate a localized depositional halt (see p. 137).

Discussion

Despite their closely similar sedimentology, the trace fossils of the Pendleian and Lower Kinderscoutian assemblages contrast most strikingly. The restricted assemblages of the latter with the absence of forms with obvious marine affinities suggests that, irrespective of local fluctuations, the basin of the central southern Pennines was generally less saline during the advance of the Kinderscoutian delta than in Pendleian time. Only the occurrence of *Sanguinolites* in the shallow-water sediments towards the top of the Pennine deltaic sequence argues against this; and these bivalves, with the evidence of *Pelecypodichnus*, which fits shells referable to cf. *Sanguinolites* (see the following section and Fig. 6), probably reflect short-lived periods of higher, but less than fully marine salinities. One may compare the well-known tolerance of lowered salinities shown by modern "estuarine" assemblages.

During intervals between the deposition of marine bands with goniatites, periods possibly associated with episodes of rising sea-level, it would be increasingly easy throughout the Namurian to dilute the basin as a result of the high freshwater discharges from delta-building rivers. If one assumes no great tectonic enlargement of the basin through the Namurian, each successive delta advance which filled and shallowed the deep basin would lead to a progressively smaller volume of sea water to be displaced. The trace fossils therefore suggest that the basin may have changed its character in a subtle way. The evidence for this salinity change depends entirely on the trace fossils and cannot be detected through the application of other sedimentological evidence.

THE UPPER KINDERSCOUTIAN, MARSDENIAN AND YEADONIAN STAGES

Nature of the Deltaic Sequences

In contrast to the Lower Kinderscoutian, Upper Kinderscoutian to Yeadonian strata of the Central Pennine Basin consist mainly of shallow-water sheet delta sequences. However, a thick turbidite unit occurs in the Marsdenian west of Blackburn (Collinson et al., 1977) and thin turbidites have been found locally in deltaic slope deposits at different stratigraphical levels. A summary of the full sedimentary development in the center of the Basin is shown in Figure 5 (left). Throughout the succession of sheet deltas paleocurrents are dominantly from the north and east, in the direction of progradation. An exception to this pattern is provided by the Haslingden Flags of the Rossendale area (Fig. 1), which are interpreted as elongate delta sequences and in which the regional paleocurrent direction is from the west (Collinson and Banks, 1975). The succession is completed by the relatively uniform Rough Rock, deposited over a very wide area on a broad plain (Shackleton, 1962), the sediment being distributed by a complex of probably braided distributary channels flowing in general from the northeast and establishing the broad paralic basis of the subsequent Westphalian Coal Measures (Reading, 1964). Seven major marine bands, yielding mainly goniatites and

EXPLANATION OF PLATE 3
All figured specimens are from Baines (1977).

Fig. A, B, D, E.— *Lophoctenium* spp. *A*, *L.* cf. *L. comosum* Richter, irregularly shaped isolate spreiten fields preserved as full relief on a parting plane within parallel-bedded fine-grained sandstone. Pendleian, Slope Association. Low Snaygill, Skipton, Yorkshire. U.K.JGB.148. *B*, *L.* cf. *L. comosum*, endichnia having "cock's tail"- like pattern made of chevron-like spreiten and a marginal burrow. Provenance as for Figure A. U.K.JGB.148. *D*, *L.* cf. *L. haudimmineri* Chamberlain, small spreiten field composed of curved, non-overlapping chevron spreiten on a parting plane. Provenance as for Figure A. U.K.JGB.150. *E*, *L.* cf. *L. haudimmineri*; parting plane in fine-grained sandstone with isolated patches of non-overlapping chevron spreiten. Pendleian, Slope Association. Pickles Gill, Hazlewood, 11 km ENE of Skipton Moor. U.K.JGB.143.

C.— *Cochlichnus*; positive and negative epichnial expression of sinuous burrows on linguoid, ripple-marked top surface of a thin turbidite sandstone. Pendleian, Turbidite Association. Vicar's Allotments, Skipton Moor. U.K.JGB.264.

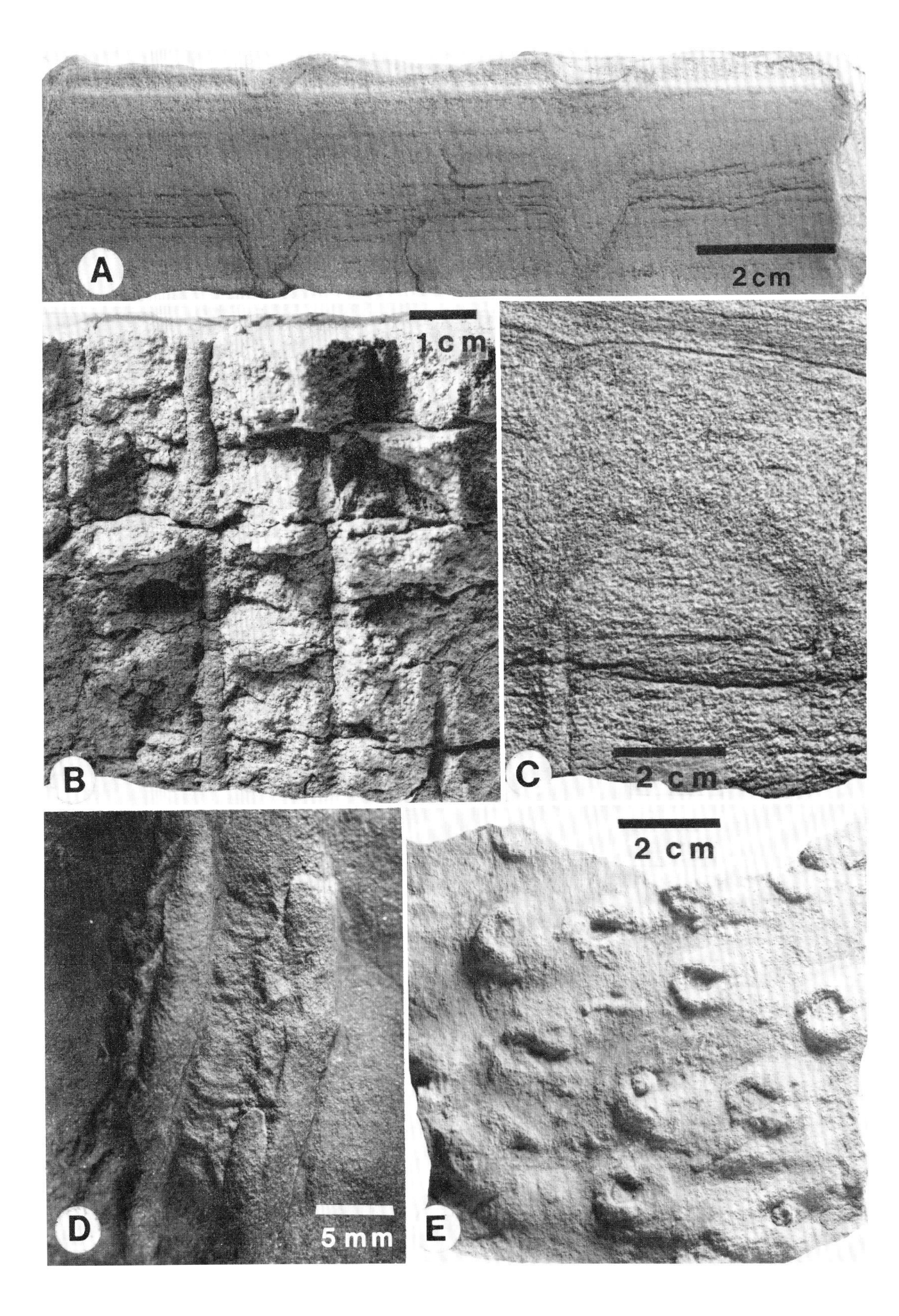
A
2 cm
1 cm
B
C
2 cm
D
5 mm
2 cm
E

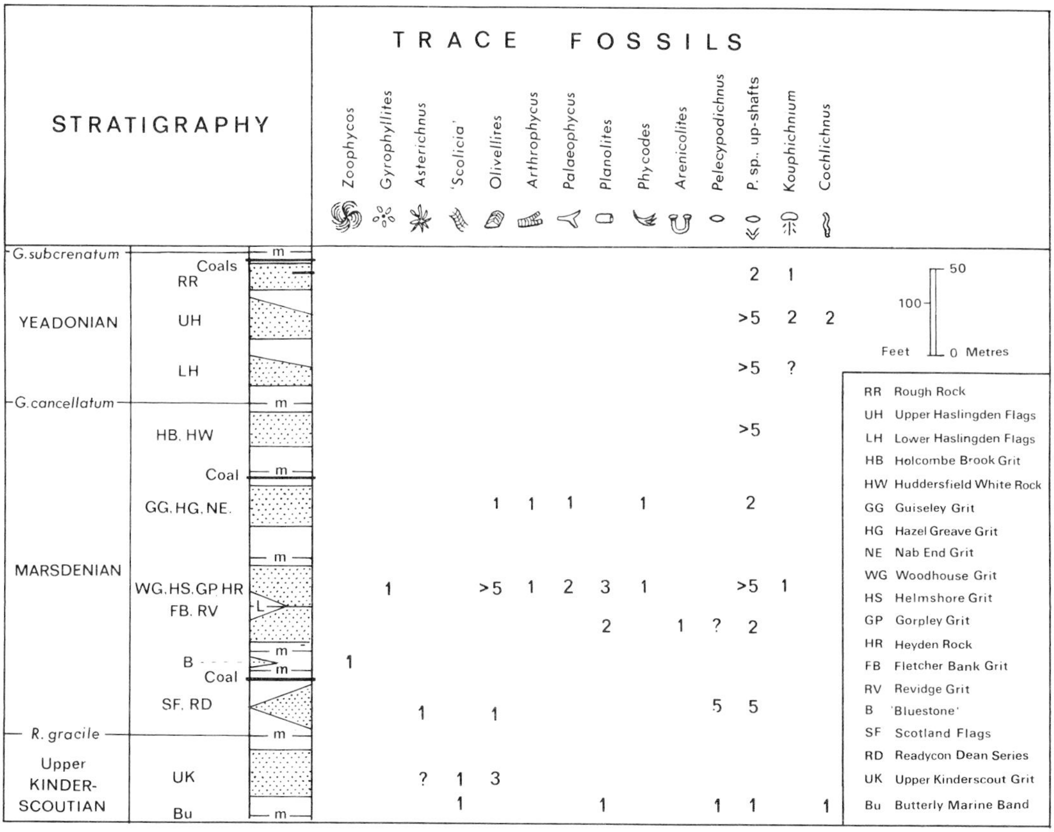

STRATIGRAPHY		Zoophycos	Gyrophyllites	Asterichnus	'Scolicia'	Olivellites	Arthrophycus	Palaeophycus	Planolites	Phycodes	Arenicolites	Pelecypodichnus	P. sp. up-shafts	Kouphichnium	Cochlichnus
G. subcrenatum	m														
YEADONIAN	Coals RR												2	1	
	UH												>5	2	2
	LH												>5	?	
G. cancellatum	m														
MARSDENIAN	HB. HW												>5		
	Coal m														
	GG. HG. NE.					1	1	1		1			2		
	m														
	WG. HS. GP HR		1			>5	1	2	3	1			>5	1	
	FB. RV (L)								2		1	?	2		
	m														
	B (m)	1													
	Coal														
	SF. RD			1		1						5	5		
R. gracile	m														
Upper KINDER-SCOUTIAN	UK			?	1	3									
	Bu (m)				1				1			1	1		1

Scale: Feet 100; 50, 0 Metres

RR Rough Rock
UH Upper Haslingden Flags
LH Lower Haslingden Flags
HB Holcombe Brook Grit
HW Huddersfield White Rock
GG Guiseley Grit
HG Hazel Greave Grit
NE Nab End Grit
WG Woodhouse Grit
HS Helmshore Grit
GP Gorpley Grit
HR Heyden Rock
FB Fletcher Bank Grit
RV Revidge Grit
B 'Bluestone'
SF Scotland Flags
RD Readycon Dean Series
UK Upper Kinderscout Grit
Bu Butterly Marine Band

Fig. 5.—Schematic section of the Upper Kinderscoutian to the top of the Namurian. Trace fossil horizons are broadly indicated by the number of localities at which they have been seen. Data for the Huddersfield White Rock from A. C. Benfield (personal communication, 1981). L=*Lingula*. Other symbols as for Fig. 2.

EXPLANATION OF PLATE 4

All figured specimens are from Baines (1977).

Fig. A, B, C, E.— *Monocraterion (Skolithos)*. *A*, Endichnial expression of funnel structure of *Monocraterion* shown on cut vertical section through parallel-bedded fine-grained sandstone. Pendleian, Delta Top Association, Threapland Gill, Cracoe Fell, 8.5 km north of Skipton, Yorkshire. U.K.JGB.370. *B*, *Skolithos*, full relief endichnial expression of regularly spaced vertical pipe burrows without funnels. Pendleian, Delta Top Association. Lowburn Gill, Embsay Fell, 6 km north of Skipton. U.K.JGB.264. *C*, *Monocraterion*; endichnial expression in parallel-bedded sandstone, showing upward passage of vertical burrows into funnels. Provenance as for Figure B. U.K.JGB.266. *E*, *Monocraterion*, positive hypichnial molds of circular current-scour effect around resistant vertical burrow. In this example the funnel lies immediately above the scour within the sandstone slab. Pendleian, Delta Top Association. Hesker Gill, Thorpe, 14 km NNE of Skipton. U.K.JGB. Negative 87.

D.— *Rhizocorallium* type A (*R*. cf. *R. irregulare* Mayer). Enlargement of part of the long, horizontal U-burrow shown in Plate 5B, indicating positive epichnial expression of parallel, marginal burrows with intervening zone of protrusive spreiten. Pendleian, lower part of the Slope Association. Bareshaw Beck, Carleton, near Skipton. U.K.JGB.67.

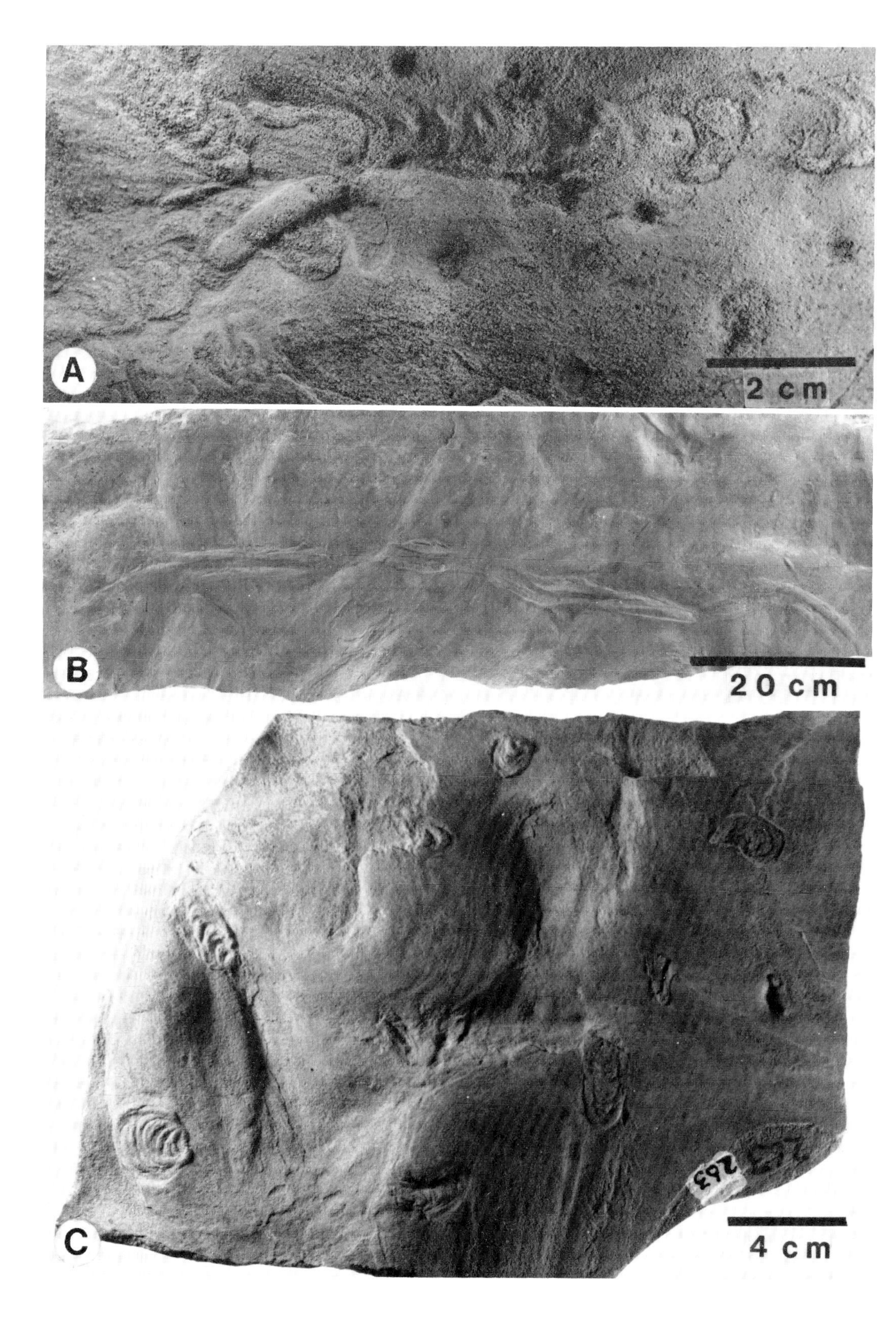
A
2 cm
B
20 cm
C
263
4 cm

less commonly shallow-water marine bivalves and gastropods, were each laid down over time intervals of the order of 10,000 years within a succession estimated to have lasted about 4 million years (Ramsbottom, 1979, 1981). Paleoenvironments were thus of much repeated, often long durations and were highly variable, ranging from interdistributary bay sediments to subaqueous levées of the mouth-bar area, channel sandstones, crevasse-splay deposits and occasionally forest swamps.

Distribution and Sequence of Trace Fossils

The trace fossils are divided, for discussion, into three main assemblages, that of *Pelecypodichnus*, which occurs through much of the succession (Fig. 5) and "evolves", with expansion in its range of paleoenvironments, "*Scolicia*" - *Olivellites* assemblages, which appear confined to these sediments, and *Phycodes-Arthrophycus-Kouphichnium* assemblages. The last group is given brief consideration here, and *Kouphichnium* (*Limulicubichnus*) is more conveniently discussed in the account of the lower Westphalian ichnofaunas. Other single occurrences are mentioned in their context and are discussed under Systematic Ichnology (p. 136).

Pelecypodichnus *and* Pelecypodichnus *escape shafts*.—Evidence of small upward movement of bivalves following downward burrowing into sands and silts has already been described from the upper part of the Lower Kinderscoutian. In the succeeding sediments of the Upper Kinderscoutian and lower Marsdenian, increasing vertical extent, or height of escape shafts, in terms of mean, minimum and maximum values, indicated a consistent overall increase in upward movement of bivalves on progressively higher horizons. However, there is still the trend, already noted, for shafts to be longest in the more coarse-grained beds at any one broad horizon (Eagar, 1977; Eagar et al., 1983). Moreover, there is wide overlapping in the range of heights of shafts on successive horizons. In the Upper Kinderscoutian bivalve burrows without evidence of upward movement, that is resting places only, are reduced to between 5 and 10 percent of the total structures, and bivalve trails may locally be present (Fig. 5, basal trace fossil horizon; Pl. 7C, E, D, B). In the succeeding lower Marsdenian no resting places (without also evidence of upward movement) were observed, and escape shafts, although commonly short (Pl. 10E), were generally more abundant—often prolifically so—than on lower horizons. Their mean observed height (20 mm) is double that of the Lower Kinderscoutian shafts (Eagar, 1977). In the Marsdenian there is wide variation in both cross-section of the shaft (reflecting size of shell) and also in vertical extent, which may reach 120 mm, for instance, in the coarser grained beds of the Scotland Flags (Fig. 5). Whereas obliquity of the shafts without obvious directional alignment (Pl. 7E, C) is usual in the Lower Kinderscoutian, in the Upper Kinderscoutian, and particularly in the lower Marsdenian, shafts are more nearly vertical and an oblique shaft straightening upward is by no means uncommon (Eagar et al., 1983). On higher horizons in the Marsdenian, for instance, in the Ashover Grit (Figs. 2, 5) and upwards through the Huddersfield White Rock to the Yeadonian, field observation shows progressive increase in the vertical extent or height of escape shafts, which in the latter beds commonly reach a meter or more (Hardy and Broadhurst, 1978).

In the Upper Kinderscoutian, sections of escape shafts fit dimensionally an assemblage of cf. *Sanguinolites* Hind *non* M'Coy (Fig. 6) which has been described from the top of the Upper Kinderscout Grit, where the shells are preserved in calcareous siltstone and the faunal assemblage was marginally marine (Eagar, 1977). Sedimentology and associated trace fossils (Pl. 8C) suggest a quiet interdistributary bay, marginally marine to brackish (Eagar et al., 1983). On the other hand, escape shafts in the Yeadonian and overlying Westphalian strata were undoubtedly made by elongate *Carbonicola*, a non-marine genus (Hardy, 1970b; Eagar, 1971; Hardy and Broadhurst, 1978; Broadhurst et al., 1980). The earliest *Carbonicola* has been found in sandstone of the Heyden Rock, middle Marsdenian (Fig. 5). On the evidence of its internal

EXPLANATION OF PLATE 5

All figured specimens are from Baines (1977).

Fig. A,
B.— *Rhizocorallium* type A (*R.* cf. *R. irregulare* Mayer). *A*, Elongate zone of protrusive spreiten with only partial preservation of lateral burrows. Although the burrow is preserved epichnially on the top surface of a thin turbidite sandstone, it was formed interfacially, below the overlying mudstone. Pendleian, lower part of the Slope Association. Bareshaw Brook, Carleton, near Skipton, Yorkshire. U.K.JGB.83. *B*, Linguoid rippled top surface of thin turbidite sandstone with very long parallel burrows with intervening spreiten (see Pl. 4D) preserved as positive epichnia. Provenance as for Figure A. U.K.JGB.67.

C.— ?*Rhizocorallium* type B (?*R.* cf. *R. jenense* Zenker). Short horizontal U-shaped burrows in positive epirelief on linguoid rippled top surface of thin turbidite sandstone. Globular (lower left) and parallel-sided types occur, both apparently in random orientation. Pendleian, Turbidite Association. New Intake Farm, Embsay, 2.5 km north of Skipton. U.K.JGB.263.

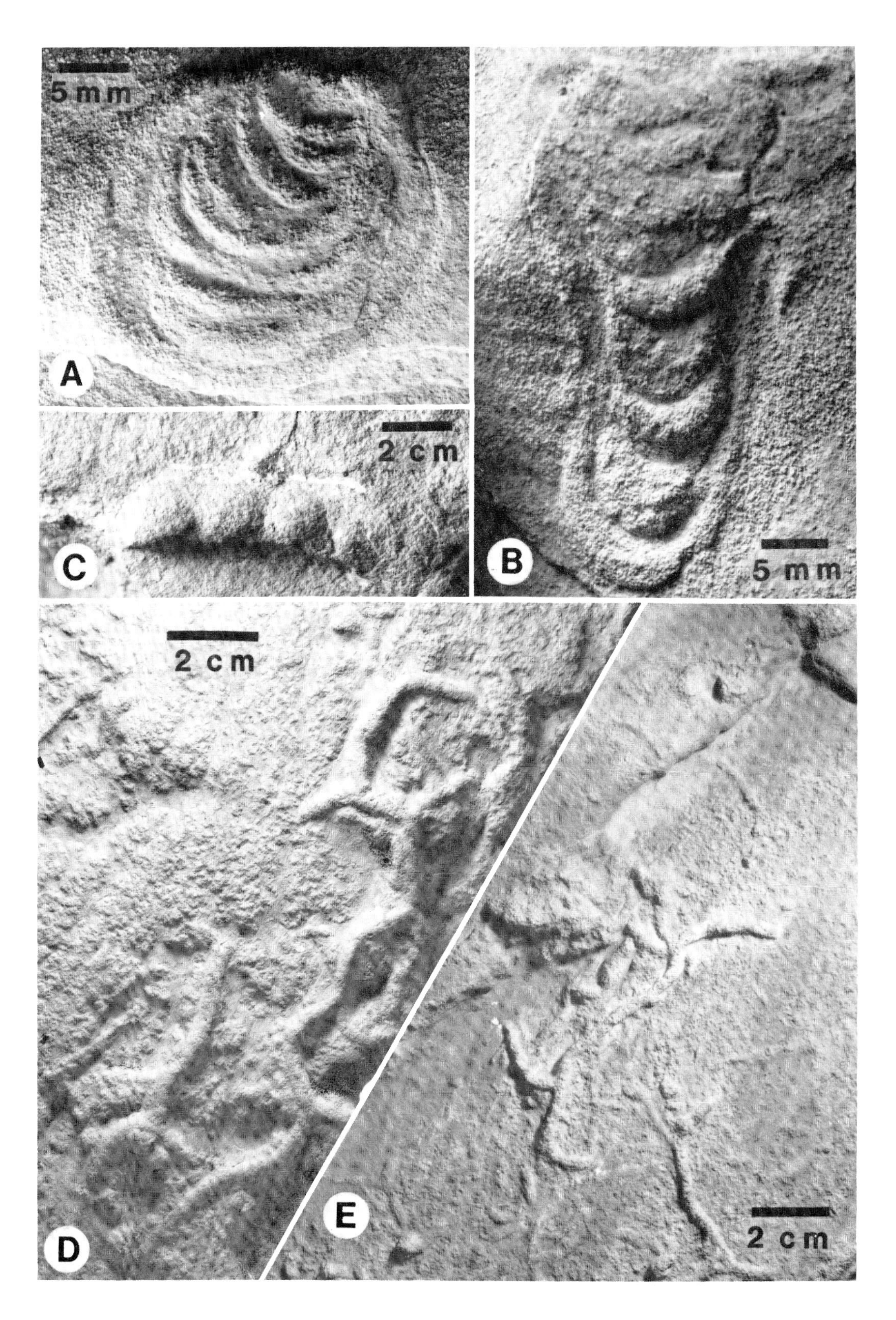
5 mm
A
2 cm
C
B
5 mm
2 cm
E
2 cm
D

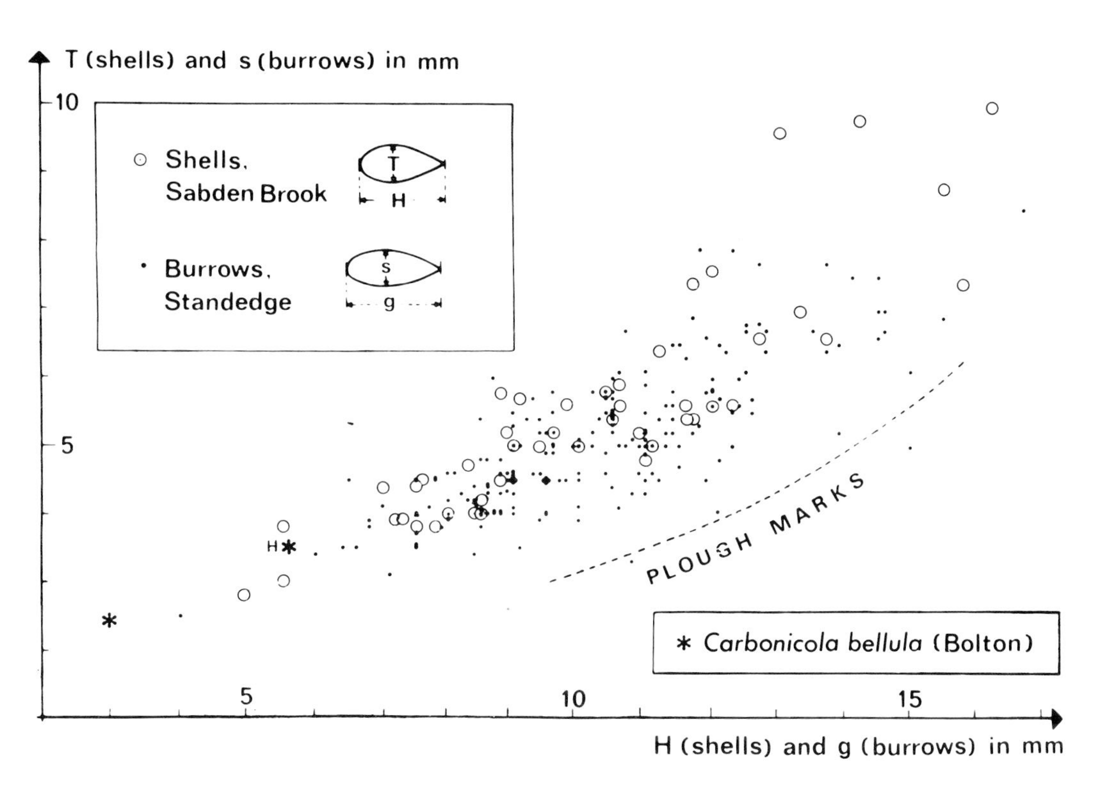

Fig. 6.—Measurements of burrows and escape shafts taken parallel to bedding planes (g, s) from above the Butterly Marine Band near Marsden (Figs. 1, 2) plotted with shell heights and obesities (H, T respectively) of an assemblage of shallow-water bivalves referred to cf. *Sanguinolites* Hind *non* M'Coy. The latter lie in calcareous siltstone 1 meter below the marker marine band, *Reticuloceras gracile* Bisat at Sabden Brook, south of Clitheroe. 'Plough marks' are elongate oblique infillings seen on the bedding planes and having one side straight or slightly reflected. From Eagar et al. (1983), by permission of the Council of the Yorkshire Geological Society.

EXPLANATION OF PLATE 6

All figured specimens are from Baines (1977).

Fig. A,

B.— *?Rhizocorallium* type B (*?R.* cf. *R. jenense* Zenker). *A*, Globular form (cf. Pl. 5C, lower left) showing asymmetrical nature of the spreiten similar to the probing lamellae of *Zoophycos circinnatus* (Brongniart). The circular marginal burrow is only poorly preserved (for other details see Pl. 5C). *B*, Parallel-sided or proximally divergent form with regular protrusive spreiten and negative epichnial expression of marginal tube. This form closely approaches *R.* cf. *jenense* from the Lower Westphalian (Pl. 14B, E). Provenances as for Plate 5C. U.K.JGB 262.

C.— *Muensteria*; positive hypichnial preservation of a short horizontal burrow with both transverse annulations and faint meniscus backfilling (see also Pl. 1B). Pendleian, Turbidite Association, Catlow Gill, Carleton, near Skipton. U.K.JGB.96.

D.— *?Protopalaeodictyon*; horizontal burrows branching in a dichotomous and "zigzag" manner to form a crude hexagonal network, preserved on the base of a thin turbidite sandstone. This sole surface bears chevron marks and other evidence of erosion suggesting that the burrows are probably predepositional in origin. Pendleian, Turbidite Association. Bareshaw Beck, Carleton, near Skipton. U.K.JGB.167.

E.— *Palaeophycus*; irregular horizontal burrows with simple dichotomous branching preserved as variable depth hypichnia associated with groove molds on the base of a thin turbidite sandstone. Pendleian, Turbidite Association. Catlow Gill, Carleton, near Skipton. U.K.JGB.156.

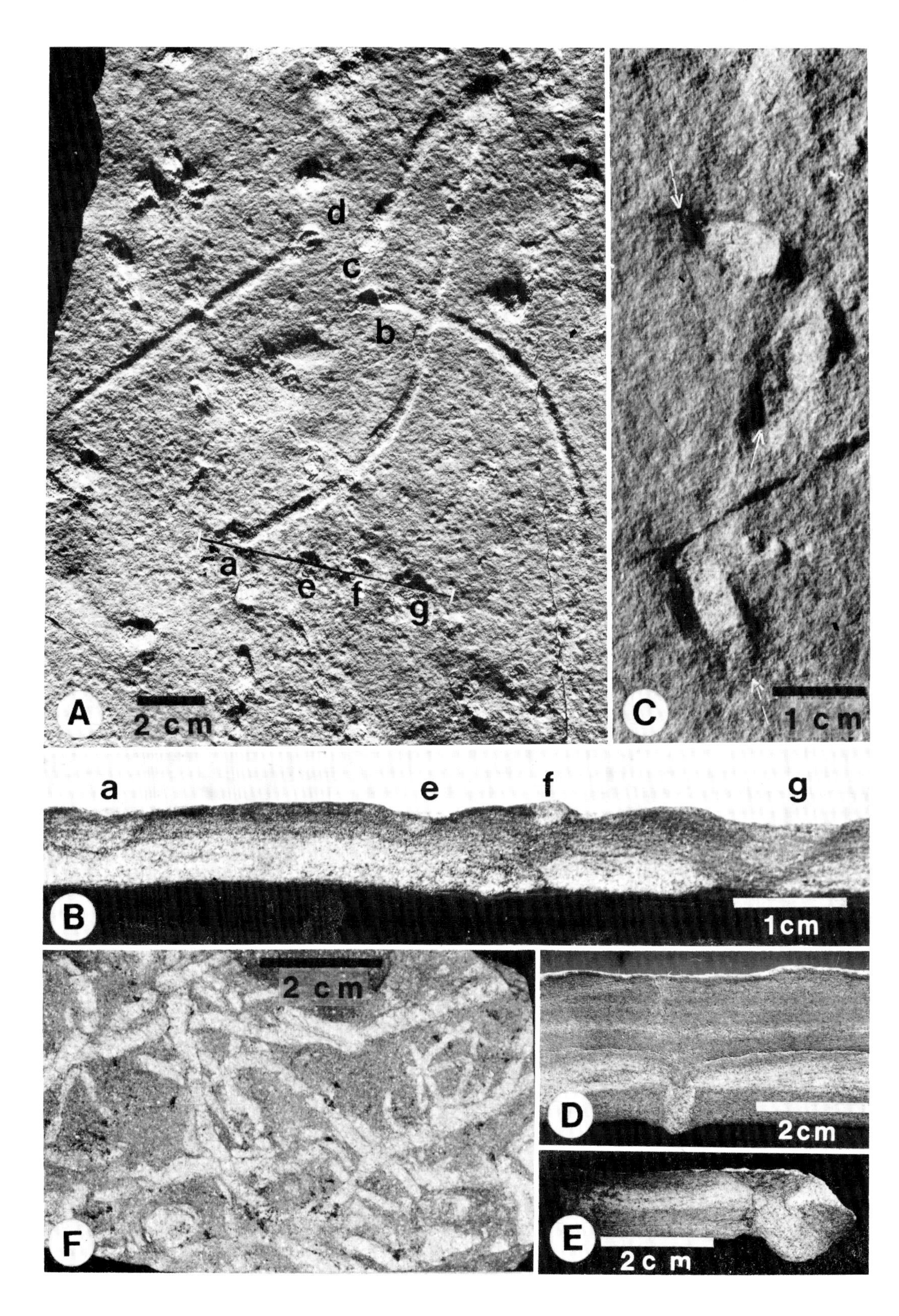
d
c
b
a
e
f
g
A
2 cm
C
1 cm
a
e
f
g
B
1cm
F
2 cm
D
2cm
E
2 c m

and external morphology it appears that the genus descended from the closely similar cf. *Sanguinolites* Hind *non* M'Coy of the Upper Kinderscoutian (Eagar, 1977; Eagar et al., 1983). The pattern of escape shafts from the Lower Kinderscoutian to the Yeadonian is continuous. It appears therefore significant that Okolo (1982) has found a very marked increase in the range of paleoenvironments characterized by *Pelecypodichnus* escape shafts in the Marsdenian compared with those in the Kinderscoutian. His examples all come from the mid-Pennine area. In addition to their continued occurrence in horizontally or parallel-bedded sandstones below the lowest fluvial channel in each sequence and, as such, interpreted as mouth bars, escape shafts have also been found in other sediments of the delta top. These include small- to medium-scale trough cross-bedding, regarded as subaqueous levées of the mouth-bar area, trough cross-laminated sandstone, interpreted as crevasse-splay deposits, and in what appears to be an abandoned channel. In all the latter cases occurrences are associated with relative quiescence or a low-energy regime. Where the flow regime appears to have been higher, as in parallel-laminated sandstones of a channel in the Fletcher Bank Grit (Fig. 5), shafts are inclined in the same direction (as contrasted with the more irregular inclination of earlier shafts on the delta slope) and suggest alignment with current flow as noted by Berg (1977) in Devonian *Archanodon* burrows (see also p.13 of this volume). There is therefore evidence from trace fossils and sedimentary facies to suggest that the salinity range of periodically upward moving bivalves was broadened, particularly in the Marsdenian and primarily in the direction of fresher water. As stressed by Eagar (1974a, 1977) and confirmed by Hardy and Broadhurst (1978), elongate shells with low umbones and low obesity (showing "streamlining") characterized the highest sandy sediments with escape shafts. This shape appears to be the result of selection operating on the more successful "risers", the "refugee communities" of Hardy and Broadhurst (1978).

Studies of the long axes of *Pelecypodichnus* escape shafts as measured on bedding planes (for example, Pl. 7C) suggest that preferred orientation parallel to paleocurrents improved with increase in grain size of the host sediment (Hardy and Broadhust, 1978). Orientation also appears to have been better in shafts attributable to *Carbonicola* than in those of its apparent ancestor (Fig. 7). Collinson and Banks (1975, pl. 26, fig. 1) used orientation of *Pelecypodichnus* as a paleocurrent indicator where depositional structures were lacking, the sharp end of the bedding section of the shaft indicating the commissure of the valves with the inhalent aperture pointing 'up-current' (compare again with the results of Thoms and Berg this volume, p.19). We stress, however, that *Pelecypodichnus* is not necessarily an indicator of non-marine conditions (see its sequence in Fig. 5). Moreover, some of the increase in the height of escape shafts progressively from Lower Kinderscoutian to Yeadonian sediments probably reflects merely a reaction to faster rates of deposition (see Schäfer, 1962). Nevertheless, between the sandstones of the Scotland Flags and the Haslingden Flags there is a difference in maximum height of shafts of one order of magnitude; and it seems reasonable to suppose that this difference, which was maintained into the lower Westphalian (p.127), primarily indicates evolution within early *Carbonicola* and its antecedents (Eagar, 1977).

EXPLANATION OF PLATE 7

Fig. A.— Four hypichnial ridges, bivalve trails, each terminating at a *Pelecypodichnus* which forms the base of an escape shaft (a-d). Lighting from bottom right. The line (a,e,f,g) shows the position of the section seen in Figure B.

B.— Section, cut vertical to the bedding, on the counterpart of the slab shown in Figure A, indicating sections of *Pelecypodichnus*, partly infilled by fine-grained sandstone. From Upper Kinderscoutian siltstones and sandstones (interdistributary bay) above the Butterly Marine Band in Standedge cutting, near Marsden, Yorkshire. From Eagar et al. (1983), by permission of the Council of the Yorkshire Geological Society. M.Mus. LL.6064A, B respectively.

C.— Epichnial surface of siltstone slab with three burrows, *Pelecypodichnus*, showing infill by fine-grained lighter colored sandstone. White arrows point to the sharpened ends of the burrows, indicating the commissures of the valves. Lighting from the left. Manch.Mus.M.3182.

D.— Section cut vertical to the bedding showing siltstone and sandstone laminae and a single *Pelecypodichnus*. The line of the section makes an angle of about 45° with the long axis of the burrow on the bedding plane. M.Mus.M.3183.

E.— Section cut vertical to the bedding and transverse to the long axis of *Pelecypodichnus* on the bedding plane. M.Mus.M.3185. Figures C, E from interdistributary, possibly crevasse-splay deposits in the Lower Kinderscout Grit at Buckden Castle Quarry, Mossley. Reproduced from Eagar (1977) by permission of the Council of the Royal Society of London.

F.— Silty mudstone with crossing *Planolites*, infilled with lighter colored fine-grained sandstone, from about the top of the Grindslow Shales, Blackden Book, Kinderscout; interpreted as uppermost delta slope or subaqueous delta top. M.Mus.LL.6208.

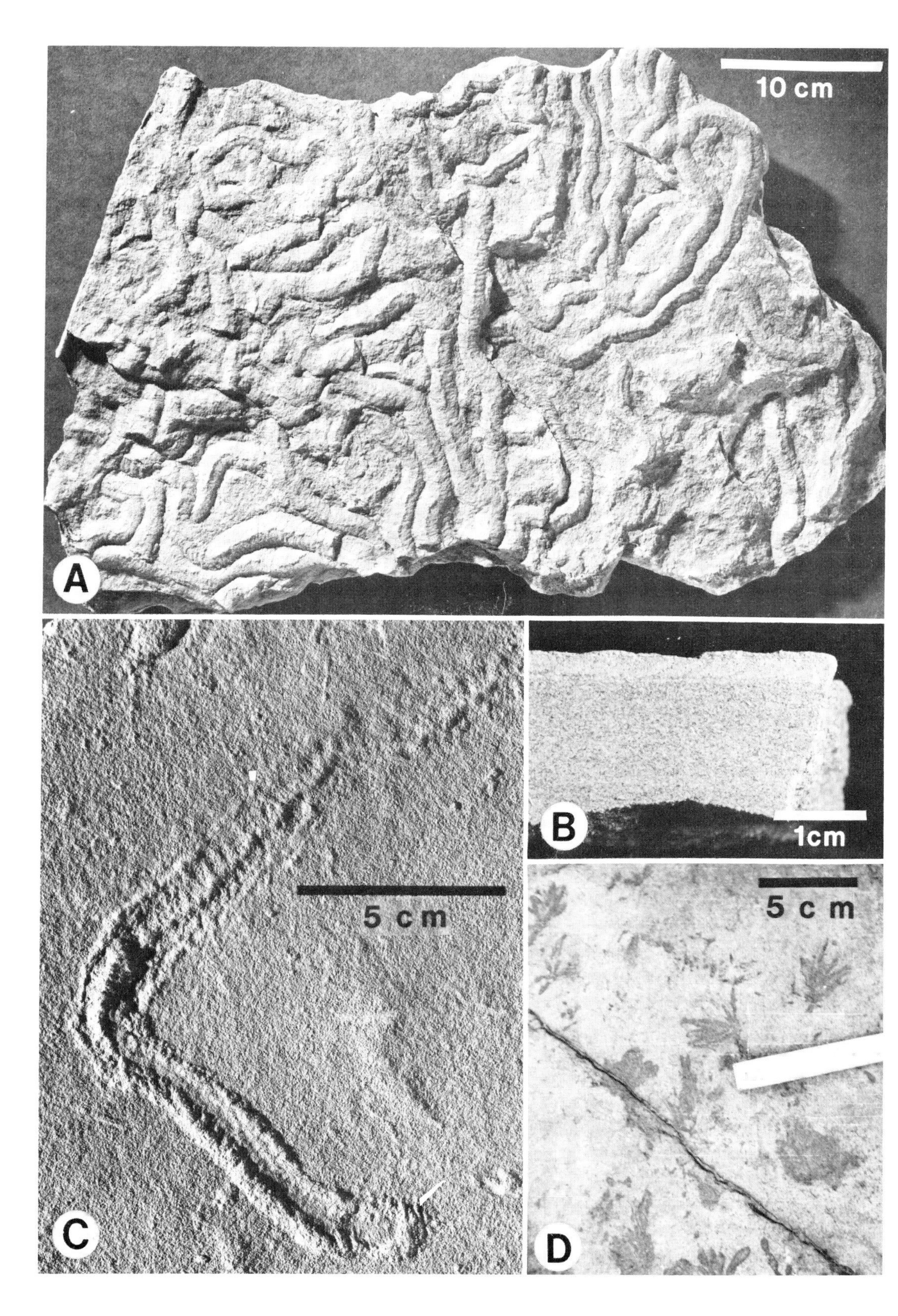
10 cm
A
5 cm
C
B
1cm
5 cm
D

Zoophycos *and* "Scolicia"–Olivellites *assemblages*.— At a single locality, near Keighley, *Zoophycos* (Pl. 10D) has been found in the "Bluestone" of the Marsdenian (Figs. 2, 5), a silty mudstone rich in sponge spicules (Stephens et al., 1953). The "Bluestone" develops over a limited area mainly in the region of Keighley and is likely to reflect clearness of marine water and absence of disturbance rather than considerable depth (p. 144). A very similar *Zoophycos*-sponge spicule association is developed in the Richmond Cherts (Pendleian) of north Yorkshire (Wells, 1955 and p.145 of this paper).

"*Scolicia*" (Pl. 8C) has been found *in situ* only above the Butterly Marine Band (Fig. 5; Eagar et al., 1983). *Olivellites* is apparently confined to the Upper Kinderscoutian and Marsdenian, where it is well developed and widespread (Pls. 8A, B; 9A, B). It occurs in horizontally bedded sandstone and in parallel laminated fine- to medium-grained well-sorted sandstone. The sandstones, which are sometimes interbedded with thin siltstone and mudstone layers, are persistent in outcrops. They commonly occur below the lowest fluvial channel and appreciably above the highest goniatite-bearing marine strata and are considered characteristic of mouth-bar sediments. With one exception, a turbidite near Holme Moss, the paleoenvironment is the same in both the Upper Kinderscoutian and the Marsdenian. Examples from the Marsdenian tend to have finer and more frequent transverse ribs (Pl. 9B) than those from the Upper Kinderscoutian, and the latter in turn are "ribbed" more frequently than "*Scolicia*", which appears to be linked to *Olivellites* in features of the endichnial structure (p.144). Thus the sequence of these assemblages within the Central Pennine Basin may have stratigraphical significance. On the other hand, there are parallels with similar trails of *Olivellites* (= *Crossopodia*) from "Yoredale" sheet deltas of Pendleian age of the Northumberland Trough and Alston-Askrigg Block areas to the north of this Basin (Hancock, 1858; Tate, 1859) (Fig. 1).

Phycodes-Arthrophycus-Kouphichnium *assemblages*.—The essential elements of these assemblages, interstratal feeding burrows of *Arthrophycus* (Pl. 10A), *Palaeophycus* (Pl. 10C), and *Phycodes* (Pl. 10B, F), have been recorded from two major stratigraphical levels in the Marsdenian (Fig. 5) but with a variety of associated traces at different localities. In the trough cross-bedded Woodhouse Grit *Phycodes* cf. *palmatum* occurs with *Gyrophyllites*, previously found in the Turbidite Association of the Skipton Moor Grits, and with *Kouphichnium* (Pl. 10B, a, b), whereas in the Fletcher Bank Grit *Arenicolites carbonarius* also occurs. The environmental interpretation of sediments containing these burrows also varies from mouth-bar or distributary channels (e.g., Woodhead and Hazel Greave Grits) to crevasse-splay sands within an interdistributary bay (Fletcher Bank Grit, Fig. 5). Species of *Phycodes* may characterize slightly different paleoenvironments (*P.* cf. *palmatum*—mouth bars, and *P.* cf. *curvipalmatum*—distributary channels). A single occurrence of cf. *Asterichnus* (Pl. 9C) in the Scotland Flags compares closely with mode of occurrence of this ichnogenus in Lower Kinderscoutian (p.111) and "Yoredale" deltaic sandstones of Weardale, County Durham (p.137), but since it was found in scree material its precise sedimentary association is uncertain. Although the earliest Silesian record of *Kouphichnium* is from the Lower Marsdenian (Fletcher Bank Grit), it is most abundant in the Yeadonian (Upper Haslingden Flags) and Westphalian (Fig. 8), where it is characteristic of shallow non-marine conditions (see below, p.129).

Discussion

Although the Upper Kinderscoutian-Yeadonian succession consists principally of sheet delta sequences and two elongate shallow-water deltas, many of the smaller-scale paleoenvironments from a sedimentological point of view are essentially the same as those of the preceding Lower Kinderscoutian. There is a proportionately smaller representation of delta slope deposits and a greater proportion of delta top, mouth-bar, interdistributary bay, channels and associated paleoenvironments.

The most obvious difference in the trace fossils is the gradual replacement of *Pelecypodichnus* resting

Explanation of Plate 8

Fig. A.— *Olivellites* cf. *O. plummeri* Fenton and Fenton in laminated siltstone-sandstone, which is interpreted as of marginal mouth-bar origin; it lies below sandstone overlain by *Reticuloceras gracile* Bisat, Upper Kinderscoutian. Grinding Stone Hole, Oxenhope, near Bradford. Lighting from top left. M.Mus.LL.6173.

B.— Section cut vertical to the bedding plane, along the center of an approximately straight part of the burrow of *Olivellites* cf. *O. plummeri*, showing imbricate structure in light colored fine-grained sandstone. Provenance as for Figure A. A transversely cut section of a similar burrow is shown in Figure 9. M.Mus.LL.6206.

C.— "*Scolicia*", seen in epichnial view, with lighting from the left, from about 3 m above the Butterly Marine Band. Provenance as for Plate 7B. From Eagar et al., 1983. Reproduced by permission of the Council of the Yorkshire Geological Society. M.Mus.LL.6191.

D.— cf. *Asterichnus*, in bedding plane view, photographed *in situ*, from about the base of the Kinderscout Grit in Heyden Brook, Woodhead. From Collinson, 1967.

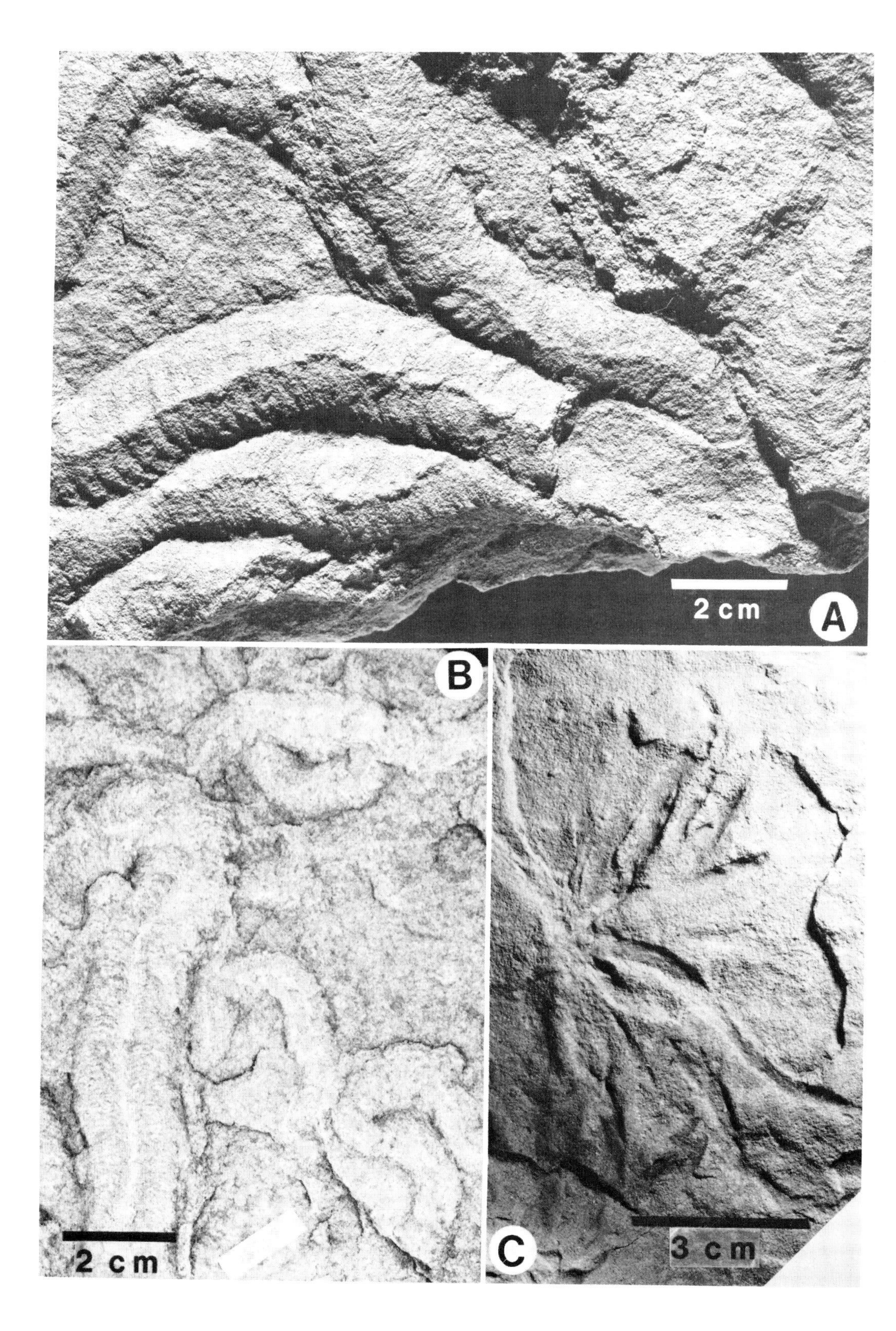
2 cm
A
B
2 cm
C
3 cm

places by escape shafts with a steady increase in their vertical extent, or height, and in their paleoenvironmental range, particularly in the Marsdenian, to include non-marine and probably freshwater settings. It is significant that this colonization of the delta top took place immediately before and contemporaneously with the first appearance of the non-marine genus *Carbonicola*, which morphological evidence strongly suggests evolved from a marine ancestor.

Among the most striking differences in the trace fossils is the replacement of the very rich *Planolites* on the delta slopes of the Lower Kinderscoutian with thin bands of sparse *Planolites* in muddy sediments evidently of an interdistributary bay. The *Olivellites* assemblages, which locally reach prolific abundance, may well represent the feeding trails of arthropods which were able to exploit the fluctuating salinities of mouth bars.

By contrast with the Lower Kinderscoutian, the "Middle Grits" of the Marsdenian have revealed, within partings, interstratal feeding systems, notably *Arthrophycus* and *Phycodes*, the latter recurring on higher horizons in the lower Westphalian with an unquestionable non-marine fauna (Fig. 8). Finally, the advent of *Kouphichnium*, in the lower Marsdenian, heralds the appearance of non-marine, probably freshwater conditions.

In essence, the diversity and frequency of the trace fossil assemblages of the Upper Kinderscoutian-Yeadonian succession enhances interpretation of the smaller-scale sedimentological features of the deltaic sequences and suggests faunal colonization of the delta top during this time. The reappearance of such trace fossils as *Zoophycos*, *Olivellites* and *Asterichnus*, common in marine-dominated sheet deltas of late Lower Carboniferous-early Namurian of the northern Pennines, may indicate that marine influence persisted for some time locally in delta-top environments following periods of major marine incursion during the Marsdenian-Yeadonian interval. The evidence, primarily ichnofaunal, suggests that in general conditions were shallower and salinities lower or more fluctuating during this long period than in Lower Kinderscoutian time. Locally or regionally for short periods conditions were similar to those of the succeeding Coal Measures floodplain.

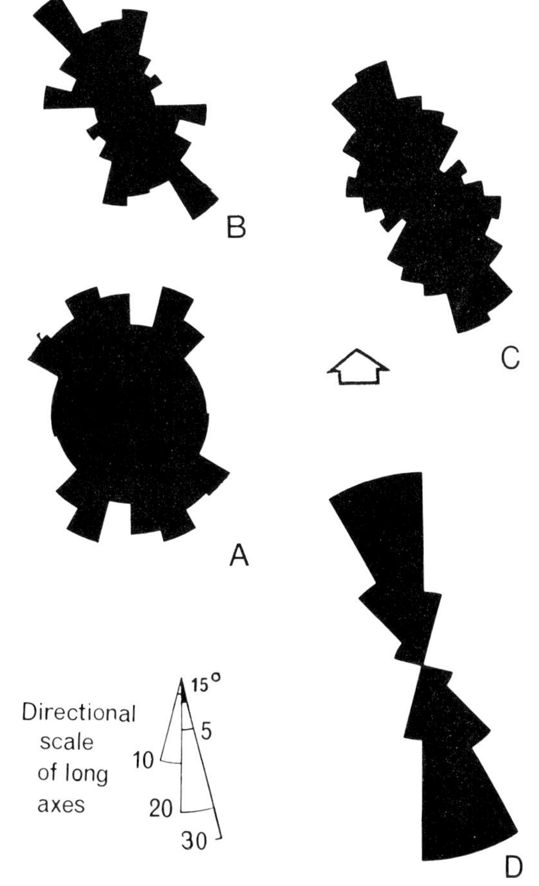

Fig. 7.—Orientations of long axes of *Pelecypodichnus* burrows and escape shafts (g of Fig. 6) plotted in 15° arc segments: A—Through 0.2 m at 1 m above the Butterly Marine Band, and B—On a single horizon 1.30 m above the Band. Both A and B are located 2 km west of Marsden, Yorkshire (Fig. 2, Marsden and Woodhead). C—At the base of a parting in the Milnrow (=Crutchman) Sandstone of the Lower Westphalian (Fig. 8, left hand section). D—In sandstone 0.1 m higher than at C in the Milnrow Sandstone, both 12 km WSW of Chapel en le Frith. (A) and (B) apparently show orientations of escape shafts of cf. *Sanguinolites* Hind, (C) and (D) of *Carbonicola*. (A) and (B) after Eagar et al. (1983) (C) and (D) after Hardy and Broadhurst (1978). The open arrow indicates north.

EXPLANATION OF PLATE 9

Fig. A.— *Olivellites* cf. *O. plummeri* Fenton and Fenton. Part of the slab shown in Plate 8A (see bottom left), photographed under strongly oblique lighting from the top right. Note the epichnial impression of the burrow (seen above the marked scale). Note also the fine parallel longitudinal grooving just discernible between the transverse ribbing on the epichnial convex surface of the burrow.

B.— Epichnial view of *Olivellites* sp. coated in ammonium chloride, from the lower part of the Woodhouse Grit, Middle Marsdenian. Ponden Clough, near Keighley. B.G.S.JS.597.

C.— cf. *Asterichnus*. A specimen showing sinuous "rays" from scree from the Scotland Flags at Midgley Quarries, southwest end, about 13 km west of Elland, Yorkshire. M.Mus.LL.6210.

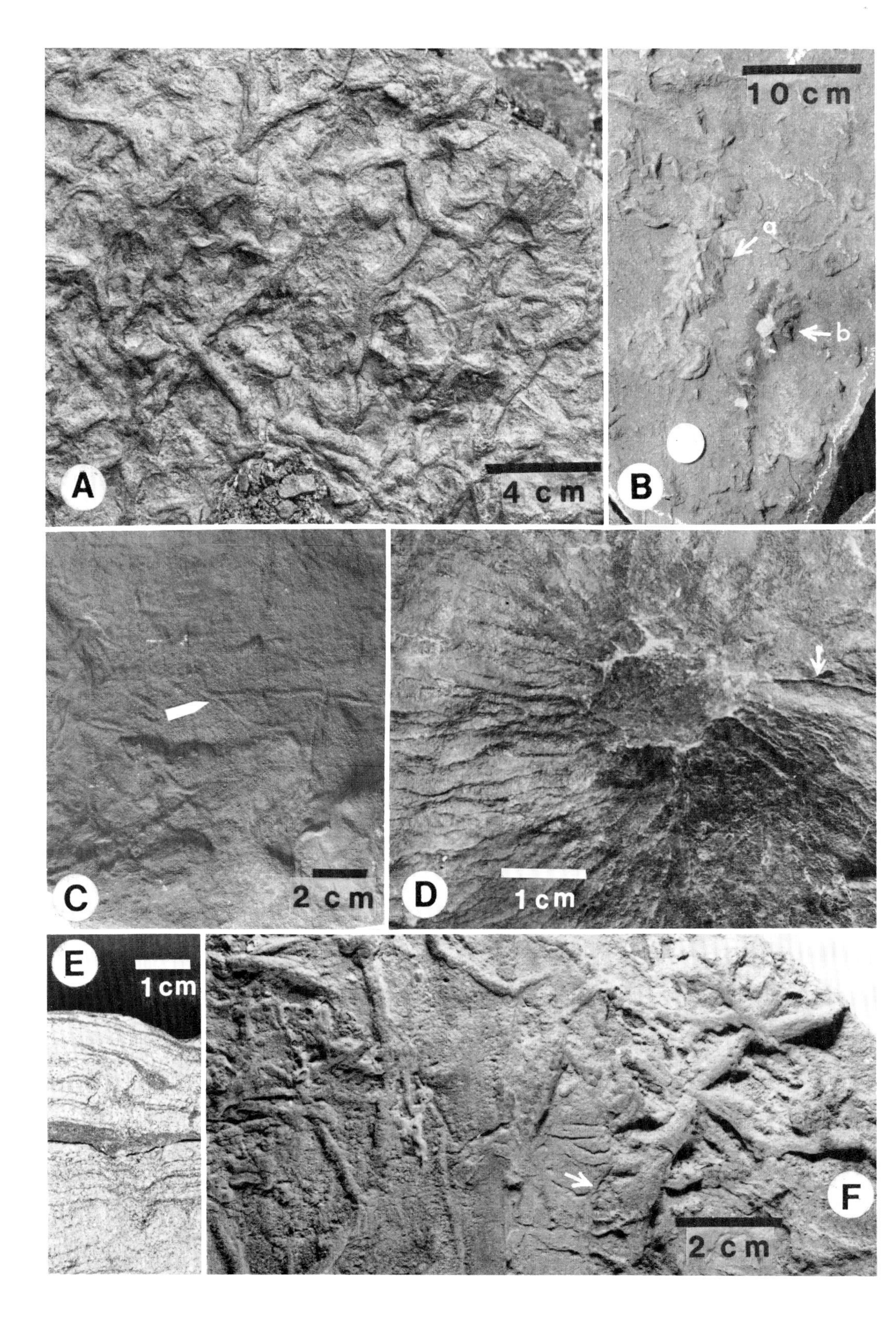
A
4 cm
B
10 cm
a
b
C
2 cm
D
1 cm
E
1 cm
F
2 cm

LOWER WESTPHALIAN OF THE PENNINE COALFIELDS

Sedimentation and Trace Fossil Assemblages

There is no major change in the sediments, faunas or environments at the base of the Westphalian; but rather a gradual increase upwards in the succession in the fluvial and swamp dominance over the earlier deltaic facies, as outlined in the preceding account of the Upper Namurian. The lowest Westphalian strata are similar in sediment types, marine and non-marine body fossils, and trace fossils to those of the underlying Yeadonian sheet deltas. However, within the lower Westphalian, Turbidite Associations are unknown, Delta Slopes are less common, and the Delta Top becomes dominated by swamps, interdistributary bays and lagoons with frequent crevasse-splay sedimentation from migrating fluvial channel systems. Marine influence is reduced to intermittent, basin-wide marine incursions forming marine bands with depth-salinity related faunal phases, including some trace fossils (Calver, 1968a, 1969). In upper Westphalian A strata, the marine bands are replaced by a non-marine mud facies with bivalve and ostracod faunas, which often form bands showing distinct faunal phases (Eagar, 1961; Pollard, 1969) but lacking trace fossils.

Trace fossils are poorly known from lower Westphalian sequences (Fig. 8) when compared with marine band and non-marine bivalve body fossil faunas. Their apparent scarcity is attributed to lack of detailed sedimentological analysis comparable to that carried out on Namurian sequences rather than to rarity of ichnofauna. Assemblages are best known from Lancashire (Fig. 8) from the work of Binney (1852), Hardy (1970a, b), Eagar (1971, 1974a) and the study of unpublished collections, which together form the basis for this analysis. Records from west Yorkshire are comparatively rare, being largely from Eagar (1974a, b). More details are available from east Yorkshire, but mainly in higher Westphalian sediments (Smith et al., 1967; Elliott, 1968). In all these areas, trace fossils occur and in any abundance in lower Westphalian strata only where there is a frequency of marine bands. Despite the preliminary nature of trace fossil examination in this and other areas, it is apparent that at least five environmentally related trace fossil assemblages can be recognized in Westphalian strata, some of which have provided very detailed information about depositional environments, as discussed below. The assemblages may be defined respectively as (1) *Planolites ophthalmoides* assemblage from marine bands; (2) *Rhizocorallium* assemblage from non-marine to possibly marine transitional strata; (3) *Pelecypodichnus-Kouphichnium-Arenicolites* assemblage from arenaceous non-marine strata; (4) *Planolites montanus* assemblage from argillaceous non-marine strata; (5) *Scoyenia* association from freshwater fluvial or lacustrine strata.

Planolites ophthalmoides *Assemblage*

This assemblage, dominated by *Planolites ophthalmoides* Jessen in silty mudstones, was first recognized from Westphalian rocks of the Ruhr area of West Germany (Jessen, 1949; Seilacher, 1963, 1964), but it has been shown by Calver (1968b) to be a distinct phase early in a marine transgression. Although unknown from the Lancashire and West Yorkshire coalfields, it is widely recorded in lower Westphalian strata of the East Midlands (Smith et al., 1967), South Wales (Woodland et al., 1957) and southeast Ireland (Eagar, 1964). Calver (1968a, b) considers that this assemblage represents a community of "worms" of uncertain affinities, living in marginal marine muds, perhaps analogous to a living *Arenicola* community.

Rhizocorallium jenense *Assemblage*

This assemblage, described here for the first time, was originally recorded by Hardy (1970b) from two localities at the same stratigraphical horizon (Fig. 8)

EXPLANATION OF PLATE 10

Fig. A.— cf. *Arthrophycus*. Hypichnial branching burrows with poor segmentation. Lower Marsdenian, Fletcher Bank Grit. Ramsbottom, Lancashire. U.K.SAO.1.

B.— *a, Phycodes* cf. *P. palmatum* Hall. Hypichnial mound of palmately branching burrows. Woodhouse Grit, Marsdenian, Parkwood Quarry, near Keighley, Yorkshire. U.K.SAO.2. *b, Gyrophyllites*. Hypichnial rosetted structure of lobes around a central protuberance (broken). Other details as for Figure Ba.

C.— *Palaeophycus*. Hypichnial horizontal burrows with rare dichotomous branching (arrowed). Marsdenian, Hazel Greave Grit. Scout End, 7.5 km northeast of Rochdale. U.K.SAO.3.

D.— *Zoophycos (Spirophyton)* cf. *cauda-galli* Hall. Endichnial conical mound with broken apex and spirally radiating and branching lamellae (poorly seen). The marginal tunnel is indicated (arrow). From the "Bluestone", Ponden Clough, SSW of Keighley, Yorkshire. U.K.SAO.4.

E.— *Pelecypodichnus* escape shaft, seen in vertical section, but with view from slightly upward to show the smooth hypichnial surface of the terminal stratal lamina. Scree from the Scotland Flags, Midgley Quarry, Yorkshire. M.Mus.LL.6209.

F.— *Phycodes* cf. *curvipalmatum* Pollard. Hypichnial horizontal branching burrows with recurved palmate terminations (arrowed). Guisley Grit, Wicking Crag, near Keighley, Yorkshire. U.K.SAO.5.

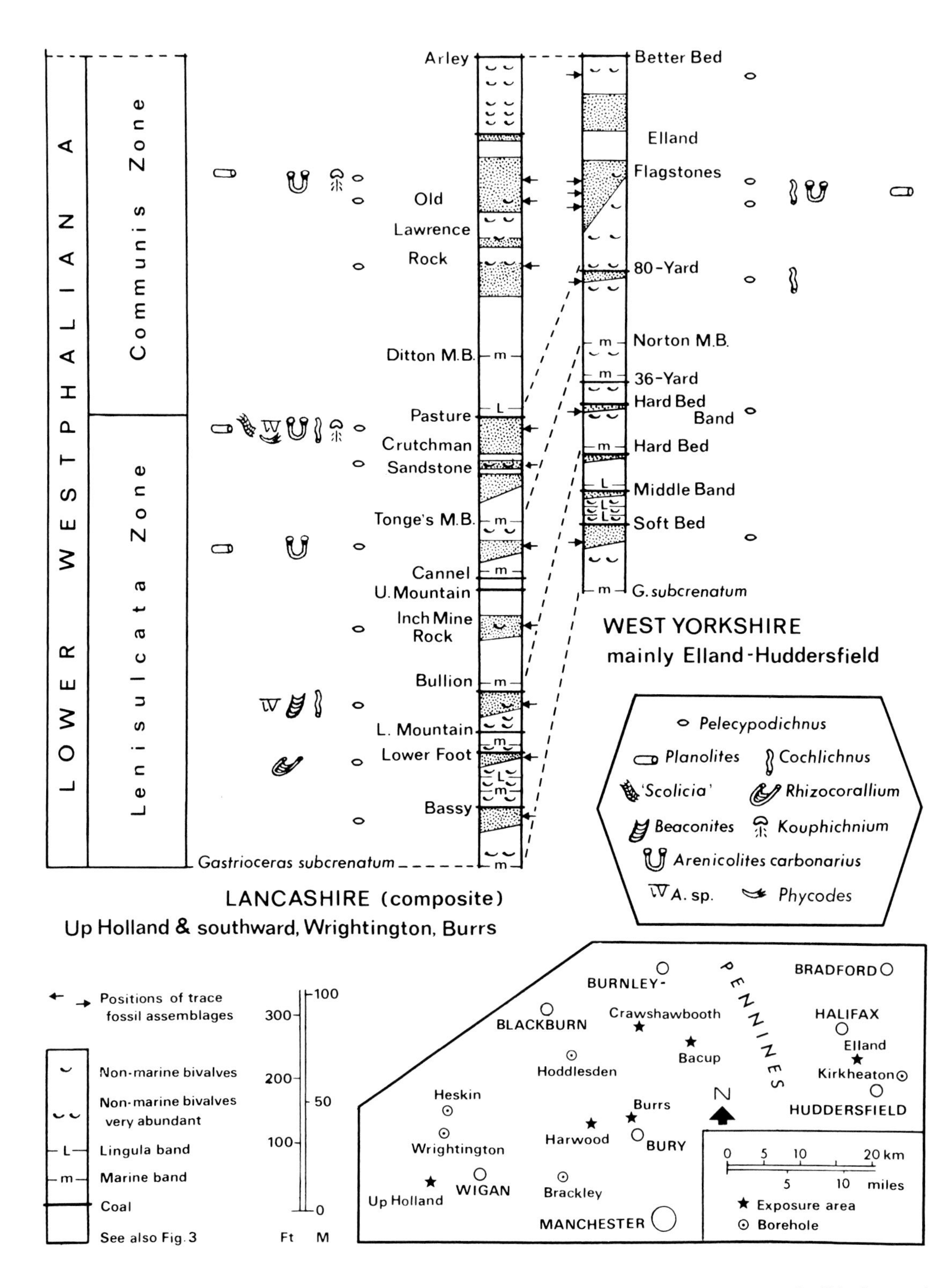

Fig. 8.—Schematic sections of lower Westphalian A sequences in Lancashire and Yorkshire with trace fossil horizons and their diversity indicated.

from lower Westphalian rocks of Lancashire and Derbyshire. At both localities, densely packed horizontal or oblique U-shaped burrows or tongue-like excavations (see below, and Pl. 14B-E) occur as hypichnia on the base of poorly sorted sandstones of variable thickness. The more oblique burrows show preferred orientation of the plane of the U-form, either in one or two opposing directions on the same axis, suggesting current response alignment. Such orientated specimens also frequently show retrusive movement of the burrow. The recorded occurrences of this trace fossil at the cross-cutting base of ill-sorted sandstones, some containing mud flakes, suggest that the animals were transported into the area by unusually turbulent water, and colonized a recently eroded surface of stiffened mud. The most likely environmental explanation is that the normally quiet "non-marine" environment of the upper Silesian delta, as detailed by the three widespread minor "cyclothems" between the Bassy and Lower Foot coals (Fig. 8), was interrupted by local turbulent erosional conditions (see Eagar, 1956, pl. 27, sections II, V, VI) and the water, which may have been marine, carried in alien animals which briefly colonized the scoured surface before being inundated by thick deposits of recently eroded material mixed with sand. The possibility of marine influence is supported by the more usual occurrence of *Rhizocorallium* in marine sediments.

Very similar environmental interpretations of *Rhizocorallium jenense* burrow communities at erosion or omission surfaces, marking a transgression of marine over non-marine conditions, have been recorded from several other localities differing widely in age (e.g., Lower Carboniferous, Scotland, by Chisholm, 1970; Lower Triassic, Vosges, by Gall, 1975; Tertiary, Hungary, by Fürsich, 1974a). Such occurrences confirm the presence of the *Glossifungites* ichnofacies (Seilacher, 1967; Frey and Seilacher, 1980) characterized by the abundance of low diversity burrows, predominantly of suspension feeders. However, Fürsich and Mayr (1981) have described recently a *R. jenense* ichnocoenosis within Miocene fluvial sediments in southern Germany. They interpret these parallel-orientated spreiten burrows as being produced by ephemerid larvae in mud-floored pools in abandoned river channels and stress extreme caution in the use of this ichnospecies alone as an environmental indicator.

Pelecypodichnus-Kouphichnium-Arenicolites *Assemblage*

This composite assemblage has its precursors in ichnofaunas of the Marsdenian Fletcher Bank Grit, of crevasse-splay origin, and the Yeadonian Haslingden Flags (Collinson and Banks, 1975) of interdistributary facies (see previous section). However, in the lower Westphalian A of Lancashire at least two components of this assemblage occur in the same formation on at least three horizons and in several localities (Fig. 8). The respective ichnogenera have not been found in intimate bedding plane association, although all can occur in such manner with *Cochlichnus* or *Planolites*, but in beds or bedding planes separated by as little as 20 mm within the same sedimentary unit. Detailed analysis of a few ichnocoenoses have provided different types of information about sedimentary environments. The significance of such independent sub-assemblages is discussed below.

Pelecypodichnus-Carbonicola *associations.*—As in sandy sediments of the Yeadonian, escape burrows, locally in prolific quantity, overlie the familiar resting impression of *Pelecypodichnus amygdaloides* Seilacher (Pl. 12A). Shafts are seen in vertical sections as downturnings of the sedimentary laminae in which the central region is perforated (Pl. 12D) or, often in the upper part, merely as sharply downbent V-structures, typically asymmetrical and without perforation (Pl. 12C). Vertical extent of the shafts is very variable, but is often in excess of 0.5 m and commonly over 1 m. Shaft origin by upward movement of *Carbonicola* is attested by association—shells lie above shafts where preservational conditions permit—by dimensional correspondence and, occasionally, as first demonstrated by Hardy (1970b), by the presence of *Carbonicola* in vertical attitude at their tops (Pl. 12B). More often, shells may be found lying on their sides in shales or mudstones between sandier beds. These "refugee communities" of Hardy and Broadhurst (1978) have the same trends in shell form as those of the Yeadonian (Eagar, 1974a), being very elongate with relatively low umbones and short anterior ends; e.g., *Carbonicola extenuata* Eagar.

Pelecypodichnus-Cochlichnus *and seasonal sedimentation.*—In some instances it is possible to count the density of shafts per unit area of sediment and to observe a progressive fall in density at higher levels. Routine observations of *Pelecypodichnus* in size frequency studies (Hardy 1970b; Hardy and Broadhurst, 1978) make it clear that in many cases communities of *Carbonicola* represented were of unimodal distribution; they thus imply that they were single generations, all individuals having grown to much the same size. Where successive beds of sandstone are penetrated and the density of shafts per unit area diminishes upwards, it is suggested that there was a progressive loss of bivalves as they attempted to escape from the accumulating sand. Recent work by Broadhurst et al. (1980) has centered on one particularly well displayed sequence of rhythmically banded sediments at Up Holland, west Lancashire (Fig. 1), between the Lower Mountain and Bullion Coals (Fig. 8). In these sandstone-siltstone units there are many horizons penetrated by bivalve escape shafts, but in some groups of successive sand beds there is a distinct fall in density of escape shafts. It is clear that at some silt horizons a spat fall gave rise to a community which, having become established, then suffered progressive losses. As

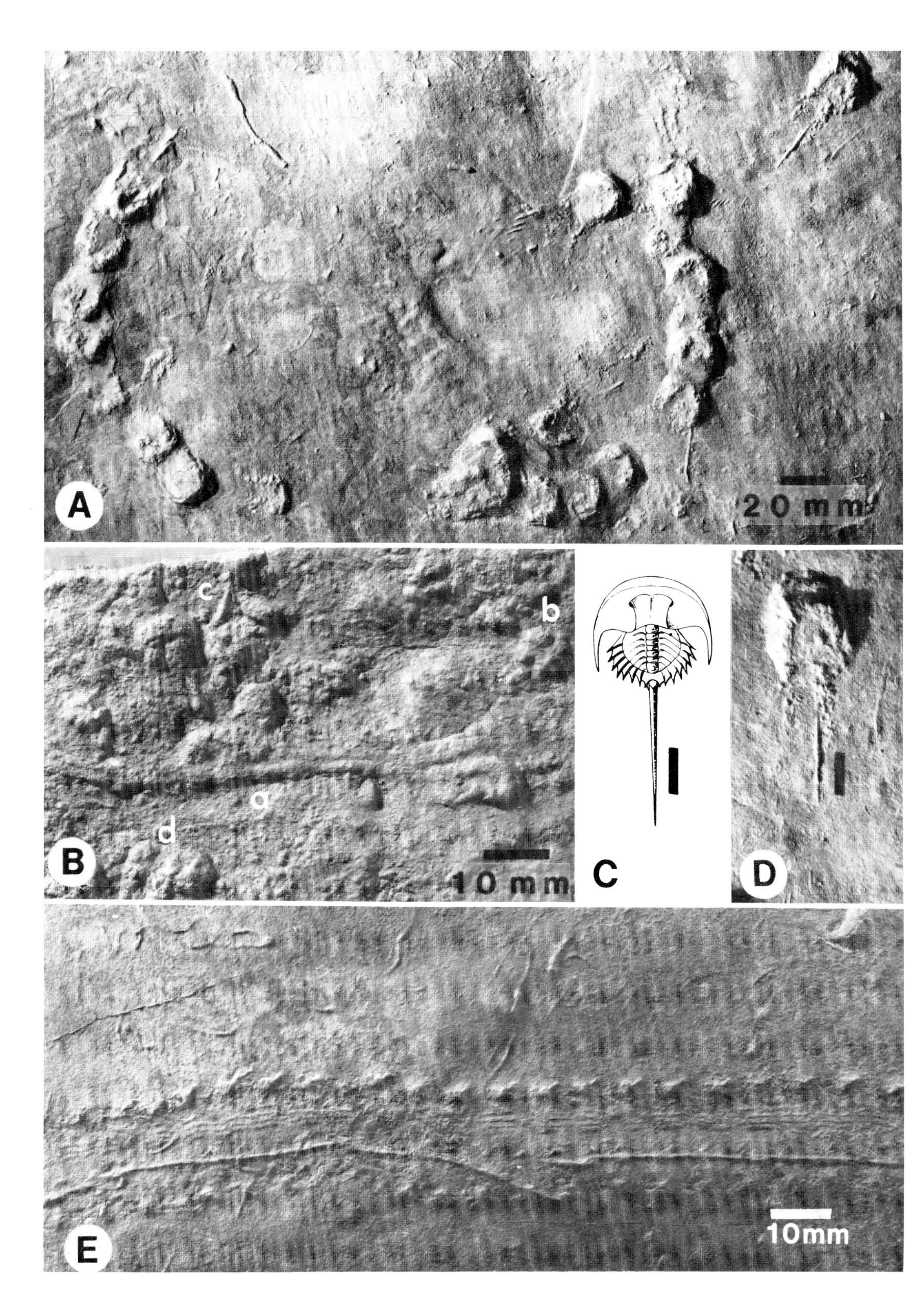
A
20 mm
B
c
b
a
d
10 mm
C
D
E
10mm

several successive sand beds exhibit structures which were made by the same community of bivalves, it follows that the rate of accumulation of these rhythmic sand-silt units must have been rapid, that is fast enough to allow several units to develop within the lifetime of a single generation of bivalves. On this evidence it was concluded that the regular alternating sand-silt sequences represent seasonal conditions of alternating floods and dry spells. With a tropical climate established for the British Upper Silesian (Scotese et al., 1979) it seems probable that the sand beds represent seasonal monsoon flood deposits which overflowed the river channel banks, covered the muddy sediments of the shallow interdistributary bays, and buried the resident bivalve community. The fate of those individuals which failed to escape upwards through the accumulating sand was, of course, death. Some of these animals were then scavenged by other residents of the sediment, probably annelid worms. These left characteristic sinuous burrows, *Cochlichnus* (Pl. 12E), commonly found associated with *Pelecypodichnus* (see previous section). Hardy (1970b) described an unusual occurrence of *Cochlichnus*, apparently associated with resting impression of *Carbonicola*. In this case the normally random orientation of *Cochlichnus* is replaced by a radial orientation around a center which, although confused by disturbance, has the shape of *Pelecypodichnus amygdaloides* (Pl. 12F). It appears from this association that after these bivalves died they were the focus for considerable numbers of small annelid worms which were attracted to the decaying bodies.

Limulicubichus [Kouphichnium] rossendalensis *Hardy—A non-marine shoreline indicator.*—Trace fossils probably made by the xiphosurid *Belinurus*, or perhaps *Euproops*, were described by Hardy (1970a) from Yeadonian sandstones (Pl. 11A, C, D) and have since been recorded from at least two horizons in lower Westphalian sediments (Fig. 8; Pl. 11B). These traces were characterized by two distinct modes of locomotion. They frequently commence with drag marks made by telson or limbs and must therefore have been made by an animal supported in water. The main part of each trace is deeply excavated with sunken areas corresponding to the position of the prosoma (head shield) and exhibiting disturbance by the limbs. These deeply disturbed sections result from the animal's forward progress through the sediment rather than through the water (cf. Pl. 11B). A single specimen of a walking trackway has been found by J. I. Chisholm and is figured here (Pl. 11E). This specimen shows light walking and pusher limb impressions and telson drag mark and was made by an animal whose body weight was largely supported by water. This is the type of trackway to be expected of animals walking over the surface of the sediment and contrasts with the deeply excavated traces previously described (see Chisholm, 1983).

Studies of the activities of the modern horseshoe or king crab, *Limulus polyphemus*, suggest that burial into the sediment and attempts at forward progress can result from feeding or from stranding of the animals on mud flats when dehydration becomes a danger (K. E. Caster, personal communication 1969; Fisher 1979). Whilst the walking type of trackway is of purely locomotory purpose, the heavily excavated trace could be accounted for either as a deposit feeding structure, or as resting traces made by animals in danger of dehydration. These traces have been found in sediments which include other trace fossils (*Pelecypodichnus*) and whose lithology and horizon is compatible with a non-marine environment. As recent work has suggested (Goldring and Seilacher, 1971; Goldring *in* McKerrow, 1978), it seems probable that the trails were made by animals living in relatively shallow non-marine water. In these circumstances it is possible that changing water levels, due to seasonal climatic fluctuations, or other sedimentological causes, resulted in some animals being stranded. The situation would sat-

EXPLANATION OF PLATE 11

Fig. A.— *Limulicubichnus [Kouphichnium] rossendalensis* Hardy. Holotype, hypichnial resting places of xiphosurids, both isolated and in curved series, Upper Haslingden Flags, Yeadonian. Whitworth, Rossendale, Lancashire. Reproduced by permission of the Editor, Palaeontology. MGSF.3.

B.— *Kouphichnium* trails and *Limulicubichnus* resting places. The hypichnial bilobed trail (a) ends in a resting trace (b). Other resting traces show telson molds (c) and digging action of the limbs (d). Milnrow Sandstone, Lower Westphalian. Kerridge, 13 km WSW of Chapel en le Frith. MGSF.86.

C.— *Belinurus*; restoration of body form for comparison with Figure D below. Black bar equals 10 mm.

D.— *Limulicubichnus [Kouphichnium] rossendalensis*. Isolated resting trace from Figure A (top right) showing outline of prosoma, two rows of appendage marks and telson impression. Note similarity to the body form of *Belinurus* (Fig. C). Provenance as for Figure A. Black bar equals 10 mm.

E.— *Kouphichnium* aff. *K. variablis* (Linck) walking trackway. Hypichnial preservation of surface trackway showing outer pusher impressions, inner walking limbs and central telson groove. Upper Haslingden Flags; provenance as for Figure A. Photograph reproduced by permission of the Director, British Geological Survey, N.E.R.C. Copyright. B.G.S.LZA.8874 (see also Chisholm, 1983, Pl. 1 and Fig. 1).

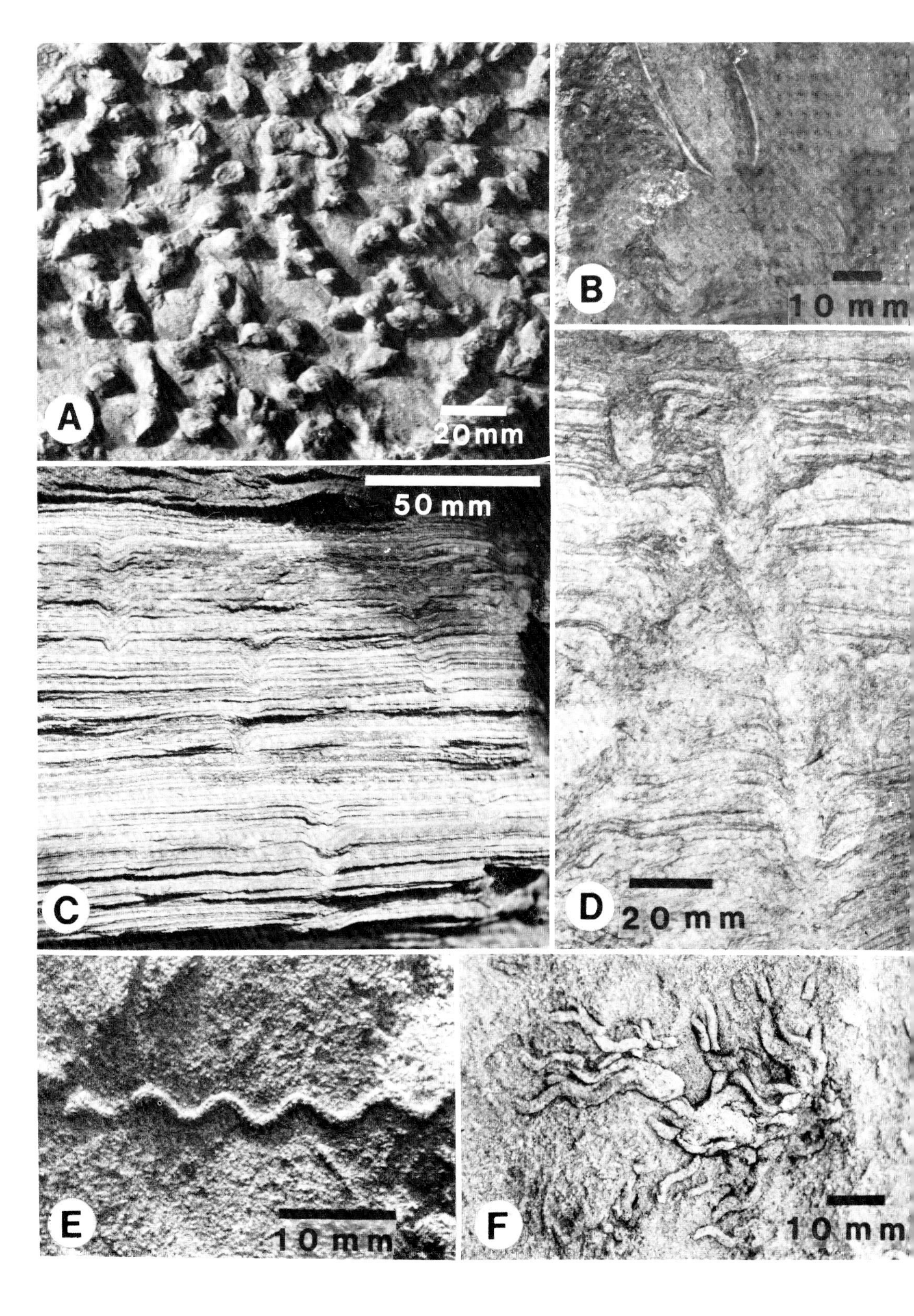
A
20mm
B
10 mm
C
50 mm
D
20 mm
E
10 mm
F
10 mm

isfy the requirements of the holotype of *Limulicubichnus [Kouphichnium] rossendalensis* (Pl. 11A) on which drag marks are seen on the commencement of each section of the excavated trail; the interpretation being that the animals were stranded on the mud and floundered through it until the water level rose, lifting them up. The holotype exhibits several short series of linked resting places which are apparently of the same individual. The resting traces are interpreted as the product of activity during subaerial exposure. The gaps between series of resting traces represent periods when water covered the animal sufficiently long for it to emerge from the sediment. Very similar behavior by juveniles of Recent *Limulus polyphemus* on drying intertidal mudflats has been observed by Fisher (1979, p. 387) and discussed by him in relationship to possible subaerial activity of *Euproops* in Silesian time.

The interpretation of paleoenvironment represented by *Kouphichnium* is one of mud banks in a shallow-water non-marine sedimentary basin in which fluctuations in water-level caused animals to be stranded on mud flats. The close association of *Kouphichnium* and *Pelecypodichnus* and the similarity of the sediments in which they occur suggest that in the lower Westphalian of the Pennine area at least, the environment was frequently one of extremely shallow water, predominantly non-marine, and subject to fluctuating water levels; that it was possibly controlled by seasonal climatic changes, with flooding a regular phenomenon.

Arenicolites *bands—Tidal flat or omission surface indicators ?*.—At several horizons in the lower Westphalian rocks of Lancashire and in the Elland Flagstones of Yorkshire (Fig. 8), thin sandstone beds occur which are rich in the trace fossil *Arenicolites*. Two forms of this ichnogenus are present, the distinctive *A. carbonarius* (Binney, 1852; Hardy, 1970b) with U-burrow form, occasional branching and funnel-shaped apertures (Pl. 13A, B), and a smaller *Arenicolites* sp., often with J- or Y-shaped variations (Pl. 13C, D). The size of burrow populations of *A. carbonarius* varies between the horizons (see below) but structures all terminate upwards at major sedimentary interfaces or partings. A brief examination of the type horizon of *A. carbonarius*, the Old Lawrence Rock of the Wigan area by P. Unsworth (personal communication 1980), shows that these burrows occur in the upper 30 mm of a grey, fine-grained quartzose sandstone, 20 mm above a parting with hypichnia of *Pelecypodichnus*. The bedding plane expression of the burrow apertures below the overlying shale consists of concentrically filled circles with radiating grooves on one side and rarely associated burrows of *Planolites* and perhaps *Palaeophycus*. In the Elland Flags (Eagar, 1974a, p. 231), a smaller variety of *A. carbonarius* in white rippled, cross-stratified sandstone (Pl. 13F) shows retrusive funnel infilling (Pl. 13E). *Arenicolites* sp. burrows appear to originate from thin muddy micaceous partings within white rippled cross-laminated sandstone (Pl. 13C, D) believed to be of levée origin (Broadhurst et al., 1980). The *Arenicolites* bands appear to represent communities of U-shaped dwelling burrows of suspension-, or perhaps partial surface deposit-feeding invertebrates, probably worms which flourished temporarily at times of reduced sedimentation. The environmental significance of these bands is somewhat ambiguous, as similar *Arenicolites* associations have been recorded both from intertidal (e.g., Corbo, 1979; Miller, 1979) and lacustrine (e.g., Bromley and Asgaard, 1979) situations. Since they largely occur in laminated or thin sandstones with *Pelecypodichnus*, and are interbedded with mudstones interpreted as of interdistributary bay origin, a non-marine environment—possibly crevasse-splay sheet sands—seems to be the most likely interpretation. As is usual elsewhere in the Silesian, there is no evidence of tides.

EXPLANATION OF PLATE 12

Fig. A.— *Pelecypodichnus amygdaloides* Seilacher. Hypichnial expression of bivalve resting burrows at the base of escape shafts, showing orientation to the direction of the paleocurrent. Lower Haslingden Flags, Yeadonian. Rossendale, Lancashire. MGSF.83.

B.— *Carbonicola*; shell preserved *in situ* above short escape shaft. The vertical section shows gaping of the valves and infilling of the shell above a zone of sandstone much bioturbated by escape shafts. Bullion Rock, Lower Westphalian. Up Holland, Lancashire. MGSF.82.

C.— Bivalve escape shafts in laminated fine-grained sandstone interpreted as seasonal flood deposits formed in an interdistributary bay environment. Elland Flags, Lower Westphalian. Elland Upper Edge, Yorkshire. Reproduced from Eagar (1974a) by permission of the Editor, Lethaia.

D.— Bivalve escape shafts in sandstone with siltstone partings, showing downbending and perforation of sedimentary laminae. Strata between the Lower Mountain and Bullion Coals, Lower Westphalian, Up Holland. MGSF.81.

E.— *Cochlichnus kochi* (Ludwig). Hypichnial expression of an interstratal sinuous burrow, Rough Rock Flags, Yeadonian, Billinge Hill, near Bollington, Cheshire, 25 km south of Bolton (Fig. 1). MGSF.85.

F.— *Cochlichnus* burrows in hypichnial expression, showing radial arrangement around *Pelecypodichnus* (see p. 129). Rough Rock Flags, as for Figure E.

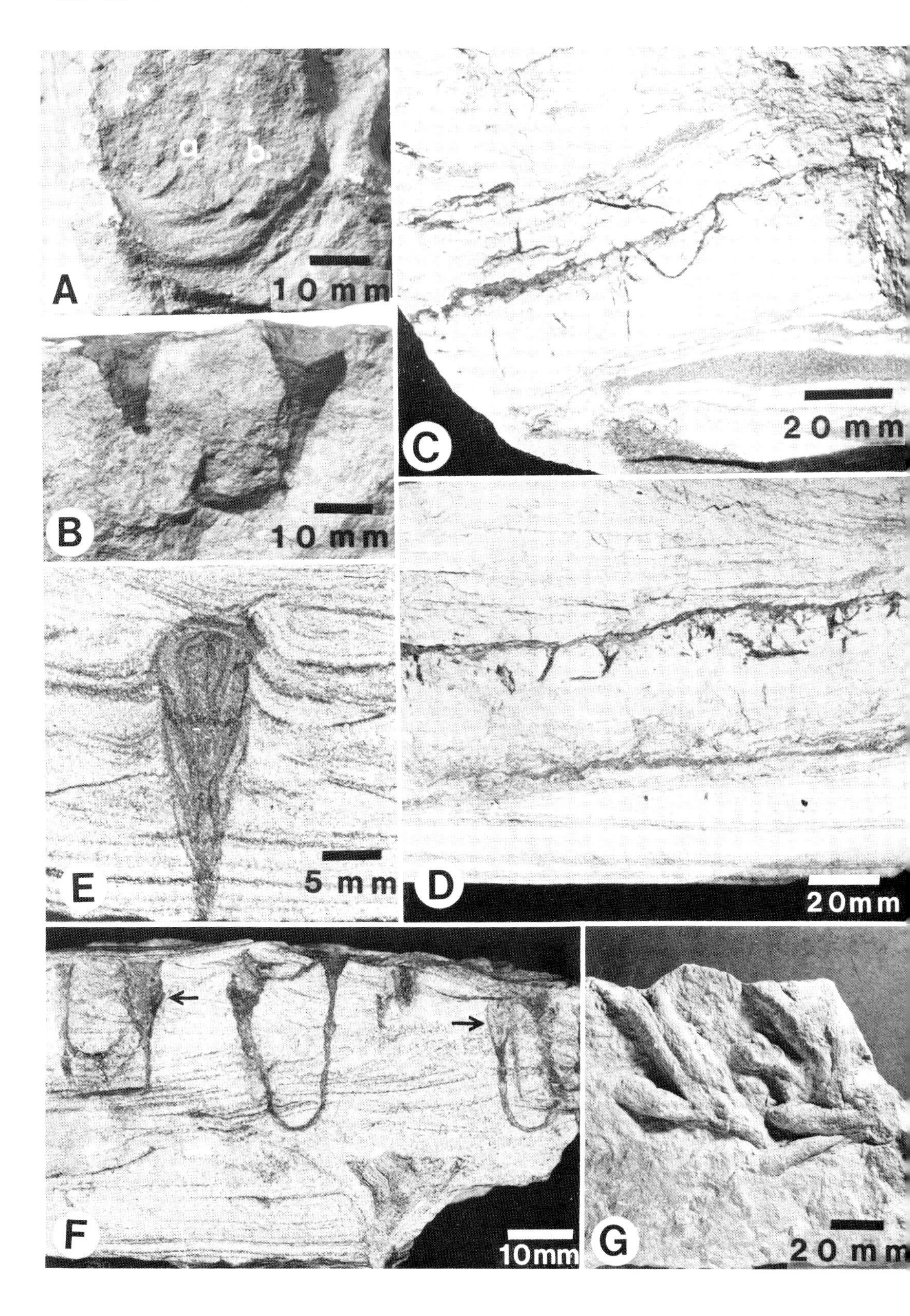
a
b
10 mm
A
10 mm
B
20 mm
C
5 mm
E
D
20mm
F
10mm
G
20 mm

Planolites montanus *Assemblage*

Although this assemblage has not yet been recorded from the Lancashire and West Yorkshire Coalfields (Fig. 8), it is well documented by Calver (in Smith et al., 1967) from the equivalent Westphalian A strata in the East Midlands. He describes *Planolites montanus* associated with *Gyrochorte carbonaria* and *Cochlichnus* in dark grey silty mudstones and ironstones of non-marine origin, very similar to their original definition from Westphalian strata of the Ruhr (Seilacher, 1963, 1964). This absence can be explained by the dominance of flood-generated sandy beds (Broadhurst et al., 1980) in the interdistributary facies of these western coalfields—notably in the lowest Westphalian, where they have been examined more closely—in contrast to the generally more muddy conditions, with infaunal mud feeding, to the east (contrast the thickness of the columns in Fig. 8).

The Scoyenia *Association*

A distinct assemblage dominated by tracks and burrows of freshwater arthropods (Seilacher, 1963, 1967, 1978) has not been clearly recognized in the Pennine coalfields, although isolated occurrences of component trace fossils appear to exist. Hardy (1970b) describes a burrow illustrated here as *Beaconites* cf. *B. antarcticus* (Pl. 14A) from strata he interpreted as of fluvial channel bank (levée) or overbank origin. Such burrows have been extensively recorded from fluvial sediments of Devonian (Allen and Williams, 1981) and Lower Carboniferous ages (Chisholm, 1977). One as yet undescribed specimen of *Diplichnites* sp. has recently been recovered from lacustrine or lagoonal sediment above the Arley Seam of Lancashire.

These limited records contrast with the diversity of the *Scoyenia* trace fossil record of Nova Scotian Coal Measures of Canada (Dawson, 1868; Seilacher, 1963, 1978), although there is a major common element in *Kouphichnium* traces. In North America such traces have been interpreted as formed in small lacustrine areas interspersed with fluvial facies (Goldring and Seilacher, 1971).

Conclusions on the Westphalian Assemblages

From this brief review it is apparent that, despite the very limited detailed sedimentological analysis as yet carried out on lower Westphalian strata of the Pennine coalfields, trace fossils permit elucidation both of major sedimentary environments and of some ephemeral sedimentary processes within this paralic delta facies. Moreover, the trace fossils so far discovered help to shed light on regional sedimentologic differences within the central Pennine coalfields; for instance, the contrasted development of the lower Westphalian to the west (Lancashire) and to the east (Yorkshire—Fig. 8). Ichnofaunal observations, which are as yet very limited, such as the southerly distribution of the *Planolites ophthalmoides* assemblage, may well have paleogeographical significance. Larger-scale paleogeographical considerations need further investigation in so far as they may bear indirectly on areas, including those of continental dimensions, where attempts have been made, using various groups of body fossils, to achieve ultimately long-range correlations.

SUMMARY AND CONCLUSIONS

The Silesian of the Central Pennine Basin provides paleoenvironments known for many years to range from basinal marine to freshwater swamp. Recent sedimentological investigations, supplemented by the recording, and occasionally the use of trace fossils, have established usually more precise paleoenvironments in Namurian sequences, where an estimated 95 percent of the sediments are without body fossils. Lower Westphalian trace fossils are less well known, although the west Pennine succession, which contains com-

EXPLANATION OF PLATE 13

Fig. A, B, E, F.— *Arenicolites carbonarius* (Binney). *A*, Negative endichnial expression with funnel-shaped apertures and branches (a, b). Crutchman Sandstone, Lower Westphalian. Bradshaw Quarry, Bolton, Lancashire. MGSF.7. *B*, Positive endichnial expression of mud-filled simple U-burrow with funnels. Provenance as for Figure A. *E*, Enlargement of complex funnel infilling of small U-burrow in Figure F. Concentric cones suggest retrusive infilling, and sedimentary laminae appear to be bent around burrow aperture rim. Elland Flags, Lower Westphalian. Elland Upper Edge, Elland, Yorkshire. *F*, Small, irregular U-shaped burrows with complex funnel infillings and rare branching, preserved in ripple cross-stratified sandstone below mud interface. M.Mus.LL.6211, 6212.

C, D.— *Arenicolites* sp. *C*, Small U-shaped, vertical or oblique burrows preserved as mud-filled endichnia within a laminated sandstone. Bullion Rock, Lower Westphalian, Up Holland, Lancashire. MGSF.8. *D*, Oblique, small Y-shaped or J-shaped burrows, mud-filled endichnia below a muddy parting layer within ripple cross-stratified fine-grained sandstone. Provenance as for Figure C. MGSF.9.

G.— *Phycodes* cf. *P. palmatum* Hall. Positive hypichnial expression of short, palmately branching horizontal burrows which overlap slightly. Crutchman Sandstone, Lower Westphalian. Bradshaw Quarry, Bolton. Bn Mus.163.1981.

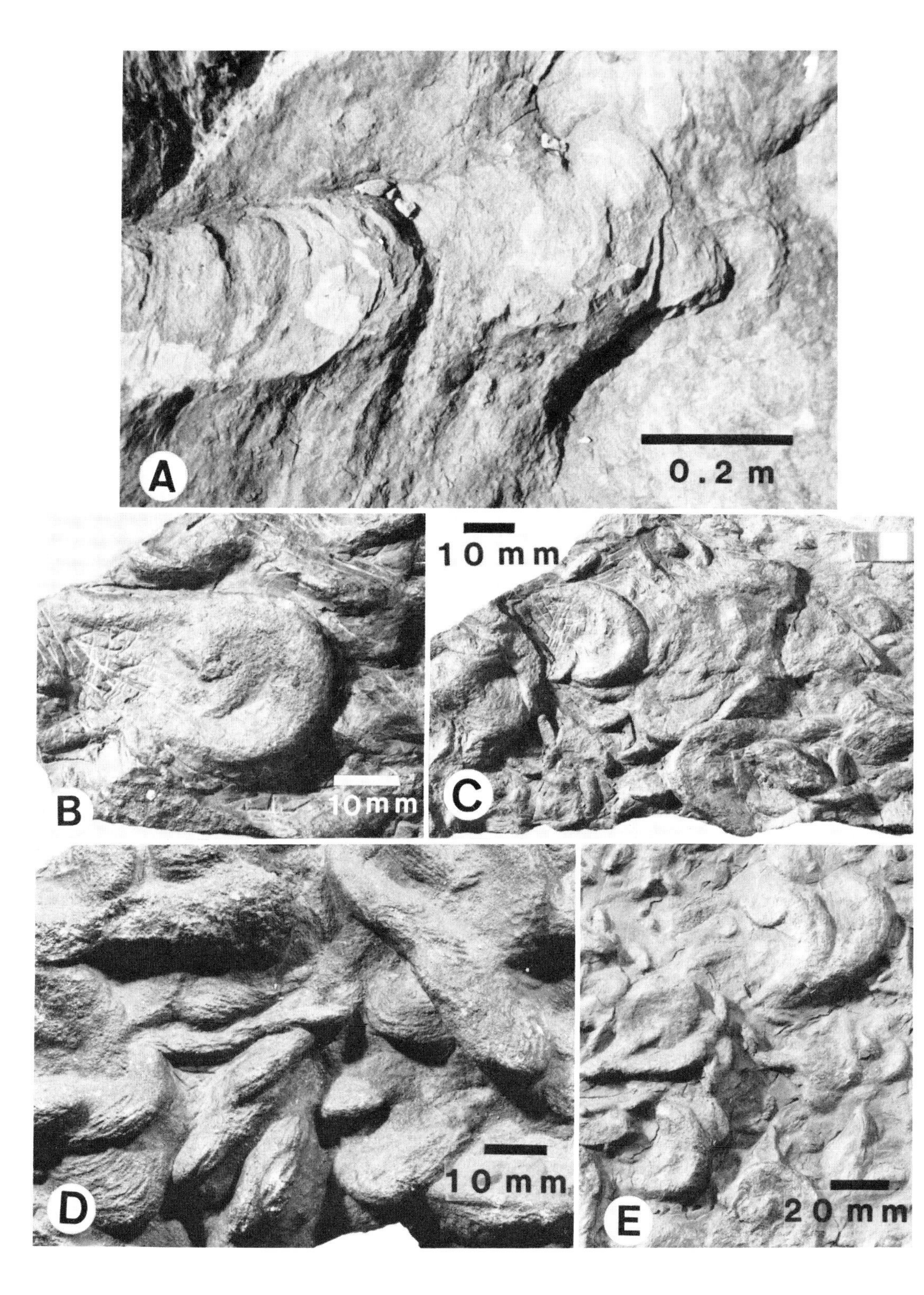
A
0.2 m
B
10mm
C
10 mm
D
10 mm
E
20 mm

paratively frequent bands of marine and non-marine body fossils, is one of the best documented sequences in the British Coal Measures (Ramsbottom et al., 1978).

Turbidite-fronted deltaic successions are contrasted from the Pendleian in the north of the Basin and from the Lower Kinderscoutian in the south. The northern Pendleian, of the Skipton area, can be divided into three sedimentary associations: the Turbidite, the Delta Slope, and the Delta Top Association. These are broadly reflected by intergrading trace fossil assemblages, although there are several facies-crossing ichnogenera (Fig. 4). In the Turbidite Association the *Bergaueria-Planolites-Rhizocorallium* assemblage appears to be related more to food distribution and preservational factors than to bathymetry. In the lower part of the Slope Association, where the ichnofauna is more diverse, the presence of *Lophoctenium* is suggestive of the *Zoophycos* ichnofacies of Seilacher (1963) rather than the bathyal *Nereites* facies, since true flysch-type trace fossils, such as *Helminthoides-Palaeodictyon* are absent (e.g., Chamberlain, 1971b). The upper Delta Slope and Delta Top Association shows a very limited trace fossil diversity, being dominated by *Pelecypodichnus* and *Monocraterion,* both characteristic of Seilacher's (1967) *Cruziana* and *Skolithos* ichnofacies and indicating shallow marine or shoreline environments. In fact, the trace fossils themselves provide very little evidence that the salinity of the Skipton basin was reduced in Pendleian time (but see p. 109).

By contrast, in the Lower Kinderscoutian of the Peak District (Fig. 2, Mam Tor Sandstones and Shale Grit), the turbidites have yielded no trace fossils; and their apparent absence is not readily explained. On the assumption that rates of sedimentation and the extent of turbidite flows were no different in the Skipton and Peak District areas, it is difficult to escape the conclusion that between marine inundations (which left bands of goniatites) salinity in the Lower Kinderscoutian of the Peak area was appreciably reduced by freshwater discharge. The *Planolites-Pelecypodichnus* assemblage of the Lower Kinderscoutian uppermost slope and delta top sediments is a precursor of the succeeding upper Namurian sediments.

Compared with the lower Namurian turbidite deltas, the ichnofaunas of the shallow-water sheet deltas of the Upper Kinderscoutian, Marsdenian, and Yeadonian Stages show two important new features. In the first place, there is an increase in diversity of trace fossils of shallow marine type in delta slope and delta top situations. Secondly, the assemblages reflect increasing faunal colonization of the diverse non-marine delta-top paleoenvironments. The first change is reflected by the recognition of both discrete *Zoophycos caudagalli*-type assemblages in clearer and possibly deeper water and also by "*Scolicia*"-*Olivellites* (-*Psammichnites*)-*Pelecypodichnus* (=*Lockeia*) assemblages of *Cruziana* ichnofacies. Similar assemblages have been described from the Pennsylvanian of north Texas (Fenton and Fenton, 1937; Yochelson and Schindel, 1978), and from the Mississippian-Pennsylvanian of the midwestern United States (Chamberlain, 1971a, b; Chamberlain and Clark, 1973; Archer and Maples, 1984).

The second new feature is shown both by *Pelecypodichnus* and the *Phycodes-Arthrophycus-Kouphichnium* assemblages. From the Upper Kinderscoutian to the Yeadonian, the size and form of *Pelecypodichnus* resting places and escape shafts appear to reflect change in form and habit of the marine bivalve cf. *Sanguinolites* in its evolution into the non-marine bivalve *Carbonicola,* a change already postulated on body fossil evidence (Eagar, 1977). The distribution of *Pelecypodichnus* within sediments during this longer time interval shows enlargement of ecological station from upper delta slope to distributary mouth bars, channels and interdistributary bays. The increasing height of escape shafts reflects the ability of these "refugee communities," or invading assemblages, to cope successfully with the increased rate of flood sedi-

EXPLANATION OF PLATE 14

Fig. A.— *Beaconites* cf. *B. antarcticus* Vyalov. Large endichnial horizontal burrow with parallel sides and retrusive meniscus infilling. Bullion Rock, Lower Westphalian. Up Holland, Lancashire. Photographied *in situ*. Negative MGSF.

B, C, D, E.— *Rhizocorallium* cf. *R. jenense* Zenker. *B*, Hypichnial expression of type A horizontal U-burrow with median constriction and distal expansion of the marginal burrow. Base of unconformable sandstone below the horizon of the Lower Foot Coal, Lower Westphalian, Goyts Moss, Derbyshire, 10 km SSW of Chapel en le Frith. MGSF.87. *C*, Hypichnial expression of short, wide U-burrows with U-bends facing in opposed directions but on the same axis. Provenance as for Figure B. MGSF.88. *D*, Enlargement of type B, tongue- or pouch-like burrows with molding of fine scratch marks crudely parallel to the burrow margin. Provenance as for Figure B. MGSF.89. *E*, Hypichnial expression of variably sized, often overlapping, short wide U-burrows with U-bends facing the same direction (right). Some of these burrows approach *Fuersichnus communis* Bromley and Asgaard, but lack true J-shape and transverse burrow sculpture. Provenance as for Figure B. MGSF.90.

mentation encountered in the newly colonized environments. However, this enlargement of range, which took place mainly in Marsdenian time, renders *Pelecypodichnus* an environmental indicator to be interpreted with due regard to stratigraphic horizon and associations.

The *Phycodes-Kouphichnium* assemblage, with subsidiary *Arthrophycus* and *Arenicolites*, first appears in the Marsdenian in distributary mouth bars but was soon successfully established in interdistributary bays and crevasse-splay environments (e.g., in the Fletcher Bank Grit of Rossendale ([Figs. 1, 2, 5]; and Collinson and Banks, 1975).

The fluvially dominated swamp deltas of the Lower Westphalian of the Pennine coalfields show further development of the environmentally related trace fossils assemblages originating in the Marsdenian. The basin-wide marine marginal mud facies constituting the *Planolites ophthalmoides* assemblage has not been seen in the sand and silt dominated inter-seam strata of the Lancashire and west Yorkshire coalfields (Fig. 8). However, within the non-marine delta-top environments, a *Rhizocorallium* assemblage, *Pelecypodichnus-Kouphichnium-Arenicolites* and interstratal feeding burrows appear to reflect environmental differences of substrate, rate of sedimentation, degree of exposure, and possibly salinity. Trace fossils attributable to freshwater or semi-terrestrial arthropods and vertebrates of the *Scoyenia* ichnofacies (Seilacher, 1963) are only poorly represented in British lower Westphalian sediments when compared with the approximately contemporaneous Nova Scotian Coal Measures of Canada. In Britain and Europe, this ichnofacies develops more fully in the uppermost Silesian (Westphalian C-Stephanian) and in the Lower Permian (Haubold and Sargeant, 1973).

Finally, the trace fossils of the Delta Top Association of the Upper Namurian and Westphalian contrast strikingly with those of the Delta Top of the Pendleian in the Skipton area, where *Monocraterion-Skolithos* assemblages are associated with *Pelecypodichnus*. There is an equal contrast with earlier Scottish deltaic sequences, including one of the Dinantian (Chisholm, 1968, 1970, 1977; Belt, 1975) where unquestionably marine trace fossils characterize shallow water of the delta top. Within the later strata of the delta top of the Namurian and Westphalian, the *Monocraterion-Skolithos* assemblages are replaced by abundant trace fossil evidence indicative or suggestive of an evolved deltaic fauna, essentially non-marine and probably mainly freshwater. Ichnological evidence is thus largely in conformity with views recently expressed on the basis of other, mainly sedimentological evidence (e.g., Ridd *in* Ramsbottom, 1979) that Carboniferous deltas of the Central Pennine Basin were built out into bodies of fresh water rather than into the sea. The evidence we present in this paper qualifies this statement with respect to earlier Namurian time but broadly substantiates it in the lower Westphalian. Irrespective of short, periodic and probably eustatic rises in sea-level (Ramsbottom, 1981), the salinity of the basinal area probably decreased in early Namurian time and had decreased markedly by the Kinderscoutian. With emergence of the delta floodplain in the Westphalian, it seems very likely that the delta did, in fact, prograde into fresh water. In brief, trace fossils suggest that the Central Pennine Basin had a unique history in Silesian time; it appears to have changed its character sufficiently slowly to become the spreading centre of the largest paralic area with evolved deltaic life (Trueman, 1946) that the geological record has so far revealed. There are no comparable modern examples.

SYSTEMATIC ICHNOLOGY

The trace fossils which make up assemblages discussed in the previous sections are now briefly described and their occurrences recorded in terms of stratigraphic horizons, lithologic, and ichnofaunal associations.

ARENICOLITES Salter, 1857

Arenicolites carbonarius Binney, 1852

Pl. 13A, B, E, F

Description.—Vertical to slightly oblique U-tubes without spreiten. A single tube on one arm may divide towards the base to form up to seven tubes on the other arm (Pl. 13A, a, b; F). Tubes are cylindrical, smooth walled with mud lining or mud-sand infilling. U-bend on one limb may show slight vertical or lateral migration but no true spreiten are formed. Dimensions vary between different burrow populations, e.g., tube diameter (Lancashire 2–6 mm, mode 5 mm; Yorkshire 1–2.5 mm), U-burrow width (Lancashire 25–80 mm; Yorkshire 10–18), U-burrow depth (Lancashire 25–50 mm; Yorkshire 22–30 mm) for Westphalian specimens. Marsdenian specimens are smaller in all dimensions. Frequently the tubes possess funnel-shaped apertures (Pl. 13B) which may show concentric infilling and sometimes dendritic grooves radiating from one side of the aperture. Sections of the funnel of specimens from the Elland Flags, Yorkshire (Pl. 13E) show irregular conical infilling, retrusive upwards, and bending of sedimentary laminae around the funnel, suggesting that the producer attempted to keep the aperture both open and above the sediment surface as sedimentation proceeded.

Occurrence.—Marsdenian examples from the Fletcher Bank Grit (Fig. 2, Rossendale area) at the top of a thin, muddy sandstone bed, are believed to be sediments of crevasse-splay origin. In the lower Westphalian, these burrows are known from at least three horizons (Fig. 8), where they occur in association with *Planolites* in laminated sandstones of crevasse-splay or interdistributary origin, near but not directly associated with *Pelecypodichnus* (Hardy, 1970b; Eagar, 1974a).

Remarks.—This ichnospecies has been regarded as the type ichnospecies of *Arenicolites* (Häntzschel,

1975) and is at present under redescription and further analysis by J.E. Pollard and P.G. Hardy. Burrows are interpreted as dwelling structures of either suspension or deposit feeding invertebrates, some burrows having been periodically enlarged to accommodate growth.

Arenicolites sp.
Pl. 13C, D

Description.—Small, vertical to oblique U-shaped, J-shaped, or straight tubular burrows, some apparently branched, all without spreiten and which together form thin burrow horizons in sandstone. Burrow diameter 0.1–1.5 mm (mode about 1.0 mm); depth variable, usually 5–15 mm, rarely 20 mm. Some burrows show flared, funnel-shaped apertures. Burrows filled with dark grey muddy sediment, sometimes around a clean sand core.

Occurrence.—Lower Westphalian A on at least two horizons (Fig. 8). In the Bullion Rock of the Lancashire succession these burrows occur repeatedly along thin muddy laminae within cross-laminated fine sandstone (Pl. 13C, D) which may be of levée origin (Broadhurst *et al.*, 1980).

Remarks.—Small burrows of this type have been recorded by several workers from non-marine beds of Carboniferous (e.g., Chisholm, 1968) or Triassic (Bromley and Asgaard, 1979) age.

ARTHROPHYCUS Hall, 1852
Pl. 10A

Description.—Horizontal, cross-cutting, branching burrows preserved in convex hyporelief. The burrows are parallel or sub-parallel sided with tapering ends and a sub-quadrate cross section. Burrow diameter 3–12 mm (mode 7 mm); vertical depth is about 8–10 mm. Branching is commonly at angles of up to 80°. On the base, the burrows are often bilobed longitudinally and show transverse annulations spaced 1–2 mm apart, although such annulation is frequently faint or absent.

Occurrence.—Fletcher Bank Grit, Marsdenian (Figs. 2, 5). Burrows occur on the base of a lenticular trough cross-laminated quartzose sandstone which is interbedded with interlaminated siltstones and mudstones. Environment of origin is interpreted as crevasse-splay sands within an interdistributary bay. The burrows occur in the same lithological unit as *Palaeophycus, Planolites,* and *Pelecypodichnus.*

Remarks.—Interstratal feeding burrow system of a worm or arthropod (Häntzschel, 1975).

cf. ASTERICHNUS Bandel, 1967
Pl. 8D; Pl. 9C

Description.—Irregular rosette structure consisting of up to 12 unbranched "rays" which are expressed as straight or slightly sinuous grooves, wider than deep, in sandstone. Characteristically they are filled with soft, richly carbonaceous, micaceous, fine-grained sediment. In epichnial expression, the "rays" run either parallel to the bedding or slightly downwards towards a center where a round perforation is sometimes discernible, having a diameter of about 3 mm. Filled, root-like cavities have been seen to branch radially and obliquely downwards from the center in some cut sections.

Occurrence.—"Rays" up to 30 mm in length were noted on a single horizon near the base of the Kinderscout Grit north of Woodhead (Fig. 2; Pl. 8D). Slightly larger structures, with a maximum radius of 50 mm, have been noted in scree from the Scotland Flags near Ripponden, Yorkshire, about 8 km west of Elland (Fig. 1) where there is again an abundance of these trace fossils on a single horizon, specimens being closely spaced in a delta top environment (Pl. 9C). If these structures represent traces at or near the surface, then a localized area of slow deposition is suggested.

Remarks.—Comparable structures are known from sandstone above the Upper Fell Top Limestone, possibly of Arnsbergian age, in Weardale, County Durham, northern England (G. A. L. Johnson, personal communication, 1975). Specimens are difficult to locate *in situ* unless they are weathered out, and very few specimens have been reported from the Central Pennine Basin. Their origin remains problematic (Bandel, 1967).

cf. AULICHNITES Fenton and Fenton, 1937
Pl. 1E, F

Description.—Curved, sinuous to meandering trails showing two convex ridges separated by a median furrow preserved in epirelief. Trail lengths up to 75 mm; some specimens show two width classes, 5 mm and about 12–15 mm, respectively, associated. Surfaces of the ridges are flattened slightly and either smooth or with transverse bilobed meniscus packing of sediment at a spacing of 0.5 mm.

Occurrence.—Pendleian, Grassington Grit in at least two localities. This rare form occurs in the flaggy-bedded, fine-grained sandstones of the Delta Top Association, although it has not been recorded with other trace fossils. These sandstones may be of inter-channel origin.

Remarks.—In sinuosity, form, and size the trail closely approaches *Aulichnites* as described by Fenton and Fenton (1937) and Hakes (1976), although trails of this type previously described from Britain have frequently been referred to *Crossopodia* (= *Olivellites* in part). This ichnogenus has previously been interpreted as the crawling or grazing trail, or burrow of a gastropod.

BEACONITES Vyalov, 1962
Beaconites cf. *B. antarcticus* Vyalov, 1962
Pl. 14A

Description.—Large horizontal U-shaped burrows lacking marginal tubes but with meniscus packed, back-fill structures of retrusive type. Lengths of two burrows observed were 1.05 m and 1.65 m, and widths were 0.18 and 0.20 m. The burrows had a semi-circular cross section and rounded distal ends and were outlined in shale or mud-flakes within white sandstone.

Occurrence.—Two specimens of *Beaconites* were observed in the lower Westphalian Bullion Mine Rock (Fig. 8; Hardy, 1970b). The horizon lies below the *Gastrioceras listeri* Marine Band and immediately above the rhythmic sandstone/siltstone units of the Lower Mountain/Bullion parting. The enclosing sediments are medium- to coarse-grained sands with large-scale cross-bedding and with enclosed macroplant debris. They appear to be of channel-bank or overbank origin.

Remarks.—The interpretation of *Beaconites antarcticus* has been a matter of considerable controversy in recent years. Although Gevers et al. (1971) favored origin as a back-filled feeding burrow of a marine polychaete worm, most recent workers favor either a vertebrate locomotory origin (Ridgeway, 1975; Pollard, 1976; Archer, 1976) or an arthropod back-filled burrow (Rolfe, 1980) for structures of this type in non-marine sediments. Large *Beaconites* burrows have been widely recorded in fluvial channels and overbank facies of both the Devonian (Old Red Sandstone, Allen and Williams, 1981) and the Lower Carboniferous (Chisholm, 1977).

BERGAUERIA Prantl, 1945
Pl. 1A, B

Description.—Small, sub-hemispherical positive hyporeliefs randomly distributed on the bases of thin turbidite sandstone beds. The traces are up to 20 mm in diameter and range between 6 and 10 mm in relief. Individual forms are circular in plan and their external surface is smooth or slightly granular. The sediment infill is identical with the overlying bed.

Occurrence.—Pendleian, at two localities. Common throughout the Turbidite Association and the lower part of the Slope Association, where *Bergaueria* occurs with *Muensteria, Mammilichnis,* and ?*Protopalaeodictyon.*

Remarks.—This structure closely resembles *Bergaueria* Prantl as figured by Häntzschel (1962, fig. 112, 4a,b; 1975), Lessertisseur (1955), Radwański and Roniewicz (1963, pl. 9), Alpert (1973, pl. 1, fig. 9) and Hakes (1976, pl. 2, figs. 1a–d). *Bergaueria* has been interpreted by various authors (see Alpert, 1973, for review) as the casts of burrows made by sedentary organisms such as coelenterates, anthozoans, or even as the infilling of decomposition hollows. If these traces are indeed the work of coelenterates, a fairly high salinity would be implied. The sudden incursion of turbidity currents depositing sands may have been responsible for the deaths of at least some of the organisms. Seilacher (1964) and Crimes et al. (1977) consider *Bergaueria* to be more common in shallow-water deposits.

COCHLICHNUS Hitchcock, 1858
Pl. 3C; Pl. 12E, F

Description.—Winding to sinuous traces, often in sine curve form, showing regularity, normally preserved as convex or concave epirelief, but also as positive hyporelief. Forms vary from 1–2 mm wide, with a wavelength of about 16 mm, to those 4–5 mm wide with a wavelength of about 45 mm.

Occurence.—Pendleian—lower Westphalian; widespread in many localities and in sediments ranging from the Turbidite Association, Delta Slope, and delta front to interdistributary bay of the Delta Top. At various levels it is associated with almost all other elements of the ichnofaunas (Figs. 4, 5, 8), being a facies-crossing ichnogenus.

Remarks.—Crawling traces and probably feeding structures of a small worm or worm-like animal. In the Upper Carboniferous of Germany and the U.S.A., as well as in Britain, *Cochlichnus* has been recorded in sediments of supposed low-salinity paleoenvironments (Michelau, 1956; Hakes, 1976).

?CURVOLITHUS Fritsch, 1908
Pl. 2A, B

Description.—Simple, ribbon-like trails preserved as half-relief epichnial molds. The trails are 10–16 mm wide and have a relief of 1–2 mm. They consist of two concave furrows on either side of a slightly wider convex ridge. Absence of any observed full cross section makes the identification tentative. Individual trails are unbranched but commonly intersect. They are straight to gently curved except where they turn fairly suddenly through angles commonly between 5 degrees and 25 degrees, and less commonly 90 degrees.

Occurrence.—Pendleian; six localities in parallel-bedded and striped siltstones most commonly occurring in the Slope Association, but also in the upper part of the Turbidite Association. ?*Curvolithus* is associated with *Cochlichnus* and *Lophoctenium.*

Remarks.—Ribbon-like trails have been diagnosed as *Curvolithus* by Häntzschel (1962, 1964), Heinberg (1970, 1973), Chamberlain (1971a), and Hakes (1976). Hakes discusses the variants which may occur on the trilobate form.

DIDYMAULICHNUS Young, 1972
Pl. 2D

Description.—An irregular curved to straight trail preserved as a positive hyporelief. The trace is 4–5 mm wide, smooth and bilobate with a very narrow median depression. Trails cut across sole structures and are therefore post-depositional and interstratal.

Occurence.—Pendleian (see p. 105). Characterizes thin-bedded turbidite sandstones of the Turbidite Association where it occurs with *Bergaueria.*

Remarks.—The forms seen are rather small for *Didymaulichnus* as originally described. Those which cut across erosional sole marks are also anomalous, since

Didymaulichnus is generally regarded as a surface trail, probably of a mollusc (Hakes, 1976). It is also regarded as of shallow-water origin, but see Occurrence and Description above.

GYROPHYLLITES Glocker, 1841
Pl. 1C; Pl. 10B, b

Description.—Positive hypichnial rosette structure consisting of a circular or slightly ellipsoidal radial pattern of petal-like lobes around a central protuberance. Pendleian examples are 16 mm in diameter with a relief of 3–4 mm and an average of 9 "petals" or lobes (Pl. 1C). Marsdenian examples are larger and more ellipsoidal, ranging from 19–60 mm in the longest dimension and 28–39 in the shortest; and the lobes, which enlarge distally, are only poorly defined, being at least 5 to 7 in number (Pl. 10B, b). These latter specimens may show a central polygonal structure 4–14 mm in diameter.

Occurrence.—This trace fossil is recorded from two horizons. In the Pendleian it occurs in the Turbidite Association, at the base of turbidites associated with *Planolites* and vertical tubes. In the Marsdenian it was found in a trough cross-bedded sandstone, the Woodhouse Grit (Figs. 2, 5), interpreted as of distributary mouth-bar origin. In the same lithological unit *Phycodes palmatum, Planolites, Kouphichnium rossendalensis,* and *Pelecypodichnus* have been recorded.

Remarks.—*Gyrophyllites* is usually defined as a vertical stem from which lobate offshoots radiate at different levels, the whole structure being conical in form (Gregory, 1969; Häntzschel, 1975). The specimens from both the Pendleian and Marsdenian appear to have only one rosette of lobes, at a sand-mud interface. This could be explained as the only preserved part of a three-dimensional feeding system and is a common mode of preservation for structures of this type in both flysch (Gregory, 1969) and shallow-water situations (Fürsich, 1974b, Fürsich and Kennedy, 1975). The rosetted lobes are due to radial feeding, and the central terminal projection, where preserved, is explained as an abandoned feeding shaft.

KOUPHICHNIUM Nopsca, 1923
Kouphichnium aff. *K. variabilis* (Linck) 1949
Pl. 11E

Description.—Hypichnial trackway consisting of a central ridge (telson groove mold) flanked by rows of markings of different form corresponding to the outer (pusher) and inner (walking) appendages of a xiphosurid arthropod. Trackway width about 20mm. The figured specimen has been described by Chisholm (1983).

Occurence.—See under *Limulicubichnus rossendalensis* below.

Remarks.—Short lengths of xiphosurid walking trackways, *Kouphichnium* in the restricted sense of Miller (1982), commonly occurs in association with resting traces, *Limulicubichnus rossendalensis*. However, in the main text and in the figures of this paper the two types of trace have been included under the name *Kouphichnium* in the broad sense of Häntzschel (1975), except where indicated.

LIMULICUBICHNUS Miller, 1982
Limulicubichnus [*Kouphichnium*]
rossendalensis Hardy, 1970
Pl. 11A, B, D

Description.—Hypichnial trails consisting of lunate casts corresponding in outline to the shape of a xiphosurid prosoma, often in a series and associated with appendage and telson marks. Width of casts 16–17 mm (more rarely 9–11 mm), deepest (2–3 mm) at the anterior convex end, shallowing to the posterior. Prosoma casts may be either isolated (Pl. 11D), with or without two rows of appendage marks and a telson cast, or may occur in a linear or curved series (Pl. 11A). Occasionally prosoma casts may occur at the end of a bilobed hypichnial trail mold with faint transverse sediment packing (Pl. 11B). Some sandstone sole surfaces may show a crude parallelism of prosoma casts suggestive of rheotaxis, but such observations are untested statistically. Biserial rows of oblique hypichnial protuberances similar to limulid digging traces are present on some thin sandstone slabs associated with these xiphosurid resting places (Pl. 11B).

Occurrence.—Marsdenian to lower Westphalian (see above); the Marsdenian includes the Woodhouse Grit and the Yeadonian the Haslingden Flags (Hardy, 1970a). There are several horizons and localities in the Westphalian (Fig. 8 and see p. 129).

Remarks.—Recently Miller (1982) has proposed the separation of limulid trackways (ichnogenus *Kouphichnium*) from limulid resting places which she assigns to a new ichnogenus, *Limulicubichnus* Miller, 1982. Consequently she renames traces described here as *L. rossendalensis* (Hardy, 1970). Such ichnogeneric reassignment appears acceptable. The ichnospecies *L. rossendalensis* differs from the ichnogenotype *L. serratus* by its smaller size, lunate rather than teardrop prosoma cast, greatest relief at the anterior border and lack of serrated posterior margins. Both *L. rossendalensis* and *Kouphichnium* traces from the Silesian of the Central Pennines are believed to have been produced by non-marine arthropods like *Belinurus* or *Euproops* rather than *Limulus*. Likewise, although the behavior patterns represented by our specimens show similarity to *Limulus polyphemus* described by Miller (1982, p. 431–432), the environmental implications are quite different, as discussed on p. 181 above.

LOPHOCTENIUM Richter, 1850
Lophoctenium cf. *L. comosum* Richter, 1851
Pl. 3A, B

Description.—Irregularly shaped horizontal spreiten fields preserved as full relief endichnia. Often the spreiten have no well-defined boundary and are formed by an overlapping series of small chevron-like marks arranged concentrically (Pl. 3B) in a "cock's tail"-like pattern. One specimen shows spreiten fields connected by a thin burrow preserved in full relief (Pl. 3A).

Lophoctenium cf. *L. haudimmineri*
Chamberlain, 1971a
Pl. 3D, E

Description.—Small isolated spreiten fields with a distinct peripheral rim (Pl. 3E) showing poor overlap of the spreiten, which are concavo-convex and protrusive to the outer margin.

Occurrence.—Pendleian; at several localities in the Skipton area in parallel-bedded and striped siltstones in the lower part of the Slope Association where it occurs with *Cochlichnus* and *Curvolithus*.

Remarks.—Chamberlain (1971a) believes that this trace was made by a worm-like organism which systematically mined the sediment. The chevron pattern is made as the animal pushes itself forward. With successive feeding passes, the animal overlaps earlier feeding passes and a field of concentric spreiten is built. The interconnecting tubes compare with those illustrated by Chamberlain (1971a). There is a similarity between *Lophoctenium* and *Zoophycos* (Simpson, 1970; Chamberlain, 1971a), suggesting a similar mode of behavior.

MAMMILICHNIS Chamberlain, 1971
Pl. 1D

Description.—A positive hyporelief consisting of smooth, sub-hemispherical protuberances with circular or elliptical terminal projections. Structures are 9 to 12 mm in diameter and relief is up to 7 mm. The terminal projections are 3 to 5 mm in diameter. Thin sections and x-radiography failed to reveal any disturbance in the sediment immediately above the structure.

Occurrence.—Pendleian, in the Skipton Moor region, in thin-bedded turbidite sandstone. Turbidite Association and lower part of the Slope Association, occurring with *Planolites*, *Cochlichnus*, and *Bergaueria*.

Remarks.—The structures compare closely with *Mammilichnis aggeris* Chamberlain (1971a). Chamberlain acknowledges that the exact nature of the form is difficult to establish but suggest that it was produced by an organism resting or hiding in the sediment. Crimes et al. (1981, p. 976) suggest that specimens of this form from the Polish Carpathian and Swiss flysch deposits may be resting traces of an anthozoan. *Mammilichnis* and *Bergaueria* are probably related and may represent preservational end members of a continuum of forms.

MONOCRATERION (SKOLITHOS) Torrell, 1970
Pl. 4A, B, C, E

Description.—Straight, cylindrical, isolated tubes that pass vertically upwards into circular, steep-sided funnels (Pl. 4A, C). Tubes are 2 to 8 mm in diameter and are up to 13 cm long. Funnel dimensions are variable with diameters averaging 15 mm at the uppermost, widest part. Funnels vary in height from 15 to 30 mm. Several specimens show funnels with raised rims, which may reflect linings to the funnels. Within this apparent lining, the central section is organized in light and dark laminae which are gently concave upwards (Pl. 4A). In some specimens a diffuse zone occurs outside the lining, and, outside that, the laminae of the host sediment are deflected downwards (Pl. 4A, C). The diffuse zone maintains a constant diameter downwards whilst the funnel diameter decreases. Some tubes show a thin lining, usually of very fine, light colored sediment which contrasts with the darker infill. Around the tubes on some parting planes are dark (?carbonaceous) flecks, generally randomly oriented but, in some cases, showing a hint of radial distribution around the tube. Rare examples show current crescent scours associated with the vertical tubes where they pass upwards from silty mudstone into medium-grained sandstone (Pl. 4E). Immediately below the funnel, there is a marked scour around the tube, preserved as hyporelief on the sandstone. Scours are parallel with one another and are consistent with regional paleocurrent direction.

Occurrence.—Pendleian; in parallel-bedded and striped siltstones and cross-laminated sandstones; top of the Slope Association and base of the Delta Top Association where *Monocraterion* occurs with *?Pelecypodichnus*, *Cochlichnus*, and *Curvolithus*.

Remarks.—*Monocraterion* is considered to be the dwelling structure of a small worm-like organism, possibly a polychaete, for which the tube of *Diopatra cuprea* may be a modern analog (see Myers, 1970; Barwis, this volume).

Cerianthus lloydi (Schäfer, 1962) has also been considered as a modern analog (Hallam and Swett, 1966). This animal lives in a burrow lined by a mucilaginous sheath. Specimens with hyporelief crescent scours suggest that the tubes were able temporarily to resist erosion. The raised rim of the funnels suggests that the animal also lined the sides of the funnels. Similar, resistant funnel rims were figured by Bruun-Petersen (1973) and Horne and Gardiner (1973). Bruun-Petersen suggests that the funnel was formed and kept open by tentacles whilst Hallam and Swett (1966) consider the size and shape of the funnels to reflect the physical stability of the sediment combined with the speed and extent of upwards migration of the animal.

Both *Skolithos* and *Monocraterion* are associated with shallow-water conditions. Seilacher's (1967) *Skolithos* ichnofacies is the shallowest of the marine assemblages in his scheme of bathymetric zonation.

MUENSTERIA Sternberg, 1833
Pl. 6C

Description.—A horizontal burrow preserved as a positive hyporelief. The burrow, which varies in width between 3 and 4 mm, is characterized by a series of transverse annulations spaced at 3 mm intervals.

Occurrence.—Pendleian, in thin-bedded turbidite sandstone. Turbidite Association, occurring with *Bergaueria*.

Remarks.—Annulations suggest that the burrows may have a meniscus backfill within them. Trace fossils with backfill structures have been called *Planolites* (Richter, 1937; Webby, 1970), *Taenidium* Heer, 1877 (Hakes, 1976), *Keckia* Glocker, 1841, and *Muensteria* Sternberg, 1833. Fürsich (1974b) considers there to be no distinctive differences between these ichnogenera. Burrows of this type are considered to be facies-independent, since they have been reported from flysch facies (Heer, 1876) to non-marine sediments (Seilacher, 1963; Stanley and Fagerstrom, 1974).

OLIVELLITES Fenton and Fenton, 1937
Olivellites cf. *O. plummeri* Fenton and Fenton, 1937
Pl 8A, B; Pl. 9A, B; Fig. 9

Description.—Curved to sharply sinuous meandering trails, 15–16 mm wide, with median, ribbon-like crest, and transverse epichnial ribbing with forward curvature. The trail is infilled with fine-grained sandstone, lighter in color than the surrounding rock host (Fig. 9). Sections cut longitudinally along the center show an imbricate structure outlined by carbonaceous debris (Pl. 8B), which appears also in transverse section but with less definition. Both cut sections suggest the structure of the etched-out filling of *Olivellites plummeri* as redescribed by Yochelson and Schindel (1978). Base of the trail shows gentle transverse curvature and regular radial undulations (Pl. 9A). Trails suggest phobotaxis, but crossing takes place above an earlier trail on the interpretation of forward movement assumed by Yochelson and Schindel (1978).

Occurrence.—Siltstone to fine-grained sandstone alternations and silty sandstone in the Upper Kinderscoutian to Marsdenian Stages of the Skipton-Bradford, Huddersfield-Halifax, and Rossendale areas

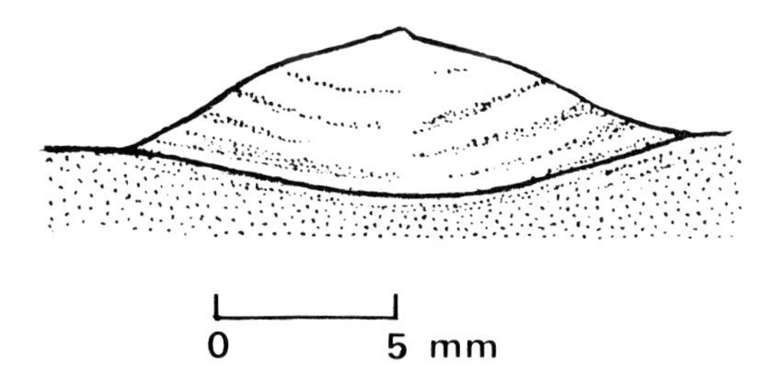

Fig. 9.—Transverse section of *Olivellites* cf. *plummeri* Fenton and Fenton, showing epichnial crest and proportions. Infilling is semi-diagrammatic, white indicating light colored fine-grained sandstone and dots carbonaceous debris. Drawn from two sections of a specimen below *Reticuloceras gracile*, Upper Kinderscoutian, near Oxenhope, Bradford: M. Mus. LL. 6205.

(Fig. 1). Where very abundant, *Olivellites* is believed to be characteristic of a proximal or marginal mouth-bar paleoenvironment. Usually it occurs without other trace fossils, but a single bedding-plane association with *Pelecypodichnus* of the cf. *Sanguinolites* group in dark siltstone of the Marsdenian (Readycon Dean Series, Figs. 2, 5) south of Holmfirth, suggests marginally marine, perhaps with fluctuating salinity.

Remarks.—*Olivellites* has long been recognized by the British Geological Survey as *Crossopodia* cf. *C. embletoni* Tate and *Crossopodia* sp., being regarded as the trail of an annelid (Stephens et al., 1953) and as a worm-track (Bromehead et al., 1933). Material from higher horizons in the Marsdenian shows finer, more numerous transverse ribs (Pl. 9B). It was recorded as *Olivellites* from the Guiseley Grit at Cornholme, near Guiseley (Fig. 1) by Williamson and Williamson (1968, p. 562). In view of the internal structure described here, we consider a feeding origin by a small arthropod as more likely than that of an annelid (Yochelson and Schindel, 1978).

PALAEOPHYCUS Hall, 1847
Pl. 6E; Pl. 10C

Description.—Horizontal cylindrical burrows preserved as positive hypichnia which show occasional dichotomous branching and smooth or striated wall sculpture. Burrows may be short and straight, or longer, slightly sinuous and intersecting. Pendleian examples are generally small, 2–4 mm in width, but in Marsdenian specimens (Pl. 10C) width may vary from 1–20 mm and length from 10–235 mm at different localities and horizons.

Occurrence.—Pendleian; in thin-bedded turbidite sandstones of the Turbidite Association where it is found also with *Planolites*. Marsdenian; in the Woodhouse and Fletcher Bank Grits (Fig. 2, left; Fig. 5) and in the overlying Hazel Greave Grit, on the bases of trough cross-bedded sandstones which are interpreted to be of distributary channel or mouth-bar origin. Associated trace fossils include *Planolites,* cf. *Arthrophycus,* and *Limulicubichus* cf. *L. rossendalensis*.

Remarks.—These traces appear to represent locomotion or feeding burrows of a worm-like organism and were formed at mud-sandstone interfaces, post-depositionally. They are distinguished from *Planolites* by their branching, although Pendleian forms may intergrade with ?*Protopalaeodictyon*. Burrows of this type have been recorded in a wide variety of environments from flysch (Ksiazkiewicz, 1970), shallow marine (Hall, 1847; Hallam, 1970) to fluvial floodplains (Ratcliffe and Fagerstrom, 1980) and so must be regarded as facies-crossing trace fossils.

PELECYPODICHNUS Seilacher, 1953 [= LOCKEIA James]
Pl. 7A-E; Pl. 10E; Pl. 12A-D

Description.—Small ovoid or almond-shaped protuberances (convex hypichnia) or hollows in bedding planes (negative epichnia) which frequently occur in great abundance and may show parallel to subparallel alignment (Fig. 7). Surface of hypichnia usually smooth, rarely with ridge basally, but one or both ends may be pointed. Hypichnia often occur at the bases of vertically retrusive structures showing downbending and/or breakages of sedimentary laminae. Vertical sectioning (Eagar, 1977) shows *Pelecypodichnus* merges gradually into bivalve upward escape shafts which consist of a succession of these structures (Pl. 10E; Pl. 12B-D). Length and width variable (see above).

Occurrence.—Pendleian (uncommon); Kinderscoutian-Westphalian (extremely abundant). For details of occurrence, origin, associated trace fossils, and interpretation, see above.

Remarks.—In accordance with the suggestions of Häntzschel (1975) and Hakes (1977), the name *Pelecypodichnus* is preferred for these trace fossils and *Lockeia* James, 1879 is regarded as a *nomem oblitum*. These trace fossils, interpreted as resting burrows of nestling, semi-infaunal bivalves, have proved to be of considerable behavioral, sedimentological, and paleoenvironmental significance in the Silesian assemblages discussed above.

PHYCODES Richter, 1850
Phycodes cf. *P. curvipalmatum* Pollard, 1981
Pl. 10F

Description.—Horizontal, ramifying and branching interstratal burrows preserved as complex hypichnial burrow systems. Individual burrows possess a tube-like form (common burrow) proximally but distally split into finger-like curved branches. Common burrows may be unilobed, bilobed, or rarely trilobed, 3–5 mm in diameter, 1–7 mm in depth, and from 40–130 mm or more in length. Cross-cutting of the burrows may form a wide net-like pattern. Distal termination may take the form of either short radiate inflated "lobes" or longer, first-order dichotomous branches which recurve both laterally and upwards into the overlying bed in a "palmate" manner. The branched terminations may occur in isolation or may overlie one another.

Occurrence.—Marsdenian, Guiseley Grit (Figs. 2, 5) at one locality south-south-east of Keighley (Fig. 1). Burrows occur at the base of a trough cross-bedded sandstone interpreted as being of interdistributary channel origin. Associated with cf. *Arthrophycus*.

Remarks.—These highly variable burrows appear to be similar to *P. curvipalmatum* recently described from the Triassic (Pollard, 1981), although the unbranched parts of the burrow system show resemblance to cf. *Fucusopsis* where unpacked, to cf. *Arthrophycus* where internal packing approaches an annulation, and even to cf. *Nereites* where compression of packing has produced an apparent series of lateral lobes. An origin as feeding burrows, exploiting the food richness of the sediment interface in a systematic manner, seems to be indicated.

Phycodes cf. *P. palmatum* Hall, 1852
PL. 10B, a; Pl. 13G

Description.—Conical or elongate hypichnia consisting of radiating or palmately branching burrows which are parallel-sided and are rounded or bluntly pointed where they curve up into the overlying bedding plane. Marsdenian examples show more radial arrangement in mound-like structures up to 75 mm long by 56 mm wide (Pl. 10B, a) with individual burrows being about 5–8 mm in width; whereas lower Westphalian examples branch palmately from a common burrow, being up to 90 mm in length, with overlapping branches about 12 mm in width and up to 35 mm in length from divergence (Pl. 13G). Externally the burrows are smooth and they may have a bioturbated infilling.

Occurrence.—Marsdenian; Woodhouse Grit, immediately east of Keighley (Figs. 1, 2, 5). Traces lie at the base of a trough cross-bedded sandstone interpreted as of distributary mouth-bar origin, intimately associated with cf. *Gyrophyllites* and from the same beds as *Pelecypodichnus*, *Limulicubichnus rossendalensis*, and *Planolites*. Lower Westphalian; Crutchman Sandstone (Fig. 8). Base of coarse carbonaceous sandstone, perhaps of crevasse-splay origin. *Arenicolites carbonarius* has been recorded from the upper parts of the same sandstone at the same locality (near Bolton, Fig. 1).

Remarks.—The Westphalian examples closely approach *P. palmatum* Hall as figured by Osgood (1970, pl. 67, fig. 7), whereas the Marsdenian specimens are morphologically more varied. This form differs from *P. curvipalmatum* by its larger size, straight burrows and lack of recurvature. Interstratal burrows of this type have been recorded in both marine paleoenvironments (Osgood, 1970; Crimes et al., 1974) and nonmarine ones (Pollard, 1981).

PLANOLITES Nicholson, 1873
Pl. 2C; Pl. 7F

Description.—Horizontal, simple interface burrows preserved as both positive and negative epichnia and positive hypichnia. They are simple, unbranched, straight or curved with smooth external surfaces and without obvious internal back-filling. Burrow length and diameter varies considerably at different horizons and localities. For instance, Pendleian examples are 2–10 mm in diameter, whereas Marsdenian specimens show two size classes, width 1–2.5 mm and 3–5 mm, and length 5–25 mm and 26–80 mm respectively. Some more sinuous examples may show terminal enlargements (7–9 mm) or end at depressions or vertical burrows up to 12 mm in diameter (Pl. 2C). Some hypichnial burrows clearly cut across erosional sole marks. Many are elliptical in cross section due to compaction.

Occurrence.—Pendleian-Westphalian. Described from a wide variety of sediment types and inferred paleoenvironments from turbidites to non-marine interdistributary bays and occurring with most other elements of ichnofaunas (see above).

Remarks.—Regarded as feeding burrows of infaunal worms. Detailed analysis of different types of *Planolites*, which may reflect environmental preferences (e.g., Chisholm, 1968), has not been possible here.

?PROTOPALAEODICTYON Ksiazkiewicz, 1970
Pl. 6D

Description.—Horizontal burrow system preserved as positive hyporeliefs. The burrows vary in size and complexity. The tubes are straight to slightly curved over distances of 20–30 mm and then branch in a zigzag manner to give an irregular and incomplete hexagonal pattern. The largest trace is over 260 mm long. Burrows range in diameter from 4–18 mm and form crude hexagons up to 25 mm in width (Pl. 6D) for a 5-mm diameter burrow.

Occurrence.—Pendleian, one locality, in thin-bedded turbidite sandstones at the top of the Turbidite Association, associated with *Bergaueria* and *Planolites*.

Remarks.—This burrow resembles *Palaeodictyon* in showing a tendency to form a hexagonal network. In this case, however, the network is not fully and regularly established and the name ?*Protopalaeodictyon* is tentatively preferred. This form was described by Ksiazkiewicz (1970) as "a meandering sole trail with ramifications on the apices of the meanders." Similar structures are also figured by Crimes et al. (1974). Seilacher (1977) points out that any direct relationship between *Protopalaeodictyon* and *Palaeodictyon* has yet to be established. It is not certain that the forms described here are graphoglyptid burrows. They could possibly be post-depositional, interface burrows related to *Palaeophycus* or *Cochlichnus* (see p.141).

RHIZOCORALLIUM Zenker, 1836
Rhizocorallium type A (*R.* cf. *R. irregulare* Mayer, 1954)
Pl. 4D; Pl. 5A, B

Description.—Positive epirelief traces, either straight or slightly sinuous (Pl. 5B). The structure consists of two parallel tubes separated by a furrow carrying numerous, closely spaced, irregular protrusive spreiten (Pl. 4D). Tubes are oval in cross section with a maximum diameter of 6 mm. The furrow is between 50 and 100 mm wide. Maximum observed length is 1.5 m. The material compares closely with specimens collected by N. H. Trewin from the Pendleian of the Staffordshire Basin to the south. That material shows more clearly developed spreiten, a U-shaped termination to the traces and striations on the upper surfaces of the lateral tubes, parallel to the general trend of the tubes. Other specimens from the Pendleian of Skipton are preserved as elongate parallel-sided zones of protrusive spreiten lacking a marginal tube (Pl. 5A).

Rhizocorallium type B (?*R.* cf. *R. jenense* Zenker, 1836)
Pl. 5C; Pl. 6A, B

Description.—Short, positive epirelief traces which are sub-rectangular or globular in shape (Pl. 5C). The former type has an abrupt, square proximal end with protrusive spreiten distally and is about 15 mm in width. The distal end is rounded with a peripheral tube which, in some examples, extends along the sides of the trace (Pl. 6B). The globular form (Pl. 6A) is 27 mm long by 25 mm wide and shows strongly asymmetrical spreiten diverging from one proximal point of burrow origin, much in the manner of *Zoophycos circinnatus* (Brongniart). Later spreiten are more concentric distally and are cut off by the almost circular peripheral tube.

Occurrence.—*Rhizocorallium* type A and B are Pendleian only, both commonly occurring on the rippled upper surfaces of thin-bedded turbidite sandstones of the Turbidite Association and the lower part of the Slope Association. Associated trace fossils on rippled bedding planes are *Planolites* and *Cochlichnus*.

Remarks.—Type A, by having a long and somewhat sinuous form, exclusively protrusive spreiten, and by following a rippled sand-mud interface, compares with *R. irregulare* Mayer. This form was interpreted by Fürsich (1974b) as the burrow of a deposit-feeding animal mining nutrient-rich sediment on the rippled top. Unlike examples from the Jurassic of southern England, figured by Fürsich, the traces from the Pendleian cut across ripple crests and are not preferentially aligned along the ripple troughs. Scratch marks on the tube walls suggest that the animal may have been a crustacean (cf. *R. jenense*, below). Type B, with its much smaller horizontal extent, compares at first sight with *R. jenense* Zenker, although its mode of preservation (compressed epichnial spreiten) and its *Zoophycos* type of spreite formation suggests a deposit-feeding rather than suspension-feeding producer. Straight-sided variants of type B (Pl. 6B) compare very closely with proximal parts of type A and so could simply represent very short-lived initiation and abandonment of type A burrows.

Rhizocorallium cf. *R. jenense* Zenker, 1836
Pl. 14B-E

Description.—Short, oblique or horizontal U-shaped burrows preserved as hypichnia in two intergrading forms; type A—horizontal marginal U-tubes enclosing a spreiten zone (Pl. 14B), and type B— an infilled tongue or pouch-shaped excavation with scratch-marked walls. Type A may show proximal expansion or median constriction and distal expansion of the marginal burrow (Pl. 14B), whereas type B may show considerable overlapping or cross-cutting of the burrow excavations, which at times are nearly vertical in attitude (Pl. 14E). Width of U-burrow 10–47 mm; length (depth) up to 77 mm; marginal tube diameter

4–5 mm. Length to width ratio rarely more than 1:1, the burrows appearing as crescent-shaped hypichnia. Burrows in any single population may vary in size (Pl. 14E) and may mutually interfere or overlap and show preferred orientation (see above). Spreite may be either protrusive or retrusive. External surface of the burrows either smooth (eroded?) or with molds of fine criss-crossing or bifurcating scratch marks arranged crudely parallel to the burrow margin (Pl. 14D).

Occurrence.—Lower Westphalian; sandstone below the Lower Foot Coal in west Lancashire and north Derbyshire (p. 125). Burrows have been made as excavations into mudstone below an erosion surface at the base of the casting sandstone, which is pebbly, highly carbonaceous and possibly of tidal (p. 127) or distributary channel origin. No associated trace fossils are known.

Remarks.—The variable U-shaped burrows appear to be similar to *R. jenense* in terms of shape, attitude, width-length ratio, protrusive-retrusive spreiten and scratch-mark pattern. They also show some similarity to "oblique striated burrows" recently described from Triassic sediments (Bromley and Asgaard, 1979; Pollard, 1981). Although densely packed, tongue-like variations of this burrow show some similarity to *Fuersichnus communis* Bromley and Asgaard (1979); more isolated examples clearly show a U-shaped rather than a J-shaped horizontal burrow (Pl. 14C). The surface sculpture is of fine scratch marks broadly parallel to the burrow margin (Pl. 14D), not a transverse striation or knobbly ornament as in *F. communis*. The ichnofacies significance of this burrow and comparable occurrences is discussed above (p. 126).

"SCOLICIA" de Quatrefages, 1849
Pl. 8C

Description.—In this paper, "*Scolicia*" has been used as a group name for irregularly meandering, sandstone-filled trails of low relief, typically 11–14 mm wide, but ranging up to 18 mm, which are crossed by transverse curved "ribs" or ridges. Up to three epichnial longitudinal ridges may be seen on some specimens, but a central one is always present, except at terminations (or initiations ?) of the trail, where the organism moved out of (or into ?) the sediment (Pl. 8C, bottom), where transverse elements only are preserved. The infilling, lighter in color than the host sandstone, is considerably flatter in cross section than that of *Olivellites* (Fig. 9), possibly in part due to compaction, but shows a similar pattern of carbonaceous inclusions.

Occurrence.—"*Scolicia*" has been found in thin fine-grained sandstones and siltstones above the Butterly Marine Band in the Marsden region (Fig. 1), interpreted as delta front mouth-bar deposits, where it is commonly associated with *Pelecypodichnus* (of cf. *Sanguinolites* origin) and *Cochlichnus* (Eagar et al., 1983). Closely similar trails have been described by Sheldon (1968) from the Todmorden district (Fig. 1, 15 km west of Halifax) and are thought to have come from about the same horizon or from the Upper Kinderscout Grit.

Remarks.—Sheldon (1968) compared the Todmorden burrows with trails of recent intertidal gastropods but did not section his material. Our endichnial evidence suggests affinity of origin with *Olivellites* (Eagar et al., 1983).

ZOOPHYCOS Massalongo, 1855
Zoophycos (Spirophyton) cf. *Z. cauda-galli* Hall, 1863
Pl. 10D

Description.—This trace consists of a large conical radiating structure at least 200 mm in diameter (edge of specimen not seen), preserved in positive endichnial relief above layers of spreiten laminae which vary considerably in direction. The central apex is polygonal with a longer diameter of 20 mm, a shorter diameter of 15 mm, and an elevation of 5 mm above the laterally extending lamellae. About 35 lamellae radiate downslope from the apex; each lamella flattens, bends in helical spiral, and bifurcates distally. Width of each lamella ranges from 2–4 mm. A prominent radial band (Pl. 10D, arrow) which curves from apex to periphery, appears to be a vertically compressed marginal tunnel of Simpson (1970, fig. 1a). The total depth of *Zoophycos* structure is greater than 50 mm, although this is hard to assess as it is composed of superimposed layers of spreiten lamellae varying in both thickness and direction.

Occurrence.—Marsdenian; a single locality 10 km west-southwest of Keighley (Fig. 1) in the "Bluestone" (Fig. 2, Skipton-Bradford area, Fig. 5) of Stephens et al. (1953, p. 103). The bioturbated sediment is bluish grey, micaceous silty mudstone, rich in *Hyalostelia* sponge debris indicative of a low energy marine environment, subject to occasional current activity. No other trace fossils were found with this form.

Remarks.—These specimens of *Zoophycos* show a central cone, helical spreiten curvature, and a compressed axial marginal tube, all characters of the *Spirophyton* type of Simpson (1970, p. 510). The thorough bioturbation of the matrix suggests *Z. (Spirophyton) cauda-galli* type, although the total structure is small for this form. This trace fossil is regarded as a complex feeding structure, involving successive tunnelling of a series of feeding probes of an unknown worm-like animal (Simpson, 1970, p. 511) which probably lived in a shallow marine environment. Although *Zoophycos* is usually regarded as characteristic of deep-water facies (Seilacher, 1967; Ekdale, 1977), recent studies have shown that this ichnogenus appears to have a considerable bathymetric range and may occur wherever the sea floor was undisturbed by waves, currents or other forms of bioturbation (Hallam, 1975). In several other studies in the Carboniferous it has been recorded from shallow marine sediments found above wave base (Wells, 1955; Taylor, 1967; Hecker, 1970; Osgood and Smzuc, 1972). Closely comparable occurrences of *Zoophycos* in Yoredale (deltaic) facies are recorded from the sponge-rich Richmond Cherts of the Pendleian of north Yorkshire (Wells, 1955) and from siltstones alternating with coals and rootlet beds (Kennedy, 1975), confirming that the environmental interpretation given here is not unique. Moreover, Miller and Johnson (1981) explain the presence of dense assemblages of small *Spirophyton* burrows in alluvial-tidal facies of the Devonian Catskill deltaic complex of New York State, in terms of opportunistic deposit-feeding animals whose distribution was controlled more by biological competition than by bathymetry, substrate type, or energy level of the environment.

ACKNOWLEDGMENTS

The paper has drawn on both published and unpublished information, the extent of previous work by the authors being indicated in the list of references. Collinson is responsible for the sedimentary introduction; Baines and Collinson for details of the Namurian turbidite deltas; Okolo and Eagar for Namurian sheet deltas; Hardy, Eagar, and Pollard for lower Westphalian ichnofaunas. Eagar has acted as coordinating editor and Pollard as advisor on trace fossil taxonomy.

The authors gratefully acknowledge field guidance locally from Philip Holroyd, Graham Miller, and Philip Unsworth. We are also indebted to Peter Douglas, David Kelsall, Susan Maher, and Wilfred Thomas for photography, and to Philip Stubley and John Pepper for some of the drafting. We finally thank Edward S. Belt and Roland Goldring for helpfully reviewing the paper in manuscript.

REFERENCES

ALLEN, J.R.L. AND WILLIAMS, B.P.J., 1981, *Beaconites antarcticus*—a giant channel-associated trace fossil from the Lower Old Red Sandstone of South Wales and the Welsh Borders: Geol. Jour., v. 16, p. 155–169.

ALPERT, S., 1973, *Bergaueria* Prantl (Cambrian and Ordovician), a probable actinian trace fossil: Jour. Paleontology, v. 47, p. 919–924.

ARCHER, A.W. AND MAPLES, C.G., 1984, Trace-fossil distribution across a marine-to-nonmarine gradient in the Pennsylvanian of southwestern Indiana: Jour. Paleontology, v. 58, p. 448–466.

ARCHER, R., 1976, *in* Scott, A.C., Edwards, D., and Rolfe, W.D.I., Fossiliferous Lower Old Red Sandstone near Cardross, Dunbartonshire: Proc. Geol. Soc. Glasgow, v. 117, p. 4–5.

BAINES, J.G., 1977, The stratigraphy and sedimentology of the Skipton Moor Grits (Namurian E_{1c}) and their lateral equivalents: [Ph.D. Thesis]: England, University of Keele.

BANDEL, K., 1967, Trace fossils from two upper Pennsylvanian sandstones in Kansas: Paleont. Contributions Univ. Kansas, Paper 18, p. 1–13.

BELT, E.S., 1975, Scottish Carboniferous cyclothem patterns and their paleoenvironmental significance, *in* Broussard, M.L.S. (ed.), Deltas, Models for Exploration: Houston, Texas, Houston Geol. Soc., p. 427–449.

BENFIELD, A.C., 1969. The Huddersfield White Rock cyclothem in the central Pennines: Report of Field Meeting: Proc. Yorkshire Geol. Soc., v. 37, p. 181–187.

BERG, T.M., 1977, Bivalve burrow structures in the Bellvale Sandstone, New Jersey and New York: Bull. New Jersey Acad. Sci., v. 22, p. 1–5.

BINNEY, E.W., 1852, On some trails and holes found in rocks of the Carboniferous strata, with remarks on the *Microconchus carbonarius*: Mem. Lit. and Phil. Soc. Manchester, 2nd Ser., v. 10, p. 181–201.

BISAT, W.S., 1924, The Carboniferous goniatites of the north of England and their zones: Proc. Yorkshire Geol. Soc., v. 20, p. 40–124.

BROADHURST, F.M., SIMPSON, I.M., AND HARDY, P.G., 1980, Seasonal sedimentation in the Upper Carboniferous of England: Jour. Geology, v. 88, p. 639–651.

BROMEHEAD, C.E.N., EDWARDS, W., WRAY, D.A., AND STEPHENS, J.V., 1933, The geology of the country around Holmfirth and Glossop: London, Mem. Geol. Surv. England and Wales, 109 p.

BROMLEY, R.G., AND ASGAARD, U., 1979, Triassic freshwater ichnocoenoses from Carlsberg Fjord, east Greenland: Palaeogeogr., Palaeoclimatol., Palaeoecol., v. 28, p. 703–741.

BRUUN-PETERSEN, J., 1973, "Conical structures" in the Lower Cambrian Balka Sandstone, Bornholm (Denmark), and in the Lower Devonian Coblenz Sandstone, Marburg (West Germany): Neues Jahrbuch für Geologie und Paläontologie Monatshefte, 1973(9), p. 515–528.

CALVER, M.A., 1968a, Coal Measures invertebrate faunas, *in* Murchison, D.G. and Westoll, T.S. (eds.), Coal and Coal-Bearing Strata: Edinburgh, Oliver and Boyd, p. 147–177.

_____, 1968b, Distribution of Westphalian marine faunas in northern England and adjoining areas: Proc. Yorkshire Geol. Soc., v. 37, p. 1–72.

_____, 1969, Westphalian of Britain: Comptes Rendus Sixième Congrès International de Stratigraphie et de Géologie du Carbonifère (Sheffield 1967), v. 1, p. 233–254.

CHAMBERLAIN, C.K., 1971a, Morphology and ethology of trace fossils from the Ouachita Mountains, southeastern Oklahoma: Jour. Paleontology, v. 45, p. 212–246.

______, 1971b, Bathymetry and paleoecology of Ouachita geosyncline of southeastern Oklahoma as determined by trace fossils: Amer. Assoc. Petroleum Geologists Bull., v. 55, p. 34–50.

______, AND CLARK, D.L., 1973, Trace fossils and conodonts as evidence for deep water deposits in the Oquirrh Basin of central Utah: Jour. Paleontology, v. 47, p. 663–682.

CHISHOLM, J.I., 1968, Lower Carboniferous trace fossils from the Geological Survey boreholes in East Fife (1965–6): Bull. Geol. Surv. Great Britain, no. 31, p. 19–35.

______, 1970, *Teichichnus* and related trace fossils in the Lower Carboniferous of St. Monance, Scotland: Bull. Geol. Surv. Great Britain, no. 32, p. 21–51.

______, 1977, *in* Forsyth, T.H. and Chisholm, J.I., The Geology of East Fife: Mem. Geol. Surv. Scotland, Edinburgh, 284 p.

______, Xiphosurid traces, *Kouphichnium* aff. *variabilis* (Linck), from the Namuriam Upper Haslingden Flags of Whitworth, Lancashire: Rep. Inst. Geol. Sci., No. 83/10, p. 37–44.

COLLINSON, J.D., 1967, Sedimentation of the southern part of the Grindslow Shales and the Kinderscout Grit [D. Phil. Thesis]: England, University of Oxford.

______, 1969, The sedimentology of the Grindslow Shales and the Kinderscout Grit: a deltaic complex in the Namurian of northern England: Jour. Sed. Petrology, v. 39, p. 194–221.

______, 1970, Deep channels, massive beds and turbidity current genesis in the Central Pennine Basin: Proc. Yorkshire Geol. Soc., v. 37, p. 495–519.

______, 1976, Deltaic evolution during basin fill—Namurian of the Central Pennines, England (Abstr.): Amer. Assoc. Petroleum Geologists Bull., v. 60, p. 52.

______, AND BANKS, N.L., 1975, The Haslingden Flags (Namurian G_1) of southeast Lancashire: bar finger sands in the Pennine Basin: Proc. Yorkshire Geol. Soc., v. 40, p. 431–458.

______, JONES, C.M., AND WILSON, A.A., 1977, The Marsdenian (Namurian R_2) succession west of Blackburn: implications for the evolution of Pennine delta systems: Geol. Jour., v. 12, p. 59–76.

CORBO, S., 1979, Vertical distribution of trace fossils in a turbidite sequence, Upper Devonian, New York State: Palaeogeogr., Palaeoclimatol., Palaeoecol., v. 15, p. 169–184.

CRIMES, T.P., 1977, Trace fossils of an Eocene deep-sea sand fan, northern Spain, *in* Crimes, T.P., and Harper, J.C. (eds.), Trace Fossils 2: Geol. Jour. Spec. Issue No. 9, Liverpool, Seel House Press, p. 71–90.

______, GOLDRING, R., HOMEWOOD, P., VAN STUIJVENBERG, J., AND WINKLER, W., 1981, Trace fossil assemblages of deep-sea fan deposits, Gurnigel and Schlieren flysch (Cretaceous-Eocene), Switzerland: Ecologae Geologicae Helvetiae, v. 74, p. 953–995.

______, LEGG, I., MARCOS, A., AND ARBONEYA, M., 1977, ?Late Precambrian-low Lower Cambrian trace fossils from Spain, *in* Crimes, T.P., and Harper, J.C. (eds.), Trace Fossils 2: Geol. Jour. Spec. Issue No. 9, Liverpool, Seel House Press, p. 91–138.

______, MARCOS, M., AND PEREZ-ESTAUNA, J., 1974, Upper Ordovician turbidites in western Asturias; a facies analysis with particular reference to vertical and lateral variations: Palaeogeogr., Palaeoclimatol., Palaeoecol., v. 15, p. 169–184.

DAWSON, J.W., 1868, Acadian Geology: London, Macmillan and Co., 694 p.

EAGAR, R.M.C., 1948, Variation in shape of shell with respect to ecological station. A review dealing with Recent Unionidae and certain species of the Anthracosiidae in Upper Carboniferous times: Trans. Royal Soc. Edinburgh, Ser. B., v. 63, p. 130–147.

______, 1952, The succession above the Soft Bed and Bassy Mine in the Pennine region: Geol. Jour., v. 1, p. 23–56.

______, 1956, Additions to the fauna of the Lower Coal Measures of the North-Midlands Coalfields: Geol. Jour., v. 1, p. 328–369.

______, 1961, A summary of the results of recent work on the palaeoecology of Carboniferous non-marine lamellibranchs: Comptes Rendus Quatrième Congrès pour l'Avancement des Études de Stratigraphie et de Géologie du Carbonifère (Heerlen, Netherlands 1958), v. 1, p. 133–149.

______, 1964, The succession and correlation of the Coal Measures of south-eastern Ireland: Comptes Rendus du Cinquième Congrès de la Stratigraphie et de la Geologie Carbonifère (Paris, 1963), v. 1, p. 359–374.

______, 1971, A new section in the Lower Coal Measures (Westphalian A) of Up Holland, near Wigan: Proc. Geol. Assoc., v. 82, p. 71–85.

______, 1974a, Shape of shell of *Carbonicola* in relation to burrowing: Lethaia, v. 7, p. 219–238.

______, 1974b, New non-marine bands and *Carbonicola* burrows in lower Westphalian A measures of the British Pennines: Bulletin de la Société Belge de Géologie, v. 83, p. 205–213.

______, 1975, Neuere Arbeiten über das Westfal im Irland: Zentralblatt für Geologie und Paläontologie, v. 1, p. 555–572.

______, 1977, Some new Namurian bivalve faunas and their significance in the origin of *Carbonicola* and in the colonization of Carboniferous deltaic environments: Phil. Trans. Royal Soc. London, Ser. B, v. 280, p. 535–570.

______, OKOLO, S.A., AND WALTERS, G.F., 1983, Trace fossils as evidence in the evolution of *Carbonicola*: Proc. Yorkshire Geol. Assoc., v. 44, p. 283–303.

EKDALE, A.A., 1977, Abyssal trace fossils in world-wide Deep Sea Drilling Project cores, *in* Crimes, T.P., and Harper, J.C. (eds.), Trace Fossils 2: Geol. Jour. Spec. Issue No. 9, Liverpool, Seel House Press, p. 163–182.

ELLIOTT, R.E., 1968, Facies, sedimentation and cyclothems in Productive Coal Measures in the East Midlands, Great Britain: Mercian Geologist, v. 2, p. 351–371.

FENTON, C.L., AND FENTON, M.A., 1937, *Olivellites*, a Pennsylvanian snail burrow: Amer. Midland Naturalist, v. 18, p. 1079–1084.

FISHER, D.C., 1979, Evidence for subaerial activity of *Euproops danai* (Merostomata, Xiphosurida) *in* Nitecki, M. H. (ed.), Mazon Creek Fossils: Chicago, Academic Press, p. 379–447.

FREY, R.W., AND SEILACHER, A., 1980, Uniformity in marine invertebrate ichnology: Lethaia, v. 13, p. 183–207.

FÜRSICH, F.T., 1974a, Ichnogenus *Rhizocorallium*: Paläont. Zeitschr., v. 48, p. 12–28.

______, 1974b, Corallian (upper Jurassic) trace fossils from England and Normandy: Stuttgarter Beiträger zur naturkunde, Ser. B, no. 13, p. 1–52.

______, AND KENNEDY, W.J., 1975, *Kirklandia texana* Carter—Cretaceous hydrozoan medusoid or trace fossil Chimaera?: Palaeontology, v. 18, p. 665–679.

______, AND MAYR, H., 1981, Non-marine *Rhizocorallium* (trace fossil) from the Upper Freshwater Molasse (Upper Miocene) of southern Germany: Neues Jahrbuch Geologie und Paläontologie Monatschefte, 1981 (6), p. 321–333.

GALL, J.C., 1975, *in* Gall, J.C. and Perriaux, J., La sedimentation continentale du Buntsandstein et les modalités de la transgression marine du Muschelkalk: Neuvième Congrès International de Sédimentologie, Nice. Excursion 8, p. 1–23.

GEVERS, T.W., FRAKES, L.A., EDWARDS, L.N., AND MARZOLF, J.E., 1971, Trace fossils in lower Beacon sediments (Devonian), Darwin Mountains, South Victoria Land, Antarctica: Jour. Paleontology, v. 45, p. 81–94.

GILLIGAN, A., 1920, The petrography of the Millstone Grit of Yorkshire: Quart. Jour. Geol. Soc. London, v. 75, p. 251–294.

GOLDRING, R., AND SEILACHER, A., 1971, Limulid undertracks and their sedimentological implications: Neues Jahrbuch Geologie, Paläontologie Abhandlungen, v. 137, p. 422–442.

GREGORY, M.R., 1969, Trace fossils from the Turbidite Facies of the Waitemata Group, Whangaparoa Peninsula, Auckland: Trans. Roy. Soc. New Zealand: Earth Sciences, v. 7, p. 1–20.

HAKES, W.G., 1976, Trace fossils and depositional environments of four clastic units, Upper Pennsylvanian megacyclothems, northeast Kansas: Univ. Kansas Paleont. Contributions, Art 63, p. 1–46.

______, 1977, Trace fossils in late Pennsylvanian cyclothems, Kansas, *in* Crimes, T.P., and Harper, J.C. (eds.), Trace Fossils 2: Geol. Jour. Spec. Issue No. 9, Liverpool, Seel House Press, p. 209–226.

HALL, J., 1847, Natural History of New York, Paleontology: Albany, New York, v. 1, 338 p.

HALLAM, A., 1970, *Gyrochorte* and other trace fossils in the Forest Marble (Bathonian) of Dorset, England, *in* Crimes, T. P., and Harper, J.C. (eds), Trace Fossils: Geol. Jour. Spec. Issue No. 3, Liverpool, Seel House Press, p. 189–200.

______, 1975, Preservation of trace fossils, *in* Frey, R.W. (ed.), The Study of Trace Fossils: New York, Springer-Verlag, p. 55–64.

______, AND SWETT, K., 1966, Trace fossils from the Lower Cambrian pipe rocks of the north-west Highlands: Scottish Jour. Geol., v. 2, p. 101–106.

HANCOCK, A., 1858, Remarks on certain vermiform fossils found in the Mountain Limestone districts of the north of England: Tyneside Naturalists' Field Club (Transactions), v. 4, p. 17–33.

HÄNTZSCHEL, W., 1962, Trace fossils and problematica, *in* Moore, R.C. (ed.), Treatise on Invertebrate Paleontology, Part W: Lawrence, Kansas, Geol. Soc. Amer. and Univ. Kansas Press, p. W177–245.

______, 1964, Spuren fossilien und Problematica im Campan von Backum: Fortschritte in der Geologie von Rheinland und Westfalen, v. 7, p. 295–308.

______, 1975, Trace fossils and problematica, 2nd revised edit., *in* Teichert, C. (ed.), Treatise on Invertebrate Paleontology, Part W: Lawrence, Kansas, Geol. Soc. Amer. and Univ. Kansas Press, p. W1–269.

HARDY, P.G., 1970a, New xiphosurid trails from the Upper Carboniferous of northern England: Palaeontology, v. 13, p. 188–190.

______, 1970b, Aspects of palaeoecology in arenaceous sediments of Upper Carboniferous age in the area around Manchester [Ph.D. Thesis]: England, University of Manchester.

______, AND BROADHURST, F.M., 1978, Refugee communities of *Carbonicola*: Lethaia, v. 11, p. 175–178.

HAUBOLD, H., AND SARJEANT, W.A.S., 1973, Tetrapodenfährten aus dem Keele und Enville Groups (Permokarbon: Stephan und Autun) von Shropshire und south Staffordshire, England: Zeitschift für Geologisch Wissenschaften, v. 1, p. 895–933.

HECKER, R.T., 1970. Palaeoichnological research in the Palaeontological Institute of the Academy of Sciences of the U.S.S.R., *in* Crimes, T.P., and Harper, J.C. (ed.), Trace Fossils, Geol. Jour. Spec. Issue 3, Liverpool, Seel House Press, p. 215–226.

HEER, O., 1876, Flora fossilis Helvetiae. Die vorwetliche Flora der Schweiz: Zurich, Verlag J. Wurster, 182 p.

HEINBERG, C., 1970, Some Jurassic trace fossils from Jameson Land (East Greenland), *in* Crimes, T.P., and Harper, J.C. (eds.), Trace Fossils, Geol. Jour. Spec. Issue No. 3, Liverpool, Seel House Press, p. 227–234.

______, 1973, The internal structure of the trace fossils *Gyrochorte* and *Curvolithus*: Lethaia, v. 5, p. 227–238.

HORNE, R.R., AND GARDINER, P.R.R., 1973, A new trace fossil from non-marine Upper Palaeozoic red beds in County Wexford and County Kerry, Ireland: Geologie en Mijnbouw, v. 47, p. 131–148.

JESSEN, W., 1949, "Augenschiefer"-Grabgänge, ein Merkmal für Faunenschiefer-Nähe im westfalischen Oberkarbon: Zeitschrift der Deutschen geologischen Gesellschaft, v. 101, p. 23–43.

KELLING, G., AND COLLINSON, J.D., in press, Silesian, *in* Duff, P. McL., and Smith, A.J., The Geology of England and Wales: Edinburgh, Scottish Academic Press.

KENNEDY, W.J., 1975, Trace fossils in carbonate rocks, *in* Frey, R.W., The Study of Trace Fossils: New York, Springer-Verlag, p. 377–398.

KOLB, C.R., AND VAN LOPIK, J.R., 1966, Depositional environments of the Mississippi River deltaic plain, southeastern Louisiana, *in* Shirley, M.L., and Ragsdale, J.E. (eds.), Deltas: Houston, Houston Geol. Soc., p. 17–62.

KSIAZKIEWICZ, M., 1970, Observations on the ichnofauna of the Polish Carpathians, *in* Crimes, T.P., and Harper, J.C. (eds.), Trace Fossils: Geol. Jour. Spec. Issue No. 3, Liverpool, Seel House Press, p. 283–322.

LESSERTISSEUR, J., 1955, Traces fossils d'activité animale et leur significance palaeobiologique: Mémoires de la Société Géologique de France. Paléontologie, Nouvelle Série, v. 74, p. 1–150.

MAYHEW, R.W., 1967, A sedimentological investigation of the Marsdenian grits and associated measures in north-east Derbyshire [Ph.D Thesis]: England, University of Sheffield.
MCCABE, P.J., 1975, The sedimentology and stratigraphy of the Kinderscout Grit group (Namurian R_1) between Wharfedale and Longdendale [Ph.D. Thesis]: England, University of Sheffield.
______, 1978, The Kinderscoutian Delta (Carboniferous) of Northern England: a slope influenced by density currents, *in* Stanley, D.J., and Kelling, G. (eds.), Sedimentation in Submarine Canyons, Fans and Trenches: Stroudsburg, Pa., Hutchinson and Ross, p. 116–126.
MCKERROW, W.S., 1978, ed., The Ecology of Fossils: London, Duckworth, 384 p.
MICHELAU, P., 1956, *Belorhaphe kochi* (Ludwig 1869), Eine Wurmspur im europäischen Karbon: Geologisch Jahrbuch, v. 71, p. 299–330.
MILLER, M.F., 1979, Palaeoenvironmental distribution of trace fossils in the Catskill deltaic complex, New York State: Palaeogeogr., Palaeoclimatol., Palaeoecol., v. 28, p. 117–141.
______, 1982, *Limulicubichnus*: a new ichnogenus of limulid resting places: Jour. Paleontology, v. 56, p. 429–433.
______, AND JOHNSON, K.G., 1981, *Spirophyton* in alluvial-tidal facies of the Catskill deltaic complex: possible biological control of ichnofaunal distribution: Jour. Paleontology, v. 55, p. 1016–1027.
MYERS, A.C., 1970, Some palaeoichnological observations on the tube of *Diopatra cuprea* (Bosc): Polychaeta, Onuphidae, *in* Crimes, T.P., and Harper, J.C. (eds.), Trace Fossils: Geol. Jour. Spec. Issue No. 3, Liverpool, Seel House Press, p. 331–334.
OKOLO, S.A., 1982, A sedimentologic-stratigraphic investigation of Marsdenian (Namurian R_{2a-b}) sediments in the Central Pennines: [Ph.D. Thesis]: England, University of Keele.
OSGOOD, R.G., 1970, Trace fossils of the Cincinnati area: Paleont. Americana, v. 6, p. 281–444.
______, AND SZMUC, E.J., 1972, The trace fossil *Zoophycos* as indicator of water depth: Bull. Amer. Paleont., v. 62, p. 1–22.
POLLARD, J.E., 1969, Three ostracod-mussel bands in the Coal Measures (Westphalian) of Northumberland and Durham: Proc. Yorkshire Geol. Soc., v. 37, p. 239–276.
______, 1976, A problematic trace fossil from Tor Bay Breccias of south Devon: Proc. Geol. Assoc., v. 87, p. 105–108.
______, 1981, A comparison between Triassic trace fossils of Cheshire and south Germany: Palaeontology, v. 24, p. 555–588.
RADWÁNSKY, A., AND RONIEWICZ, P., 1963, Upper Cambrian trilobite ichnocoenoses from Wielka Wissniowka (Holy Cross Mountains, Poland): Acta Palaeontologica Polonica, v. 8, p. 259–280.
RAMSBOTTOM, W.H.C., 1966, A pictorial diagram of the Namurian rocks of the Pennines: Trans. Leeds Geol. Assoc., v. 7, p. 181–184.
______, 1969, The Namurian of Britain: Comptes Rendus Sixième Congrès International de Stratigraphie et de Géologie du Carbonifère (Sheffield 1967), v. 1, p. 219–232.
______, 1979, Rates of transgression and regression in the Carboniferous of NW Europe: Quart. Jour. Geol. Soc. London, v. 136, p. 147–154.
______, 1981, Eustacy, sea-level and local tectonism, with examples from the British Carboniferous: Proc. Yorkshire Geol. Soc., v. 143, p. 473–482.
______, CALVER, M.A., EAGAR, R.M.C., HODSON, F., HOLLIDAY, D.W., STUBBLEFIELD, C.J., AND WILSON, R.B., 1978, Correlation of Silesian rocks in the British Isles: Geol. Soc. Spec. Report, no. 10, p. 1–81.
RATCLIFFE, B.C., AND FAGERSTROM, J.A., 1980, Invertebrate lebensspuren of Holocene flood plains; their morphology, origin and paleoecological significance: Jour. Paleontology, v. 54, p. 614–630.
READING, H.G., 1964, A review of the factors affecting the sedimentation of the Millstone Grit (Namurian) in the central Pennines, *in* Van Straaten, L. M. J. U. (ed.), Deltaic and Shallow Marine Deposits: Amsterdam, Elsevier, p. 340–346.
RICHTER, R., 1937, Marken und Spuren aus allen Zeiten, I-II: Senckenbergiana, v. 19, p. 150–169.
RIDGEWAY, J.M., 1975, A problematical trace fossil from the New Red Sandstone of south Devon: Proc. Geol. Assoc., v. 85, p. 511–517.
ROLFE, W.D.I., 1980, Early invertebrate terrestial faunas, *in* Panchen, A.L. (ed.), The Terrestrial Environment and the Origin of the Land Vertebrates: London, Academic Press, p. 117–157.
SCHÄFER, W., 1962, Actuo-Paläontologie nach studien in der Nordsee: Frankfurt a. M., Verlag Waldemar Kramer, 666 p.
SCOTESE, C.R., BAMBACH, R.K., BARTON, C., VAN DER LOO, R., AND ZIEGLER, A.M., 1979, Paleozoic basemaps: Jour. Geol., v. 87, p. 217–233.
SCOTT, A.C., 1980, Sedimentological and ecological control of Westphalian B plant assemblages from West Yorkshire: Proc. Yorkshire Geol. Soc., v. 41, p. 461–508.
SEILACHER, A., 1963, Lebensspuren und Salinitätsfazies: Fortschritte in der Geologie von Rheinland und Westfalen, v. 10, p. 81–94.
______, 1964, Biogenic sedimentary structures, *in* Imbrie, J. and Newell, N.D. (eds.), Approaches to Paleoecology: New York, John Wiley & Sons, p. 296–316.
______, 1967, Bathymetry of trace fossils: Marine Geology, v. 5, p. 413–428.
______, 1977, Pattern analysis of *Palaeodictyon* and related trace fossils, *in* Crimes, T.P., and Harper, J.C. (eds.), Trace Fossils 2: Geol. Jour. Spec. Issue No. 9, Liverpool, Seel House Press, p. 289–354.
______, 1978, Use of trace fossil assemblages for recognizing depositional environments, *in* Basan, P.B. (ed.), Trace Fossil Concepts, S.E.P.M. Short Course No. 5, p. 167–181.
SHACKLETON, J.S., 1962, Cross strata from the Rough Rock (Millstone Grit Series) in the Pennines: Geol. Jour., v. 3, p. 109–118.
SHELDON, R.W., 1968, Probable gastropod tracks from the Kinderscout Grit of Soyland Moor, Yorkshire: Geol. Mag., v. 105, p. 365–366.

SIMPSON, S., 1970, Notes on *Zoophycos* and *Spirophyton, in* Crimes, T.P., and Harper, J.C. (eds.), Trace Fossils: Geol. Jour. Spec. Issue No. 3, Liverpool, Seel House Press, p. 505–514.

SMITH, E.G., RHYS, G.L., AND EDEN, R.A., 1967, Geology of the Country around Chesterfield, Matlock and Mansfield: London, Mem. Geol. Surv. England and Wales, 433 p.

SORBY, H.C., 1859, On the structure and origin of the Millstone Grit in south Yorkshire: Proc. Yorkshire Geol. and Poly. Soc., v. 3, p. 669–675.

STANLEY, K.O. AND FAGERSTROM, J.A., 1974, Miocene invertebrate trace fossils from a braided river environment, western Nebraska, U.S.A: Palaeogeogr., Palaeoclimatol., Palaeoecol., v. 15, p. 63–82.

STEPHENS, J.V., MITCHELL, G.H., AND EDWARDS, W., 1953, The Geology of the Country between Bradford and Skipton: London, Mem. Geol. Surv. Great Britain, 180 p.

TATE, G., 1859, The geology of Beadnell, in the County of Northumberland, with the description of some annelids of the Carboniferous formation: The Geologist, 1859, p. 59–70.

TAYLOR, B.J., 1967, Trace fossils from the Fossil Bluff Series of Alexander Island: British Antarctic Survey Bulletin, no. 13, p. 1–30.

TRUEMAN, A.E., 1946, Stratigraphical problems in the Coal Measures of Europe and North America. Anniversary Address of the President: Quart. Jour. Geol. Soc. London, v. 102, p. xlix-xciii.

WALKER, R.G., 1964, Some aspects of the sedimentology of the Shale Grit and Grindslow Shales (Namurian R_{1c} of north Derbyshire) and the Westward Ho! and Northam Formations (Westphalian, north Devon) [D.Phil. Thesis]: England, University of Oxford.

______, 1966a, Shale Grit and Grindslow Shales; transition from turbidite to shallow water sediments in the Upper Carboniferous of northern England: Jour. Sed. Petrology, v. 36, p. 90–114.

______, 1966b, Deep channels in turbidite-bearing formations: Am. Assoc. Petroleum Geologists Bull., v. 50, p. 1899–1917.

WEBBY, B.D., 1970, Late Pre-Cambrian trace fossils from New South Wales: Lethaia, v. 6, p. 79–109.

WELLS, A.J., 1955, The development of chert between the Main and Crow Limestones in north Yorkshire: Proc. Yorkshire Geol. Soc., v. 30, p. 177–196.

WILLIAMSON, I.A., AND WILLIAMSON, R.I.H., 1968, Trace fossils from Namurian sandstone, Yorkshire: Geol. Mag., v. 105, p. 562.

WOODLAND, A.W., ARCHER, A.A., AND EVANS, W.B., 1957, Recent boreholes in the Lower Coal Measures below the Gellideg-Lower Pumpquart Coal horizon in South Wales: Bull. Geol. Surv. Great Britain, no. 13, p. 39–60.

WRIGHT, W.B., SHERLOCK, R.L., WRAY, D.A., LLOYD, W., AND TONKS, L.H., 1927, The Geology of the Rossendale Anticline: Mem. Geol. Surv. England and Wales, London, H.M.S.O., 182 p.

YOCHELSON, E.L., AND SCHINDEL, D.E., 1978, A re-examination of the Pennsylvanian trace fossil *Olivellites*: Jour. Res. U. S. Geol. Survey, v. 6, p. 789–796.

TRACE FOSSILS FROM A MIDDLE CAMBRIAN DELTAIC SEQUENCE, NORTH SPAIN

IAIN C. LEGG
Geological Survey of Northern Ireland, Belfast BT9 6BS, Northern Ireland

ABSTRACT

An analysis of the lithofacies and trace fossils from a Middle Cambrian deltaic sequence in the Cantabrian Mountains, northwest Spain, indicates the importance of energy conditions, as evidenced by bed thickness and the presence of sand/mud alternations, in determining the distribution of trace fossils. This facies control of trace fossils is particularly strong for trilobite tracks and U-tubes.

An analysis of *Arenicolites* and *Diplocraterion* demonstrates that the size of the trace fossil can be taken as an indicator of onshore/offshore conditions.

Thirteen ichnogenera are described, including the new ichnospecies *Monomorphichnus pectenensis*, *Rusophycus ramellensis*, and *Teichichnus ovillus*.

INTRODUCTION

This paper describes the lithofacies and the contained trace fossils from the best exposed section at Villar del Puerto of the Middle Cambrian Oville Sandstones and Shales in the Cantabrian Mountains of North Spain (Fig. 1). As such, it forms part of a more detailed study (Legg, 1980) on the sedimentology and ichnology of Spanish Cambrian sediments.

Trace fossils from the Lower Cambrian Herreria Sandstone of North Spain have been previously described by Crimes et al. (1977); those from the overlying Barrios Quartzite were described by Baldwin (1977a). The age of the Oville Sandstones and Shales has been given as late Middle Cambrian to latest Late Cambrian by Lotze and Sdzuy (1961) which confirmed the findings of Comte (1959) who placed the upper limit at the Tremadoc. Baldwin (1976; 1977a) gave an age range for the Barrios Quartzite, based on ichnostratigraphy, of Late Cambrian to Arenig. During the course of this study, no Upper Cambrian trace fossils were found in the Oville Sandstones and Shales; however, Middle Cambrian trace fossils such as *Cruziana barbata* were found throughout the sections, so a Middle Cambrian age is demonstrated.

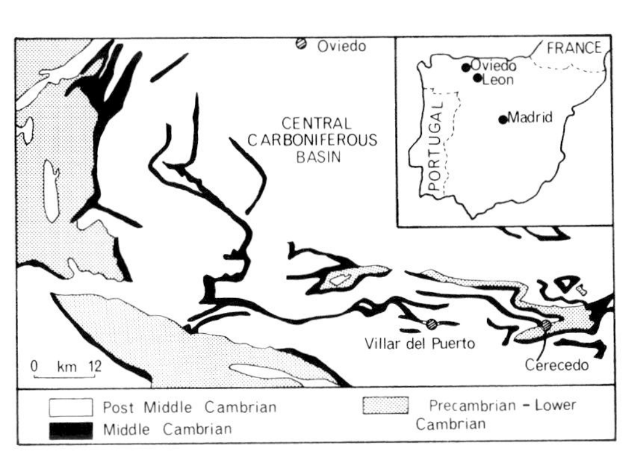

FIG. 1.—Simplified geological map of the southern Cantabrian Mountains with location map (inset).

The Oville Sandstones and Shales at Villar del Puerto overlie the Lancara Formation, which consists of nodular limestones and dolomites, with a transition zone between them. The formations are continuously exposed through 230 m. Only the top 20 m of the sections is poorly exposed.

LITHOFACIES

Eight major lithofacies are recognized in the Villar del Puerto section of the Oville Sandstones and Shales and the characteristics are briefly described below. Recognition of the lithofacies is based on the associations of inorganic sedimentary structures and no account is taken of the contained trace fossils.

Tidal Channel Facies

This facies consists of medium to thickly bedded (11 to 55 cm) laminated and cross-bedded sands which are coarse to medium grained. Commonly mud clasts are found in the lowest few centimeters of the laterally discontinuous beds, with fining upwards sequences poorly developed. Massive bedding, parallel lamination, and planar cross-bedding with herringbone patterns are the dominant sedimentary structures, while bimodal-bipolar paleocurrents are common.

Sand Flat Facies

This facies is characterized by thinly to thickly bedded (5 to 54 cm) sands with minor silts and muds. Planar cross-bedding with herringbone cross-stratification, parallel laminations, abundant small scale wave and current ripples, bimodal paleocurrents and mudcracks indicating exposure, all suggest an intertidal sand flat setting for this facies.

Mixed Flat Facies

Alternations of sand and mud characterize this medium energy intertidal flat facies. These alternations are either coarse (2–5 cm thick) or fine (1–2 mm thick). The coarsely interlayered bedding, much of which can be classified as wavy bedding, or lenticular bedding with connected lenses (Reineck and Wunderlich, 1968), contains ripple bedding and abundant asymmetric current ripples. The finely interlayered bedding contains no structures.

Paleocurrent data demonstrate quadrimodal currents, which have been considered to be indicative of tidal flats (Klein, 1970; 1971).

The juxtaposition of these small scale units (mixed flats) with large scale units (tidal channels) is one of the criteria listed by de Raaf and Boersma (1971) as diagnostic of tidal environments. The presence of mudcracks indicates a probable intertidal situation and the other criteria demonstrate that an intertidal mixed flat origin is likely.

Beach Facies

This facies consists of clean, well-sorted, and well-rounded, fine and medium sands with subordinate silty shales. It is found interdigitating with the intertidal sand and mixed flat facies.

Sedimentary structures include low angled cross-bedding, planar cross-bedding, and parallel lamination, all characteristic of beach sediments. Paleocurrent data indicate trimodal currents.

Parallel-laminated sand is generally abundantly distributed on beaches or other sandy areas exposed to wave action. Reineck (1963) attributed the laminated sand of beaches to the swash and backwash activity of waves.

The low angled cross-bedding and parallel lamination with heavy mineral bands are typical of beach sediments and particularly of the foreshore. Planar cross-bedding can also develop in such a setting with the shoreward migration of longshore bars and ridges.

A mainland beach origin for this facies is unlikely due to the inclusion in the sequences of muds with evidence of tidal deposition and to the close relationship with intertidal flat facies. Gietelink (1973) concluded that this facies could be similar to the shoals in the entries of estuaries, as described by Dörjes et al. (1970) or could represent beaches lying upon tidal flats, since salt marshes were not developed in the Cambrian.

Barrier Beach Facies

This medium to thickly bedded (11 to 55 cm) facies is composed of clean, well-sorted and well-rounded sands. Sedimentary structures include low angled cross-bedding and parallel lamination with trimodal or quadrimodal paleocurrents.

With quadrimodal paleocurrents described from beaches (Davis et al., 1972) and attributed to ridge and runnel activity, a beach origin seems likely. The position of this facies between delta slope and intertidal flat deposits makes a barrier beach origin probable.

Bar Facies

This facies is composed of fine to medium sand and is found intedigitating with shelf and pro-delta deposits. Low angled cross-bedding dipping seawards and high angled planar cross-bedding dipping landwards are the characteristic sedimentary structures. Parallel lamination is only rarely developed. Davidson-Arnott and Greenwood (1974; 1976) described low angled cross-bedding from the seaward slopes of nearshore bars and high angled cross-bedding from landward facing slopes; a feature also common in longshore bars and beach ridges (McKee and Sterrett, 1961; Werner, 1963; Fraser and Hester, 1977). Bar crests may develop parallel lamination during periods of near exposure (Davis et al., 1972), although the preservation potential is not high (Davidson-Arnott and Greenwood, 1974; 1976).

This facies is thus interpreted as representing a nearshore bar. Gietelink (1973) interpreted these deposits as sub-beach, i.e., forming in deeper water than the barrier beach proper.

Delta Slope Facies

The characteristic bedding feature of this facies is the regular alternation of sands and muds. Grain size of the sands is from fine sand at the base of the section to medium sand at the top. The sand percentage varies from 18 to 90 percent and generally increases in an upward direction. The sandstones are very thinly to medium bedded (1 to 15 cm) and become thicker up the sequence. Sands are poorly sorted in general and pass upwards from quartzwackes to quartzites.

The lower part of the sequence has sand percentages of 18 to 70 percent increasing in an upwards direction. Most of the sands have small scale current ripple lamination with some parallel lamination interpreted as forming by settlement from decelerating current flows. Groove casts are commonly developed in this part of the section and reflect the erosive nature of the current flows. Load casts and small scale slumps are also common. Paleocurrent data demonstrate the unimodal nature of the currents.

The upper part of the section tends to have thicker sands (10 to 40 cm) and sand percentages of 70 to 90

FIG. 2.—Distribution of trace fossils and vertical facies relationships in the Villar del Puerto section, Leon.

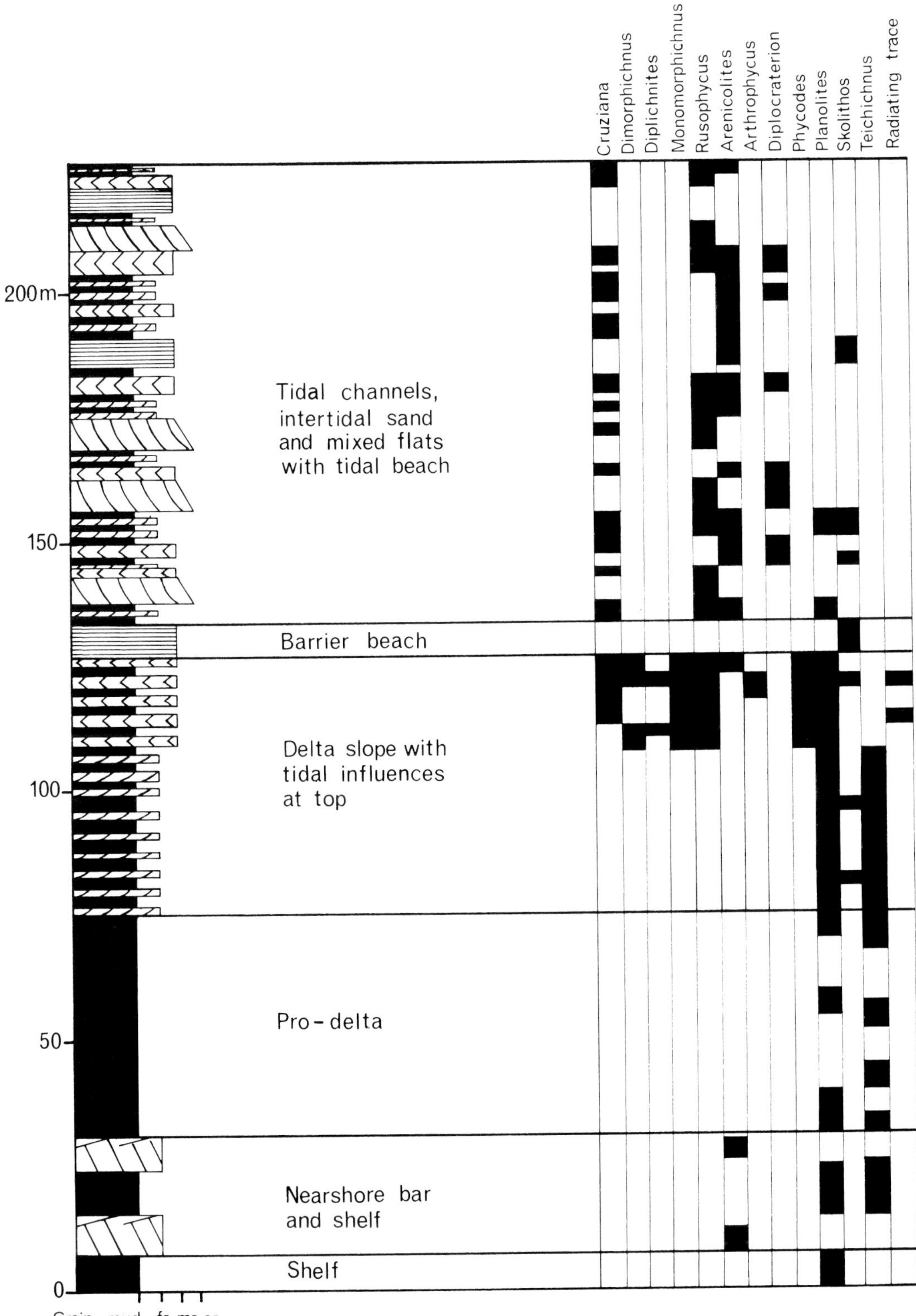
Cruziana
Dimorphichnus
Diplichnites
Monomorphichnus
Rusophycus
Arenicolites
Arthrophycus
Diplocraterion
Phycodes
Planolites
Skolithos
Teichichnus
Radiating trace
200m
150
100
50
0
Tidal channels, intertidal sand and mixed flats with tidal beach
Barrier beach
Delta slope with tidal influences at top
Pro-delta
Nearshore bar and shelf
Shelf
Grain size
mud
fs
ms
cs

percent. Parallel lamination is more common and ripples, both wave- and current-formed, are well developed. Herringbone cross-stratification is found in the upper part of the section.

The generally coarsening upwards nature of this sequence and the unimodal paleocurrents suggest a deltaic origin.

The lower part of the sequence was probably deposited below wave base, and the presence of sand beds suggests a delta slope environment (cf. Allen, 1965; Oomkens, 1967; Vos, 1977; Stanley and Surdam, 1978). Small scale slumps and load casts which indicate rapid deposition on a relatively steep slope are also described from delta slope deposits (Vos, 1977).

The upper part of the sequence contains many wave ripples, and together with current ripples, indicates deposition above wave base. The relative maturity of the sediment, the presence of herringbone cross-bedding and the position between delta slope and barrier beach facies, suggests that deposition occurred on a tidally influenced upper delta slope.

Shelf and Pro-Delta Facies

This facies is composed of silty muds and muds with some small scale current ripple-bedded sands.

The stratigraphic position of these deposits, below the delta slope facies, suggests that a pro-delta environment is likely. The boundary between pro-delta and delta slope deposits is gradual, and near the delta slope, deposits are more silty and may contain ripple bedding (Cherven, 1978). The differentiation between shelf and pro-delta deposits is not easy because the two environments merge into each other.

The lithofacies distribution in the Villar del Puerto section is shown in Figure 2, together with the contained trace fossils.

TRACE FOSSILS

All figured specimens are deposited with the Institute of Geological Sciences, London.

Ichnogenus Cruziana d'Orbigny, 1842
Cruziana barbata Seilacher, 1970 (partim)
Pls. 1A, B; 4F

Lectotype.—As originally defined, the ichnospecies *Cruziana barbata* contains forms which can be assigned to both *Cruziana* and *Rusophycus* (Seilacher, 1970). An occurrence in the Lower Boñar Beds at Cerecedo, Boñar, Spain was selected by Seilacher as the type locality, but no holotype was selected. *C. barbata* is here restricted to the *Cruziana* form, following the precedent set by most authors after 1970 retaining *Rusophycus* for the resting excavations and using *Cruziana* only for the furrows. *C. barbata* specimen number 7 of Figure 7 of Seilacher (1970) is selected as lectotype (Tübingen catalogue number 1392/8).

Description.—Shallow bilobed furrows, preserved on the soles of beds, 3.6 to 8.7 cm wide, with coarse, more or less equal scratches, in bundles of at least three, and making a high V-angle, typically 120° to 160°. Genal spine marks are absent. The trails rarely occur as straight-sided, constant-width trace fossils, and they tend to broaden out in the anterior direction.

Occurrence.—In all occurrence lists the order of lithofacies is in decreasing numbers of trace fossils. *Cruziana* occur in mixed flat, sand flat, tidal delta slope, and tidal channel facies.

Ichnogenus Dimorphichnus Seilacher, 1955
Dimorphichnus sp.
Pl. 2C

Description.—The trail is 8 cm long and 6.8 cm wide, and consists of two parallel series of well-developed ridges 2.5 to 3 cm apart preserved on the sole of a bed. One series consists of short (1.2 to 1.5 cm long) ridges, and the other of long (up to 3.4 cm) ridges. The short ridges are parallel and straight, at an angle of about 50° to the central axis of the trace, and may be arranged in pairs. The long ridges are less regular and are gently curved through 50° towards the center of the trace.

Discussion.—Although much coarser with more widely spaced ridges than most figured specimens of *Dimorphichnus*, this specimen is assigned to that ichnogenus. An alternative interpretation as *Diplichnites* is rejected on the grounds that it is the asymmetry of the trace that is diagnostic. The long imprints (*harksiegel* of Seilacher, 1955) represent raking movements made by the limbs, while the short imprints (*stemmsiegel*) represent support marks, formed during sideways movement of the trilobite.

Occurrence.—Tidal delta slope facies.

Ichnogenus Diplichnites Dawson, 1873
Diplichnites sp.
Pl. 4D

Description.—Two series of ridges preserved on a sandstone sole. The single specimen is 8.5 cm long

EXPLANATION OF PLATE 1

Fig. A.— *Cruziana barbata*. IGS/FOR 4144.
B.— *Cruziana barbata*. Note the grouping of scratches in bundles of four or five in the center of the trace fossil. IGS/FOR 4145.
C.— *Phycodes pedum*. IGS/FOR 4153.
D.— Radiating trace. IGS/FOR 4161.

A
2 cm

B
2 cm

3 cm
C

1 cm
D

A
B
C
2 cm
D
2 cm
E
5 cm

with the series 1 to 2 cm apart. The ridges are up to 2.5 cm long and 0.8 cm wide and tend to be more deeply developed on the outside of the trace fossil with a bulbous portion passing inwards to a narrow ridge.

Discussion.—The specimen has imprints made by only one claw. The bulbous portion represents a deep excavation by the limb which is gradually raised through the sediment before withdrawal. This pattern contrasts with other descriptions, e.g., *Diplichnites* sp. described by Crimes et al. (1977, fig. 5g) in which the bulbous portion is present on the inside of the track. Such a distinction could be the reflection of higher energy conditions in this case, with a deep excavation necessary for the animal to gain a "foothold".

Occurrence.—Tidal channel, sand flat, mixed flat, and tidal delta slope facies.

Ichnogenus MONOMORPHICHNUS Crimes, 1970
Monomorphichnus bilinearis Crimes, 1970
Pl. 2A, B

Description.—The trace fossils, which are preserved on the soles of sandstone beds, comprise a series of eight or nine paired ridges, 2.4 to 5.7 cm long, which are parallel or sub-parallel. Ridges are 0.7 to 1.6 cm apart and may be gently curved. Total lengths of the trace fossils are 5.5 to 11.0 cm and the pattern may be repeated across bedding plane surfaces.

Discussion.—The form of the tracks suggests excavation by animals with eight or nine limbs and which had at least two claws; one claw may be larger than the other to produce the high and low ridges, or the claw could have been turned to give a shallower impression. The repeated forms of *M. bilinearis* indicate that the trace fossil was produced by movement parallel to the current, with the same animal producing all four examples seen on one bedding plane, in the swimming-grazing style, as envisaged by Crimes (1970).

Occurrence.—Mixed flat, tidal delta slope, tidal channel, and sand flat facies.

Monomorphichnus pectenensis n. ichnosp.
Pl. 2C

Name.—From the latin pecten = comb, after the regular nature of the fine scratches.

Holotype.—Institute of Geological Sciences, London (collection no. IGS/FOR 4146).

Type Locality.—From the Villar del Puerto section, 120 m above the base of the section.

Diagnosis.—A set of paired straight ridges with intervening fine, comb-like striations.

Description.—The holotype consists of a single set of three paired ridges, up to 4.3 cm long preserved on the sole of a sandstone bed; each set of ridges is 0.3 cm wide and separated by 0.9 to 1.1 cm. The ridges occur as couplets, with one ridge much larger than the other. Between the ridges are sets of fine parallel striations, up to five in number, less than 1 mm apart, which approach the main ridges at a low angle.

Discussion.—This species differs from the type species *M. bilinearis*, which has no comb-like striations between the sets of paired ridges. *M. multilineatus* Alpert, 1976 consists of a number of striations or ridges, but these are parallel. *M. pectenensis* differs in that the fine striations are parallel to each other but not to the main ridges. This configuration suggests excavation by different appendages. Bergstrom (1973) illustrates a specimen of the trilobite *Cryptolithus* with teleopodites and rake-like exite branches which would be capable of producing the comb-like striations. An alternative explanation that the fine striations were produced by cilia cannot be discounted.

Occurrence.—Tidal delta slope facies.

Ichnogenus RUSOPHYCUS Hall, 1852
Rusophycus radwanskii Alpert, 1976
Pl. 2E

Description.—Specimens are preserved on the soles of sandstone beds and are bilobed trace fossils, 12.7 to 14.5 cm wide, and up to 20.5 cm long. Scratches are highly irregular, but a series of fine scratches in bundles of eight runs parallel to the side of the trace fossil. Other, coarser scratches run across the major depression. The anterior part of the trace, approximately circular, is 4.2 cm deep, and the posterior part, which is weakly bilobed is 1.4 cm deep in the best-preserved specimen.

Discussion.—This trace fossil is comparable to the forms of *Cruziana rusoformis* described by Orlowski, Radwanski, and Roniewicz (1970), being transitional in nature between *Cruziana* and *Rusophycus* and the *R. radwanskii* described by Alpert (1976). It reflects the behavior of the trilobite in first furrowing, forming the cruzianid portion of the trace, and then forming a resting impression. There must have been considerable turning during the production of the resting trace to give such irregular scratches.

EXPLANATION OF PLATE 2

Fig. A.— *Monomorphichnus bilinearis*. Note repitition of the trace fossils across the total bedding plane exposure. Field photograph. Diameter of coin is 1 cm.
B.— *Monomorphichnus bilinearis*. Close-up of part of Figure A. Field photograph.
C.— *Monomorphichnus pectenensis* n. ichnosp. on left, with *Dimorphichnus* sp. on right. IGS/FOR 4146.
D.— *Skolithos linearis*. Burrow tops in a thin sandstone bed. IGS/FOR 4155.
E.— *Rusophycus radwanskii*. IGS/FOR 4148.

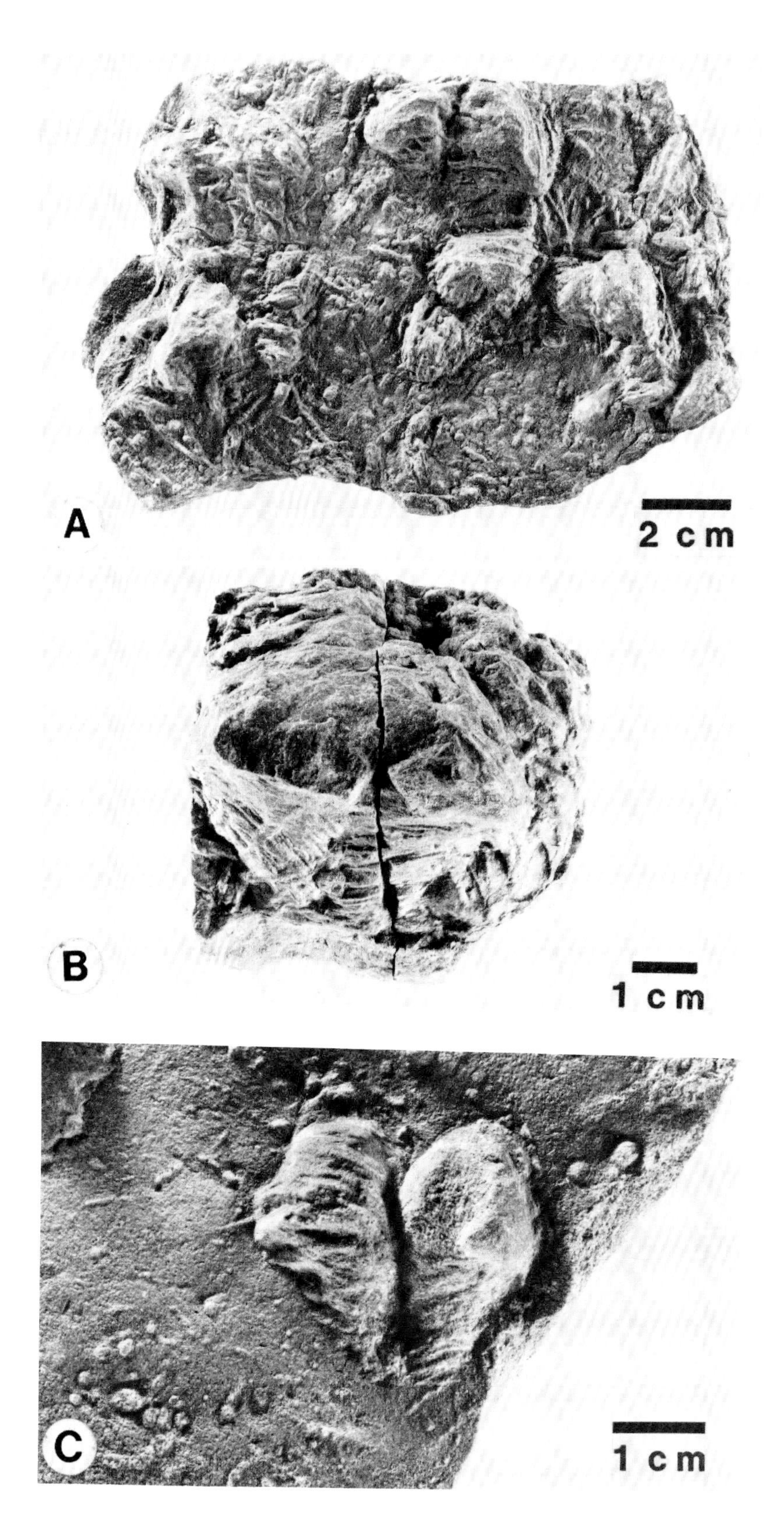
A
2 cm
B
1 cm
C
1 cm

Occurrence.—Tidal channel facies.

Rusophycus ramellensis n. ichnosp.
Pl. 3A, B, C

Cruziana barbata Seilacher, 1970, p. 457, Fig. 7, 6-7 (partim)

Name.—From the type locality which is near the village of Ramellan.

Lectotype.—*Cruziana barbata* Seilacher, 1970 contains both *Cruziana* and *Rusophycus* forms. An occurrence at Cerecedo near Boñar, Leon, was selected as type locality but no holotype was chosen. The rusophycid forms are here removed from the ichnospecies and named *Rusophycus ramellensis* n. ichnosp. The specimen number 6 of Figure 7 of Seilacher (1970) is here selected as lectotype (Tübingen catalogue number 1392/7).

Description.—Shallow to deep, well-defined, bilobed traces, 2.7 to 8.4 cm wide preserved on the soles of sandstone beds. The anterior part of the trace is of two deep lobes meeting medially to give a moustache shape. Each lobe has shallow scratches running parallel to the margin, in bundles of up to 5. Commonly, this is the only part of the trace which is preserved. Where found, the posterior portion of the trace consists of a shallow bilobed excavation, with fine, parallel, curved scratches in groups of 4 or 5, meeting medially at a V angle of about 100°.

Discussion.—The traces conform to Seilacher's figured specimens of *C. barbata* (rusophycid form) with short rear leg markings only commonly preserved under a prominent beard of front leg markings. "Scoop marks" consisting of only part of *R. ramellensis* are commonly found.

Where only the anterior part of the trace if found, *R. ramellensis* could be confused with *R. bonnarensis* Crimes, Legg, Marcos, and Arboleya, 1977, but the latter are generally wider and the scratches coarser.

Occurrence.—Tidal channel, mixed flat, sand flat, and tidal delta slope facies.

Ichnogenus ARENICOLITES Salter, 1857
Arenicolites sp.

Description.—Simple U-tubes, perpendicular to bedding and without spreite. Normally infilled with sand. Burrows are usually only seen at their intersections with the top of sandstone beds where they appear as paired circular outlines. Burrows vary in diameter from 0.8 to 3.9 cm (mean 2.12 cm) and burrow spacings range from 11 to 43 mm with a mean of 31 mm (341 specimens). These traces generally occur crowded on bedding planes.

Occurrence.—Sand flat, mixed flat, tidal channel, bar, beach, and tidal delta slope facies.

Ichnogenus ARTHROPHYCUS Hall, 1852
Arthrophycus sp.
Pl. 4E

Description.—These trace fossils are vertical, straight-walled, unbranched, infilled with fine to coarse sand and developed in beds of silty sandstone. The best-preserved specimen is 2.0 cm wide, 4.8 cm deep, and 12.5 cm long. The trace fossils are strongly annulated with the individual rings up to 2 mm wide separated by 1 to 2 mm. A faint median groove can be traced along 6 cm of the deepest part of the specimen, and along the rest, the rings can be seen to be displaced about the median line. The trace fossil has a series of parallel or near parallel striations on the near vertical walls of the specimen, reflecting internal spreite.

Discussion.—The trace fossils partly resemble *Teichichnus ovillus* but differ in their annular nature. The median line and groove probably indicate production by an animal with bilateral symmetry. *Arthrophycus*, sometimes considered as a junior synonym of *Phycodes* Richter, 1842 (e.g., Seilacher, 1955) typically contains regularly spaced transverse ridges (Häntzschel, 1975). Specimens have been described from the Silurian of New York State by Saarle (1906) where the organism was forced to raise the level of the burrow in response to sedimentation. Such activity in these specimens resulted in the production of the internal spreite.

Occurrence.—Tidal delta slope facies.

Ichnogenus DIPLOCRATERION Torell, 1870
Diplocroterion sp.

Description.—U-burrows, sand infilled within beds of sand, oriented perpendicular to bedding, containing spreite and with funnel-shaped apertures. The burrows are usually only seen at their intersections with bedding planes where they appear as dumbbell shapes with paired circular openings joined by a slit-shaped area of sediment corresponding to the spreite.

Occurrence.—Sand flat, mixed flat, and tidal channel facies.

Ichnogenus PHYCODES Richter, 1842
Phycodes pedum Seilacher, 1955
Pl. 1C; 4C

Description.—Straight to gently curving, sand infilled burrows. The main burrow is preserved on the base of a sandstone bed and has frequent bifurcations which pass as minor branches around the main burrow, and then upwards into the overlying sandstone bed. Burrow widths are 2 to 9 mm.

EXPLANATION OF PLATE 3

Fig. A.— *Rusophycus ramellensis* n. ichnosp. Note that only the anterior scratch marks are preserved. IGS/FOR 4149.
B.— *Rusophycus ramellensis* n. ichnosp. IGS/FOR 4150.
C.— *Rusophycus ramellensis* n. ichnosp. IGS/FOR 4151.

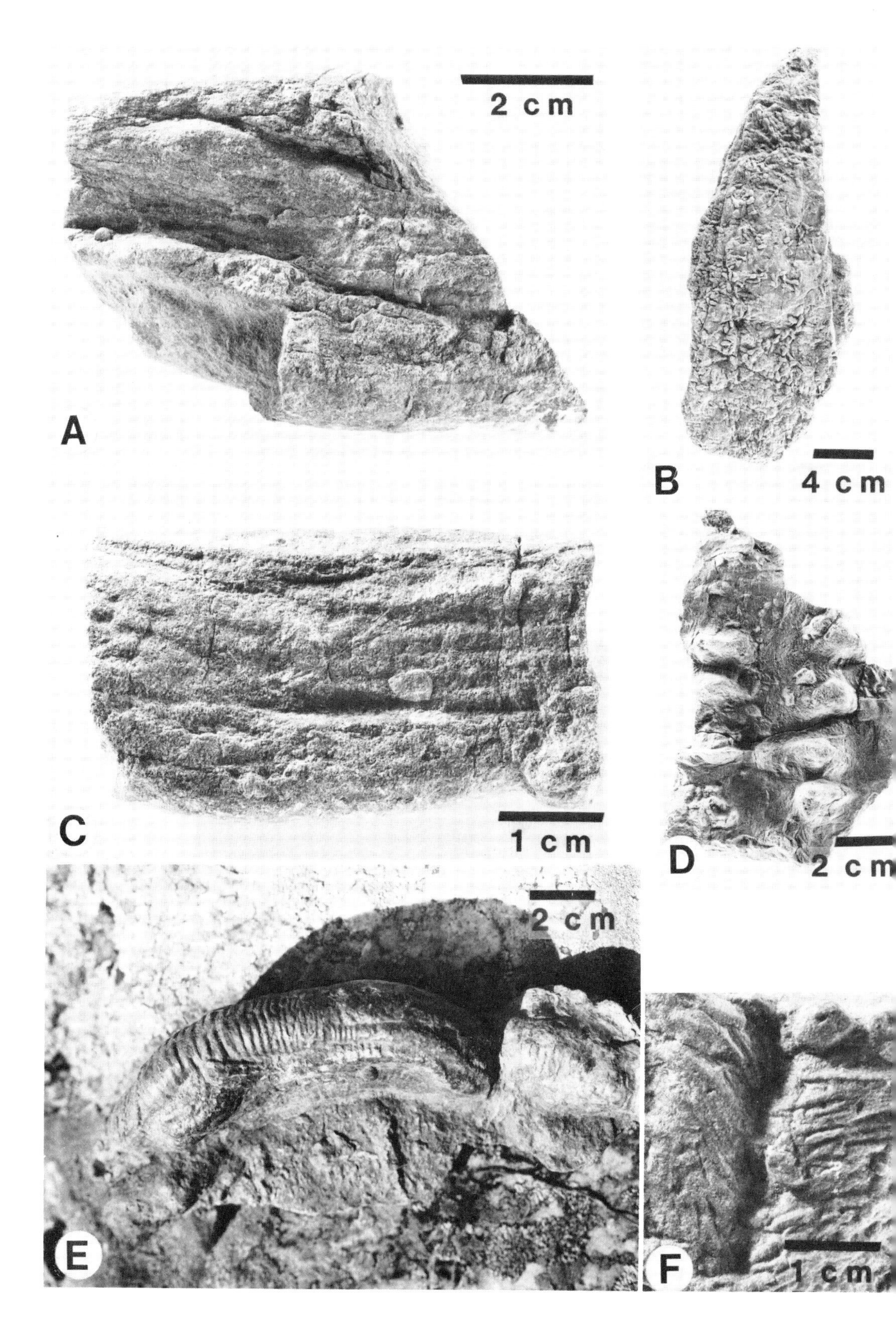
2 cm
A
B
4 cm
C
1 cm
D
2 cm
E
2 cm
F
1 cm

Occurrence.—Mixed flat and tidal delta slope facies.

Ichnogenus PLANOLITES Nicholson, 1873
Planolites sp.

Description.—Simple, unbranched, unornamented, cylindrical burrows, infilled with sand and up to 16 mm in diameter. Parallel to bedding.

Occurrence.—Mixed flat, pro-delta, lower delta slope, tidal delta slope, sand flat, tidal channel, and bar facies.

Ichnogenus SKOLITHOS Haldeman, 1840
Skolithos linearis
Pl. 2D

Description.—Simple vertical burrows which are unbranched, unornamented, and infilled with sand. They occur either closely spaced or isolated on bedding planes. Mostly seen as circular or subcircular outlines on the top surfaces of sandstone beds. Burrow diameters range from 3 to 11 mm.

Occurrence.—Mixed flat, sand flat, beach, barrier beach, tidal channel, lower delta slope, and tidal delta slope facies.

Ichnogenus TEICHICHNUS Seilacher, 1955
Teichichnus rectus Seilacher, 1955
Pl. 4B

Description.—Straight, unbranched, cylindrical, wall-like structure, infilled with medium sand and preserved in silty mudstone. The trace fossils are 1.4 to 2.0 cm wide, 3.0 to 3.2 cm deep and 5.8 to 9.5 cm long. Spreite of a retrusive nature are seen in transverse sections and are reflected as a series of sub-parallel lineations in the straight-sided walls of the specimens.

Discussion.—The morphology of these traces is sufficient to distinguish them from the teichichnids described below, in that they have steep vertical walls and a relatively low width:depth ratio (approximately 1 to 2.1), appearing as a series of stacked vertical burrows.

Occurrence.—Lower delta slope facies.

Teichichnus ovillus n. ichnosp.
Pl. 4A

Name.—From the village of Oville which gives its name to the Oville Sandstones and Shales, and from which the specimens were collected.

Holotype.—Institute of Geological Sciences, London (collection number IGS/FOR 4157).

Diagnosis.—Vertical, wall-like structures, straight to slightly sinuous, unbranched with protrusive or retrusive spreite.

Description.—The specimens are vertical, straight to slightly sinuous, unbranched and wall-like. They are 2.4 to 4.0 cm wide and 2.0 to 5.8 cm deep, and collected specimens are up to 23 cm long, although material examined in the field was up to 80 cm long. Internal spreite are reflected in laminations where they meet the outside wall. Spreite are both protrusive and retrusive.

Discussion.—These teichichnids are infilled with fine to coarse sand and are excavated in beds of silty sandstone and mudstone. They differ from the type species *T. rectus* in being less straight-walled and generally having a higher width:depth ratio (approximately 1 to 1) and the burrows are more circular in outline. As such they have similarities with the large teichichnid burrows described from the Middle Cambrian of Oland by Martinsson (1965) and from the Cambrian Bray Group of Ireland by Crimes (1976). They are also similar to the *T. rectus* described by Baldwin (1977a) from North Spain. Baldwin noted that the spreite are both retrusive and protrusive, whereas the type species has only retrusive spreite. Martinsson (1965, p. 216) described both forms of spreite, and the overall dimensions and structures of the trace fossils, compared to the Spanish material, makes them analogous.

Occurrence.—Lower delta slope, pro-delta, and mixed flat facies.

RADIATING TRACE FOSSIL
Pl. 1D

Description.—Radiating trace fossil, up to 5 cm diameter, preserved on the base of a sandstone bed. It consists of a series of burrows, 0.6 to 0.8 cm wide, radiating outwards from a central area. Up to 15 radiating burrows can be seen, and they appear to overlap, but indifferent preservation prevents detailed examination.

Discussion.—These irregular burrows were probably formed by an animal systematically foraging through the sediment from the central area. As such, it has some similarities with the feeding burrow *Phy-*

EXPLANATION OF PLATE 4

Fig. A.— *Teichichnus ovillus* n. ichnosp. IGS/FOR 4157.
B.— *Phycodes pedum*. IGS/FOR 4154.
C.— *Teichichnus rectus*. IGS/FOR 4156.
D.— *Diplichnites* sp. IGS/FOR 4147.
E.— *Arthrophycus* sp. Field photograph. IGS/FOR 4152.
F.— *Cruziana barbata*. IGS/FOR 4162.

codes and may simply reflect a different response by the same animal. The beds contain an abundance of other traces, mostly *Planolites*, and crowding could have resulted in an animal choosing this method of feeding.

Occurrence.—Tidal delta slope facies.

TRACE FOSSILS AND LITHOFACIES

The relationship between trace fossils and lithofacies is shown in Table 1; the data are supplemented by material from other nearby measured sections in the Cantabrian Mountains. Two major factors should be considered in explaining the distributions of trace fossils-mud content and bed thickness.

Mud Content

The amount of mud found in each of the lithofacies is an important control on the distribution of ichnogenera. The amount of mud can be considered firstly as indicating the energy of the environment. All mud or dominantly mud environments such as the shelf facies can be considered a less active environment than the beach facies with only little mud, and it might be expected that organisms would find it easier to colonize the lower energy area. Secondly, if many of the animals are sediment feeders on organic-rich muds, or suspension feeders on suspended fine grained material, then appreciable quantities of mud will encourage such organisms to thrive. It is important to note that most of the trace fossils described herein are interpreted to have been formed at the sediment-water interface. An alternation of sand and mud would give the best preservation potential (cf. Baldwin, 1976), with the trace fossils usually forming in mud with sand acting as the casting medium.

Monolithologic facies (those consisting only of sand or mud) such as the shelf, bar, and sand flat facies contain fewer numbers of individuals and ichnogenera than the closely related heterolithologic facies composed of alternations of sand and mud, such as the delta slope, mixed flats, and tidal channels.

Bed Thickness

There is a tendency for very thin beds (less than 10 cm) to contain many trace fossils, while the thickest beds (over 90 cm) contain none. In all measured Lower and Middle Cambrian sections, thin beds contain the most trace fossils with a tendency for the percentage of beds with trace fossils to decrease as bed thickness increases (cf. Baldwin, 1977b; Crimes et al., 1977). Thus, facies consisting dominantly of thin-bedded material, such as the mixed flat facies, contain more ichnogenera and individuals than facies with thick beds such as beach and tidal channel facies. This is presumably because thinner beds indicate lower energy where conditions are more favorable for colonization.

Tidal Lithofacies Association

Closely associated lithofacies can best be considered in terms of their association and the relationship between style of sedimentation and trace fossil occurrence. Thus the heterolithologic facies and most thinly bedded facies, the mixed flat, contains more ichnogenera (ten) than the other facies which are essentially monolithologic. Sand flats and tidal channels are transitional with mixed flats but have less mud, thicker beds, and fewer ichnogenera (eight).

Considering the individual ichnogenera, the sand flats are dominated by *Diplocraterion* and *Cruziana*, while mixed flats are dominated by *Arenicolites* and

TABLE 1.—LITHOFACIES AND ASSOCIATED TRACE FOSSILS FROM THE OVILLE SANDSTONES AND SHALES. NUMBERS REFER TO THE NUMBER OF BEDS WITH TRACE FOSSILS, NOT THE NUMBER OF INDIVIDUAL SPECIMENS.

	Tidal Channel	Sand Flat	Mixed Flat	Bar/Beach	Tidal Delta Slope	Lower Delta Slope	Pro-Delta/Shelf
Cruziana	6	20	35	—	6	—	—
Dimorphichnus	—	—	—	—	3	—	—
Diplichnites	2	2	2	—	2	—	—
Monomorphichnus	3	2	12	—	5	—	—
Rusophycus	85	12	57	—	11	—	—
Arenicolites	14	190	129	6	2	—	—
Arthrophycus	—	—	—	—	2	—	—
Diplocraterion	25	348	28	—	—	—	—
Phycodes	—	—	18	—	7	—	—
Planolites	7	16	70	3	18	32	60
Skolithos	2	36	43	26	1	2	—
Teichichnus	—	—	4	—	—	323	50
Radiating trace	—	—	—	—	3	—	—

Rusophycus, with *Rusophycus* and *Diplocraterion* important in the tidal channels. Thus *Diplocraterion* is most important in the higher energy facies and is replaced by *Arenicolites* in the lower energy facies. This would be because in the highest energy environments the animals could retreat into their burrows and build deeper ones when necessary and thus were more adaptable to periods of erosion and deposition. The animals forming *Arenicolites* burrows were presumably less adaptable and would simply vacate their burrows during periods of desication and/or erosion.

The trilobite traces show a similar pattern. The highest energy tidal channels and lowest energy mixed flats are dominated by *Rusophycus*, with the intermediate energy sand flats dominated by *Cruziana*. This implies that the trilobite resting trace is produced in a number of environments, while the furrowing trace is restricted. *Cruziana* declines in numbers from mixed flat, through sand flat to tidal channels, and thus favors the lower energy end of the spectrum, while *Rusophycus* is found in all environments but is concentrated in the highest energy tidal channels. Trilobite swimming traces (*Monomorphichnus*) are more prevalent in the mixed flat facies. Other ichnogenera found in the tidal association (*Phycodes* and *Planolites*) also occur preferentially in lower energy settings.

Deltaic Lithofacies Association

This association comprises the tidal delta slope, lower delta slope, and pro-delta facies. The shallowest water tidal delta slope, which is heterolithologic, contains eleven ichnogenera while the deepest water shelf facies, which is essentially monolithologic, has two ichnogenera; the intermediate lower delta slope has three ichnogenera.

Planolites and *Teichichnus* present in the pro-delta facies are found together with *Skolithos* in the lower delta slope. These *Skolithos* are only found as isolated

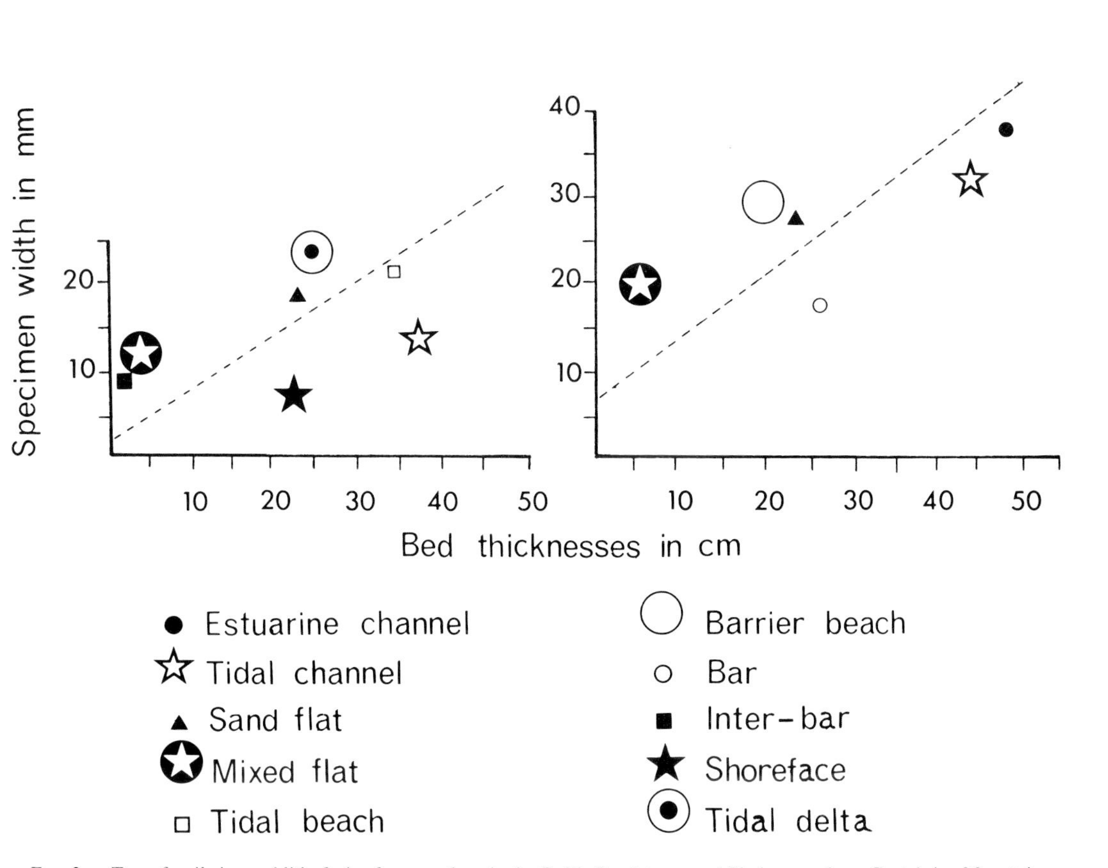

FIG. 3.—Trace fossil size and lithofacies from sections in the Oville Sandstones and Shales, southern Cantabrian Mountains, Leon.

burrows and never occur crowded on bedding planes. The widest ichnospectrum is developed in the tidally influenced delta slope. Here there is a dominance of trilobite traces and complex feeding burrows (*Phycodes*) and the "radiating trace".

TRACE FOSSIL SIZE AND LITHOFACIES

Myers (1970) suggested from a study of modern trace fossils found in the littoral and sublittoral zones, that the depth of burrowing can be taken as an indicator of water depth, with the depth of burrowing decreasing in an offshore direction. Little attention has been paid to the application of this hypothesis to ancient deposits and trace fossils.

To test this hypothesis the U-tubes, *Arenicolites* and *Diplocraterion*, were used in this study. The size of these U-tubes is taken as the distance between the two limbs as exposed on bedding planes. Little data is available to confirm that as the distance between the two limbs increases, the depth of the U-tube also increases, but data presented by Knox (1973) on the U-tube ichnogenus *Corophioides* indicated that this is the case.

For this study, additional information is taken from other Lower and Middle Cambrian sections in the Cantabrian Mountains. In particular, this includes information from other facies, i.e., estuarine channel, shoreface, and interbar facies.

Arenicolites

In the tidal environments, there is a decrease in the width of the U-tube and presumably in the depth of burrowing, from tidal beach through sand flat to mixed flat. This relationship could be related to water depth, although energy levels in general, and bed thickness in particular, are more directly related, with the decrease in specimen width corresponding to a decrease in bed thickness. The data, as shown in Figure 3, show that the onshore tidal facies (tidal channel, sand flat, mixed flat, and tidal beach) have wider traces and deeper burrows than the further offshore facies (interbar and shoreface).

Diplocraterion

Diplocraterion is found in a much narrower range of facies than *Arenicolites* and is most important only in the shallowest water facies. In the tidal facies, the widest traces (mean = 38.0 mm) are found in the shallowest water estuarine channel facies; the traces are 31.1 mm wide in the tidal channel facies, and 27.2 mm wide in the sand flat facies, with the narrowest specimens (20.3 mm) found in the mixed flat facies, which is developed in the shallowest water.

Such a relationship also exists within the wave-dominated facies. The traces are wider in the barrier beach facies and narrower in the more offshore bar facies.

This study therefore illustrates that the depth of burrowing, as reflected in the width of the U-tubes can be taken as an indicator of water depth, and as a guide to onshore or offshore environments.

CONCLUSIONS

This paper illustrates the importance of sedimentological parameters in determining the occurrence of trace fossils in a well-exposed deltaic sequence. Water depth alone is insufficient to explain the observed distributions but other factors, namely mud content and bed thickness, both of which can be used as general indicators of energy conditions, must also be considered. Medium energy environments with medium thick sandstone beds and alternations of sandstone and mudstone contain the most trace fossils, both in terms of diversity and numbers. Such environments would be most conducive to colonization and would also provide the best preservation potential.

ACKNOWLEDGMENTS

Part of this work was funded by the award of a Natural Environment Research Council Research studentship, which is gratefully acknowledged. At all stages, considerable assistance was afforded by T. P. Crimes (University of Liverpool). My thanks also to W. G. Hakes for his critical review of the manuscript. This paper is published by permission of Director, British Geological Survey (NERC).

REFERENCES

ALLEN, J.R.L., 1965, The sedimentation and palaeogeography of the Old Red Sandstone of Anglesey, North Wales: Proc. Yorkshire Geological Society, v. 35, p. 139–185.

ALPERT, S.P., 1976, Trilobite and star-like trace fossils from the White-Inyo Mountains, California: Jour. Paleontology, v. 50, p. 226–239.

BALDWIN, C.T., 1976, The trace fossil stratigraphy of some shallow marine Cambro-Ordovician rocks from Brittany, N.W. Spain and the U.K. [unpub. Ph.D. thesis]: Liverpool, Liverpool University.

______, 1977a, The stratigraphy and facies associations of trace fossils in some Cambrian and Ordovician rocks of northwestern Spain, *in* Crimes, T.P., and Harper, J.C. (eds.), Trace Fossils 2: Geol. Jour. Special Issue No. 9, Liverpool, Seel House Press, p. 9–40.

______, 1977b, Internal structures of trilobite trace fossils indicative of an open surface furrow origin: Palaeogeogr., Palaeoclimatol., and Palaeoecol., v. 21, p. 273–284.

BERGSTRÖM, J., 1973, Organization, life and systematics of trilobites: Fossils and Strata no. 2, Oslo, 69 p.

CHERVEN, V.B., 1978, Fluvial and deltaic facies in the Sentinel Butte Formation, Central Williston Basin: Jour. Sed. Petrology, v. 48, p. 159–170.

COMTE, P., 1959, Recherches sur les terrains anciens de la Cordillere Cantabrique: Mem. Inst. Geol. y Esp., v. 60, p. 1–440.

CRIMES, T.P. 1970, Trilobite tracks and other trace fossils from the Upper Cambrian of North Wales: Geol. Jour., v. 7, p. 47–68.

______, 1976, Trace fossils from the Bray Group (Cambrian) at Howth, Co. Dublin: Geol. Surv. Ireland Bull., v. 2, p. 53–67.

______, LEGG, I.C., MARCOS, A., AND ARBOLEYA, M., 1977, ?Late Precambrian—low Lower Cambrian trace fossils from Spain, *in* Crimes, T.P., and Harper, J.C. (eds.), Trace Fossils 2: Geol. Jour. Special Issue No. 9, Liverpool, Seel House Press, p. 91–138.

DAVIDSON-ARNOTT, R.G.D., AND GREENWOOD, B., 1974, Bedforms and structures associated with bar topography in the shallow water wave environment, Kouchibouguac Bay, New Brunswick, Canada: Jour. Sed. Petrology, v. 44, p. 698–704.

______, AND ______,, 1976, Facies relationships on a barred coast, Kouchibouguac Bay, New Brunswick, Canada, in Davis, R.A., and Ethington, R.L. (eds.), Beach and Nearshore Sedimentation: Soc. Econ. Paleontologists Mineralogists Spec. Pub. No. 24, p. 149–168.

DAVIS, R.A., FOX, W.T., HAYES, M.O., AND BOOTHROYD, J.C., 1972, Comparison of ridge and runnel systems in tidal and non-tidal environments: Jour. Sed. Petrology, v. 42, p. 413–421.

DE RAAF, J.F.M., AND BOERSMA, J.R., 1971, Tidal deposits and their sedimentary structures: Geol. en Mijn., v. 50, p. 479–504.

DÖRJES, J., GADOW, S., REINECK, H.E., AND SINGH, I.B., 1970, Sedimentologie und Makrobenthos der Nordergrunde und der Ausenjade (Norsdsee): Senckenbergiana Maritima, v. 2, p. 31–59.

FRASER, G.S., AND HESTER, N.C., 1977, Sediments and sedimentary structures of a beach-ridge complex, south western shore of Lake Michigan: Jour. Sed. Petrology, v. 47, p. 1187–1200.

GIETELINK, G., 1973, Sedimentology of a linear prograding coastline followed by three high destructive delta complexes (Cambro-Ordovician, Cantabrian Mountains, N.W. Spain): Leidse Geol. Meded., v. 49, p. 125–144.

HÄNTZSCHEL, W., 1975, Trace fossils and problematica, *in* Teichert, C., ed., Treatise on Invertebrate Paleontology: Lawrence, Kansas, Geol. Soc. America and Univ. Kansas Press, Part W, 269 p.

KLEIN, G. DE V., 1970, Depositional processes and dispersal dynamics of intertidal sand bars; Jour. Sed. Petrology, v. 40, p. 1095–1127.

______, 1971, A sedimentary model for determining paleotidal range: Geol. Soc. America Bull., v. 82, p. 2585–2592.

KNOX, R.W.O'B., 1973, Ichnogenus *Corophioides*: Lethaia, v. 6, p. 133–146.

LEGG, I.C., 1980, Sedimentology and trace fossils of Lower and Middle Cambrian clastic deposits of North Spain [unpub. Ph.D. thesis]: Liverpool, Liverpool University.

LOTZE, F., AND SDZUY, K., 1961, Das Kambrium Spaniens: Akad. Wiss. Lit. Abh. Math. Nat. Kl. v. 6, p. 283–693.

MARTINSSON, A., 1965, Aspects of a Middle Cambrian thanatotope on Oland: Geol. For. Stockh. Fohr., v. 87, p. 181–230.

MCKEE, E.D., AND STERRETT, T.S., 1961, Laboratory experiments on form and structure of longshore bars and beaches, *in* Peterson, J.A., and Osmond, J.C. (eds.), Geometry of Sandstone Bodies: Am. Assoc. Petroleum Geologists, p. 13–28.

MYERS, A.C., 1970, Some palaeoichnological observations on the tube of *Diopatra cuprea* (Bosc): Polychaeta, Onuphidae, *in* Crimes, T.P., and Harper, J.C. (eds.), Trace Fossils: Geol. Jour. Special Issue No. 3, Liverpool, Seel House Press, p. 331–334.

OOMKENS, E., 1967, Depositional sequences and sand distribution in a deltaic complex: Geol. en Mijn., v. 46, p. 265–278.

ORLOWSKI, S., RADWANSKI, A., AND RONIEWICZ, P., 1970, The trilobite ichnocoenoses in the Cambrian sequence of the Holy Cross Mountains, *in* Crimes, T.P., and Harper, J.C. (eds.), Trace Fossils: Geol. Jour. Special Issue No. 3, Liverpool, Seel House Press, p. 345–360.

REINECK, H.E., 1963, Sedimentgefuge im Bereich der sudliche Nordsee: Senckenberg. Nat. Ges. Abh., v. 505, p. 1–138.

______, AND WUNDERLICH, F., 1968, Classification and origin of flaser and lenticular bedding: Sedimentology, v. 11, p. 99–104.

SARLE, C.T., 1906, *Arthrophycus* and *Daedalus* of burrow origin: Rochester Acad. Sci. Proc., v. 4, p. 203–210.

SEILACHER, A., 1955, Spuren und Lebensweiss der Trilobiten, *in* Schindewolf, O., and Seilacher, A. (eds.), Beitrage zur Kenntnis des Kambriums in der Salt Range (Pakistan): Jb. Akad. Wiss. Lit. Mainz, no. 10, p. 11–143.

______, 1970, *Cruziana* stratigraphy of non-fossiliferous sandstones, *in* Crimes, T.P., and Harper, J.C. (eds.), Trace Fossils: Geol. Jour. Special Issue, No. 3, Liverpool, Seel House Press, p. 447–476.

STANLEY, K.O., AND SURDAM, R.C., 1978, Sedimentation on the front of Eocene Gilbert-type deltas, Washabie Basin, Wyoming: Jour. Sed. Petrology, v. 48, p. 557–573.

VOS, R.G., 1977, Sedimentology of an upper Palaeozoic river, wave and tide influenced delta system in Southern Morocco: Jour. Sed. Petrology, v. 47, p. 1242–1260.

WERNER, F., 1963, Uber den inneren Aufbau von Strandwallen an einen Kustenabschnitt der Eckernforder Bucht: Meyniana, v. 13, p. 108–121.

ENVIRONMENTAL CONTROL OF TRACE FOSSIL MORPHOLOGY

LARRY W. KNOX AND MOLLY F. MILLER
Department of Earth Sciences, Tennessee Technological University, Cookeville, Tennessee 38505; and Geology Department, Vanderbilt University, Nashville, Tennessee 37235

ABSTRACT

Marked differences in the morphology of modern gastropod trails can be related to sediment consistency. In uncompacted sediment of tidal channels, tidal channel banks, and low tidal flats in Barnstable Harbor, Massachusetts, the gastropod *Polinices duplicatus* produces deep V-shaped grooves as the animal moves several centimeters below the sediment surface. Bilobed, transversely ridged trails are formed in firmer substrate of the high tidal flat where *P. duplicatus* moves nearer the surface. Results of laboratory experiments on *P. duplicatus* moving through various sediment mixtures suggest that grain size and substrate consistency are two controlling factors of trail morphology.

Trails, presumably formed by snails, occur in Pennsylvanian sandstones of tidal origin in Tennessee. Substrate characteristics of the rocks were determined qualitatively by examination of sedimentary features. Comparison of trail morphology with syn-depositional sediment mass properties suggests a close relationship between substrate consistency and trail morphology. In sandstones likely deposited on higher parts of a tidal flat, single trails change from (1) V-shaped furrows to (2) bilobed, transversely ridged trails to (3) longitudinal rows of tiny knobs. These morphologies apparently resulted from the snail moving at decreasing depth in a uniformly firm substrate. In sandstones representing looser sand of lower tidal flat origin the snails produced V-shaped and lobed trails that lack prominent transverse ridges.

Recognition that different trace morphologies can be produced by the same organism is useful for paleoecological interpretation for at least three reasons. (1) The variety of trails produced may be a clue to syn-depositional sediment mass properties. (2) Variety (or degree of variation) of gastropod trail morphology may help identify the tidal flat environment of deposition. (3) Recognition of variation will result in a truer picture of trace producer diversity and development of more accurate compilations of ichnogeneric paleoenvironmental distributions.

INTRODUCTION

For more than a decade trace fossils and associations of trace fossils have been successfully employed in the determination of marine paleoenvironments. For example, Seilacher (1967) has demonstrated the value of trace fossil assemblages as depth indicators. However, as Whittington (1964) has pointed out, the usefulness of fossils for interpretation of any kind depends on sound taxonomy. One problem with trace fossils is that a single animal can produce traces that vary a great deal in morphology. Several authors (e.g., Seilacher, 1953; Osgood, 1975) have noted that the causes for this variability include (among others) behavior of the organism, environmental conditions, and mechanics of preservation. Examples of trace variability due to these factors are common in the literature. Trewin (1976) illustrated the range in variation of *Isopodichnus*, which he ascribed to differences in behavior of differently-sized individuals. Rhoads (1970) pointed out differences in burrows produced in sediments of different mass properties. Osgood (1970) illustrated surface tracks of arthropods that are quite different from imprints of arthropod appendages preserved at depth on cleavage planes. A sound taxonomic framework for trace fossil interpretation can often be made by observation of similar modern organisms and the factors that control the morphology of the traces they produce. This technique has been employed in this paper.

The purpose of this paper is twofold. One objective is to document the variability of trails made by modern gastropods in sediment of different tidal flat subenvironments. Relationships between trail variation and sediment cohesiveness, texture, and composition were determined. Field observations of gastropods in intertidal environments and laboratory experiments with different sediments were utilized in order to accomplish this objective.

The second objective is to describe one type of gastropod trail occurring in Pennsylvanian rocks of Tennessee and interpret the trail variability attributed to (1) organism behavior, and (2) substrate characteristics related to depositional environment. In order to accomplish this objective, syn-depositional mass characteristics of the sediment were determined by sedimentary features, and trails of modern snails were used as analogues for interpretation.

TRAILS OF MODERN SNAILS IN INTERTIDAL ENVIRONMENTS

General

Gastropods are abundant on modern tidal flats. Along the central California coast 23 species of shelled gastropods live on intertidal sand and mudflats (Smith and Carlton, 1975) and 16 species live on soft substrates in intertidal areas of the Woods Hole, Massachusetts, area (Smith, 1964). Many intertidal snails (e.g., *Bulla gouldiana* and *Cerithidea* on the West Coast and *Littorina irrorata* on the East Coast) are grazers and surface deposit feeders. Alexander (1979) found that the grazer *L. irrorata* ingests a variety of

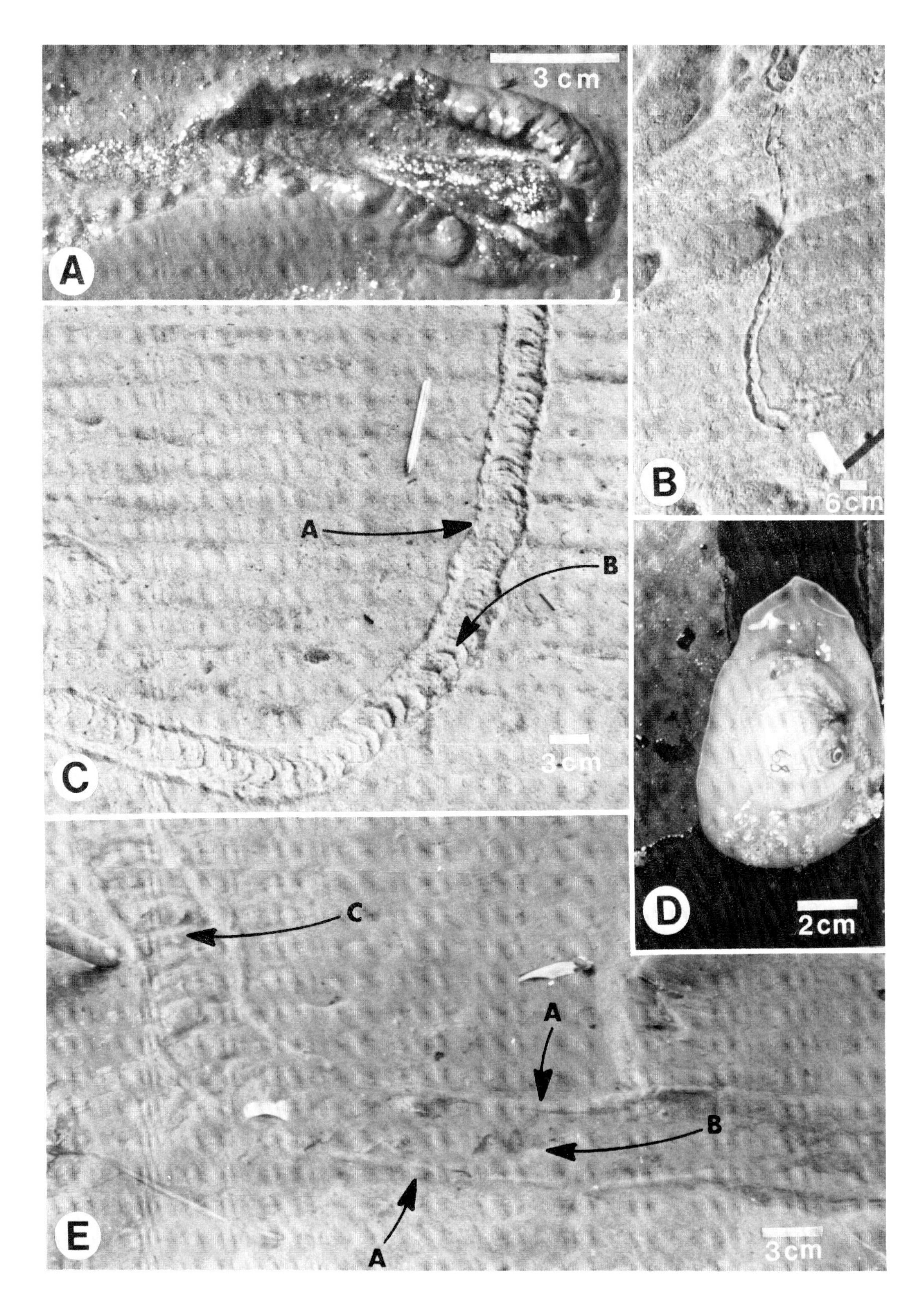
3 cm
A
B
6cm
A
B
3cm
C
D
2cm
C
A
B
A
E
3cm

foods, including organic and inorganic detritus, filamentous algae and diatoms, and interstitial organisms. Others such as *Polinices* are predatory on clams. *Polinices lewisii* moves as much as 30 cm below the surface in search of prey (MacGinitie and MacGinitie, 1949, p. 81) and less commonly travels across the sediment surface. Morphology of traces resulting from snail movement varies as a function of snail behavior and of substrate consistency. Substrate consistency may affect behavior as well as the post-formational response of the sediment to the trace produced. Variations in gastropod trail morphology observed at Barnstable Harbor, Massachusetts and Mugu Lagoon, California are described below.

Barnstable Harbor

Trails of the predatory snail *Polinices duplicatus* (Pl. 1B-E) have been observed at Woods Hole, Barnstable Harbor, a shallow bay in the northern part of Cape Cod, Massachusetts. Approximately 80 percent of the bay is covered by salt marsh, consisting of peat marsh islands, marsh margins with tidal channels, and sand flats (Redfield, 1972). The morphology of trails made by *P. duplicatus* in tidal channel and sand flat sediments is highly variable.

Sand in tidal channels, channel banks, and lower tidal flats is saturated with water and loose in consistency. *P. duplicatus* produces deep (1–2 cm), generally smooth, V-shaped trails (Pl. 1B) as it moves several centimeters below the substrate/water interface searching for prey. Loose sediment on the lateral margins of the trail collapses inward after passage of the snail. This post-formational collapse results in a narrowing of the bottom of the trail, producing a V-shaped transverse profile. The trail morphology is caused both by the gastropod moving beneath the sediment and by the character of the substrate.

On the high sand flat the sediment is not water-saturated and is firmer than in the tidal channels and their banks. The morphology of the snail trails reflects this difference. Here, trails of *P. duplicatus* are shallow (<1 cm), wide (4 to 6 cm), and bilobed, with prominent, evenly spaced, transverse ridges (Pl. 1C). Because the substrate is firm, the gastropod probes the sediment, thereby producing transverse ridges. The trail is not affected by collapse in this firm substrate. Trails with indistinct transverse ridges were observed in tidal channels and lower parts of the sand flat (Pl. 1E, left part). It is unclear whether the absence of distinct ridges in the looser sediment in and near the tidal channels is due to partial obliteration by post-formational sediment collapse or whether the probes by the gastropods produced initially less distinct ridges.

The most common trail variety consists of two lateral ridges (<1 cm in height) separated by a broad (1 to 6 cm wide) valley. The resulting transverse profile is U-shaped (Pl. 1E, right side). Trails with this morphology are found in the channels, channel banks, and less commonly on the sand flat. The U-shaped morphology is intermediate in form between the V-shaped and the transversely ridged varieties.

Some trails change morphology along their length (i.e., from U-shaped to transversely ridged) (Pl. 1E). Changes in morphology that occur in close proximity are more likely caused by differences in behavior than by substrate differences because there is probably little difference in physical character of the substrate within distances of a few centimeters.

Transversely ridged and V-shaped trails produced by *P. duplicatus* are end members of a morphological continuum. The end members represent trail morphologies made in sediment types differing in degree of water saturation and degree of compaction. In firm sediment the snails move near the surface and probe the sediment to produce broad furrows with clearly preserved transverse ridges, whereas in loose, water-saturated sediment they move deeper in the sediment and produce unornamented furrows whose outlines are further modified by post-formational collapse of the loose sediment. Here, sediment consistency affects trace morphology by influencing behavior and by partially controlling trace preservation.

Mugu Lagoon

Snails are abundant in Mugu Lagoon, a small, tidally dominated barrier system in southern California. The physiographic and sedimentologic features,

EXPLANATION OF PLATE 1

Fig. A.—Bilobed trail of gastropod *Cerithidea californica* on water-saturated sediment in artificial enclosure on tidal sand flat, Mugu Lagoon, California.

B.—V-shaped trail of gastropod *Polinices duplicatus* on loose sediment of lower part of tidal sand flat, Barnstable Harbor, Massachusetts.

C.—Bilobed trail of *Polinices duplicatus* on firm sediment of higher part of tidal sand flat, Barnstable Harbor. A = lateral ridge; B = transverse ridge.

D.—*Polinices duplicatus* in motion, foot is 80 percent extended.

E.—U-shaped trail of *Polinices duplicatus* on sediment of intermediate compaction from middle part of tidal sand flat, Barnstable Harbor. A = lateral ridge; B = U-shaped valley; C = indintinct transverse ridge.

and faunal communities of Mugu Lagoon have been described by Warme (1971). The epifaunal deposit-feeding snail *Cerithidea californica* lives in profusion on the sediment surface of the lower marsh; it is also common in muddy areas of the sand flat and channel margins. It produces a barely discernible trail on the compacted muddy sand of the marsh. In uncompacted muddy sediment of the channel margin its trail is a poorly defined groove. Although its behavior was no different, *C. californica* produced a strikingly different trail (Pl. 1A) in water-saturated sediment of approximately the same texture as that deposited on the sand flat in an enclosure being used to test the escape response of infaunal thalassinid shrimp. The trail was deep (approximately 1 cm), V-shaped, and marked by prominent lateral lobes (Pl. 1A), which are lacking in trails produced in similarly-sized sediment of the marsh or channel margin. For this trail, substrate consistency clearly was the primary control of morphology.

EXPERIMENTAL STUDIES

Purpose and Procedure

Aquaria experiments on *Polinices duplicatus* were conducted at the Marine Biological Laboratory, Woods Hole, Massachusetts. Snails were placed on different sediment mixtures of varying grain size and composition, and the variation in morphology of their trails was documented under controlled conditions.

Each of four mixtures of sediment was placed in a separate aquarium tank. Each tank was filled with sea water and, with one exception, allowed to sit overnight. Water temperature just above the sediment, percent oxygen saturation of water within one centimeter above the sediment, and percent oxygen saturation of the interstitial water were monitored continuously. Temperature and oxygen saturation values varied little during the experiment. Therefore, no behavior differences could be attributed to these parameters.

Three sediment types were used in the experiments. One sediment type (MBL) was predominantly medium-grained quartz sand from the beach at the Marine Biological Laboratory (Table 1). A second sediment type (BH) was predominantly fine-grained quartz sand from a tidal channel at Barnstable Harbor (Table 1). The third sediment type (LSM) was muddy sand from Little Sippewissett Marsh, located about 13 kilometers (8 miles) north of Woods Hole. LSM consisted of sediment smaller than 0.25 mm, contained a large percent of organic material, emitted a strong odor of hydrogen sulfide, and had a large water content.

Using the three sediment types, four different sediment mixtures of varying composition, texture, and degree of compaction were placed in the aquaria (Table 2). A snail was placed in each aquarium for approximately 15 minutes, and its trail was measured after removal; a total of ten snails was placed in each aquarium. To the extent possible, the same ten snails were used in each tank. The following parameters (Fig. 1) of the trails were measured: A) total width of trail, B) minimum width of bottom of trail, and C) depth from middle of trail to top of lateral ridge (maximum depth = MD). If the trails varied in shape the extremes of shape were measured.

Results and Discussion

Two parameters of the trails—maximum depth (MD) and the ratio of minimum bottom width to total width (MBW/TW) (Table 2)—are most significant because they appear to be independent of snail size. The deepest trails were produced in the softest sediment (type 1). Maximum depth decreased with increasing firmness of substrate, and the shallowest trails were produced in the firmest substrate (type 4). V-shaped trails have small MBW/TW ratios; U-shaped trails have large MBW/TW ratios. The most V-shaped trails were produced in firm, type 4 sediment. Trails of intermediate depth and shape were produced in types 2 and 3 sediments, which were of intermediate degree of compaction. The smaller MBW/TW ratio in the softer (and coarser) sand is a function of the greater depth at which the snails moved and of the increased tendency of the lateral ridges to collapse toward the central portion of the trail after passage of the snail.

Five of the ten trails made in type 4 sediment had transverse ridges, which apparently resulted from the

TABLE 1.—GRAIN SIZE DISTRIBUTION (PERCENT OF TOTAL WEIGHT) FOR BARNSTABLE HARBOR (BH) AND MARINE BIOLOGICAL LABORATORY (MBL) SANDS USED IN AQUARIA EXPERIMENTS.

Phi (ϕ) class interval	Wentworth size class	Percent of total weight	
		BH sand	MBL sand
0–1	coarse sand	3.1	9.3
1–2	medium sand	7.3	83.2
2–3	fine sand	79.8	7.4
3–4	very fine sand	6.9	0.1
>4	silt and clay	2.9	trace
		100.0	100.0

snails' attempt to move more deeply into the compacted sediment. Transverse ridges were never produced in the softer types 1, 2, or 3 sediments.

The experiment suggests that relative differences in substrate, which affect trail morphology, are directly related to sediment size. It is reasonable to expect variation in both modern and ancient gastropod trail morphology due to sediment size variation.

SIMILAR TRAILS OF PENNSYLVANIAN AGE

Introduction

Trails identified as *Palaeobullia* have been found in Pennsylvanian sandstones and mudstones of the northern Cumberland Plateau, Tennessee. These rocks represent deposition in a tidal flat environment. The morphology of the trails varies considerably. Differences in morphology that occur along a single trail resulted primarily from differences in behavior of the producer. Other variation, however, can be attributed to differences in sediment characteristics related to subenvironments within the tidal flat depositional environment. Interpretations of the effects of behavior and environmental control of morphology of the fossil trails have been enhanced significantly by comparison with trails of modern snails in intertidal environments.

Stratigraphic Setting

The trails occur on large bedding plane surfaces (Pl. 2G) of the uppermost part of the Fentress Formation (Lower Pennsylvanian) of Tennessee (see Miller and Knox, Fig. 2, this volume, for stratigraphic nomenclature). The trails were exposed by quarry operations at four localities (3a, 3b, 4, and 5) between 2½ and 4¼ kilometers west of Fairview, Tennessee (see Miller and Knox, Fig. 1 and Appendix, this volume, for index map and outcrop locations). The Fentress Formation and the overlying Rockcastle Conglomerate belong to the Crab Orchard Mountains Group, which represents

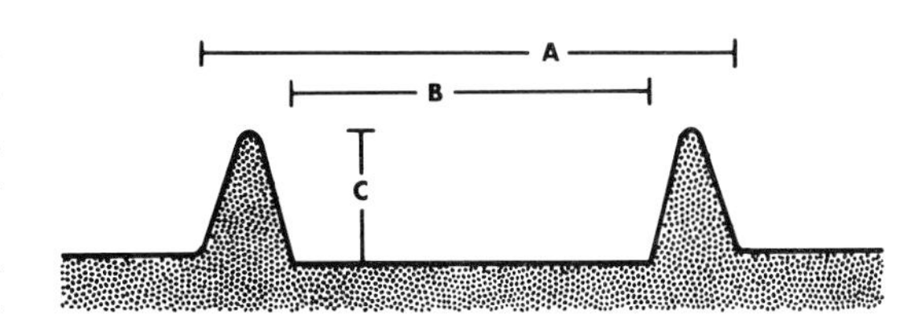

Fig. 1.—Line drawing of transverse vertical section showing orientation of measurements on surface of *Polinices duplicatus* trails. A = width of entire trail; B = width of bottom of trail; C = depth from middle of trail to top of lateral ridge.

TABLE 2.—SEDIMENT TYPE, SEDIMENT MIXTURE, SEDIMENT COMPOSITION AND TEXTURE, RELATIVE DEGREE OF COMPACTION, MEAN ($\bar{x}$) AND STANDARD DEVIATION (σ) FOR MAXIMUM DEPTH (MD), MEAN ($\bar{x}$) AND STANDARD DEVIATION (σ) FOR MINIMUM BOTTOM WIDTH/TOTAL WIDTH (MBW/TW), AND MORPHOLOGICAL VARIETY FOR TRAILS OF *POLINICES DUPLICATUS* FORMED IN AQUARIA.

AQUARIA SEDIMENTS				TRAIL CHARACTERS				
				Parameters				
Type	Mixture	Composition and texture	Relative degree of compaction	MD (mm) $\bar{x}$	MD (mm) σ	MBW/TW (ratio) $\bar{x}$	MBW/TW (ratio) σ	Morphological variety
1	100% MBL	medium quartz sand	loosely compacted	9.3[1]	3.0	0.34[2]	0.26	deep, V-shaped, no transverse ridges
2	75% MBL 25% LSM	medium quartz sand and organic mud	intermediate	6.1[1]	2.2	0.36[3]	0.24	intermediate, no transverse ridges
3	50% MBL 50% LSM	medium quartz sand and organic mud	intermediate	5.3[1]	3.7	0.57[3]	0.30	intermediate, no transverse ridges
4	100% BH	fine quartz sand	firm	2.9[1]	2.2	0.62[3]	0.22	shallow, U-shaped, often with transverse ridges

[1]Based on 10 measurements
[2]Based on 9 measurements
[3]Based on 8 measurements

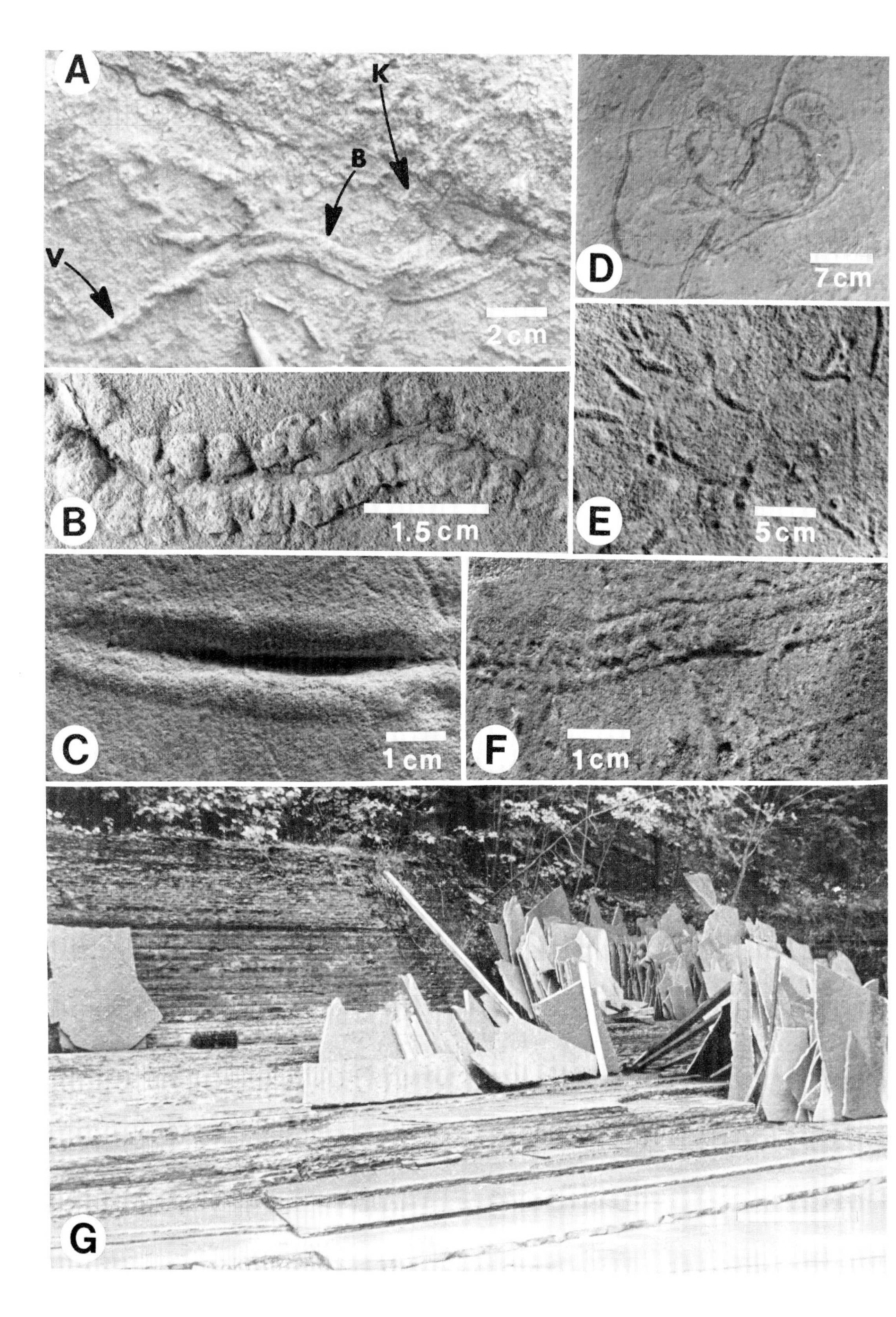
A
K
B
V
2 cm
D
7 cm
B
1.5 cm
E
5 cm
C
1 cm
F
1 cm
G

a suite of regressive littoral sedimentary environments, including barrier and associated tidal delta, tidal flat, and lagoon paleoenvironments (Milici, 1974; Miller and Knox, this volume).

Depositional Environment

Description.—Exceptionally well-preserved specimens of *Palaeobullia* are found within a 4-meter sequence (very thinly bedded sandstone facies of Miller and Knox, this volume) of interbedded quartz sandstones and dark mudstones with sharp contacts (localities 3a and 3b). The sandstones (0.3 to 10 cm thick) are very fine to fine grained, contain as much as 23 percent matrix (based on thin section analysis of three samples), and often contain abundant comminuted plant debris. The sandstone beds are composed of horizontal laminae.

The most prominent sedimentary structure of the thinly bedded sandstone facies is parallel horizontal stratification (Pl. 2G). Individual sandstone beds at locality 3a can be followed for at least 50 meters. Other sedimentary structures include low amplitude (2 to 3 mm) interference ripples and occasional mud cracks. Neither scour marks nor tool marks are present on the mudstones.

At locality 4 (about 100 meters from 3b) the mudstone interbeds are usually absent, interference ripples are more common, oscillation ripples occur rarely, and flaser bedding occurs rarely. There is less matrix in the sandstones.

At locality 5, about 1½ kilometers west of locality 3a, the upper part of the Fentress Formation consists of thinly to very thinly bedded (mostly 1 to 4 cm thick), fine to very fine grained sandstones and coarse siltstones (rippled sandstone-siltstone facies of Miller and Knox, this volume). Mudstone interbeds occur only rarely, and the sandstones contain less matrix than sandstones of localities 3a and 3b. Flat topped and straight, sharp-crested oscillation ripples are abundant. Interference ripples are common.

Interpretation.—Both rocks of the thinly bedded sandstone facies (localities 3a, 3b, and 4) and the sandstone-siltstone facies (locality 5) were deposited in a tidal flat paleoenvironment (Miller and Knox, this volume). Rocks of these four localities represent deposition under different intensities of wave and current energy within the tidal flat.

The interbedded sandstones and mudstones with sharp contacts (localities 3a and 3b) likely represent episodic bed load and suspension deposition. Such mixed lithologies are characteristic of the mid-tidal flat (Klein, 1977). The presence of mud cracks indicates occasional desiccation of these beds.

At locality 4 the reduced thickness or absence of mudstone interbeds, less matrix in the sandstones, common occurrence of interference ripples, rare oscillation ripples, and rare flaser bedding indicate higher energy of deposition. These beds were apparently deposited lower on the tidal flat than those of localities 3a and 3b. The decreased mud content suggests that these sediments were less firm during deposition than those of localities 3a and 3b.

The sandstones of locality 5 reflect deposition under the highest energy conditions of the four localities. The sandstones are clean, mudtone interbeds occur only rarely, and ripples are more abundant than at the other localities. These sandstones were apparently deposited lower on the tidal flat than the sandstones of the other localities. Because of their location they likely had a higher water content. The lack of mud and probable higher water content imply that these sediments formed the softest substrates of the four localities.

Trails

Sandstones exposed at the quarries contain tracks, trails and burrows (Miller and Knox, this volume). Vertical burrows are rare and few laminae have been disturbed by burrowing, but surface trails are common. One of the most common traces is a meandering and looping surface trail preserved in concave epirelief on the mudstones and convex hyporelief on the sandstones. The trails (Pl. 2A, D-F) most closely resemble the *Palaeobullia* type of the *Scolicia* group (Häntzschel, 1975). These trails are usually about 10 mm wide, but a few are as small as 5 mm in width. We are confident that they represent trails produced at or near the surface because there is no disruption of sandstone laminae within or above the trail, which would have

EXPLANATION OF PLATE 2

All figures from the Fentress Formation (Pennsylvanian)

Fig. A.— *Palaeobullia* showing characteristic change in morphology along trail length. Sole of bed, locality 3a. V = V-shaped portion (deepest movement); B = bilobed, transversely ridged portion; K = rows of tiny knobs (shallowest movement).

B, C.— Bilobed and V-shaped trails that may have been formed by snails that produced *Palaeobullia*. Tops of beds, locality 4.

D.— Meandering *Palaeobullia* on sole of bed, locality 3a.

E.— V-shaped form of *Palaeobullia* on top of bed, locality 5.

F.— Flat-bottomed form of *Palaeobullia* on top of bed, locality 4.

G.— Lower part of quarry sequence at locality 3b showing thin, dark mudstone interbeds on surfaces of laminated sandstone beds. *Palaeobullia* preserved in concave epirelief on mudstones and convex hyporelief on sandstones.

occurred if the traces were endogenic. It is doubtful that the traces were preserved by subsequent erosion and filling of endogenic burrows because it is highly unlikely that such a process would have preserved thin, parallel, horizontal beds of large lateral dimensions and not have produced scour marks or tool marks.

The Fentress Formation specimens of *Palaeobullia* likely represent gastropod trails for several reasons. Surface trails of gastropods often are bilobate with transverse markings (Schäfer, 1972, p. 266). The trails tend to cross earlier-formed loop portions, and the maker generally avoided traveling along previously formed parts of its own trail-traits characteristic of gastropods (Schäfer, 1972, p. 267). Looped, unbranched trails are typical of grazing trails of gastropods (Seilacher, 1953, p. 442).

The *Palaeobullia* is similar to traces generally believed to be traces of arthropods (i.e., *Isopodichnus* and small *Cruziana*). However, some varieties of the *Palaeobullia* (V-shaped grooves and rows of knobs) are not typical of the most similar types of arthropod traces (*Cruziana, Isopodichnus, Pteridichnites, Oniscoidichnus*). Burrowing traces of arthropods (*Cruziana, Isopodichnus*) are commonly associated with walking traces (*Diplichnites*) and resting traces (*Rusophycus*)(Häntzschel, 1975). No traces that resemble *Diplichnites* or *Rusophycus* have been found associated with *Palaeobullia*. *Cruziana* has a herringbone pattern consisting of sets of transverse scratch marks, each set overlapping and truncating the previous set (Birkenmajer and Bruton, 1971), whereas the transverse ridges of the Fentress Formation *Palaeobullia* are never truncated. Furthermore, specimens of *Palaeobullia* from the Fentress Formation are 1) wider than *Isopodichnus* (Häntzschel, 1975), 2) wider than and lack the alternating pits of *Pteridichnites* (Clark and Swartz, 1913), and 3) lack the footprints associated with *Oniscoidichnus* (Brady, 1947). Bilobed traces of small (less than 1 cm wide) trilobites and other arthropods, worms and gastropods are similar and very difficult to distinguish (Seilacher, 1970, p. 456). Trewin (1976) pointed out that where well-preserved material occurs illustrating a range of activity, distinction between *Cruziana* and *Isopodichnus* can probably be made. The well-preserved *Palaeobullia*, because of the range of activity represented by its variation, can be distinguished from *Cruziana* and *Isopodichnus*.

Variation in Palaeobullia.—Variation in *Palaeobullia* morphology occurs because of behavior and differences in substrate characteristics. Differences caused by behavior were produced in the thinly bedded sandstone facies at localities 3a and 3b. There, single trails of *Palaeobullia* sometimes change from V-shaped, smooth furrows to bilobed trails with evenly spaced, oblique, transverse ridges, and finally to four longitudinal rows of tiny knobs (Pl. 2A) within a few centimeters. The gastropod always moved near the surface, but the different morphologies were produced, apparently, by movement at slightly decreasing depth in the substrate. Single trails with these different morphologies are common. We are confident that they resulted from the activity of the gastropod and not from exposure of different bedding plane surfaces. The varieties occur so close together that it is highly unlikely they could have been caused by differences in substrate characteristics.

Other variation in *Palaeobullia* morphology was controlled by substrate consistency. The bilobed, transversely ridged trail is by far the most common variety produced on the originally firm substrate at localities 3a and 3b. The looser substrates represented by the sandstones at locality 4 resulted in *Palaeobullia* with flat bottoms and less distinct transverse ridges (Pl. 2F; Miller and Knox, Pls. 1D, 2F, this volume). Other trace varieties at locality 4 include V-shaped (1 to 2 cm wide, $\approx$0.4 cm deep) trails (Pl. 2C) and distinctly lobed (1 to 2 cm wide) trails lacking transverse ridges (Pl. 2B). These trails are very similar to modern gastropod trails of *Polinices* (Pl. 1B) and *Cerithidea* (Pl. 1A), respectively, which were produced on soft substrates. The V-shaped and lobed trails of locality 4 may represent additional varieties of trails made by the snails that produced *Palaeobullia*. *Palaeobullia* at locality 5 was formed in the softest substrate. These trails are V-shaped (about 15 mm wide) and the deepest (5 to 6 mm) of all the *Palaeobullia* form the Fentress Formation.

No body fossils of gastropods or any other invertebrate have been found in the rocks of these four localities.

Relationship of Ancient and Modern Trails

In view of our observations of both modern and ancient intertidal gastropods, the most critical factor in the control of trail morphology appears to be stability of the substrate. In firm substrate of the higher sand flat, both modern and ancient gastropods commonly produced shallow trails with prominent transverse ridges. In soft substrate of the lower sand flat, the modern and ancient trails are deeper, more V-shaped, and almost always lack transverse ridges. These shallow and deep varieties represent the extreme types of a morphological continuum. Varieties that were intermediate in depth and shape were produced in aquaria on substrates that were intermediate in consistency.

Substrate characteristics are broadly related to depositional environments. Sediment size, cohesion, water content, and degree of consolidation are critical factors in sediment stability (Crimes, 1975). Within the tidal flat, sediment size is controlled by energy and transport mechanism. Mud is more commonly deposited on the higher flat under low wave and current energy; sand is more prevalent on the lower flat where wave activity is strongest and active for the longest time (Reineck and Singh, 1973, p. 358).

Sediment of the high tidal flat is generally exposed for the longest period of time and often, but not invariably, the result is reduced water content and greater degree of compaction. Sediment of the lower flat is covered by water for longer periods and moved about by stronger waves and currents. The result is generally greater water saturation and looser consistency. Data in the literature on Recent intertidal areas are not sufficient to permit an assessment of how consistently these sediment relationships may hold. The relationships are true for some areas of Mugu Lagoon and most areas of Barnstable Harbor. However, where there is an abundance of water-saturated silt and clay high on the tidal flat the sediment is very poorly consolidated (examples include the upper reaches of Mugu Lagoon and Newport Bay, California). Because the variety of gastropod trail is in large part also controlled by substrate consistency the trail morphology can also be used to help identify subenvironments within the tidal environment.

The intertidal gastropods, both ancient and modern, produced a large variety of trail morphologies. If only small bedding-plane surfaces of the Fentress Formation had been available for study it is unlikely that the large degree of variation would have been easily recognized, if at all. Limited continuous bedding-plane exposure would likely have resulted in several ichnogenera named for different morphological varieties of the *Palaeobullia*. This would result in two problems. If diversity were used as a paleoecological index the result would be overestimation of trace producer diversity. A second problem relates to environment of deposition. Individual segments of *Palaeobullia* might resemble some portions of other ichnogenera that have different paleoenvironmental distributions. Assigning several ichnogeneric names to different morphological varieties of the Fentress Formation trails would unnecessarily broaden the reported environmental distribution of these ichnogenera, thereby reducing their value as paleoenvironmental indicators. Large morphological variation within ichnogenera may be greatest where large differences in substrate consistency occur because of subaerial exposure. A conservative approach to trace fossil taxonomy is warranted where dealing with tidal deposits or any other rocks deposited where the gradient of substrate consistency is steep.

CONCLUSIONS

1) The depth at which a snail moves below the sediment/water interface is related to sediment consistency. Both in Pennsylvanian *Palaeobullia* and in modern snail trails a deep, V-shaped groove represents movement at greatest depth in uncompacted sediment. Transversely ridged trails represent movement nearer the surface in compacted substrate.

2) Substrate consistency varies within the tidal environment. Substrates of the higher tidal flat generally are firmer than those of the lower flat. Differences in morphology of both ancient and modern gastropod trails reflect differences in substrate mass properties. Recognition of variation in gastropod trail morphology can be used in conjunction with associated sedimentary features to recognize tidal flat paleoenvironments and to distinguish higher from lower tidal flat subenvironments. Similar results could be expected with other organisms and other environments if substrate characteristics produce varying consistency.

3) Environments with large variation in substrate consistency are likely to result in significant variation in the trace morphology of gastropods and probably other organisms as well. A conservative approach to taxonomy will result in the most accurate estimates of trace producer diversity and environmental distribution in these paleoenvironments.

ACKNOWLEDGMENTS

The authors are most grateful to John Wheeler and Oakley Hinds of Jamestown, Tennessee, who permitted access to their quarries and generously donated specimens for our study collections. We extend thanks to Kelvin W. Ramsey, Department of Geology, Vanderbilt University, for helpful discussions of the Fentress trace fossils. We thank Gary Hill and Paul Basan for critically reviewing this manuscript and suggesting improvements. Studies at Barnstable Harbor were carried out while the junior author was a student in the Experimental Invertebrate Zoology class; the late Ralph Gordon Johnson made many suggestions and provided helpful discussions.

REFERENCES

ALEXANDER, S.K., 1979, Diet of the periwinkle *Littorina irrorata* in a Louisiana salt marsh: Gulf Research Reports, v. 6, p. 293–295.

BIRKENMAJER, K., AND BRUTON, D.L., 1971, Some trilobite resting and crawling traces: Lethaia, v. 4, p. 303–319.

BRADY, L.F., 1947, Invertebrate tracks from the Coconino Sandstone of northern Arizona: Jour. Paleontology, v. 21, p. 466–472.

CLARK, J.M., AND SWARTZ, C.K., 1913, Systematic paleontology of the Upper Devonian deposits of Maryland, *in* Prosser, C.S., and others, Middle and Upper Devonian: Maryland Geological Survey, p. 535–701.

CRIMES, T.P., 1975, The production and preservation of trilobite resting and furrowing traces: Lethaia, v. 8, p. 35–48.

HÄNTZSCHEL, W., 1975, Treatise on Invertebrate Paleontology, Part W, Miscellanea, Supplement 1, Trace Fossils and Problematica: Lawrence, Kansas, Geological Society of America and Univ. of Kansas Press, 269 p.

KLEIN, G. DEV., 1977, Clastic Tidal Facies: Champaign, Illinois, Continuing Education Publishing Company, 149 p.

MacGinitie, G.E., and MacGinitie, N., 1949, Natural History of Marine Animals: New York, McGraw-Hill Book Co., 473 p.
Milici, R.C., 1974, Stratigraphy and depositional environments of Upper Mississippian and Lower Pennsylvanian rocks in the Southern Cumberland Plateau of Tennessee, *in* Briggs, G. (ed.), Carboniferous of the Southeastern United States: Geological Society of America Spec. Paper 148, p. 115–133.
Osgood, R.G., 1970, Trace fossils of the Cincinnati area: Palaeontographica Americana, v. 6, no. 41, p. 281–444.
______, 1975, The paleontological significance of trace fossils, *in* Frey, R.W. (ed.), The Study of Trace Fossils: New York, Springer-Verlag, p. 87–108.
Redfield, A.C., 1972, Development of a New England salt marsh: Ecological Monographs, v. 42, p. 201–237.
Reineck, H.E., and Singh, I.B., 1973, Depositional Sedimentary Environments: New York, Springer-Verlag, 439 p.
Rhoads, D.C., 1970, Mass properties, stability, and ecology of marine muds related to burrowing activity, *in* Crimes, T.P., and Harper, J.C. (eds.), Trace Fossils: Geological Journal, Special Issue No. 3, Liverpool, Seel House Press, p. 147–160.
Schäfer, W., 1972, Ecology and Palaeoecology of Marine Environments: Edinburgh and Chicago, Oliver & Boyd and University of Chicago Press, 568 p.
Seilacher, A., 1953, Studien zur Palichnologie. 1. Über die Methoden der Palichnologie: Neues Jahrb. Geol. Paläontologie, Abhandlungen, v. 96, p. 421–452.
______, 1967, Bathymetry of trace fossils: Marine Geology, v. 5, p. 413–428.
______, 1970, Cruziana stratigraphy of "nonfossiliferous" Paleozoic sandstones, *in* Crimes, T.P., and Harper, J.C. (eds.), Trace Fossils: Geological Journal, Special Issue No. 3, Liverpool, Seel House Press, p. 447–476.
Smith, R.I., 1964, Keys to marine invertebrates of the Woods Hole region: Systematics-Ecology Program, Marine Biological Laboratory Contrib. no. 11, 208 p.
______, and Carlton, J.T., 1975, Light's Manual: Intertidal Invertebrates of the Central California Coast, 3rd ed.: Berkeley-Los Angeles, University of California Press, 716 p.
Trewin, N.H., 1976, *Isopodichnus* in a trace fossil assemblage from the Old Red Sandstone: Lethaia, v. 9, p. 29–37.
Warme, J.E., 1971, Paleoecological aspects of a modern coastal lagoon: Univ. California Publ. Geol. Sci. 87, p. 1–131.
Whittington, H.B., 1964, Taxonomic basis of paleoecology, *in* Imbrie, J., and Newell, N., Approaches to Paleoecology: New York, John Wiley and Sons, p. 19–27.

PART II
ASSEMBLAGES OF BIOGENIC STRUCTURES IN INTERTIDAL TO SUBTIDAL ENVIRONMENTS

INTRODUCTION.—This part of the volume contains papers that deal with biogenic structures found in intertidal to shallow subtidal environments. Microbial endoliths (microorganisms that live within a rock or shell substrate) can be significant agents of bioerosion, particularly within the intertidal and supratidal zones of carbonate rocky coasts. The paper by E.J. Hoffman documents a clear zonation in carbonates of microbial endoliths within the intertidal and supratidal zones of Bermuda. The zones established by Hoffman are represented in the same manner from site to site on Bermuda and appear to be very similar to those found along Mediterranean coasts and in some areas of the Caribbean. The distribution of microbial endoliths along the Bermuda coast is summarized in Hoffman's Figure 3, and Table 1 presents a carefully constructed list of criteria for the recognition of different endolith forms. Hoffman concludes by suggesting that fossil endolith assemblages could be most useful for identifying and subdividing ancient carbonate coastal sequences.

Gary Hill's study of the modern ichnofacies of the sediment size-graded shelf of the northwestern Gulf of Mexico shows that bioturbation varies significantly and systematically across the shelf. The study reveals that the diversity and density of traces decrease across the shelf as substrate grain-size becomes finer and more uniform and as infaunal assemblages become less dense and diverse. Hill concludes with some useful observations on how geologists might recognize better zonation patterns in shelf sediments based on patterns and degree of bioturbation.

Algal mats and stromatolites commonly are thought to occur only in tropical, often arid, carbonate marine environments. The paper by Barry and Diane Cameron and Richard Jones shows that algal mats formed by cyanophytes also can be common in quartz sands in the intertidal and supratidal zones of modern, higher latitude coasts. The authors clearly document zonation within algal mats along the northeastern coast of Massachusetts, and they describe a number of physical/biogenic structure forms associated with the mats. The preservation potential for quartzose algal mats may be low owing to lack of carbonate cement, but the authors give criteria for recognition and suggest that the occurrence of algal mats may be much more common in the stratigraphic record than is now known.

The polychaete *Diopatra cuprea* commonly constructs dwelling burrows in nearshore sands along the Atlantic coast of the United States. The study by John Barwis documents the occurrence of *D. cuprea* along the South Carolina coast and shows how tidal channel hydrodynamics influence the distribution and geometry of the dwelling tubes. *D. cuprea* is a good modern analog for some forms of the trace fossils *Skolithos* and *Monocraterion*. Barwis uses a clearly constructed example of comparison to show how his information on *D. cuprea* distribution can be used to interpret the occurrence and significance of *Skolithos* and *Monocraterion* in a Cambrian tidal flat setting.

The *Glossifungites* ichnofacies occurs in firm but unlithified substrates in marine, intertidal to shallow subtidal settings. George Pemberton and Robert Frey's paper reviews and clarifies the concept of the *Glossifungites* ichnofacies and describes the characteristics of the ichnofacies from Holocene examples along the Georgia coast. Table 1 presents a useful comparison of the *Trypanites*, *Glossifungites*, and *Skolithos* ichnofacies, and the many figures of the article illustrate well the distinguishing features of the *Glossifungites* ichnofacies. The authors analyze carefully the associations of organisms and traces in Holocene examples of the ichnofacies and discuss how the ichnofacies might occur and best be recognized in the rock record.

The final paper of this section by Allen Curran describes an extremely well-preserved assemblage of Late Cretaceous trace fossils from Delaware. Modern tracemaker analogs are suggested for many of the trace fossils, thus enabling further interpretation of their paleoenvironmental significance. The assemblage is shown to represent a prograding, low energy, nearshore environment with a low energy, *Ophiomorpha*-dominated, upper shoreface-lower foreshore zone similar to that found today along the Sea Isles coast of Georgia.

DISTRIBUTION PATTERNS OF RECENT MICROBIAL ENDOLITHS IN THE INTERTIDAL AND SUPRATIDAL ZONES, BERMUDA

E.J.HOFFMAN[1]
Department of Geology, Boston University, Boston, Massachusetts 02215

ABSTRACT

Previous work indicates that microbial endoliths exhibit distinct habitat preferences within the intertidal and supratidal zones of Bermuda. An analysis of species diversity and abundance for eight physically varying sites shows several trends which are typical for Bermuda. From the lower intertidal zone through the upper supratidal zone, there is a gradual reduction in the number of endolithic species and a change in the dominance of a given species. The intertidal and supratidal zones can be divided into four distinct, well-defined subzones on the basis of co-occurring assemblages of endoliths. These four subzones can be correlated with visible changes in rock surface relief and color, tidal range, and invertebrate and macro-algal associations.

Most of the endolithic species, as well as their characteristic distributional patterns in Bermuda, are essentially the same as those found along Mediterranean coasts. Preliminary work in Florida and Jamaica suggests that these patterns may be typical for the Caribbean area.

Since a given assemblage of endolithic organisms occupies a certain position with respect to mean sea level, it can be used as an indicator of tidal range of intertidal and supratidal conditions. Thus, if preserved and recognized in the rock record, the endolithic assemblages described in this study could be used to identify and interpret ancient carbonate intertidal and supratidal sequences.

INTRODUCTION

Endolithic microphytes are microorganisms which penetrate and live within carbonate substrates. They influence constructive and destructive carbonate coastal processes in a variety of ways (see Folk et al., 1971; Schneider, 1976; Golubic and Schneider, 1979; Lukas, 1979b). Endolithic organisms and the herbivores which graze on them combine to erode carbonate substrates and produce the biokarst typical of carbonate coasts. Although endolithic microorganisms (commonly referred to as endoliths) have been known to botanists for a long time (e.g., Bornet and Flahault, 1888, 1889), recent workers have been more interested in their geological effects and have directed their work in two areas:

1) the role of endoliths in bioerosion and micritization processes (Bathurst, 1966; Kobluk and Risk, 1977a and b; Schneider, 1977), and
2) their potential as paleoecological indicators of depth and environment (Swinchatt, 1969; Golubic et al., 1975; Budd and Perkins, 1980).

Realization of the utility of microbial endoliths in interpreting paleobathymetry has become possible with the development of the methodology for comparing fossil and Recent assemblages (Golubic et al., 1975, 1979). An increased understanding of modern endoliths, and new evidence for the evolutionary conservatism of the endolithic biota (Campbell, 1980), has also contributed to our understanding of fossil endoliths.

Most studies of bathymetric distribution among modern endoliths have been carried out in the subtidal photic and aphotic regions (e.g., Rooney and Perkins, 1972; Perkins and Tsentas, 1976; Lukas, 1979a; Zeff and Perkins, 1979). The few studies that have been carried out in the intertidal and supratidal zones suggest that distinct communities of microbial endoliths and epiliths (organisms which live on the surface and do not actively penetrate the carbonate substrate) occur within these areas (Ercegović, 1932, Schneider, 1976; Le Campion-Alsumard, 1979a). However, none of these studies was undertaken in a tropical carbonate regime, and the degree of regional distribution of the endolith assemblages, which is central to paleoecological application, was not evaluated.

The purposes of this study are (1) to assess the diversity and relative abundance of microbial endolith species in the intertidal and supratidal zones of the Bermuda carbonate platform, and (2) to characterize particular assemblages of endoliths within these zones. These endolith assemblages were compared to those found in studies from other areas, and the degree of regionality was estimated. The broader objective is to evaluate the potential of modern endolith associations as indicators of the intertidal and supratidal zones, and to assess their potential use for the recognition of these environments in the fossil record.

[1]Present address: Cities Service Oil and Gas Corporation, P. O. Box 1919, Midland, Texas 79702

METHODS

Study Area

Bermuda was chosen as the study area because the numerous sedimentological studies carried out there indicate that Bermuda's carbonate platform contains many analogs to the high-energy, unrestricted shallow carbonate environments seen in the fossil record (Upchurch, 1970). Bermuda's present coastline is composed of cemented eolianite and shallow water marine skeletal limestones deposited as strandline or shoreline-dune deposits during successive Pleistocene high sea level stands (Land et. al., 1967; Plummer et al., 1976).

The intertidal and supratidal zonation of macro-algae and invertebrates along Bermuda's rocky coasts has been the subject of intensive studies. The zonation seen in Bermuda can be correlated to zones established elsewhere in the world (Stephenson and Stephenson, 1972). Data on many of the physical factors known to affect endolith distribution (i.e., desiccation potential, tidal range, water and air temperature) have been monitored for some time and are readily available (Plummer et al., 1976; Morris et al., 1977).

The Bermuda Islands are located at 32° north latitude, 60° west longitude, and are approximately 1000 km off the coast of North Carolina. The warm waters of the Gulf Stream are responsible for the local subtropical marine environment and here is found the most northerly development of coral reefs. Bermuda experiences two seasons with associated differences in wave height and frequency, wind speed, gale occurrences, insolation, precipitation, and temperature (Morris et al., 1977). Winter is characterized by lower temperatures, less insolation, and frequent rains of long duration (Plummer et al., 1976). Larger waves associated with the stronger, more numerous winter storms physically abrade the coast. However, abrasion is minimized since waves break first on the shallow reef tract which surrounds Bermuda. The energy dissipation is most pronounced in areas with better developed reef barriers, such as along the north shore (Morris et al., 1977). Summer is characterized by increased insolation, which is accompanied by increased desiccation potential, fewer and weaker storms, but more numerous rainstorms of short duration (Plummer et al., 1976; Morris et al., 1977). The tides at Bermuda are semidiurnal, with a high or low phase every 6.21 hours (Morris et al., 1977). Diurnal inequality is small and occurs principally with the high tides (Stephenson and Stephenson, 1972). Tidal range varies from 0.45 to 1.2 meters with an average of 0.75 meters (Morris et al., 1977).

Samples for this study were collected from eight sites, which were chosen to include both north- and south-facing exposures with different degrees of shore slope (Fig. 1). At each site, four subzones—lower intertidal subzone through upper supratidal subzone—were designated on the basis of rock surface morphology, color, and associated macro-organisms (see Stephenson and Stephenson, 1972 for discussion of associated organisms). Several rock chips, each approximately 1.0 cm^2, were taken from the middle of each subzone and preserved in the field with buffered 4% formaldehyde.

Microscopic Methods

In the laboratory, the samples were subdivided and prepared for light microscopy or scanning electron microscopy (SEM) as described in Golubic et al. (1970, 1975) and outlined in Lukas (1979b, Fig. 1). For light microscopy, part of each sample was placed in 3% HCl to free the endolithic organisms from their carbonate substrate. Four or five slides of the isolated endolithic organisms from each subzone were made and examined under transmitted light and Nomarski interference phase contrast. The percent coverage by each species present on the slides was tabulated by visual estimation as seen under low magnification (x250) and given a semi-quantitative coverage value on the following scale (after Golubic, 1967):

+ present
1 10% coverage
2 10 to 25% coverage
3 25 to 50% coverage
4 50 to 75% coverage
5 75 to 100% coverage

These values were used to calculate the percent occurrence and percent dominance. The percent occurrence is the percentage of times a species occurs in slides from a particular subzone. Percent dominance is the percentage of times that the average coverage value for a species is greater than or equal to three. These values of percent dominance and percent occurrence were used to divide the taxa into associations of species which are unique to a subzone and to evaluate the abundance of a particular species within that subzone.

DESCRIPTION OF INTERTIDAL AND SUPRATIDAL ZONES

Introduction

The intertidal and supratidal zones in Bermuda usually have been defined on the basis of coloration of the rocks and the assemblages of macroscopic invertebrates and algae (Charlton, 1969; Hanley, 1970; Stephenson and Stephenson, 1972; Morris et al., 1977). A correlation of zones as used in this study with those of previous workers in Bermuda and elsewhere is given in Figure 2.

A predictable change in color, rock surface relief, macro-organisms, and endolith assemblages from the intertidal through the supratidal zone occurs. These trends are introduced below and discussed by subzone in the following section. The boundary between the intertidal and supratidal zones is most pronounced at sites with the steepest slope and is less distinct, al-

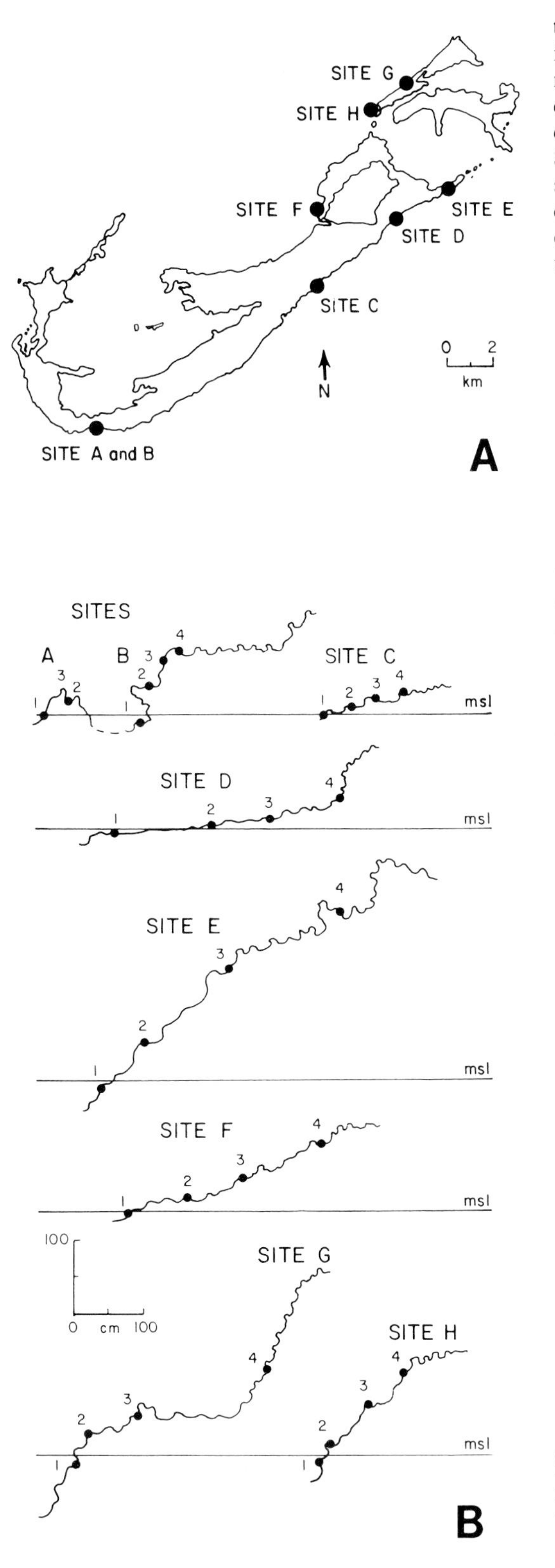

though visible, at sites with gentler slope. Zone width is controlled by a complex of factors, including coast morphology and exposure to wave action, and can be correlated directly to moisture availability, which is determined primarily by tidal range (Stephenson and Stephenson, 1972). Relative zone width varies from site to site, as is to be expected from geometric considerations (Fig. 1B). The following predictable changes occur at each site from the lower intertidal through the upper supratidal subzone:

1) a decrease in time of exposure to sea water;
2) an increase in desiccation potential;
3) an increase in effects of rainwater;
4) an increase in temperature variation;
5) an increase in the macro- and micro-relief of the rock;
6) a decrease in visible macroscopic invertebrates;
7) a decrease in grazing pressure; and
8) a change in color of the rock.

Because the greatest number of microbial endoliths from the intertidal and supratidal zones of Bermuda are cyanophytes (blue-green algae), this study concentrates on this group. Eleven species of cyanophytes representing seven different genera were identified on the basis of cell dimensions, filament length, branching pattern, and thallus color and appearance. In addition to the cyanophyte genera, three eukaryotic groups were observed: a conchocelis form of bangioid rhodophyte, the chlorophytes *Ostreobium quekettii*, and an unidentified form and several unidentified fungi. Detailed descriptions of these organisms are not given here, but Table 1 gives cell dimensions, as well as selected references to the organisms and SEM photomicrographs of their resin-cast borings. A diagrammatic sketch of individual organisms and light photomicrographs of the common organisms are given in Figures 3, 4, and 5.

The following changes in the microbial endolith assemblages were observed concurrent with changes in physical parameters from the lower intertidal through the upper supratidal subzones:

1) a reduction in the number of species present;
2) a change in the association of species present;
3) a change in the dominance of particular species.

The two intertidal subzones have the largest total and average number of species; the upper supratidal subzone has the smallest average number of species. Species diversity as calculated by the Shannon-Weiner function decreases from the lower intertidal subzone through the upper supratidal subzone (Table 2). Table 2

FIG. 1.—Site locations and site profiles. *A*, Site locations in Bermuda. *B*, Site profiles in relation to mean sea level. 1 = lower intertidal subzone; 2 = upper intertidal subzone, 3 = lower supratidal subzone; 4 = upper supratidal subzone; msl = mean sea level.

also presents the percent dominance and occurrence of endolith species by subzone. The percent dominance and occurrence of a species change from subzone to subzone. These changes define the associations of endolith species which are characteristic for particular subzones. The changes in associations across the intertidal-supratidal zone boundary are more pronounced than the differences between the subzones within each zone (Fig. 3).

Intertidal Zone

Lower Intertidal Subzone (Subzone 1)

Physical description.—The lower intertidal subzone (Subzone 1) in Bermuda is exposed only at low tide. The rocks vary from buff to light red-brown in color, and the surface has a gently undulating profile and rounded macro- and micro-relief (Fig. 4A, B). A dense algal "turf" (Stephenson and Stephenson, 1972), molluscs, and arthropods are common. Grazing marks are visible, including those of parrot fish, which are made during high tide.

Endolith assemblages (Figs. 4C-H; and 5F, H).—The lower intertidal subzone has the greatest number of endolith species and includes the largest eukaryotic component. *Plectonema terebrans*, *Mastigocoleus testarum*, and *Hyella caespitosa* are the dominant species, while *Solentia foveolarum* and *Hormathonema* spp. are common species (Table 2, Fig. 3). *Hyella* sp. A (see Lukas and Hoffman, 1984) the unidentified chlorophyte and fungi, although far less abundant, are diagnostic of both intertidal subzones since they are limited to this habitat. *Ostreobium quekettii* is even more limited in distribution since it is found only in the lower intertidal subzone. *Hyella* sp. A and *O. quekettii* may be subtidal species that are able to exist in the lower intertidal subzone in localized microenvironments typical of subtidal conditions.

Upper Intertidal Subzone (Subzone 2)

Physical description.—Rocks of the upper intertidal subzone are yellow-brown in color. Like the lower intertidal subzone, the surface has a gentle, rounded profile, but its macro- and micro-relief is slightly more irregular (Figs. 1B, 4A). The more rounded profile typical of both of the intertidal subzones is due to increased bioerosion of the rock from the physical presence of the endoliths and from grazing by fish, crabs, gastropods, and other invertebrates (Stephenson and Stephenson, 1972; Schneider, 1976; Golubic and

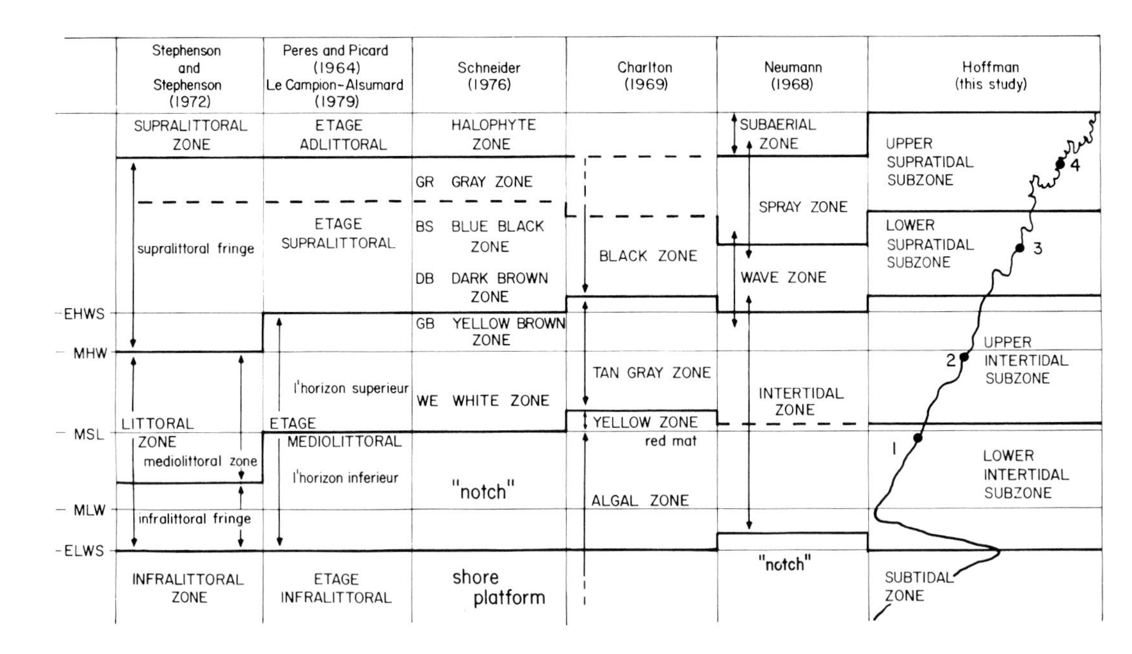

FIG. 2.— Correlation chart of selected classifications of the littoral system as seen in Bermuda and elsewhere compared to a composite intertidal and supratidal profile from this study. The classifications of Neumann (1968), Charlton (1969), Schneider (1976) and Hoffman (this study) apply directly to the Bermuda coast. The classifications of Pérès and Picard (1964) and Le Campion-Alsumard (1979a) apply to European coasts; the classification of Stephenson and Stephenson (1972) has world-wide application. Dots indicate approximate sample locations for this study in Bermuda and thickened horizontal lines indicate equivalent zones. EHWS = extreme high water, spring tides; ELWS = extreme low water, spring tides; MHW = mean high water; MLW = mean low water; MSL = mean sea level.

Schneider, 1979). The upper intertidal subzone is exposed to the air for a longer time than is the lower intertidal subzone, but it is wetted by larger waves during low tide.

Endolith assemblages (Figs. 4C-H; 5F-H).—Most of the species present in the lower intertidal subzone also occur in the upper intertidal subzone but with decreased abundance (Table 2, Fig. 3). The dominant species of the lower intertidal subzone (Subzone 1), with the addition of *Solentia foveolarum*, are also the dominant species of the upper intertidal subzone (Subzone 2) (Table 2). The presence of *Kyrtuthrix dalmatica*, a distinctively branching cyanophyte which occurs also in the lower supratidal subzone, and the absence of *Ostreobium quekettii* help to distinguish the endolith assemblages of the upper intertidal subzone from those of the lower intertidal subzone. The percent occurrence and percent dominance of the eukaryotic species also decrease from the lower to the upper intertidal subzones (Table 2).

Supratidal Zone

Lower Supratidal Zone (Subzone 3)

Physical description.—Rocks of the lower supratidal subzone are black in color, and are wetted by wave splash and by larger waves at high tide. The surface of the lower supratidal subzone has a convoluted profile with pitted and irregular macro- and micro-relief consisting of small depressions and pinnacles (Figs. 1B, 5A, B). This surface relief may be related to the intensity of bioerosive activity (Schneider, 1976; Golubic and Schneider, 1979). The gastropod *Tectarius*, which can withstand desiccation, is the only macroscopic invertebrate present.

Endolith assemblages (Figs. 4C-F; 5C, E-H).—There is a major change in the endolith assemblages across the intertidal-supratidal zone boundary. *Hyella caespitosa*, a dominant species in both intertidal subzones is absent from the lower and upper supratidal subzones. *Mastigocoleus testarum* and *Plectonema*

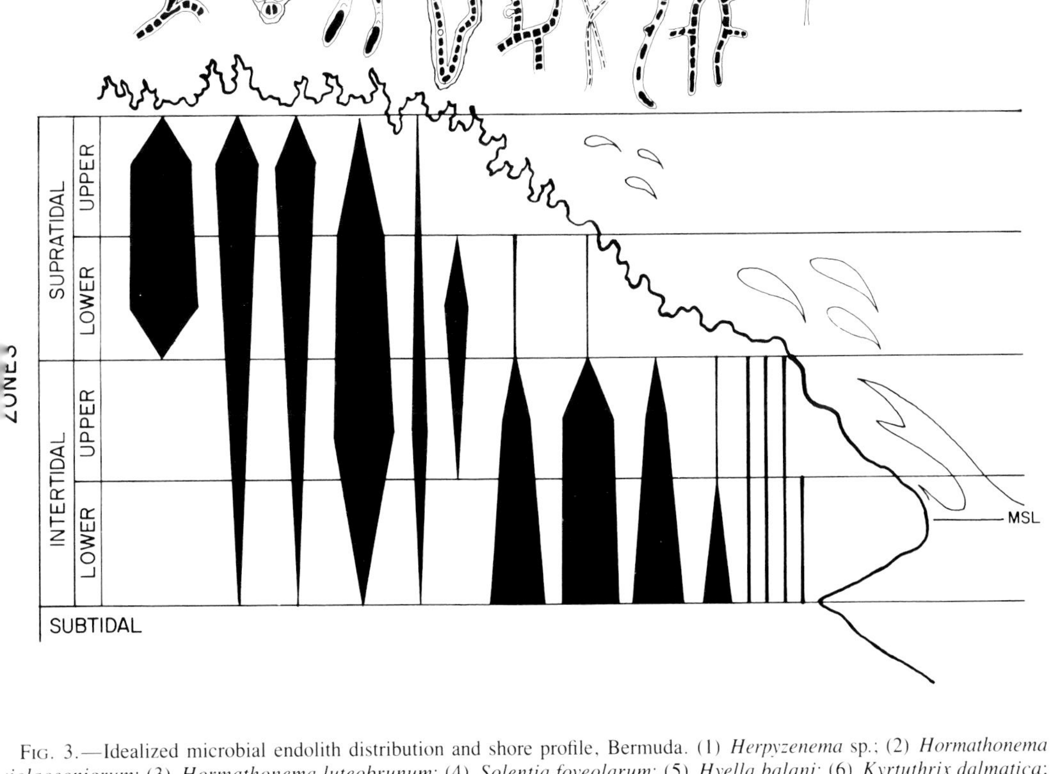

FIG. 3.—Idealized microbial endolith distribution and shore profile, Bermuda. (1) *Herpyzenema* sp.; (2) *Hormathonema violaceonigrum*; (3) *Hormathonema luteobrunum*; (4) *Solentia foveolarum*; (5) *Hyella balani*; (6) *Kyrtuthrix dalmatica*; (7) *Mastigocoleus testarum*; (8) *Plectonema terebrans*; (9) *Hyella caespitosa*; (10) *Hyella* sp. A; (11) conchocelis stage of bangioid rhodophyte; (12) unidentified fungi; (13) unidentified chlorophyte; (14) *Ostreobium quekettii*.

TABLE 1.—CRITERIA USED FOR DISTINGUISHING THE MICROBIAL ENDOLITHS OF THIS STUDY, AND SELECTED REFERENCES TO THE ENDOLITH SPECIES AND PHOTOGRAPHS OF THEIR RESIN-CAST BORINGS.

Species	Cell Dimensions Breadth/Length	Filament Length	Branching	Gel	References	
					Organism	SCAN photograph
Cyanophytes						
Hormathonema violaceonigrum	2.0-5.0μ 2.0-8.0μ	Colonies to 60μ	Not observed	Common gel with individual sheaths around cells	Ercegović, 1929 Golubic et al., 1975 Le Campion-Alsumard, 1979a, b	Le Campion-Alsumard, 1979a Budd and Perkins, 1980
H. luteobrunum	2.0-5.0μ 2.0-8.0 μ	Colonies to 60μ	Not observed	Common gel with individual sheaths around cells	Ercegović, 1929 Golubic et al., 1975 Le Campion-Alsumard, 1979a, b	Le Campion-Alsumard, 1979a Budd and Perkins, 1980
Solentia foveolarum	5.0-20.0μ to 40.0μ	to 350μ	Lateral cell protrusion	Common gel strongly layered 10-25μ between cells	Ercegović, 1930 Schneider, 1976 Le Campion-Alsumard, 1979a, b	Schneider, 1976 Le Campion-Alsumard, 1979a
Hyella caespitosa	4.0-5.0μ to 15μ	to 300μ	Occasional lateral cell protrusion	Common gel to 2.5μ thick; not layered to 6-14μ between cells	Bornet et Flahault, 1888 Schneider, 1976 Le Campion-Alsumard, 1979a, b	Schneider, 1976 Harris et al., 1979 Budd and Perkins, 1980
H. balani	4.0-6.0μ 4.0-6.0μ	Colonies to 60μ	Not observed	Common gel with individual sheaths around cells	Lehmann, 1903 Le Campion-Alsumard, 1979a, b	Le Campion-Alsumard, 1979a
H. sp. A	2.5-3.8μ 3.0-17.5μ	to 350μ	Common lateral cell protrusion	Common gel to thick up to between cells	Lukas and Hoffman, 1984	
Plectonema terebrans	0.80-1.5μ 3.6-7.5μ	Not measured	Filamentous Not observed	Thin, rarely visible	Bornet and Flahault, 1889 Lukas, 1973	Budd and Perkins, 1980

Kyrtuthrix dalmatica	2.5-8.0μ 3.5-7.0μ	Not measured	Filamentous False branching Intercalary heterocysts with tapering trichome ends	Thin	Ercegović, 1929 Golubic and Le Campion-Alsumard, 1973 Schneider, 1976 Le Campion-Alsumard, 1979a, b	Golubic and Le Campion-Alsumard, 1973 Golubic et al., 1975 Schneider, 1976
Herpyzenema sp.	3.5-6.0μ 4.5-15.0μ	to 250μ	Filamentous True V-shaped branching and false branching intercalay heterocysts.Rounded trichome ends.	Thin	Weber van Bosse, 1913 Hoffman, 1981	This study, Fig. 5D
Mastigocoleus testarum	4.0-5.5μ 5.5-13.5μ	Not measured	Filamentous True T-shaped branching	Thin, rarely visible	Laegerheim, 1886 Lukas, 1973 Golubic and Le Campion-Alsumard, 1973 Perkins and Tsentas, 1976 Schneider, 1976	Golubic and Le Campion-Alsumard, 1973 Schneider, 1976 Perkins and Tsentas, 1976 Budd and Perkins, 1980
Rhodophytes						
Conchocelis	Vegetative filaments to 3.5μ in diameter	Not measured	Common	Not applicable	Rooney and Perkins, 1972 Conway and Cole, 1977 Campbell, 1980	Rooney and Perkins, 1972 Campbell, 1980
Chlorophytes						
Ostreobium quekettii	Filaments 3.0-10.0μ in diameter	Not measured	Common	Not applicable	Lukas, 1973 and 1974	Budd and Perkins, 1980
unidentified chlorophytes	Not measured	Not measured	Occasional	Not applicable	Hoffman, 1981	
Fungi						
unidentified fungi	Filaments to 2μ in diameter	Not measured	Common; T-shaped or dichotomous	Not applicable	Hoffman, 1981	

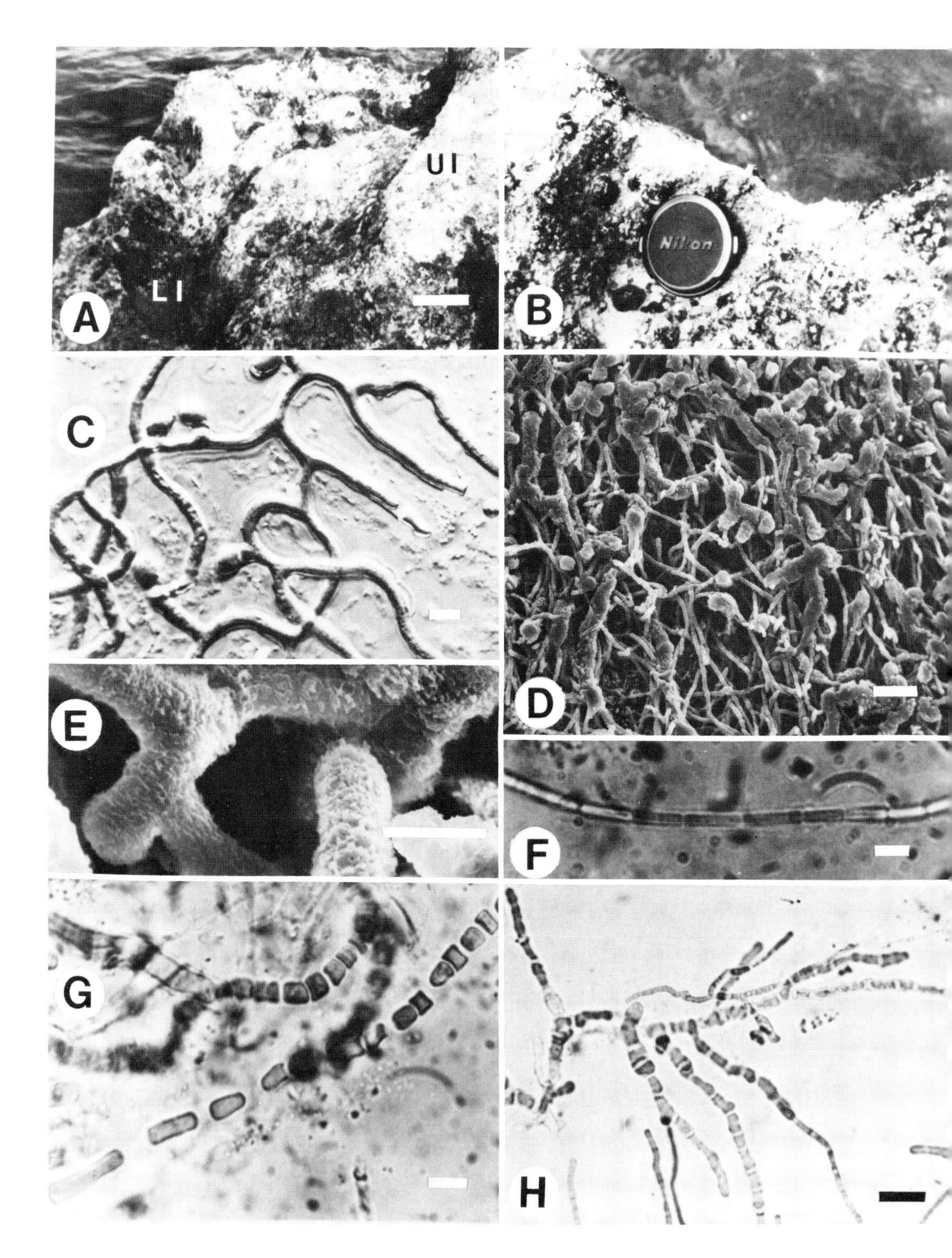
UI
LI
A
B
Nikon
C
D
E
F
G
H

terebrans, dominant species in the intertidal subzones, become uncommon or rare species in the lower supratidal subzone (Subzone 3). No eukaryotic species are found in either supratidal subzone. *Solentia foveolarum*, which was a dominant species in the upper intertidal subzone, and *Herpyzenema* sp. become dominant species in the lower supratidal subzone (Table 2, Fig. 3). The V-shaped branching of *Herpyzenema* sp. helps to distinguish it from *Kyrthuthrix dalmatica*, the other common filamentous, branching species in this subzone. *Hormathonema* spp., which are present throughout the two intertidal subzones, become increasingly abundant in the supratidal zone.

Upper Supratidal Zone (Subzone 4)

Physical description.—Rocks of the upper supratidal subzone are gray in color and are wetted only by wave spray or storm splash. The surface of the upper supratidal subzone has a sharp, jagged profile with convoluted and irregular macro- and micro-relief. Numerous pinnacles and spikes are typical of this subzone, perhaps due to decreased bioerosion (Figs. 1B, 5A)(Schneider, 1976, Golubic and Schneider, 1979). Visible macroscopic life is rare.

Endolith assemblages (Fig. 5C-F, H).—The upper supratidal subzone (Subzone 4) has the least number of endolith species. *Herpyzenema* sp. is the dominant species. *Hormathonema* spp. are very common, but are not dominant. *Solentia foveolarum* and *Hyella balani* are common but quantitatively unimportant in comparison to *Herpyzenema* sp. and *Hormathonema* spp. (Table 2, Fig. 3).

REGIONAL DISTRIBUTION

At all eight sites of varying physical properties around Bermuda, the endolith assemblages exhibited a well-defined zonation pattern. This zonation pattern is recognized by changes in dominance and abundance of species and can be correlated to differences in tidal range, macroscopic organism, and the physical appearance and color of the rock surface. The general trends of endolith distribution in Bermuda, such as the reduction of the number of species in the supratidal zones, are consistent with accepted models of the effect of increasing rigor on community composition (Krebs, 1972).

The similarity of endolith zonation patterns in other areas to that in Bermuda has not been evaluated. An evaluation is necessary because, without some consistency in the zonation patterns of endolith assemblages, recognition of global intertidal and supratidal zones and, eventually paleoecological application, is not possible.

There have been three intensive studies correlating endolith distribution with recognized intertidal zonation. Beyond the present study, Le Campion-Alsumard (1969, 1975, 1979a, b) has studied the endolith distribution around Les Calanques, Marseille, France, and Schneider (1976) has worked on the endolith and physical zonation along the carbonate coasts around Rovinj, Yugoslavia, and Bermuda, and in a preliminary manner around the Florida Keys. In addition, Budd and Perkins (1980, Table 1) present some limited data on the occurrence of endoliths from studies done in different geographical areas.

Only Schneider (1976) has drawn any comparisons among the physical and biological zonations in these different areas. He found that the endolith distribution, rock surface relief, rock color, and texture in France and Yugoslavia were identical up to his Gray Zone (= lower supratidal subzone) (see Fig. 2 for correlations). Schneider suggested that differences above the Gray Zone are due to increased climatic aridity around Marseille, France, and inhomogeneities within the rock texture and increased humidity and grazing pressure in Florida and Bermuda. Despite these differences, Schneider thought that the general trends in physical appearance could still be seen. Moreover, he found that the distribution of endolith species and their zonation was, in detail, the same for Marseille, France; Rovinj, Yugoslavia; and Bermuda, although he does not elaborate on the organisms or characteristics of their distribution patterns.

FIG. 4.—Rock surface relief and common microbial endoliths in the intertidal zone, Bermuda. *A*, Typical macro-relief of the lower and upper intertidal subzones. Scale bar is 20 cm. LI = lower intertidal subzone; UI = upper intertidal subzone. *B*, Typical surface microrelief of lower intertidal subzone. Note presence of vermetids, barnacles, and macro-algae. Scale is 5 cm in diameter. *C*, Aspect of *Mastigocoleus testarum* in decalcified material from the lower intertidal subzone. Scale bar is 10 μ. *D*, Resin-cast borings of *Mastigocoleus testarum* and *Plectonema terebrans* from the lower intertidal subzone. Scale bar is 20 μ. *E*, Detail of T-shaped branching and terminal heterocysts of *Mastigocoleus testarum* as seen in the resin-cast borings. Scale bar is 10 μ. *F*, Portion of filament of *Plectonema terebrans* in decalcified matter from the lower intertidal subzone. Scale bar is 2 μ. *G*, Portion of filament of *Hyella caespitosa* from the upper intertidal subzone; note long and short cells within filament. Scale bar is 10 μ. *H*, Aspect of *Hyella* sp. A from the lower intertidal zone. Scale bar is 20 μ.

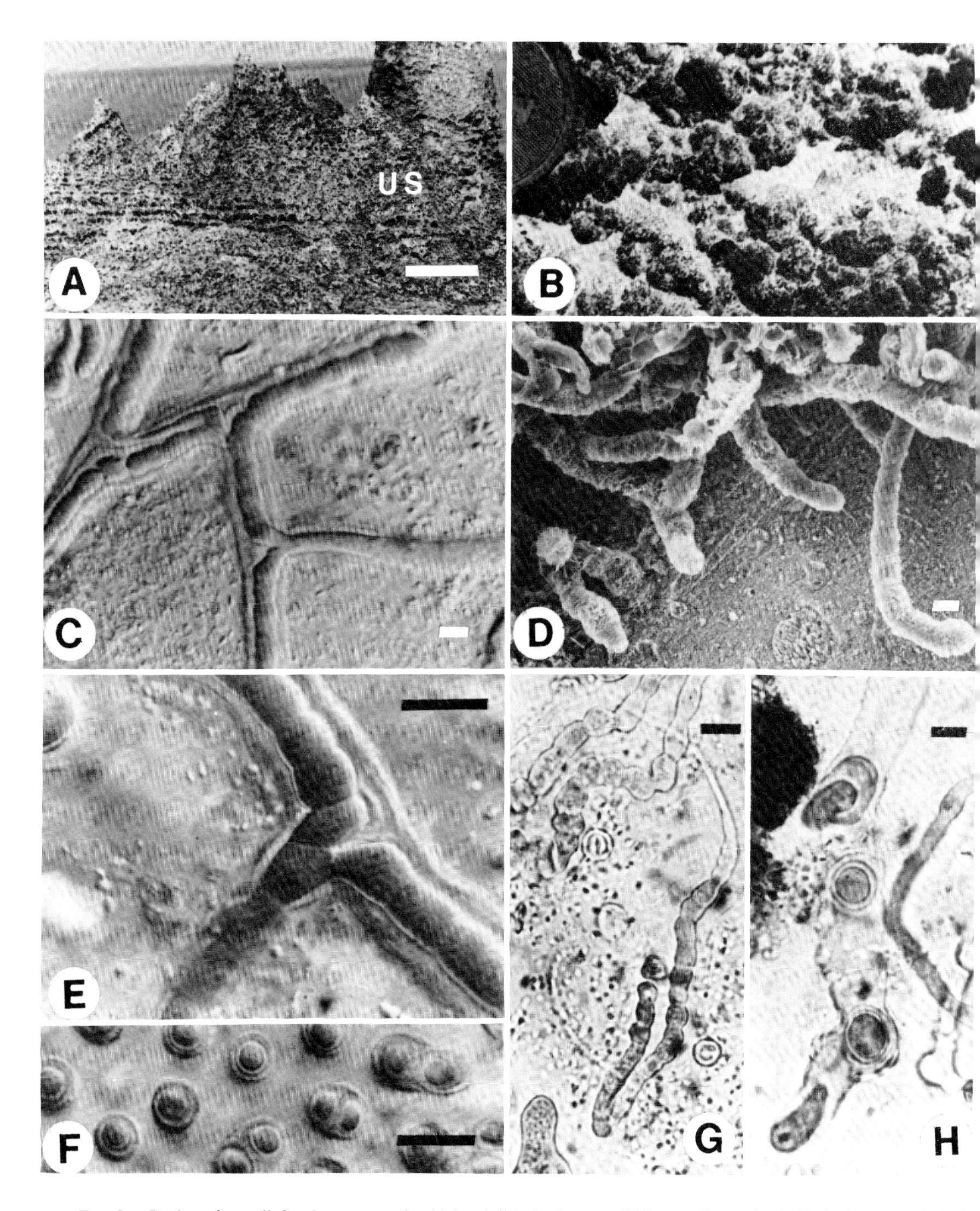

FIG. 5.—Rock surface relief and common microbial endoliths in the supratidal zone, Bermuda. *A*, Typical macro-relief of the supratidal zone, Bermuda. Lower supratidal subzone not shown; US = upper supratidal subzone. Scale is 20 cm. *B*, Typical irregular surface micro-relief of lower supratidal subzone. Scale is 5 cm in diameter. *C*, Aspect of *Herpyzenema* sp. in decalcified material from the lower supratidal zone. Scale bar is 10 μ. *D*, Resin-cast boring of *Herpyzenema* sp. Scale bar is 10 μ. *E*, Detail of V-shaped branching of *Herpyzenema* sp. Scale bar is 10 u. *F*, Portion of colony of *Hormathonema violaceonigrum* in decalcified material from the upper intertidal subzone. Scale bar is 10 μ. *G*, Aspect of *Kyrtuthrix dalmatica* in decalcified material from the lower supratidal subzone. Scale bar is 10 μ. *H*, Shorter supratidal (?) form of *Solentia fovedarum* in decalcified material from the lower supratidal subzone. Scale bar is 10 μ.

Because of the emphasis on the organisms themselves, comparisons between Bermuda and the European coasts can best be made based on the work of Le Campion-Alsumard (1979a, b). The similarities in the endolith distribution of these two areas are discussed below.

In the lower intertidal subzone, equivalent to l'étage médiolittoral inférieur (see Fig. 2), both studies found *Hyella caespitosa, Plectonema terebrans, Mastigocoleus testarum, Solentia foveolarum, Ostreobium quekettii*, and fungi to be present. In the upper intertidal subzone, equivalent to l'étage médiolittoral supérieur, both studies reported *Kyrtuthrix dalmatica, H. caespitosa, M. testarum* and *P. terebrans*. Despite the reduction in width of the supratidal zone in the more arid Mediterranean climate, both studies found *Hyella*

TABLE 2.—ABUNDANCE AND DIVERSITY DATA FOR MICROBIAL ENDOLITHS IN THE INTERTIDAL AND SUPRATIDAL ZONES, BERMUDA.[1]

	Lower			Upper		
SPECIES	Percent Occurrence		Percent Dominance	Percent Occurrence		Percent Dominance
			INTERTIDAL ZONE			
Hormathonema violaceonigrum	46	C	0	41	C	0
Hormathonema luteobrunum	42	C	0	41	C	0
Solentia foveolarum	58	C	21	85	D	74
Hyella caespitosa	58	D	64	52	D	57
Hyella balani	13	R	0	11	R	33
Hyella sp. A	29	U	71	22	U	0
Plectonema terebrans	75	D	44	82	D	41
Kyrtuthrix dalmatica	0		0	15	U	25
Herpyzenema sp.	0		0	0		0
Mastigocoleus testarum	63	D	60	93	D	48
conchocelis	17	U	25	11	R	0
Ostreobium quekettii	13	R	0	0		0
Unidentified chlorophytes	21	U	0	4	R	0
Fungi	21	U	0	15	U	0
Average number of species present at the eight sites	6.1			6.1		
Total number of slides examined	24			27		
Shannon-Weiner Diversity Index	1.9			1.8		
			SUPRATIDAL ZONE			
Hormathonema violaceonigrum	83	C	0	100	C	21
Hormathonema luteobrunum	75	C	0	100	C	17
Solentia foveolarum	75	D	44	42	C	40
Hyella caespitosa	0		0	0		0
Hyella balani	38	U	0	8	R	50
Hyella sp. A	0		0	0		0
Plectonema terebrans	13	R	0	0		0
Kyrtuthrix dalmatica	42	C	50	0		0
Herpyzenema sp.	79	D	79	92	D	59
Mastigocoleus testarum	25	U	0	0		0
conchocelis	0		0	0		0
Ostreobium quekettii	0		0	0		0
Unidentified chlorophytes	0		0	0		0
Fungi	0		0	0		0
Average number of species present at the eight sites	5.1			3.6		
Total number of slides examined	27			24		
Shannon-Weiner Diversity Index	1.6			1.2		

[1]Percent occurrence and dominance are explained in text. The percent occurrence and dominance were used to designate dominant (D), common (C), uncommon (U), and rare (R) species, as follows: A dominant species must have a percent occurrence of greater than 40% and a percent dominance of greater than 50%. A common species (C) has a percent occurrence of greater than 35%. The percent occurrences of an uncommon species (U) ranges from 15–35%, and the percent occurrence of a rare (R) species is less than 15%.

balani, *Hormathonema* spp., and a form of *Solentia foveolarum* present. There were, however, differences in the physical appearance of the rock and in the endolith distribution. The porous nature and diverse skeletal grains of Bermuda's cemented eolianite contrasts markedly with the dense, homogeneous, well-cemented limestone of the Marseille coast. The rock surface relief and pinnacled biokarst of the supratidal zone is more extensively developed in Bermuda than around Marseille. Despite these physical differences, there were relatively few differences in endolith distribution. In Bermuda, *Herpyzenema* sp. was very common in the supratidal zone, but it does not occur around Marseille. *Kyrtuthrix dalmatica* was limited to "les curvettes", small moisture-rich depressions in l'étage médiolittoral supérieur around France. In Bermuda, *K. dalmatica* was found in an equivalent zone, but it was not limited to any such depressions. Schneider's (1976) listing of intertidal organisms (White through Dark Brown Zones) is similar to that found in this study and those of Le Campion-Alsumard (1979a, b). However, Schneider's listing of *Hyella caespitosa* as an important lower supratidal species (Blue-Black Zone) and *Mastigocoleus testarum* and *Hyella caespitosa* as present in the upper supratidal zone (Gray Zone) is contrary to the findings of this study and those of Le Campion-Alsumard (1979a, b).

The findings of Schneider (1976), Le Campion-Alsumard (1969, 1975, 1979a, b), this study, and the unpublished work of K. Lukas in Florida and Jamaica suggests that certain general trends in the distribution of endolith assemblages are common to the Mediterranean and the Atlantic coast of North America. These trends include:

1) a decrease in species diversity from the intertidal through the supratidal zone;
2) a decrease in the abundance of eukaryotic species from the lower intertidal through the supratidal zone;
3) an abundance of *Hyella* spp., *Mastigocoleus testarum*, and *Plectonema terebrans* in the intertidal zone; and
4) an abundance of *Hormathonema* spp. in the supratidal zone.

In addition to these general trends, certain endolith species (*Hyella caespitosa, Plectonema terebrans, Mastigocoleus testarum, Kyrtuthrix dalmatica, Solentia foveolarum, Hormathonema* spp., and *Ostreobium quekettii*) were present in all studies and are more cosmopolitan in distribution (Lukas, 1973; Budd and Perkins, 1980, Table 1). Although most taxa are widely distributed, some, such as *Hyella* sp. A and *Herpyzenema* sp. which are found in Bermuda and not in France, may be limited regionally. Since the species identity of some forms, such as the fungi or the conchocelis stage of the bangioid rhodophyte, is not known, the regional distribution of these forms can only be evaluated by the occurrence of morphologically similar forms. Other general reports of bathymetric zonation (Golubic et al., 1975; Lukas 1979b; Budd and Perkins, 1980) support these common trends of intertidal and supratidal distribution, and the worldwide distribution of certain endolith species, but most studies do not differentiate among the subtidal, intertidal, and supratidal subzones. For instance *Ostreobium quekettii*, which Budd and Perkins (1980) cite as diagnostic of the lower photic zone (20–75 m deep), was also present in the lower intertidal areas of Bermuda and Marseille.

Regional comparisons of supratidal, intertidal, and subtidal endolith distribution based on common elements and widely distributed species can be made. However, to substantiate fully the degree of regionality an analysis of endolith zonation in physically comparable but regionally distinct areas is needed. Such studies should include a profile from the supratidal through the subtidal areas to characterize fully the vertical distribution of microbial endoliths.

PALEOECOLOGICAL APPLICATIONS

Most researchers acknowledge the wide time range (Cambrian to Holocene) of preserved endolith borings and their potential for use in making paleoecological interpretations. They also acknowledge that several factors limit such application at the present time: (1) our knowledge of the vertical and global distribution of modern and fossil endoliths; (2) the uncertain taxonomic status of modern and fossil endoliths; and (3) diagenetic changes which affect boring morphology (Golubic et al., 1975; Lukas, 1979b; Harris et al., 1979; May and Perkins, 1979).

A knowledge of the global distribution of fossil endoliths is essential to the use of fossil endoliths as paleoenvironmental indicators. If fossil endoliths were limited geographically, interpretation based on modern global assemblages may not be correct. However, the growing number of microfossil to modern organism comparisons suggests that some types of modern endolith organisms or related genera have existed from early in Earth's geologic history and lived in the same environments where their modern counterparts are found today (Knoll et al., 1975; Golubic and Hofmann, 1976; Golubic and Campbell, 1979). Published SEM photomicrographs of resin-cast borings comparing Pleistocene and Holocene *Hyella gigas* (Harris et al., 1979), Silurian and modern conchocelis stages of a bangioid rhodophyte (Campbell, 1980), and resin-cast borings of a Pleistocene endolith assemblage including *Mastigocoleus, Plectonoma, Hyella*, and *Phaeophila* (Budd and Perkins, 1980) illustrate that fossil to modern comparisons are possible with endoliths. In addition, it is likely that some stromatolite-building microorganisms might have had a global distribution. Stratigraphic correlation, based on the world-wide distribution of similar stromatolite morphologies seems possible (Walter, 1976, Chap. 7; Semikhatov et al., 1979). The stromatolites were formed by communities of cyanophytes, bacteria, and algae, and these microbial taxa have endolithic members.

However, the potential for paleoecological interpretation rests solely on our ability to recognize taxa by the morphology of their borings, and there are some uncertainties in our ability to do so. Problems in determining the influence of the substrate on the borehole, the degree of correspondence between the borehole morphology and the morphology of the organism itself, as well as the extent of spatial and temporal overlap of borings, contribute to these difficulties (Campbell and Hoffman, 1979). For example, does a single boring represent a boring filament or a single boring cell? We need to discriminate between borings made by morphologically similar members of different taxa whose size ranges overlap and to decide whether a boring morphology is the product of two different organisms or of two different periods of boring. These questions can be answered in many instances, but only with an increased ability to recognize taxa at the species level (Campbell and Hoffman, 1979).

Despite the problems, there are taxa which exhibit well-defined boring patterns; these species are readily recognizable in resin cast, SEM-scanned material. For example, in the intertidal zone of Bermuda, resin cast borings of many shapes are common, but the two dominant species are readily recognizable. The borings of *Mastigocoleus testarum* have characteristic enlarged terminal heterocysts and right-angle bends (Fig. 4C). The small, uniform diameter borings of *Plectonema terebrans* (Fig. 4D) are characteristic and can be separated from the less common but similarly sized fungal borings by the presence or absence of sporangia. In addition, the irregular and branched borings of *Ostreobium quekettii* produce an unmistakable cast and are characteristic of the lower intertidal and subtidal zones (see Budd and Perkins, 1980 for photograph of resin cast boring). A reduction in the number of species and the presence of two distinctively branching organisms which have well-defined casts characterize the supratidal zone. *Kyrtuthrix dalmatica* is typical of the lower supratidal subzone, while *Herpyzenema* sp. is more common in the supratidal zone (see Fig. 5D, and Golubic and Le Campion-Alsumard, 1973 for photographs of resin-cast borings). Although we may not be able to find modern analogs for all of these fossil casts, enough well-defined, recognizable morphologies should exist to aid in the recognition and interpretation of paleoenvironments.

Based on the presence of certain characteristic borings, the well-defined endolith zonation of the intertidal and supratidal zones can be recognized and has geological significance. This zonation can extend or refine previously suggested paleobathymetry and can aid in the recognition of ancient shoreline deposits (Perkins and Halsey, 1971) or of ancient carbonate coastlines, such as the supratidal zone interpretation Schneider (1976) made for a previously described coastline (Wendt, 1969).

Perhaps the most practical application, however, is the use of endolith zonation in discriminating between supratidal, intertidal, and subtidal environments. Assemblages of preserved endolith borings present in the lithified products of the supratidal, intertidal, and subtidal zones sould be usable for recognition of these zones. Since endolith distribution is affected by tidal range, their zonation may provide an independent means of assessing tidal range in a manner similar to Cloud's (1968) use of stromatolite morphology to estimate paleo-tidal range. It may even be possible to recognize displaced sediments and sea level changes by the co-occurrence of normally intertidal and supratidal species (Perkins and Halsey, 1971; Zeff and Perkins, 1979). Although rocky carbonate coastlines are not common in the fossil record, carbonate bioherms, including stromatolites, are common and originally formed in these environments. Living endoliths are present in the recent columnar stromatolites of Shark Bay. Australia (Golubic, 1976), and, by analogy, they should be present in such structures formed in ancient environments. Preserved endolith borings, which were contemporaneous with stromatolite build-up, could provide the key to defining the depositional environment and tidal range where the stromatolite was formed.

Paleoecological application of fossil endoliths on a limited level is possible even now. Present limitations can be overcome by an increased knowledge of the distribution and ecology of modern endoliths, and particularly of their boring morphologies. The use of preserved fossil endoliths will become more important in the future as our need to make and refine paleoenvironmental interpretations grow.

CONCLUSIONS

1) Microbial endolith assemblages vary between and within the intertidal and supratidal zones of Bermuda. The patterns of abundance and dominance of microbial endolith species are easily determined and can be used to recognize four intertidal and supratidal subzones around Bermuda. Differences in the distributional patterns of species across the intertidal and supratidal boundary are most pronounced, but significant differences occur in the abundance or dominance of endolith species in all subzones.
2) Associations of microbial endolith species can be correlated with, and may be in part responsible for, the changes in small-scale physical appearance and color of the rock surface.
3) Microbial endolith assemblages in the intertidal and supratidal zones of Bermuda are the same from place to place, and the same species are dominant at all sites, regardless of changes in the degree of slope or exposure to wave action.
4) On a regional level, the distribution of microbial endoliths of the supratidal-intertidal zones of Marseille, France; Rovinj, Yugoslavia; and Ber-

muda is similar. Differences in the physical characteristics of the rock surface, desiccation potential, insolation, and the presence of different invertebrates and macro-algae (rather than differences in geographical distribution of species) may account for the differences in the observed distributional patterns.

5) The use of fossil microbial borings as indicators of intertidal and supratidal conditions is possible and may yield important information on tidal ranges of the past and on ancient depositional environments. It seems probable that the endolith assemblages described in this study could be recognized in the rock record and used to identify and subdivide ancient carbonate supratidal and subtidal zones.

ACKNOWLEDGMENTS

Many people aided me in this research; I can acknowledge but a few personally. S. Golubic, Boston University, helped with the taxonomic identifications, and E. Seling (SCAN Laboratory, Harvard University), provided an expert eye on the SEM. I am grateful to B. W. Cameron, K. J. Lukas, J. M. Queen, and W. N. Tiffney for their helpful criticism and for reviewing the manuscript. I also wish to express my appreciation to the Bermuda Biological Station's permanent staff who helped my scientific efforts. Limited support for this research was provided by National Science Foundation Grant EAR-79-11200 to B. W. Cameron and S. Golubic, Boston University.

REFERENCES

BATHURST, R.G.C., 1966, Boring algae micrite envelopes and lithification of molluscan biosparite: Jour. Geology, v. 5, p. 15–32.

BORNET, E., AND FLAHAULT, C., 1888, Note sur deux nouveaux genres d'algues perforantes: Jour. Botanique, v. 2, p. 161–165.

______, AND ______, 1889, Sur quelques plantes vivant dans le test calcaire des Mollusques: Societé Botanique de France, v. 36, p. 148–177.

BUDD, D.A., AND PERKINS, R.D., 1980, Bathymetric zonation and paleoecological significance of microborings in Puerto Rican shelf and slope sediments: Jour. Sed. Petrology, v. 50, p. 881–904.

CAMPBELL, S.E., 1980, *Palaeoconchocelis starmachii* (Campbell, Kazmierczak & Golubic. A carbonate boring microfossil from the Upper Silurian of Poland (425 million years old): Implications for the evolution of the Bangiaceae (Rhodophyta): Phycologia, v. 19, p. 25–36.

______, AND HOFFMAN, E.J., 1979, Endoliths and their microborings: how close is the fit? (Abs.): Second International Symposium on Fossil Algae, Paris, France, April 1979.

CHARLTON, D.S., 1969, Intertidal zonation on Bermuda's rocky shores as an indicator of tidal range and wave energy, *in* Ginsburg, R.N., and Garrett, P. (eds.), Reports of Research–1968, Seminar on Organism–Sediment Interrelationships: Bermuda Biological Station Special Publication No. 2, p. 27–34.

CLOUD, P.E., 1968, Atmospheric and hydrospheric evolution on the primitive earth: Science, v. 160, p. 729–736.

CONWAY, E., AND COLE, K., 1977, Studies in the Bangiaceae: Structure and reproduction of the conchocelis phase of *Porphyra* and *Bangia* in culture (Bangiales, Rhodophyceae): Phycologia, v. 16, p. 205–216.

ERCEGOVIĆ, A., 1929, Sur quelques nouveaux types des Cyanophycées lithophytes de la côte adriatique: Archiv f. Protistenk, v. 66, p. 164–174.

______, 1930, Sur quelques types peu connus des Cyanophycées lithophytes: Archiv f. Protistenk, v. 71, p. 361–373.

______, 1932, Etudes écologiques et sociologiques des Cyanophycées lithophytes de la côte de Yugoslave de l'Adriatique: Bull. Int. Acad. Yugosl. Sci., v. 26, p. 33–56.

FOLK, R.L., ROBERTS, H.H., AND MOORE, C.H., 1971, Black phytokarst from Hell: Geol. Soc. America, Ann. Mtg. Abs. with Programs, v. 3, No. 7, p. 569–570.

GOLUBIC, S., 1967, Algen vegatation der Felsen, eine okologische Aldenstudie im dinarischen Karstegebiet: Binnengewasser: v. 23: Stuttgart, Schweizerbart, 183 p.

______, 1976, Organisms that build stromatolites, *in* Walter, M.R. (ed.), Stromatolites: Developments in Sedimentology No. 20, Amsterdam, Elsevier, p. 113–126.

______, BRENT, G., AND LE CAMPION, T., 1970, Scanning electron microscopy of endolithic algae and fungi using a multipurpose casting-embedding technique: Lethaia, v. 3, p. 203–209.

______, AND CAMPBELL, S.E., 1979, Analogous microbial forms in Recent subaerial habitats and in Precambrian cherts. *Gloeothece coerula* Geitler and *Eosynechococcus moorei* Hoffmann: Precambrian Research, v. 8, p. 201–217.

______, AND HOFMANN, H.J., 1976, Comparison of Holocene and mid-Precambrian Entophysalidaceae (Cyanophyta) in stromatolitic algal mats; cell division and degradation: Jour. Paleontology, v. 50, p. 1074–1082.

______, HOFFMAN, E.J., AND CAMPBELL, S.E., 1979, Study of fossil microborings; new approach: Am. Assoc. Petroleum Geologists Bull., v. 63, p. 458.

______, AND LE CAMPION-ALSUMARD, T., 1973, Boring behavior of marine blue-green algae *Mastigocoleus testarum* Lagerheim and *Kyrtuthrix dalmatica* Ercegovic as a taxonomic character: Schweiz. Zeitschr. Hydrol., v. 35, p. 157–161.

______, PERKINS, R.D., AND LUKAS, K.J., 1975, Boring microorganisms and microborings in carbonate substrates, *in* Frey, R.W. (ed.), The Study of Trace Fossils: New York, Springer Verlag, p. 229–259.

______, AND SCHNEIDER, J., 1979, Carbonate dissolution, *in* Trudinger, P.A., and Swaine, D.J. (eds.), Biochemical Cycling of Mineral-Forming Elements: Amsterdam, Elsevier, p. 107–129.

HANLEY, J.H., 1970, Intertidal and subtidal benthonic organisms of Shelly Bay, Bermuda, *in* Ginsburg, R. N., and Stanley, S.M. (eds.), Reports of Research–1969, Seminar on Organism-Sediment Interrelationships: Bermuda Biological Station Special Publication No. 6, p. 53–62.

HARRIS, P.M., HALLEY, R.B., AND LUKAS, K.J., 1979, Endolith microborings and their preservation in Holocene- Pleistocene (Bahama-Florida) ooids: Geology, v. 7, p. 216–220.

HOFFMAN, E.J., 1981, The distribution of microbial endoliths in the intertidal and supratidal coastal zones of Bermuda [Unpub. M.S. Thesis]: Boston, Boston University, 130 p.

KNOLL, A.H., BARGHOORN, E.S., AND GOLUBIC, S., 1975, *Paleopleurocapsa wopfnerii* gen. et sp. nov.: A late Precambrian alga and its modern counterpart: Proc. Natl. Acad. Sci. U.S.A., v. 72, p. 2488–2492.

KOBLUK, D.R. AND RISK, M.J., 1977a, Micritization and carbonate grain-binding by endolithnic algae: Am. Assoc. Petroleum Geologist Bull., v. 61, 1069–1082.

______, AND ______, 1977b, Calcification of exposed filaments of endolithic algae, micritic envelope formation and sediment production: Jour. Sed. Petrology, v. 47, p. 517–528.

KREBS, C.J., 1972, Ecology, the experimental analysis of distribution and abundance: New York, Harper and Row, 697 p.

LAEGERHEIM, G., 1886, Notes sure le *Mastigocoleus*, nouveau genre des algues marine de l'order des Phycochromacées: Notarisia, v. 1, p. 65–69.

LAND, L.S., MACKENZIE, F.T., AND GOULD, S.J., 1967, Pleistocene history of Bermuda: Geol. Soc. America Bull., v. 78, p. 933–1006.

LE CAMPION-ALSUMARD, T., 1969, Contribution a l'étude des Cyanophycées lithophytes des étages supralittoral et médiolittoral (region de Marseille): Tethys, v. 1, p. 119–172.

______, 1975, Etude experimental de la colonization de'éclats de calcite par les Cyanophycées endolithes marines: Cahiers de Biologie Marine, v. 16, p. 177, 185.

______, 1979a, Les Cyanophycées endolithes marines: systematique ultrastructure, ecologie et biodestruction. [unpub. Ph.D. thesis]: Marseille, Univ. Aix-Marseille-ll. T.I. (texte), 198 p., T. II (illustrations) 48 pl.

______, 1979b, Les Cyanophycées endolithes marines. Systematique, ultrastructure, écologie et biodestruction: Oceanol. Acta., v. 2, p. 143–156.

LEHMANN, E., 1903, Uber *Hyella balani* nov. spec.: Nyt. mag. Naturvidenskab, v. 41, p. 77–88.

LUKAS, K.J., 1973, Taxonomy and ecology of Recent endolithic microflora of reef corals with a review of the literature on endolithic microphytes [unpub. Ph.D. thesis]: Kingston, Univ. Rhode Island.

______, 1974, Two species of the chlorophyte genus *Ostreobium* from the skeletons of Atlantic and Caribbean Reef Corals: Jour. Phycology, v. 10, p. 331–335.

______, 1979a, Depth distribution and form among common microboring algae from the Florida continental shelf: Geol. Soc. America Ann. Mtg. Abs. with Programs, v. 19, p. 443.

______, 1979b, The effects of marine microphytes on carbonate substrata: Scanning Electron Microscopy II, p. 447–455.

______, AND HOFFMAN, E.J., 1984, New endolithic cyanophytes from the North Atlantic Ocean III. *Hyella pyxis* Lukas and Hoffman sp. nov.: Jour Phycology, v. 20, p.331–335.

MAY, J.A., AND PERKINS, R.D., 1979, Endolithic infestation of carbonate substrates below the sediment-water interface: Jour. Sed. Petrology, v. 49, p. 357–378.

MORRIS, B., BARNES, J., BROWN, F., AND MARKHAM, J., 1977, The Bermuda Marine Environment: Bermuda Biological Station Special Publication No. 15, 120 p.

NEUMANN, A.C., 1968, Biological erosion of limestone coasts, *in* Fairbridge, R. W. (ed.). The Encyclopedia of Geomorphology: New York, Rheinhold Book Corp., p. 75–81.

PÉRÈS, J.M., AND PICARD, J., 1964, Nouveau manuel de bionomie benthique de la mer Mediterranee: Rec. Trav. St. mar. Endoume, v. 31, p. 137.

PERKINS, R.D., AND HALSEY, S.D., 1971, Geologic significance of microboring fungi and algae in Carolina shelf sediments: Jour. Sed. Petrology, v. 41, p. 843–853.

______, AND TESNTAS, C.I., 1976, Microbial infestation of carbonate substrates planted in the St. Croix shelf, West Indies: Geol. Soc. America Bull., v. 87, p. 1615–1628.

PLUMMER, L.N., VACHER, H.L., MACKENZIE, F.T., BRICKER, O.P., AND LAND, L.S., 1976, Hydrogeochemistry of Bermuda: A case history of ground water diagenesis of biocalcarenites: Geol. Soc. America Bull., v. 87, p. 1301–1316.

ROONEY, W.S., AND PERKINS, R.D., 1972, Distribution and geological significance of microboring organisms within sediments of the Arlington Reef Complex, Australia: Geol. Soc. America Bull., v. 83, p. 1139–1150.

SCHNEIDER, J., 1976, Biological and inorganic factors in the destruction of limestone coasts: Contributions to Sedimentology, No. 6: Stuttgart, Schweizerbartsche Verlag, 112 p.

______, 1977, Carbonate construction and decomposition of epilithic and endolithic micro-organisms in salt- and freshwater, *in* Fluge, E. (ed.), Fossil Algae Results and Developments: Berlin, FDR, Springer-Verlag, p. 248–260.

SEMIKHATOV, M.A., GEBELEIN, C.D., CLOUD, P., AWRAMIK, S.M., AND BENMORE, W.C., 1979, Stromatolite morphogenesis—progress and problems: Canadian Jour. Earth Sciences, v. 16, p. 992–1015.

SWINCHATT, J.P., 1969, Algal boring: A possible depth indicator in carbonate rocks and sediments. Geol. Soc. America Bull., v. 80, p. 1391–1396.

STEPHENSON, T.A., AND STEPHENSON, A., 1972, Life Between Tide-Marks on Rocky Shores: San Francisco, W.H. Freeman, 425 p.

UPCHURCH, S.B., 1970. Sedimentation on the Bermuda platform [unpub. Ph.D. thesis]: Chicago, Northwestern Univ., 145 p.

WALTER, M.R. (ED.), 1976, Stromatolites: New York, Elsevier, 790 p.

WEBER VAN BOSSE, A., 1913, Siboga Expedition, Liste des Algues du Siboga. I. Myxophyceae, Chlorophyceae, Phaeophyceae: Monograph 59a Livr., v. 68, p. 1–136.

WENDT, J., 1969, Stratigraphie und Palaeogeographie des Roten Jurakalks im Sonnwendgebirge (Tirol, Osterreich): Neues Jahrb. Geol. Palaeont. Abh., v. 132, no. 2, p. 219–238.

ZEFF, M.L., AND PERKINS, R.D., 1979, Microbial alteration of Bahamian deep-sea carbonates: Sedimentology, v. 26, p. 175–201.

ICHNOFACIES OF A MODERN SIZE-GRADED SHELF, NORTHWESTERN GULF OF MEXICO

GARY W. HILL
U.S. Geological Survey, 12201 Sunrise Valley Dr., Reston, Virginia 22092

ABSTRACT

Biogenic sedimentary structures decrease in diversity and abundance with increasing water depth across the size-graded shelf of south-central Texas. Regional bioturbation patterns are related to macrobenthic infaunal assemblages and sediment facies. Dense and diverse assemblages of biogenic sedimentary structures are associated with shallow-water areas with relatively coarse substrates, low sedimentation rates, and dense and diverse infaunal assemblages. The outer shelf zone exhibits little bioturbation due to the presence of very few organisms, high sedimentation rates, and an underconsolidated substrate of muddy sediments of near-uniform grain size. Most traces have limited distribution ranges, high preservation potential, and vertical to subvertical orientation (or represent deep subhorizontal burrowing). In the rock record, estimating the degree of bioturbation in mud beds is the key to correctly interpreting regional bioturbation patterns for size-graded shelf deposits.

INTRODUCTION

A continental shelf where sediment cover is in equilibrium with its hydrodynamic regime, and where sediments become finer in a seaward direction, is termed a graded shelf (Stetson, 1953; Swift, 1969). Because about 70 percent of modern continental shelves are covered by relict sediments which are in at least partial disequilibrium with their present hydrodynamic environment (Emery, 1968), most modern shelves are unsuitable for study as the uniformitarian key to their ancient graded counterparts. This paper describes ichnofacies characteristics of the modern size-graded shelf off south-central Texas (Fig. 1) and discusses the nature of and balance between biogenic sedimentary structures, regional bioturbation patterns, biofacies, and grain size. Three major shelf facies are defined: (1) lower shoreface (water depth 10-30 m), (2) mid-shelf (30-120 m), and (3) outer shelf (120-200 m).

Prior to the mid-1960's, most geologic studies of the continental shelf focused on describing only the texture and mineralogy of surficial sediments (e.g., Shepard, 1932; Emery, 1952). These studies demonstrated that most shelves are composed of a complex mosaic of relict and modern sediments. This conclusion contrasted with an earlier supposition (e.g., Johnson, 1919) that the continental shelf was an equilibrium surface whose sediments decreased in grain size offshore. During the 1960's and 1970's shelf studies began to emphasize the operation and products of processes currently active in the shelf environment. It was during this period that the significance of biogenic sedimentary processes and products relative to their usefulness in paleoenvironmental interpretations and their effect on sediment dynamics was realized by the general geologic community. Applying our understanding of these processes and their products to shelf deposits has yielded distinctive sedimentary models (e.g., Swift et al., 1972) and ichnologic models (e.g., Seilacher, 1964, 1967) to classify continental shelf deposits. Most of the sedimentary models are in sharp contrast with the concept of a graded shelf (Johnson, 1978).

Today, continental shelves around the world are being extensively investigated. Few studies, however, have focused on modern graded shelves. A recent example is the sedimentological study of the Bering Sea by Sharma (1972). Pioneering ichnologic studies of shelf environments off the United States have been done by Robert W. Frey (University of Georgia) and James D. Howard (Skidaway Institute of Oceanography). Many of their studies (e.g., Frey and Howard, 1972; Howard and Reineck, 1972) are on the Georgia shelf where locally graded sediments occur out to the mid-shelf. An opportunity for ichnological studies on a fully graded shelf (beach to shelf-slope break) evolved from the need to do large-scale environmental studies associated with offshore petroleum lease sales off Texas. These studies resulted in detailed reports on the geology (Berryhill *et al.*, 1976), chemistry and biology (Parker, 1976), and physical oceanography (Angelovic, 1976).

ENVIRONMENTAL SETTING

The study area (Fig. 1) is part of the continental shelf in the western Gulf of Mexico off south-central Texas. This area extends from Corpus Christi Bay in the north, to Baffin Bay in the south, and seaward to about the 200 m isobath. The shoreward limit approximates the 10 m isobath. The study area encompasses approximately 4,650 km^2 and is the only part of the Texas continental shelf which is size-graded. Ancestral deltas of the Brazos-Colorado River and Rio Grande River lie to the north and south respectively. These Pleistocene deltas extend to the shelf-slope break and are made up of relict sediments.

The surface of the continental shelf in the study area is relatively smooth and gently sloping (Fig. 1). Average shelf width is 94.5 km, with a gradient of 1.8 m/km.

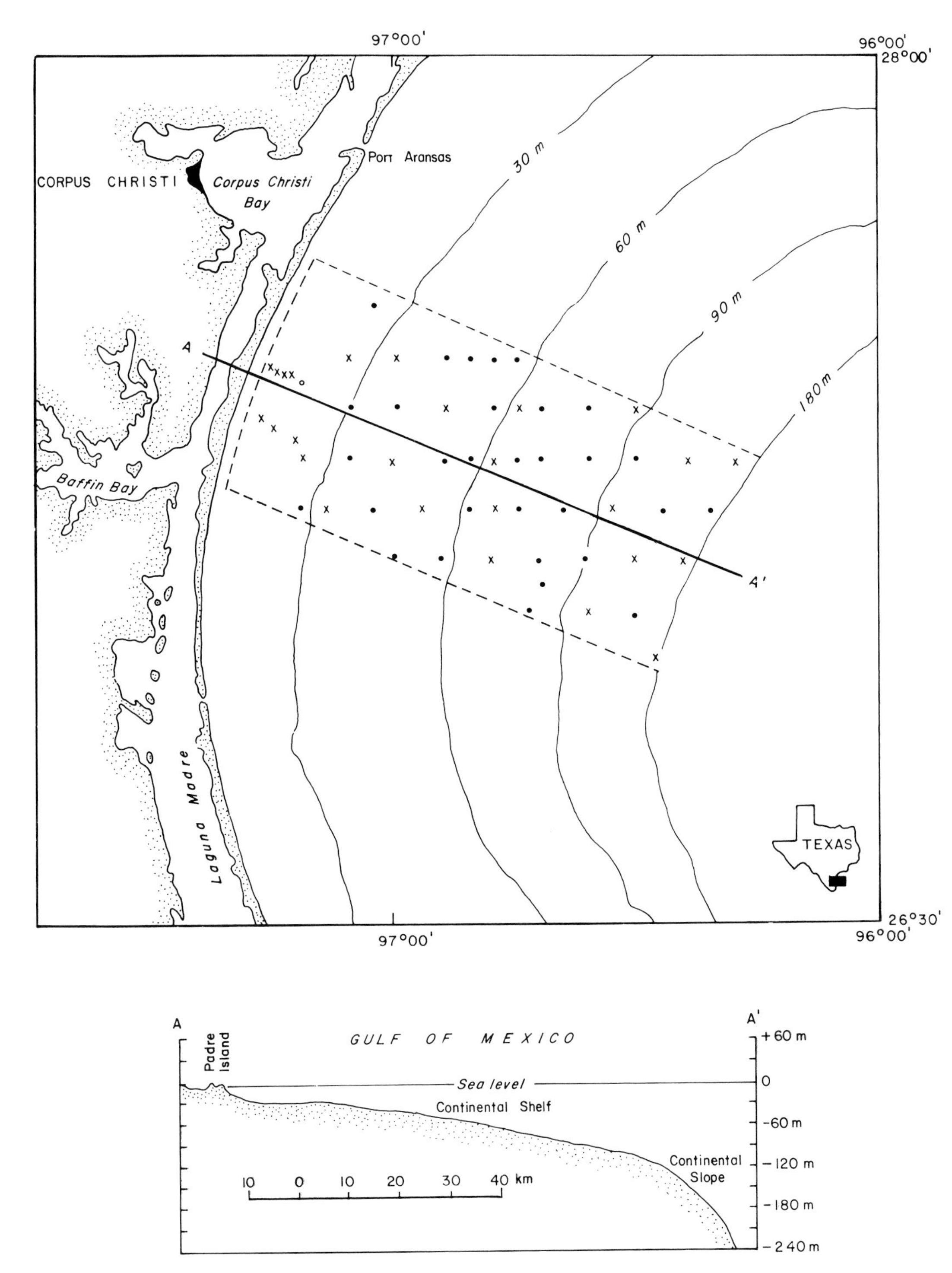

Fig. 1.—Index map of study area and sample locations. Biological and grain size analyses made of all samples; cores taken a stations indicated by crosses.

The study area is in a semiarid warm-temperate climate. Mean annual rainfall at Corpus Christi is 72.5 cm. Mean monthly temperature ranges from 14.0°C in January to 29.0°C in August, and the mean annual temperature is 22.1°C (National Oceanic and Atmospheric Administration (NOAA), 1974). Winds are predominantly southeasterly during the summer and northerly during the winter. Mean annual wind speed at Corpus Christi is 5.3 m/sec, and resultant wind direction is 121°.

Tides along the south-central Texas coast have a mean diurnal range of 51.8 cm (microtidal) at Port Aransas (NOAA, 1975). Few data are available regarding wave climate. Heights of breakers along the nearshore are normally 0.3 to 1.0 m (personal observation). Breakers higher than 2 m occur on several days each year, largely during storms in the fall, winter, and spring. Wave approach is from the southeast during summer and dominantly from the northeast during winter.

Mean monthly water temperature at Port Aransas ranges from 13.6°C in January to 30.0°C in August; mean annual water temperature is 22.7°C (NOAA, 1973). Mean monthly salinity, calculated from the water density at Port Aransas, ranges from 29.5‰ to 36‰ (NOAA, 1973).

Well-defined temperature and salinity stratification exists intermittently off the south-central Texas coast (Jones et al., 1965; Berryhill et al., 1976). Winter water temperatures in the northern Gulf of Mexico approach those found offshore in the Atlantic from North Carolina to Long Island; in the summer, temperatures rise higher than those in the Caribbean. Parker (1960) compared summer and winter average bottom-water temperatures in the northern Gulf of Mexico and noted convergence of temperature values at about 80 m and again at 120 m.

Circulation patterns of both littoral and semi-permanent shelf currents are complex in the study area. Numerous investigations have been conducted in the area to determine drift rates and patterns (Curray, 1960; Kimsey and Temple, 1963, 1964; Watson and Behrens, 1970; Hunter et al., 1974; Hill et al., 1975; Shideler, 1979). These studies suggests the existence of a yearly cycle of coastward water movement primarily controlled by seasonal winds. Uniformly southward winter drift and uniformly northward summer drift define the extremes of seasonal variation, but atypical winds may alter these conditions during some years.

During some seasons, a drift convergence may occur in the study area. The position of the convergence tends to shift northward during the spring and southward during the fall, resulting in the net development of migrating surface and bottom convergence zones along the south-central Texas coast. Convergence zones are structurally complex both in horizontal and vertical sections.

METHODS

A Smith-MacIntyre grab sampler[1] was used to collect bottom samples from fifty stations (Fig. 1). After subsamples were removed for textural analyses, the remaining sediments (about 14.5 liters per sample) were washed through a 0.5 mm mesh sieve. Organisms recovered were fixed in 10 percent formalin, preserved in 45 percent isopropyl alcohol, and later identified and counted in the laboratory. These samples were collected during the period October 25 to December 22, 1974; consequently, this study represents only a "snapshot" picture of benthic biological conditions. Only one grab sample (approximately 1,096 cm^2; 15 L) was taken at each station because of time and logistical constraints.

Eleven box cores (30 x 30 x 50 cm) and thirty pipe cores, some as much as 2.0 m long, were collected for analysis of physical and biogenic sedimentary structures (Fig. 1). In the laboratory, X-ray radiographs were made of half-round cores and slabs using radiographic scanning techniques as described by Hill et al. (1979).

Textural data are from Berryhill et al. (1976) and are part of a larger study by Schideler (1977). The silt/mud fraction was analyzed by a 16-channel TA Coulter Counter, and the sand-size fraction was analyzed by use of a Rapid Sediment Analyzer (RSA). Detailed description of analytical methods for the textural analysis is in Berryhill et al. (1976).

BIOGENIC SEDIMENTARY STRUCTURES

On the south-central Texas shelf, biogenic sedimentary structures, which are described below, vary in type, size, and orientation. Some larger biogenic structures are obvious, while others are indicated by subtle textural differences or changes in sediment color; most small traces are revealed only by X-ray radiography. Because very few organisms were collected directly from the biogenic sedimentary structures, associating organisms and structures is difficult.

Trace A
Fig. 2A

Description.—Vertical, tightly spiralling burrow; 4 to 6 mm tube diameter, 1 to 1.5 cm spiral diameter, 1 to 1.5 cm between spiral loops; burrow with mucus-lined smooth walls; observed burrow lengths up to 35 cm.

Environment.—Outer continental shelf, water depth 90 to 200 m (Table 1); clayey-silt sediment.

Tracemaker.—Unknown.

Trace B
Fig. 2B

Description.—Tube with 3 mm overall diameter, 1 to 1.5 mm tube shaft, 1 mm mucus-agglutinated, thin

[1]Any use of trade names is for descriptive purposes only and does not imply endorsement by the United States Geological Survey.

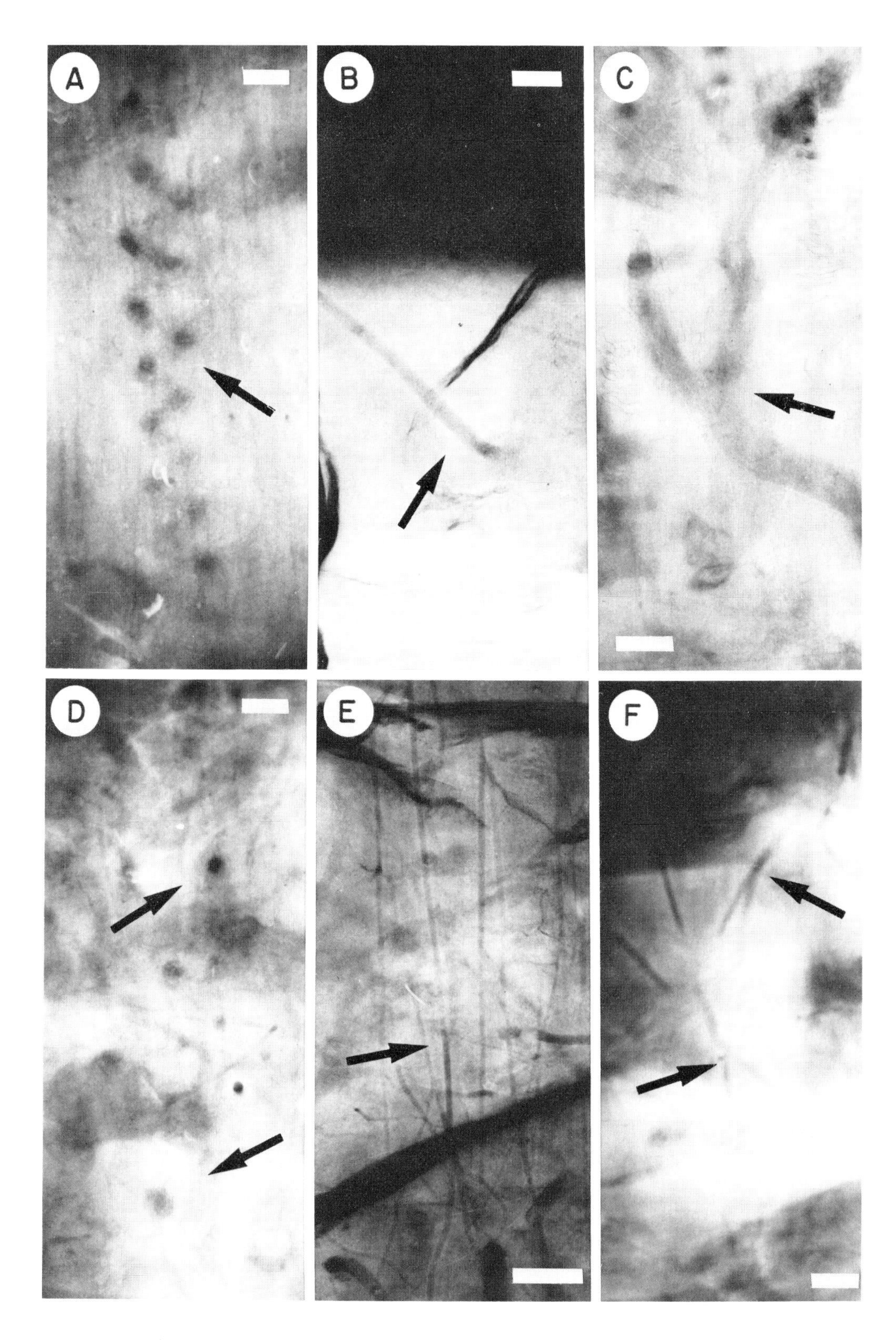
A
B
C
D
E
F

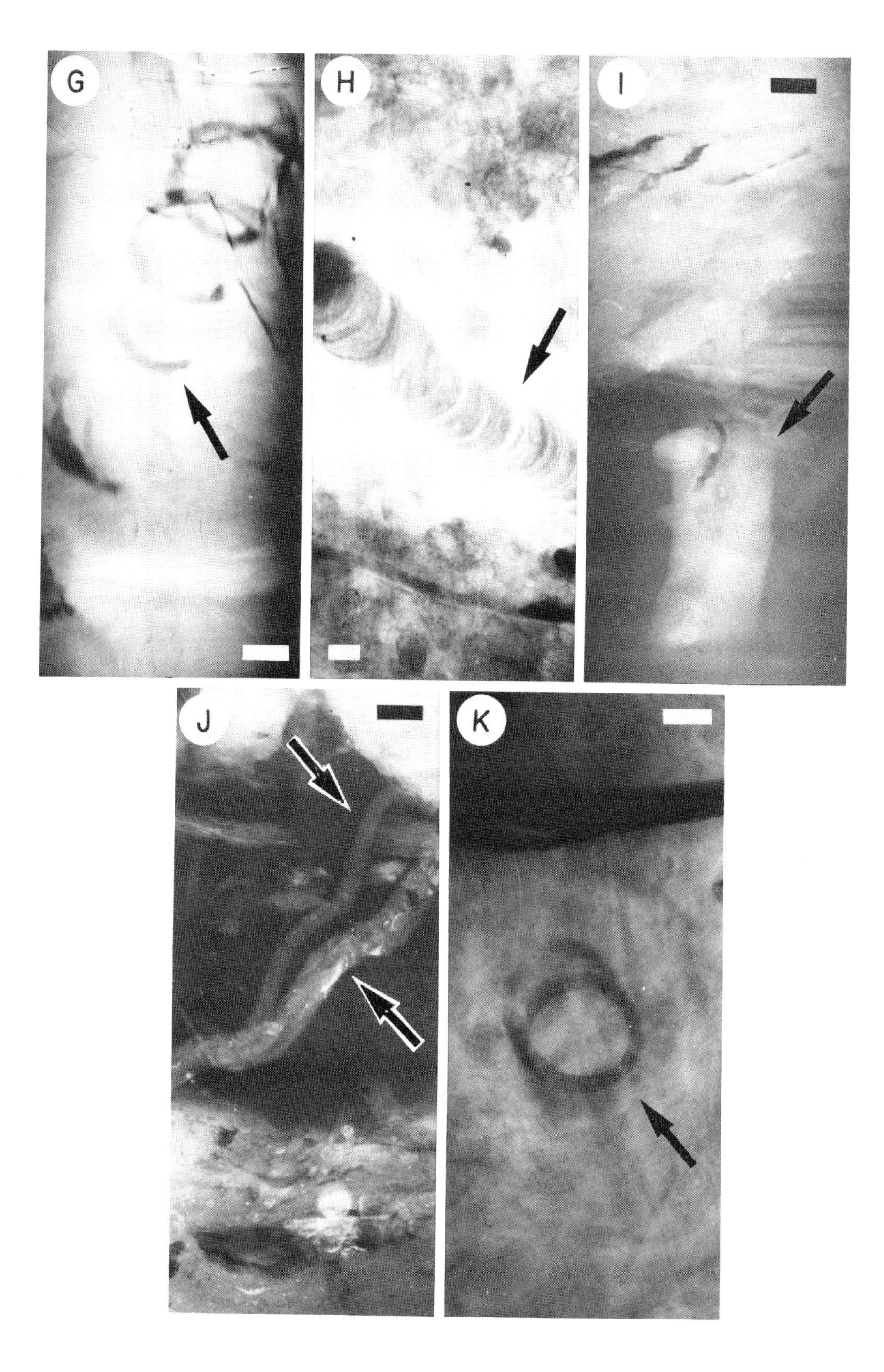
G
H
I
J
K

Fig. 2.—Examples of specific biogenic sedimentary structures observed in cores (sections perpendicular to bedding). In these X-ray negatives mud is black, sand is white, and sandy mud is a mottled gray. Arrows point to trace; see text for descriptions. (Bar scale = 1 cm)

TABLE 1.—DISTRIBUTION OF TRACES. SEE TEXT FOR DESCRIPTION OF EACH TRACE; TRACE DESIGNATION CORRESPONDS TO TEXT

Water Depth						Trace					
Interval (m)	A	B	C	D	E	F	G	H	I	J	K
<30		X	X	X	X	X	X	X	X	X	X
30–60		X	X	X	X	X					
60–90		X	X	X	X						
90–120	X	X	X	X	X						
120–150	X	X	X	X							
>150	X	X	X	X							

walls; straight to irregular and unbranched; subhorizontal to subvertical; observed tube lengths as much as 20 cm.

Environment.—All water depths throughout the study area (Table 1); common in sandy to clayey-silt sediments.

Tracemaker.—Unknown.

Trace C
Fig. 2C

Description.—Burrow with 4 to 7 mm diameter, burrow diameter constant. Occasional thin (0.1 mm) mucus lining. Configuration is variable—straight sections (a few cm long) subvertical, possibly U-shaped, rarely branched; observed lengths up to 15 cm.

Environment.—Found at all water depths (Table 1); at shallower depths in coarser sediments, this trace is confined to muddy beds. (Note: In this report, mud refers to detritus less than 63 μm in diameter; i.e., silt + clay detritus.)

Tracemaker.—*Glycera* sp., Polychaeta.

Trace D
Fig. 2D

Description.—Tube with horizontal orientation, cross to oblique sections common in radiographs; 0.2 to 1.0 cm in diameter, with agglutinated walls as much as 3 mm thick; tubes mostly filled by mud, wall material slightly coarser than fill material; observed lengths up to 12 cm.

Environment.—Commonly found at all water depths (Table 1) in coarse- to fine-grained sediments.

Tracemaker.—Unknown.

Trace E
Fig. 2E

Description.—Tube with vertical to subvertical unbranched shafts; 3 to 8 mm overall diameter, walls 1 to 3 mm thick, shaft 0.5 to 2.0 mm wide; agglutinated walls of coarse-grained material which commonly contain some sand; extends from sediment-water interface to depths of 35 cm.

Environment.—Inner to mid-shelf, water depths less than 120 m (Table 1); common in sandy to silty sediments.

Tracemaker.—Unknown.

Trace F
Fig. 2F

Description.—Burrow with branched, subvertical shafts; branching at 25 to 45° angles; 1 to 3 mm overall diameter, shaft 1 to 2 mm, thin mucus-impregnated walls (0.2 to 0.5 mm); observed lengths to about 5 mm.

Environment.—Inner shelf, water depths less than 60 m (Table 1); rare, and found only in heavily bioturbated, muddy sand or sandy mud.

Tracemaker.—Unknown.

Trace G
Fig. 2G

Description.—Loosely wound spiralling burrow; loop spacing slightly irregular (1 to 3 cm); burrow winds upward subvertically; tube diameter 5 to 7 mm, spiral diameter 1.5 cm; smooth mucus-lined walls; observed lengths up to 15 cm.

Environment.—Inner shelf at depths less than 30 m (Table 1); rare, and found only in muddy sand or sandy mud deposits.

Tracemaker.—Unknown.

Trace H
Fig. 2H

Description.—Locomotion trace composed of U-in-U backfill structures, 3 to 6 cm in diameter; cross-section tends to be semicircular but poorly defined; horizontal orientation; laminae defined by alternating sand and finer-grained layers; observed lengths as much as 30 cm.

Environment.—Distinct traces found only on the inner shelf in water less than 30 m deep (Table 1), but organism producing trace was collected in deeper water (as deep as 120 m); rare in sandy substrates.

Tracemaker.—*Moira atropos* (Lamarck), Echinodermata, Echinoidea.

Trace I
Fig. 2I

Description.—Unbranched, vertical burrow measuring 1 to 2 cm in diameter; thin mucus lining (0.1 to 0.2 mm); in places filled with shell debris in otherwise shell-free sediment.

Environment.—Inner shelf, water depths less than 30 m (Table 1); penetrates bedding types ranging from mud to sand; rare.

Tracemaker.—Unknown.

Trace J
Fig. 2J

Description.—Sinuous burrow with oblique orientation; 2 to 5 mm in diameter; no wall, no lining; filled with sand and/or shell material.

Environment.—Inner shelf, water depths less than 30 m (Table 1); common in relatively thin (a few cm) mud beds.

Tracemaker.—Unknown.

Trace K
Fig. 2K

Description.—Horizontal, tightly spiralling burrow; 3 to 4 mm burrow diameter, spiral diameter 2.5 cm; smooth walls not reinforced; observed lengths as much as 20 cm.

Environment.—Inner shelf, water depths less than 30 m (Table 1); common in intensely bioturbated sandy mud or muddy sand.

Tracemaker.—Unknown.

The preservation potential of these traces is worth noting. All of the traces were found in upper (less than 100 cm) and lower (greater than 100 cm) sections of core. Traces in lower core sections are well below erosion depth. Additionally, all types of traces were often found in a "filled" condition; that is, when the animal departed, the trace did not simply collapse and disappear. As a result, these traces have good potential for being incorporated into the rock record.

Differences in the distribution and diversity of traces across the shelf have environmental significance. Most of the traces (more than 70%) are limited to specific ranges in water depths (Table 1). Shallow water sediments tend to have a greater diversity of traces than do sediments on the outer shelf (Table 1). However, only two conspicuous traces in this study (1) occur principally in only one shelf environment, and (2) are found in most samples in some abundance. These traces (which have the same characteristics as traces described by Hertweck, 1972) are trace A (Fig. 2A), found in water depths greater than about 100 m, and trace E (Fig. 2E), found in water depths of 20 to 100 m (see Table 1).

BIOTURBATION PATTERNS

The degree to which sediments are bioturbated (Fig. 3) varies systematically across the shelf. A graded scale next to each graphic core (Fig. 4) illustrates the degree of bioturbation in the core. Comparison of many cores analyzed in this manner enables one to define the bioturbation pattern vertically within cores and regionally across the shelf.

Bioturbation varies vertically between beds within each core. These beds are components of major, cyclic bedding sequences representing fluctuating depositional conditions associated with storm versus fair-weather conditions (Hill, in prep.). On the inner shelf, for example, heavy bioturbation is normally associated with the uppermost bed (bioturbated muddy sand) in the sequences (Fig. 4). Heavily bioturbated beds represent fair-weather periods during which the rate of biogenic reworking of the sediment exceeds the rates of depositional processes.

However, complete depositional sequences (e.g., basal laminated sand grading upward into interlaminated mud-sand and then into muddy sand) are rare (Fig. 4). This is in part due to (1) bioturbation of storm-deposited sands obscuring the sharp contact between cycles, especially in pipe cores with small diameters, (2) complete bioturbation of some thin (a few cm) storm deposits over time, and (3) termination of part or all of a sequence at any point, depending upon the depth to which the sea bed may be eroded by subsequent storms.

Howard and Reineck (1972) suggested that under ideal conditions bioturbation would gradually increase with increasing water depth (i.e., greater bioturbation in environments less subject to bottom disturbances such as storms). In this study, however, bioturbation generally decreases seaward (Fig. 4). Even very thin (approximately 1 mm) laminations are clearly evident in deeper water cores (Fig. 5). Although the study was not designed specifically to identify factors that control regional bioturbation patterns, the variety of environmental aspects that were measured (biological, hydrological, and sedimentological; see Berryhill et al., 1976) can be assessed as to their role in observed regional bioturbation patterns.

FACTORS CONTROLLING BIOTURBATION PATTERNS

The observed regional bioturbation pattern is related to macrobenthic infaunal assemblages and sediment facies.

Biota

Number of species and individuals.—Samples from 50 stations yielded 952 individuals representing several

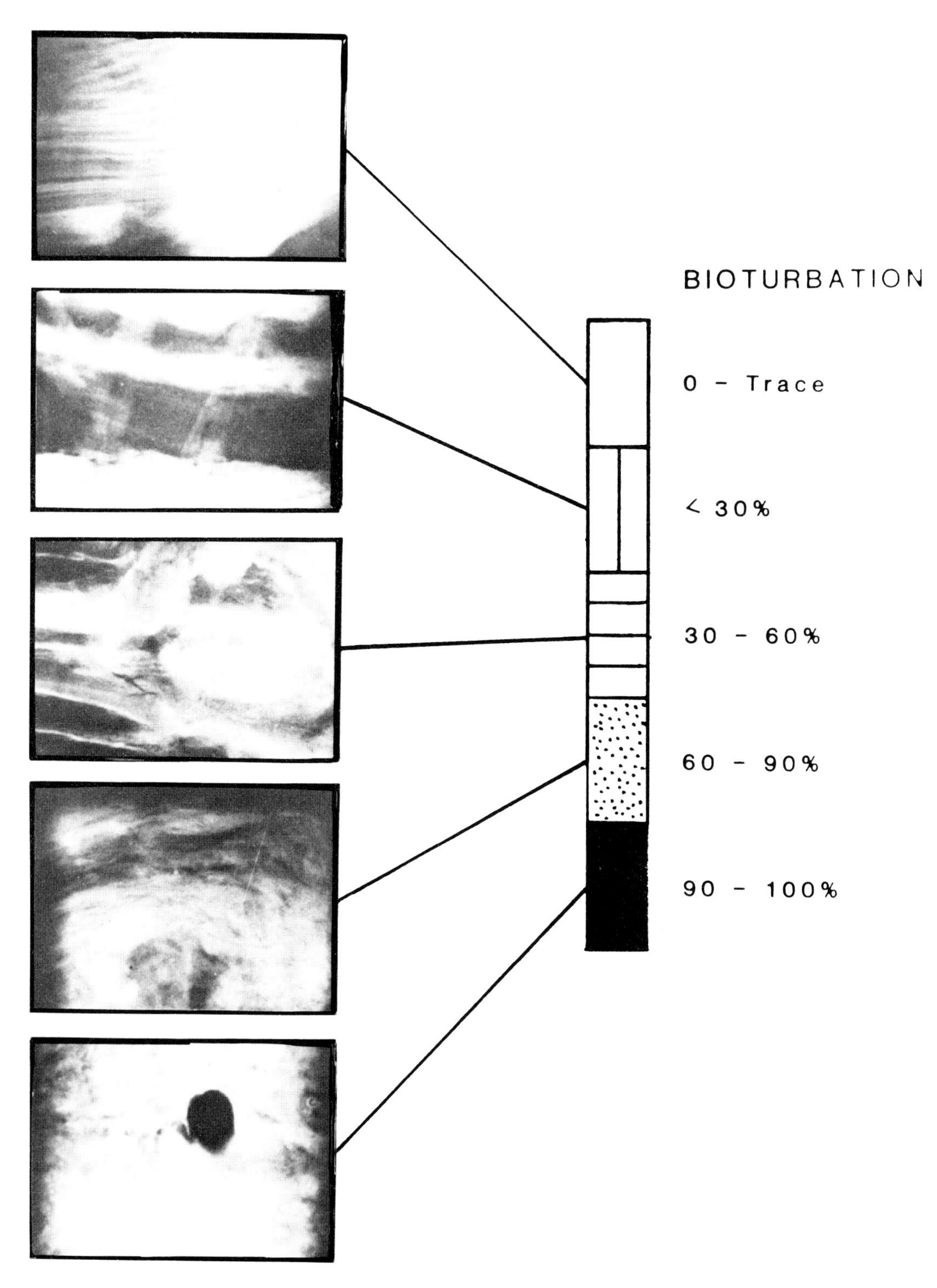

Fig. 3.—Scale showing different degrees of bioturbation. In these X-ray negatives, mud is black and sand is gray. (after Howard and Reineck, 1972)

taxonomic groups (Table 2) and 74 species (Table 3). Major taxonomic groups having the greatest number of species were Polychaeta (53%), Mollusca (23%), and Crustacea (16%). The greatest number of individuals collected belong to Crustacea (50%), Polychaeta (32%), and Mollusca (8%). Other taxonomic groups account for less than 10 percent of the total number of species and individuals.

Numbers of species per sample (species richness) ranged from zero to 13 with a mean of 6.5 (Fig. 6). Species richness generally decreases with increasing water depth to about 70 to 100 m and then gradually increases again offshore. The degree of variation in species richness is relatively small throughout the study area.

Number of individuals per sample (density) ranges from zero to 143 with a mean of 20.3 (Fig. 7). Density decreases sharply offshore to about 40 to 50 m and then, on the average, slightly increases with increasing water depth. Variation in density between samples is small on the mid-shelf and relatively larger at depths shallower than 50 m and deeper than 150 m. The greatest number of individuals occurs in water depths shallower than 30 m.

Species distribution.—Each taxonomic group exhibits its distinct patterns of diversity and density distribution with increasing depth (Figs. 6 and 7, Table 3). The number of crustacean and polychaete species per sample is about the same throughout the shelf, although both groups may decrease slightly in diversity seaward. Diversity of mollusc species increases offshore and density of crustacean individuals shows a marked decrease with increasing depth below 30 m. On average, molluscs are more numerous on the outer shelf (>150 m) than in shallower water. Polychaete density generally decreases with increasing water depth.

On the basis of distribution, the macrobenthic infauna (Table 3) can be divided into three main groups. Group one consists of several species that occur across the entire shelf. These species were very common at many stations and include nemerteans, ophiurids, echiurids, the mollusc *Corbula* sp., the *Nephtys* amphipod *Ampelosca cristoides*, the polychaetes *Nephtys picta*, *Cossura delta*, *Paraprionspio pinnata*, *Lumbrineris* sp., *Onuphis* sp., *Ninoe nigripes*, a cirratulid, and a pilargid. A second major group, including only a few species, consists of two smaller subgroups: species only at water depths of 20 to 120 m (e.g., *Ancistrocyllis papillosa*) or only at water depths of 90 to 200 m (e.g., *Glycera* sp.). The majority of infauna are in a third general group of species which is restricted to the inner, middle, or outer third of the Texas shelf. Twenty-four species (e.g., *Diopatra cuprea*) were found only on the inner shelf (20 to 60 m). *Sigambra tentaculata* is one of some 12 species collected only from the mid-shelf (60 to 120 m). Eleven species (e.g., *Terebellides stroemi*) were restricted to the outer shelf (120 to 200 m).

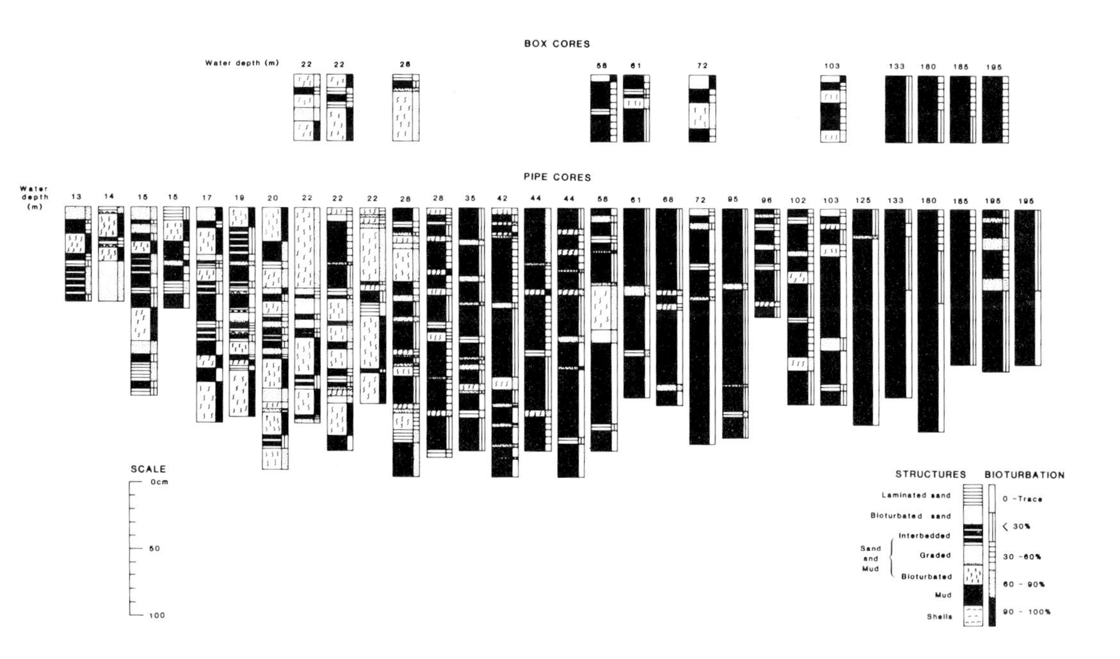

Fig. 4.—Graphic display of cores. Symbols indicate various sedimentary structures. Degree of bioturbation is shown in the right-hand column (see also Fig. 3).

Because sampling was done in the winter and only one grab sample was taken at each station, it is necessary to determine whether the sampling design biased the data and precluded valid interpretations. A study of benthic invertebrates on the south Texas shelf during 1975 (Holland, 1976) included two transects of six stations each in or immediately adjacent to this study area. Holland took four grab samples, which he estimated would collect about 60 percent of the species present, at each station during winter, spring, and summer seasons. His results (Table 4) indicate that the number of species and individuals per sample generally decreases in a seaward direction regardless of season. Quantitatively, his winter results are very similar to those reported here. Holland did not sample sediment in water more than 131 m deep, so his data do not document the gradual increase in number of species and individuals at depths greater than 150 m observed in this study. I conclude that the greater density of sample stations used in this study overcomes some of the variability caused by use of a single grab per station, therefore yielding valid regional patterns of distribution for both numbers of species and of individuals.

The values of species and individual density reported in this study are considerably lower than those reported from other nearby areas. For example, Holland et al. (1974) reported 338 benthic taxa from Corpus Christi Bay, and densities as high as 11,856 individuals/0.5 m^3. Manheim (1975) reported 190 species of polychaetes alone from the shelf off Mississippi, Alabama, and the Florida panhandle. The disparity in densities is probably the result of homogeneous fine-grained sediments characteristic of the study area (Berryhill et al., 1976).

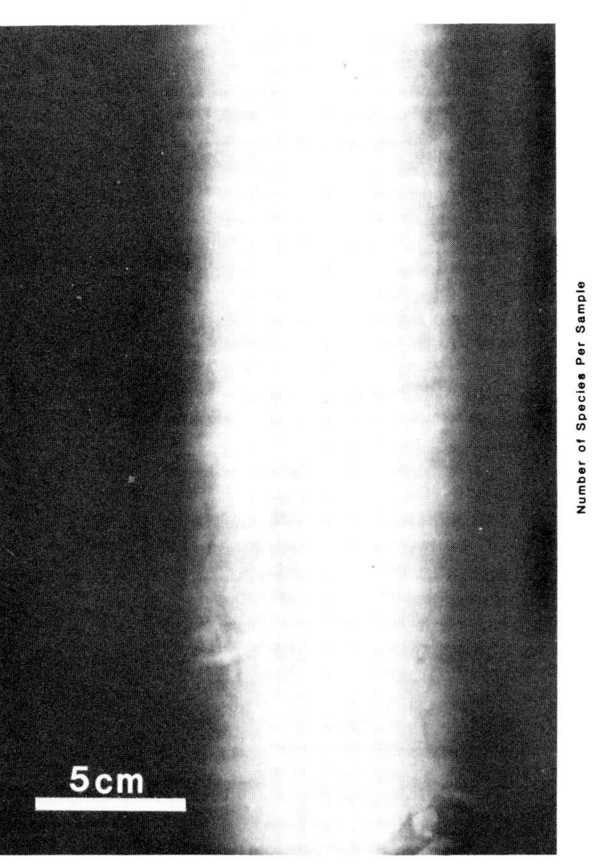

Fig. 5.—X-ray negative showing very thin parallel lamination in a mud bed cored at 195 m.

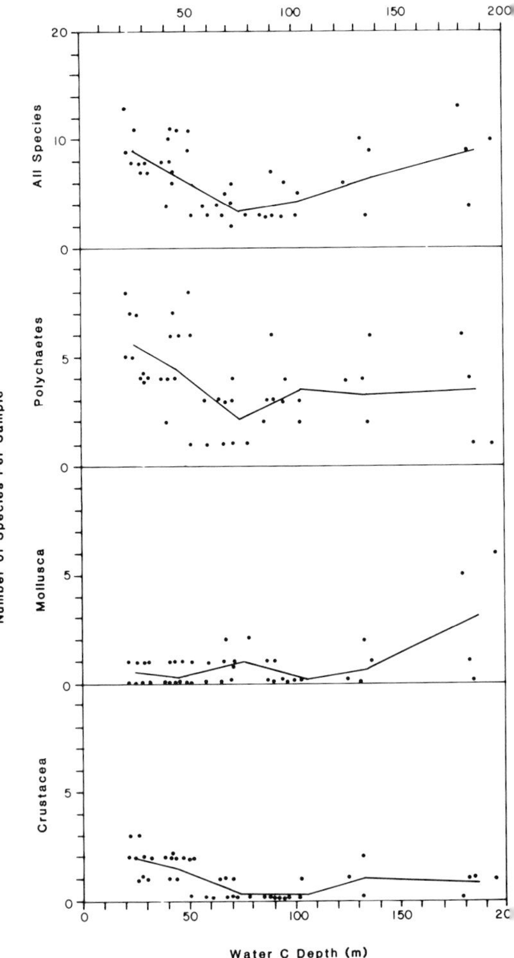

Fig. 6.—Taxonomic diversity (species richness) across the south-central Texas continental shelf. Average values represented by solid line.

Factors controlling infaunal zonation.—Based mainly on the distribution of molluscan fauna, Parker (1960) recognized several benthic assemblages on the Texas shelf. Three of the assemblages he described are of particular significance to this study: intermediate shelf assemblage (22 to 73 m), outer shelf assemblage (73 to 119 m), and upper continental slope assemblage (119 to 1,098 m). Parker concluded that the distribution of species followed certain ranges of bottom temperatures and major sediment types (to be discussed later).

As part of the total geological study of the south Texas continental shelf (Berryhill et al., 1976), water temperature data were collected from expendable bathythermographs. For the winter of 1974, bottom-water temperatures reflect seasonally changing temperatures to about mid-shelf, with the bottom of the

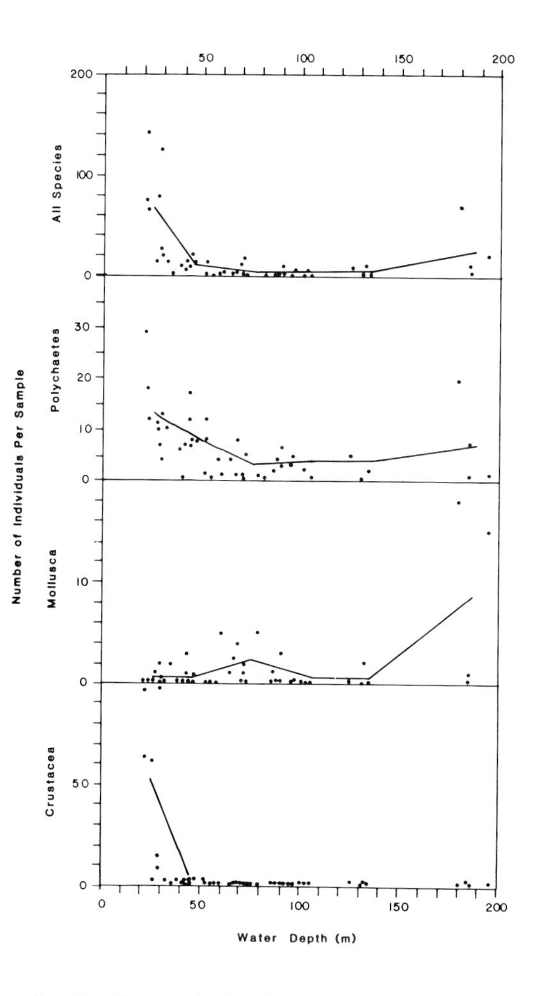

Fig. 7.—Taxonomic abundance across the south-central Texas continental shelf. Average values represented by solid line.

TABLE 2.—NUMBER OF SPECIES AND INDIVIDUALS COLLECTED FROM 50 STATIONS

Taxon	No. of Species	No. of Individuals
Polychaeta	39 (53%)	309 (32%)
Crustacea	12 (16%)	474 (50%)
Mollusca	17 (23%)	72 (8%)
Others	6 (8%)	97 (10%)
Total =	74 (100%)	952 (100%)

TABLE 3.—DISTRIBUTION OF INFAUNAL SPECIES

Species	Water Depth (m) 60	60-120	120
Cossura delta	----------	----------	----------
nemerteans	----------	----------	----------
Paraprionspio pinnata	----------	----------	----------
Nephtys picta	----------	----------	----------
ophiurid	----------	----------	----------
echiurid B	----------	----------	----------
Lumbrineris sp.	----------	----------	----------
cirratulid	----------	----------	----------
Onuphis sp.	----------	----------	----------
pilargid	----------	----------	----------
Corbula sp.	----------	----------	----------
Ampelisca cristoides	----------	----------	----------
Ninoe nigripes	----------	----------	----------
Nuculana concentrica	----------	----------	
Scolelepis sp.	----------	----------	
unidentified polychaete	----------	----------	
Aricidea sp.	----------	----------	
Magelona sp.	----------	----------	
nereid	----------	----------	
Ampelisca cf. cucullata	----------	----------	
paraonid	----------	----------	
unidentified gastropod	----------	----------	
Ancistrocyllis papillosa	----------	----------	
Glycera sp.		----------	----------
echiurid A		----------	----------
Nuculana acuta		----------	----------
Renilla mulleri	----------		
Macoma pulleyi	----------		
Squilla empusa	----------		
Gammarus sp.	----------		
Diopatra cuprea	----------		
Pseudeurythoe sp.	----------		
Nephyts sp.	----------		
paraonid	----------		
Amphinome rostrata	----------		
Malacoceros sp.	----------		
Armandia maculata	----------		
Malacoceros indicus	----------		
Pinnixa retinens	----------		
Chasmocarcinus mississippiensis	----------		
Eudorella monodon	----------		
Armandia agilis	----------		
Leptognatha gracilis	----------		
Speocarcinus sp.	----------		
Sthenelais sp.	----------		
Notomastus latericeus	----------		
Goniada sp.	----------		
spionid	----------		
Paraonis sp.	----------		
Polyophthalmus pictus	----------		
Cyclinella tenuis		----------	
Volvulella texasiana		----------	
Diplodonta sp.		----------	
Pyrunculus caelatus		----------	
Pinnixa sp.		----------	
heterospionid		----------	
Sigambra tentaculata		----------	
maldanid		----------	
Moira atropos		----------	
Dentalium sp.		----------	
tanaid		----------	
Glycinde sp.		----------	
flabelligerid			----------
Spiophanes sp.			----------
orbiniid			----------
Terebellides stroemi			----------
Tmetonyx sp.			----------
Cycolcardia sp.			----------
nuculanid			----------
Verticordia sp.			----------
Verticordia ornata			----------
unidentified bivalve			----------
Mysella sp.			----------

seasonal layer at about 70 to 80 m (Fig. 8). These results agree with Parker (1960), who showed that at a depth of approximately 75 m the average winter and summer bottom-water temperatures differ, reflecting water mixing to the bottom (Fig. 9). There is some indication in Parker's report that seasonal changes in temperatures might occur to depths as great as 150 m. In this study, the magnitude of biological parameters decreases down to about 120 m. Seaward of that depth, the biological parameters increase.

The rough correlation of benthic assemblage boundaries with the bottom of the seasonal water-temperature layer and with the increase in the magnitude of biological parameters in depths of relatively constant bottom-water temperatures indicates that water depth and bottom-water temperature are probably important in controlling macrobenthic infaunal zonation. The increase in infaunal density and diversity in deeper water may be explained in the biological concept of the ecotone or boundary effect where two habitats overlap; the boundary environment is more favorable than that of either habitat alone (Odum, 1959). In this study, the circulation of deep water from the Gulf of Mexico over the outer shelf section of the study area (Parker, 1960;

TABLE 4.—TOTAL NUMBER OF SPECIES AND INDIVIDUALS FOR THE FOUR REPLICATES AT EACH STATION OF HOLLAND (1976)

Transect Number	Station No.	Latitude	Longitude	Water Depth (m)	Winter		Spring		Summer	
					No. of Species	No. of Individuals	No. of Species	No. of Individuals	No. of Species	No. of Individuals
II	1	27°40′	96°59′	22	22	228	43	1481	27	116
II	2	27°30′	96°44.5′	49	14	29	27	66	19	33
II	3	27°17.5′	96°23′	131	7	12	13	18	11	15
III	1	26°57.5′	97°11′	25	13	133	34	301	23	116
III	2	26°57.5′	96°48′	65	7	14	25	53	19	30
III	3	26°57.5′	96°32.5′	106	11	16	13	21	26	63

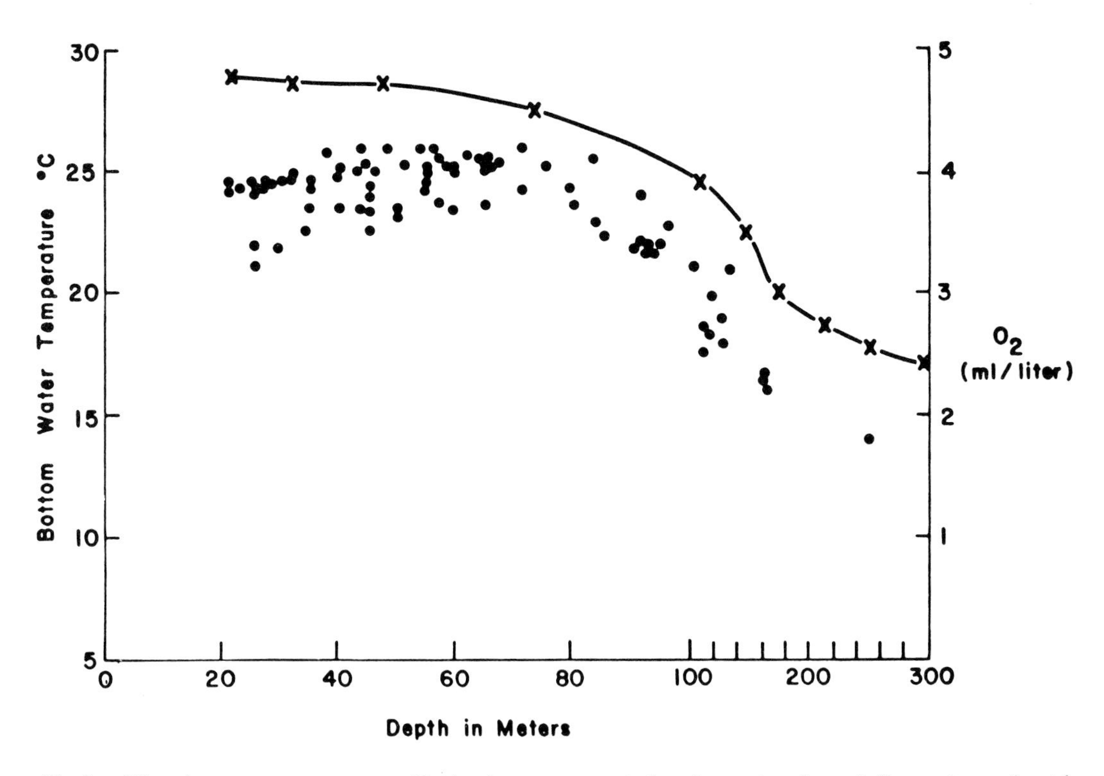

Fig. 8.—Winter bottom-water temperatures (dots) and oxygen concentrations (crosses) on the south Texas outer continental shelf. Oxygen data from Churgin and Halminski (1974).

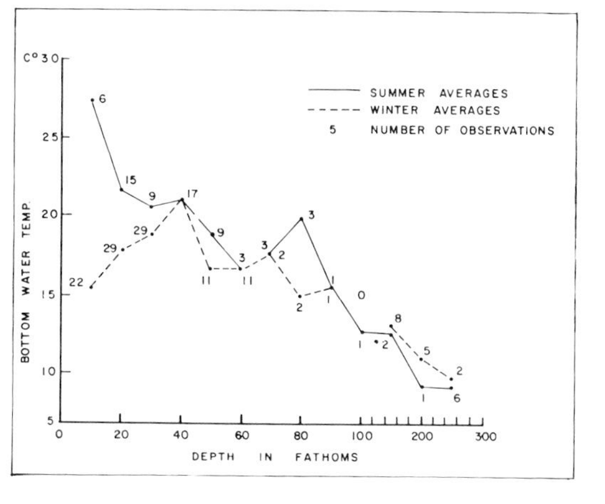

Fig. 9.—Comparison of summer and winter average bottom-water temperatures in the northern Gulf of Mexico (1951-1955) (from Parker, 1960).

Berryhill et al., 1976) causes a change in environmental conditions. The result is a mixing of outer shelf and upper slope infauna that increases both the number of species and density of the fauna on the outer shelf.

Zonation of bioturbation generally parallels macrobenthic infaunal zonation; i.e., density and diversity of both bioturbation structures and macrobenthic infauna decrease offshore. The exception to this generally parallel relation is the increase in diversity and density of organisms (and not of bioturbation structures) at the shelf's edge. Consideration of sedimentological characteristics of the study area help to explain the above parallel relationships and exceptions.

Sediments

Two types of textural relationships that illustrate the regional pattern of sediment distribution in the study area are (1) single component percentages (gravel, sand, silt, and clay), each of which has a unique distribution pattern, and (2) mean grain size (Fig. 10).

Gravel (grain size > 2 mm in diameter) is a very minor constituent, representing a maximum of 1 percent in some samples but totally absent in most. The gravel-sized material is mostly shell detritus; lithic fragments were found in only two samples. Shell material occurs throughout the area but is most common in water depths shallower than 30 m.

The sand-sized fraction (63 to 200 μm) is largely terrigenous (mainly quartz) with some biogenic detritus. Sand is the predominant sediment size in depths less than 20 m and quickly diminishes seaward to less than 10 percent. (Note: Due to problems with the analytical procedure, foraminifers, which are of sand size, were removed from the sediment sample prior to grain-size analysis. Foraminifers were extremely abundant in some samples from the shelf edge.)

The most common grain size is silt (3.9 to 63 μm). The amount of silt increases from less than 10 percent in water as shallow as 10 m to a maximum of 70 to 80 percent in water depths of 20 to 30 m. Deeper than 30 m the amount of silt gradually decreases toward the outer shelf edge.

Clay (less than 4 μm) has a distribution pattern unlike other sediment constituents. Clay detritus shows a relatively sharp but minor increase in abundance between 10 and 30 m and gradually increases with increasing water depth. Only at the shelf edge do clay-sized particles represent as much as 50 percent of the sediment. In the 3.0 to 0.45 μm size fractions, the predominant clay mineral is an expandable type, probably calcium montmorillonite (Berryhill et al., 1976). This mineral represents 40 to 90 percent of the material in the clay fraction. Illite is the second most common clay; trace amounts of a chlorite-type mineral also occur.

Mean grain size (Fig. 10) ranges from fine sand (3.0) to clay (7.9 ϕ). Most of the study area is covered by fine to very fine silt (6.0 to 7.9 ϕ). Coarser sediment (< 6.0 ϕ) is limited to water depths generally shallower than about 20 m. Overall, mean grain size decreases with increasing water depth; the most pronounced change is between 10 and 30 m. The decrease in grain size seaward largely reflects increasing distance from coastal source areas and decreasing wave energy.

Relationship Between Bioturbation and Sediment Zonation

Comparison of regional distribution maps suggests a similarity between the zonation of bioturbation structures (Fig. 4) and the physical characteristics of the sediment (Fig. 10). Diversity and density of biogenic sedimentary structures decrease as mean grain size decreases; mean grain size generally decreases as water depth increases. This parallel relationship might be explained through an understanding of the effect of sediment texture and the variation in sedimentation rate across the shelf on organism distribution.

A number of investigators (Parker, 1956, 1960; Boyer, 1970; Stanton and Evans, 1971, 1972; Hill, 1975; Holland, 1976) have noted the close relationship between the distribution of macroinvertebrates and variations in sediment texture in the northwestern Gulf of Mexico. A decreasing number of species and individuals (Figs. 6 and 7) is associated with decreasing grain size (Fig. 10). This trend might be explained in part by the dependence of species diversity on structural diversity of the habitat (Beerbower and Jordan, 1969). The shallower parts of the study area have the greatest variation in sediment grain size (habitat diversity) and consequently support a more diverse fauna than the relatively homogeneous, fine-grained sediments of deeper water. A contradiction to the general trends discussed above is the increase both in numbers of species and in individuals at the eastern edge of the shelf. Besides the ecotone concept mentioned earlier, the greater diversity and density of organisms on the outer shelf may reflect a response to a sedimentological characteristic not indicated in Figure 10. The bottom sediments on the far eastern edge of the shelf contain many foraminiferal tests, among which *Orbulina* is conspicuous. The numerous foraminifers (approaching foram oozes in some samples) in otherwise muddy sediments of near uniform grain size would effectively increase variation of the sediment grain size and thus affect the number of infauna.

The increase in density and diversity of fauna on the eastern edge of the study area is not accompanied by an increase in bioturbation. Variation in rates of sedimentation across the shelf might explain this observation.

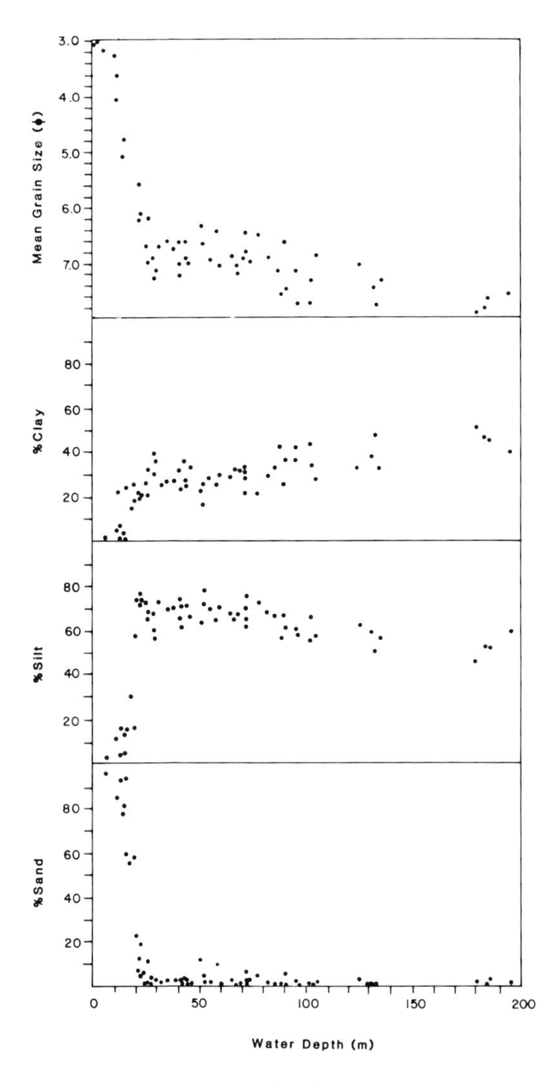

Fig. 10.—Distribution of sediment by grain size within the study area.

Studies by Holmes and Martin (1977) and Berryhill (U.S. Geological Survey, personal communication) indicate average sedimentation rates increase in a seaward direction - almost a centimeter per year on the outer shelf compared to no sedimentation for the lower shoreface. In addition to higher sedimentation rates, the underconsolidated nature (due to very high water content) of the uppermost sediment on the outer shelf apparently precludes any preservation of near-surface traces. Only "long" traces with a vertical to subvertical orientation or horizontal traces produced at depth survive both erosional events and sediment liquefaction processes near the sediment-water interface. The combination of sparse infauna, relatively high sedimentation rates, and underconsolidated sediments on the outer shelf results in sediments which exhibit little evidence of bioturbation.

Implications for Interpretation of Rock Record

A word of caution is in order concerning how these sediments might appear in the rock record. Through compaction and diagenesis, the fine parallel laminations characteristic of the muddy outer shelf sediments would probably "disappear" or be extremely difficult to detect an outcrop. The occasional sand bed in outer shelf deposits is usually highly bioturbated. It has been my experience that the occurrence of "structureless" mud beds interbedded with highly bioturbated sand beds leads to an interpretation by many field geologists that a sequence is bioturbated throughout. Consequently, it would be possible to interpret shelf deposits such as the ones described in this study as being characterized by increasing bioturbation in an offshore direction. An interpretation of this sort is reinforced by the literature, which supports such an observation on other types of shelves. The key to correctly interpreting the offshore decrease in bioturbation of deposits of a size-graded shelf as described here is in estimating the degree of bioturbation in mud beds - a very difficult job in sediments of near-uniform grain size.

SUMMARY AND CONCLUSIONS

1) The degree of bioturbation varies significantly and systematically across the sediment size-graded continental shelf off the south-central Texas coast. Diversity and density of biogenic sedimentary structures decreases seaward.

2) Most traces are limited in distribution across the shelf; only two traces, however, are conspicuous and abundant enough to be termed characteristic traces. Long, straight vertical tubes with agglutinated coarse-grained walls (trace E) are characteristic of the inner and mid-shelf (< 120 m). Outer shelf (120 to 200 m) deposits are characterized by a burrow that is best described as a tightly looping vertical spiral (trace A).

3) Zonation of biogenic sedimentary structures is related to macrobenthic infaunal assemblages and sediment facies. Diversity and density of traces generally

decrease as a) infaunal assemblages become less dense and diverse and b) the substrate grain-size becomes finer and more uniform.

4) Where deep water circulation and upper slope infaunal assemblages extend onto the outer shelf, an increased infaunal density/diversity and sediment diversity is not accompanied by increased bioturbation, apparently due to high sedimentation rates and the underconsolidated nature of the sediments.

5) The large variation in degree of bioturbation across the shelf contrasts with relatively small variations in grain size, sedimentation rates, and diversity/density of macrobenthic assemblages in the study area. The balance between physical and biogenic processes which result in observed bioturbation patterns is possibly due in large part to the overall density and diversity of macrobenthic infauna, which is very low compared with shelf areas off other parts of the United States.

ACKNOWLEDGMENTS

This paper was significantly improved through the thoughtful, critical review of Ralph Hunter, Henry Berryhill, James Howard, Robert Frey, and Allen Curran. I thank each of them for their efforts, concerns, and suggestions.

REFERENCES

Angleovic, J.W., ed., 1976, Environmental Studies of the South Texas Outer Continental Shelf, 1975–Physical Oceanography (Vol. II): NOAA, Rept. to Bureau of Land Management (Contract 08550-15A-19), 290 p.

Beerbower, J.R., and Jordan, D.S., 1969, Application of information theory to paleontologic problems: Taxonomic diversity: Jour. Paleontology, v. 43, no. 5, p. 1184–1198.

Berryhill, H.L., Shideler, G.L., Holmes, C.W., Hill, G.W., Barnes, S.S., and Martin, R.G., jr., 1976, Environmental Studies, South Texas Outer Continental Shelf, 1975: Geology: U.S. Department of Commerce, NTIS Pub. No. PB 251–341, 350 p.

Boyer, P.S., 1970, Actuopalaeontology of the large invertebrates on the coast of Louisiana [unpub. Ph.D. thesis]: Houston, Rice University, 97 p.

Churgin, James, and Halminski, S.J., 1974, Temperature, salinity, oxygen, and phosphate in waters off the United States, Volume III. Gulf of Mexico: Washington, D.C., National Oceanic and Atmospheric Administration, National Oceanographic Data Center, 117 p.

Curray, J.R., 1960, Sediments and history of Holocene transgression, continental shelf, northwest Gulf of Mexico, *in* Shepard, F.P., Phleger, F.B., and van Andel, Tj. H. (eds.), Recent Sediments, Northwest Gulf of Mexico: Tulsa, Am. Assoc. Petroleum Geologists, p. 221–266.

Emery, K.O., 1952, Continental shelf sediments off southern California: Geol. Soc. America Bull., v. 63, p. 1105–1108.

______, 1968, Relict sediments on continental shelves of the world: Am. Assoc. Petroleum Geologists Bull., v. 52, p. 445–464.

Frey, R.W., and Howard, J.D., 1972, Georgia coastal region, Sapelo Island, U.S.A.: Sedimentology and biology, VI. Radiographic study of sedimentary structures made by beach and offshore animals in aquaria: Senckenbergiana marit., v. 4, p. 169–182.

Hertweck, G., 1972, Georgia coastal region, Sapelo Island, U.S.A.: Sedimentology and biology, V. Distribution and environmental significance of lebensspuren and in-situ skeletal remains: Senckenbergiana marit., v. 4, p. 125–167.

Hill, G.W., 1975, Macrobenthic infaunal zonation on the south Texas outer continental shelf [unpub. M.S. thesis]: Corpus Christi, Texas A&I University at Corpus Christi, 81 p.

______, Dorsey, M.E., Woods, J.C., and Miller, R.J., 1979, A radiographic scanning technique for cores: Marine Geology, v. 29, p. 93–106.

______, Garrison, L.E., and Hunter, R.E., 1975, Maps showing drift patterns along the north-central Texas coast, 1973–1974: U.S. Geological Survey MF Map 714, one sheet.

Holland, J.S., 1976, Benthos project–invertebrates, *in* Parker, P.L. (ed.), Environmental Assessment of the South Texas Outer Continental Shelf–Chemical and Biological Survey Component: Rept. to Bureau of Land Management (Contract 08550-CT5-17), Univ. Texas, Texas A&M Univ., and Rice Univ., p. 200–249.

______, Maciolik, N.J., Kalke, R.D., and Oppenheimer, C.H.,1974, A benthos and plankton study of Corpus Christi, Copano and Aransas Bay system: Second Annual Report to the Texas Water Development Board, 212 p.

Holmes, C.W., and Martin, E.A., 1977, Rates of sedimentation, *in* Berryhill, H.L. (ed.), Environmental Studies, South Texas Outer Continental Shelf, 1976–Geology: Report to the Bureau of Land Management, p. 230–246.

Howard, J.D., and Reineck, H.E., 1972, Georgia coastal region, Sapelo Island, U.S.A.: Sedimentology and biology, IV. Physical and biogenic sedimentary structures of the near-shore shelf: Senckenbergiana marit., v. 4, p. 81–123.

Hunter, R.E., Hill, G.W., and Garrison, L.E., 1974, Drift patterns along the south Texas coast, 1970–1973: U.S. Geological Survey MF Map 623, two sheets.

Johnson, D.W., 1919, Shore Processes and Shoreline Development: New York, John Wiley, 584 p.

Johnson, H.D., 1978, Shallow siliciclastic seas, *in* Reading, H.G. (ed.), Sedimentary Environments and Facies: New York, Elsevier, p. 207–258.

Jones, R.S., Copeland, B.J., and Hoese, H.D., 1965, A study of the hydrography of inshore waters in the western Gulf of Mexico off Port Aransas, Texas: Publ. Institute Marine Science, v. 10, p. 22–32.

Kimsey, J.D., and Temple, R.F., 1963, Currents on the continental shelf of the northwestern Gulf of Mexico, *in* Fishery Research, Biological Laboratory, Galveston, Texas, Fiscal Year 1962: U.S. Bureau of Commercial Fisheries Circular 161, p. 23–27.

______, and ______, 1964, Currents on the continental shelf of the northwestern Gulf of Mexico, *in* Fishery Research, Biological Laboratory, Galveston, Texas, Fiscal Year 1963: U.S. Bureau of Commercial Fisheries Circular 183, p. 25–27.

Manheim, F.T., 1975, Baseline environmental survey of the Mississippi, Alabama, Florida (MAFLA) lease area: St. Petersburg, Fla., Florida State University, Institute of Oceanography, 114 p.

National Oceanic and Atmospheric Administration, 1973, Surface water temperature and density, Atlantic coast, North and South America: National Ocean Survey Publication 31-1, 12 p.

______, 1974, Local climatological data, annual summary with comparative data, 1974: Corpus Christi, Texas, National Environmental Data Service, 4 p.

______, 1975, Tide Table, 1975, East Coast of North and South America: Rockville, Md., National Ocean Survey, 288 p.

Odum, E.P., 1959, Fundamentals of Ecology: Philadelphia, W.B. Saunders, 546 p.

Parker, P.L., ed., 1976, Environmental Assessment of the South Texas Outer Continental Shelf–Chemical and Biological Survey Component: Rept. to Bureau of Land Management (Contract 08550-CT5-17), Univ. Texas A&M, and Rice Univ., 598 p.

Parker, R.H., 1956, Macro-invertebrate assemblages as indicators of sedimentary environments in east Mississippi delta region: Am. Assoc. Petroleum Geologists Bull., v. 40, no. 2, p. 295–376.

______,1960, Ecology and distributional patterns of marine macroinvertebrates, northern Gulf of Mexico, *in* Shepard, F.P., Phleger, F.B., and van Andel, Tj. H. (eds.), Recent Sediments, Northwest Gulf of Mexico: Tulsa, Am. Assoc. Petroleum Geologists, p. 307–337.

Seilacher, A., 1964, Sedimentological classification and nomenclature of trace fossils: Sedimentology, v. 3, p. 253–256.

______, 1967, Bathymetry of trace fossils: Marine Geology, v. 5, p. 413–428.

Sharma, G.D., 1972, Graded sedimentation on the Bering Shelf: Rep. 24th Int. Geol. Cong., Montreal, Section 8, p. 262–271.

Shepard, F.P., 1932, Sediments on continental shelves: Geol. Soc. America Bull., v. 43, p. 1017–1034.

Shideler, G.L., 1977, Late Holocene sedimentary provinces, South Texas Outer Continental Shelf: Am. Assoc. Petroleum Geologists Bull., v. 61, p. 708–722.

______, 1979, Regional surface turbidity and hydrographic variability on the South Texas continental shelf: Jour. Sed. Petrology, v. 49, no. 4, p. 1195–1205.

Stanton, R.J., and Evans, I.A., 1971, Environmental controls of benthic macrofaunal patterns in the Gulf of Mexico adjacent to the Mississippi delta: Gulf Coast Assoc. Geol. Soc. Trans., v. 21, p. 371–378.

______, and ______, 1972, Recognition and interpretation of modern molluscan biofacies, *in* Contributions on the geological and geophysical oceanography of the Gulf of Mexico: Texas A&M University Oceanographic Studies, v. 3, p. 203–222.

Stetson, H.C., 1953, The sediments of the western Gulf of Mexico: Part I–The continental terrace of the western Gulf of Mexico; its surface sediments, origin, and development: Mass. Inst. Tech. & Woods Hole Oceanographic Inst., v. 12, no. 4, p. 3–45.

Swift, D.J.P., 1969, Evolution of the shelf surface, and the relevance of the modern shelf studies to the rock record, *in* Stanley, D.J. (ed.), The NEW Concept of Continental Margin Sedimentation: Washington, D.C., Am. Geol. Institute, 112 p.

______, Duane, D.B., and Pilkey, O.H., 1972, Shelf Sediment Transport, Process and Pattern: Stroudsburg, Pa., Dowden, Hutchinson and Ross, 656 p.

Watson, R.L., and Behrens, E.W., 1970, Nearshore surface currents, southeastern Texas Gulf coast: Contributions to Marine Science, v. 15, p. 133–143.

MODERN ALGAL MATS IN INTERTIDAL AND SUPRATIDAL QUARTZ SANDS, NORTHEASTERN MASSACHUSETTS, U.S.A.

BARRY CAMERON, DIANE CAMERON, AND J. RICHARD JONES
Acadia University, Wolfville, Nova Scotia BOP 1XO; Brow Mountain Road, Glenmont, Nova Scotia BOP 1HO; and University of Texas, Austin, Texas 78712

ABSTRACT

Occurrences of algal mats at Plum Island along the northeastern coast of Massachusetts at 43°N latitude consist of laminated quartz silt and sand bound by mucous-secreting, filamentous blue-green algae (Cyanophyta). Recent discoveries of modern algal mats in quartz sands along cool, temperate coasts further confirms the fact that algal mat-forming organisms have wide temperature and salinity as well as substrate tolerances.

At Plum Island, algal mats occur along the high intertidal to low supratidal margins of a metahaline pond in a swale on a recurved spit at the southern end of the island. They are also found along the margin of the marsh next to the edge of the backdune area of the island's dune field.

The algal mats on Plum Island are vertically stratified into three major zones. The upper 1 mm of the mat in the spit area is dark brownish green due to filaments of *Lyngbya*, *Microcoleous*, and other smaller filamentous blue-green (?) algae. Coccoid cyanophytes (including *Enthophysalis*?), diatoms, *Euglena*, and nematode worms also are present. Below this upper green layer, there is a thinner, pinkish layer containing anaerobic, photosynthetic purple sulfur bacteria. Underlying the pinkish layer, there usually is a third layer composed of organic-rich black sand, 1-10 cm thick, indicating anaerobic conditions. This black layer contains the remains of buried older surface mats whose laminae can be recognized as alternating layers of decaying organic matter and layers of water-laid sands and wind-blown silts. Gelatinous material aiding filament-binding of sediment extends 1-7 mm below the upper surface. No carbonate cementation was found.

The larger structures associated with these algal mats include gas domes, desiccation ploygons, elongated ridges, rolled mats, mounds, and distorted mats redeposited after flotation.

Algal mat growth affects physical sedimentation in part of the interdune area of the spit by stabilizing silt and sand after erosion and/or sediment deposition. By stabilizing each new sediment surface, these mats aid upward growth of low, ponded areas of the spit. Overturned and rolled mats produce mounds that become colonized by vascular plants and thus initiate dune development on low, flat areas.

Ancient quartzose algal mats may be difficult to recognize in the stratigraphic record, but occurrences are known. Preservation potential is low because of the lack of carbonate cementation. Possible criteria for recognition include fine sand and silt laminae, desiccation cracks, elongated ridges, mounds, and overturned mats.

INTRODUCTION

Much of the literature on modern algal mats and stromatolites notes that their occurrence today is generally restricted to arid to semi-arid, tropical to subtropical carbonate marine environments (e.g., Bathurst, 1975). However, Awramik et al. (1978) pointed out in their review that modern stromatolite analogs or algal mats occur in many marine environments, not just the intertidal and supratidal zones, as well as in lakes, streams, and thermal springs. They and a number of other authors (usually in the context of other studies) also have noted that modern algal mats can occur in siliciclastic sediments, but not as much is known about these occurrences as is known about their carbonate counterparts. The purposes of this report are to (1) provide a description of the recently discovered cool temperate, quartzose algal mats at Plum Island along the northeast coast of Massachusetts (Jones and Cameron, 1978a); (2) describe their geologic environment; and (3) evaluate their geologic and paleontologic significance and fossilization potential.

A stromatolite is a laminated biogenic sedimentary structure ". . . formed by sediment trapping and binding and/or mineral precipitation with prostrate microbial communities termed algal mats . . . Algal mats form more or less cohesive fabrics of intertwined filaments and/or gelatinous matter produced by both filamentous and coccoid microorganisms" (Golubic, 1976a, p. 113). These microorganisms are principally blue-green algae (cyanophytes) (Awramik and Margulis, in Walter, 1976, p. 1; Golubic, 1976b).

The algal mats at Plum Island consist of uncemented, laminated quartz silt and sand bound by mucous-secreting, filamentous blue-green algae (Cyanophyta). These algal mats are similar in community structure and generic composition to other modern occurrences in warmer and/or more arid climates which contain a cosmopolitan flora (Golubic and Awramik, 1974).

STROMATOLITE AND ALGAL MAT OCCURRENCES

Stromatolites first appeared over three billion years ago (Golubic, 1973; Gebelein, 1976) when the only known organisms were procaryotic, such as blue-green algae and bacteria (Kingdom Monera). Although the diversity of stromatolite forms has significantly decreased since the Precambrian, possibly due to the evolution of competing eucaryotic algae and grazing metazoa such as gastropods, trilobites, and burrowers (Garrett, 1970; Awramik, 1971; Friedman et al., 1973; Kepper, 1974; Cussey and Friedman, 1976), stromatolites are well-known from Phanerozoic marine sedimentary rocks. Modern studies of stromatolites emphasize (1) microbial evolution through study of preserved cellular materials in silicified Precambrian occurrences (e.g., Golubic and Hofmann, 1976), (2) developing a stromatolite biostratigraphy for Precambrian correlations (e.g., Semikhatov, 1976), and (3) their use as paleoenvironmental indicators (see Walter, 1976). Thus, the discovery and description of modern stromatolite analogs from different sedimentary environments is significant.

A consensus from many papers (see Awramik et al., 1976, and Awramik et al., 1979, for extensive bibliographies) on modern marine stromatolite analogs indicates that most (1) form in arid to semi-arid and tropical to subtropical climates where salinities are often elevated due to high evaporation rates, (2) form in carbonate areas, and (3) are interpreted as having been deposited in intertidal to supratidal marine environments. However, exceptions to the above generalizations are becoming increasingly recognized (Leeder, 1982; Monty, 1977). Modern non-marine stromatolite analogs have been known since the 1800's (Monty, 1977). Since the first discovery of Holocene marine algal mats in the Bahamas (Black, 1933), most geologists have tended to compare intertidal to supratidal algal mats from the Bahamas, Florida, Australia (Shark Bay), and the Persian (= Arabian) Gulf (Trucial Coast) in order to determine the ecologic, biologic and geologic factors controlling their distribution today (Hoffman, 1973; Monty, 1977). However, subtidal forms have been discovered (Bathurst, 1967; Gebelein, 1969; Neumann et al., 1970; Golubic, 1973; Playford and Cockbain, 1976) and well-developed algal mats also now are known from Baja California (Horodyski and Vonder Haar, 1975). In addition, Playford and Cockbain (1969) reported deep water stromatolites of Paleozoic age.

Gebelein (1969) noted that turbulence, velocity of bottom currents, and sediment supply and texture were important factors in stromatolite formation. The algae precipitate calcium carbonate in warm climates, as well as trap and bind sedimentary particles deposited over them with their sticky filaments, thus stabilizing the sediment-water interface. Many workers have noted that the sediment layers are storm-derived (Park, 1976) and that the algal mat re-establishes itself at the new surface soon after sediments cover the former surface (Ginsburg et al., 1971). Algal mats also tend to trap the finer sand and silt grains from the surrounding sediments (Gebelein, 1969).

Hoffman (1973) noted that the aridity of an environment controls the distribution and development of algal mats along the shoreline. For example, supratidal mats are prolific in the more humid Bahamas while algal mats extend only up to the upper intertidal zone in the more arid Trucial Coast and at Shark Bay, Australia. The mats of the more humid Florida and Bahamas regions are consequently unlithified. Fluctuating and often higher salinities, as well as desiccating conditions of the intertidal and supratidal zones, limit the occurrence of competing eucaryotic algae, aquatic plants, grazing metazoa and burrowers (Awramik, 1971; Golubic, 1973; Hoffman, 1973; Friedman et al., 1973). Thus, since the late Precambrian, marine algal mats should be well-developed mainly in restricted, e.g., hypersaline environments, perhaps regardless of latitude and substrate composition.

A review of the stromatolite and algal mat literature yields a relatively small number of fairly recent papers containing reports of modern quartzose algal mats and temperate zone algal mats. In addition, such sedimentary structures are usually mentioned only in passing or in the context of other studies. The fossil record contains but few reported occurrences of algal stromatolites associated with siliciclastic sediments. Davis (1968) described early Ordovician stromatolites in Minnesota composed of quartz sandstone with thin bands of dolomite. Other occurrences were reported also by Donaldson (1963) from Labrador and by Radwanski and Szulczewski (1965) from Hungary.

High latitude, quartzose algal mats have been described briefly from southern New England (41° 31' N latitude) by Conover (1962), southwestern British Columbia (49° N latitude) by Kellerhals and Murray (1969), southeastern Massachusetts (Sippewisset marsh, 41° N latitude) by Tropper (1976), northeastern Massachusetts (43° N latitude) and northern Nova Scotia (46° N latitude) by Jones and Cameron (1978a), northeastern Massachusetts by Cameron (1979), and Cape Cod (Massachusetts, 41°30' N latitude) by Semikhatov et al. (1979). European high latitude occurrences also have been reported from the Wash in England by Evans (1965), Western Eire by Gunatilaka (1972), Normandy by Le Gall and Larsonneur (1972), North Sea by Reineck (1972), and North and Baltic Seas of Denmark and Germany by Golubic (1973). Leonard et al. (1981) also remarked in general on the high latitude occurrence of modern siliciclastic algal mats.

A number of other reports note the presence of quartzose algal mats at lower latitudes. Those from along the East Coast of the United States include Williams (1950, cited in Conover, 1962) for North Carolina; Pomeroy (1959) for Georgia; Conover (1962) for the Carolina-Georgia coastal areas; and Edwards and Frey (1977), and Frey and Basan (1978) for Georgia. Davis (1968) also reported modern quartz sand stromatolites in Laguna Madre, Texas, while Thompson (1968) reported them on the Colorado River Delta in the Gulf of California. Schwarz et al. (1975)

described modern quartzose algal mats on tidal flats in coastal embayments of Mauritania, West Africa, at 19°30' N latitude which are 20-40 percent carbonate, uncemented, and contain aeolian dune sand. Gunatilaka (1975) described modern intertidal quartzose algal mats from Ceylon at 9° N latitude which are up to 40 percent carbonate-rich, gypsum-bearing, uncemented, and contain aeolian dune sands and silts. Other references noting that predominately quartzose algal mats can occur include Awramik et al. (1978) and Walter (1977).

Some workers have referred to the presence of quartz sands in otherwise calcareous algal mats (e.g., Logan, 1961; Davis, 1966; Horodyski and Vonder Haar, 1975; Park, 1976; Gunatilaka, 1977). Friedman (1980, p. 169), for example, noted the presence of igneous rock fragments, quartz grains, mica flakes, and recycled reef particles which demonstrate the sediment-binding capabilities of algal mats.

Since most previously reported and extensively described occurrences of modern algal mats are in coastal, arid to semi-arid, low latitude, calcareous substrates, many geologists still assume that the presence of stromatolites in the stratigraphic record indicates a similar environment of deposition. By this reasoning, stromatolites should not occur in non-calcereous or quartzose, high latitude environments. Paleoenvironmental interpretations from stromatolites are biased by what is known from modern environments, or, as Hoffman (1973, p. 188) states, ". . . will continue to depend on discoveries of Recent analogs." The best known exception to this uniformitarian approach is the discovery of Devonian deep-water stromatolites of the Canning Basin in Western Australia (Playford et al., 1976). The occurrence of modern algal mats in non-carbonate sediments is actually quite widespread. Their occurrence in quartzose sediments in the cool temperate climates of coastal Massachusetts and other areas, such as Nova Scotia (Jones and Cameron, 1978a), indicates that some caution should be exercised in the interpretation of paleoenvironments for stromatolite-bearing formations. More studies of and comparisons with a variety of modern analogs are needed. The recognition and correct interpretation of more ancient quartzose stromatolites from high latitudes may lead to the development of a new paleoenvironmental indicator.

PLUM ISLAND QUARTZOSE ALGAL MATS

Introduction

The coastal quartzose algal mats at Plum Island, Massachusetts, are found today in a cool temperate climate at about 43° N latitude. They are best developed along the high intertidal to low supratidal edges of a metahaline pond between beach-dune ridges on a spit that is accreting at the southern end of the barrier island. Local occurrences elsewhere include small backbarrier areas in the marsh behind the dune field (Fig. 1). Mats found at the marsh edge contain some granule- and pebble-sized clasts in addition to silt and sand because they are near a gravel road. Because the algal mats are most extensively developed on the spit, we shall describe their occurrence there in detail.

Geologic Setting

The Plum Island barrier system (43° N, 71° W) is located approximately 40 miles northeast of Boston (Fig. 1) and was formed about 7,000 years ago as a result of the release of vast quantities of sediment coupled with a rising sealevel during the melting of the Wisconsin glacier (McIntire and Morgan, 1963; Jones and Cameron, 1976). The island consists of an oceanward linear beach, a well-developed dune field, and an extensive salt marsh (Fig. 1), and it is growing southward through spit accretion (Farrell, 1969; Jones, 1977). It is also now migrating westward by dune migration (Jones and Cameron, 1977; Cameron and Jones, 1977, 1979) as a result of decreasing sediment supply and frequent northeast storms (Jones and Cameron, 1978b, 1979a, 1979b, 1979c).

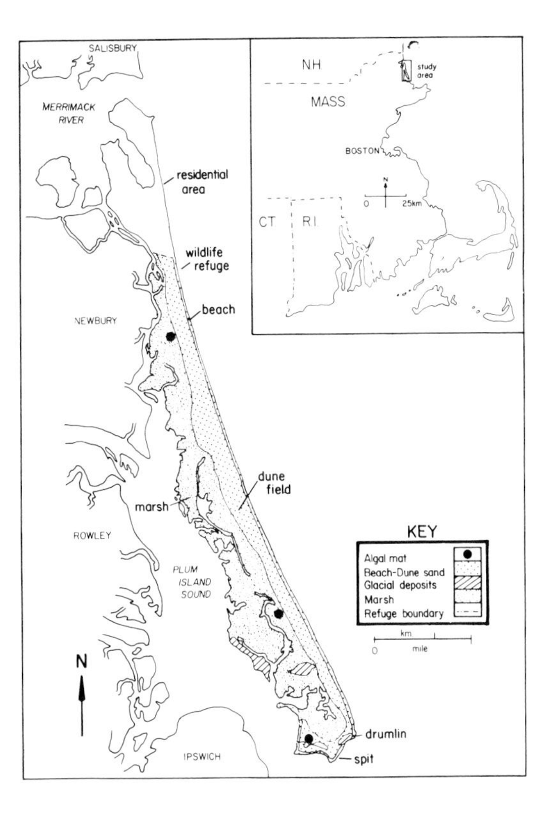

Fig. 1.—Generalized surficial geologic map of Plum Island, Massachusetts, showing localities of known algal mats. Inset map indicates geographic location of Plum Island.

A
B
C
D
E
F
G
H
I

Southward growth of the spit occurs through the addition of curved beach ridges which weld completely or incompletely to the southwest end of the island next to the salt marsh (Coastal Research Group, 1969). During incomplete bar welding, a narrow, shallow, temporary tidal channel is formed between the inland beach ridge and the seaward bar. As the end of the bar connects to the beach, a metahaline pond is formed. The algal mats reported herein are developed in a swale which is only connected to seawater during spring high tides by a shallow, meandering channel (Fig. 1; Pl. 1A-C).

The microclimatic conditions that are established in summer during low tide promote an environment which is somewhat similar to those in more arid regions. During warm sunny days the temperature of the shallow (0.2-0.5 meter) seawater pond rises, which increases the rate of evaporation leading to higher than normal salinities. The dark color of the surface mat also enhances local heating of the mat system (Stanley Awramik, personal communication, 1979). Water lost by evaporation is replenished at high tide by tidal inflow into the inter-beach ridge channel and by some marine groundwater percolation. Rainwater may dilute the brine, but the barrier ridge prevents the outflow of brine concentrated by evaporation (Heckel, 1972). Freshwater runoff is essentially zero in the low relief southern spit area of Plum Island. As a result, a potential mini-evaporite basin is formed. This increases salinity, inhibits gastropods and other grazing fauna, and permits prolific colonization of the sand by blue-green algae, with subsequent algal mat formation.

The sedimentation rate in this small basin is episodic and due to winds, storm overwash, and high tides. The small size of the pond permits little if any wind-generated water currents to move sediment. At highest tides, minor shallow tidal currents are generated at the open end of the pond, resulting in elongated narrow channels, tens of centimeters deep, containing bare sand between higher algal mat-covered areas (Pl. 1B). Older channels of this type are covered by thinner algal mats (Pl. 1D). Occasionally, coarser sediment is overwashed from the beach into the pond during storms (Pl. 3A-F), contributing to the further vertical accumulation of sediments. Multiple algal mat layers occur as the motile algae move upwards each time to recolonize the new surface (Pl. 2E). Aeolian transported silt and fine-grained sand are interlayered with these sands and also contribute to the vertical growth of the algal mats (Fig. 2).

Algal Mats

Introduction.—In the spit area of southern Plum Island, quartzose algal mats occur along the high intertidal to low supratidal margins of a metahaline pond and tidal channel between beach-dune ridges (Pls. 1A-B, 2A-B). The mats occur in an area of about 30 m by 150 m and develop up to and around the halophytic vascular plants, such as *Spartina* (grass) and *Salicornia* (saltwort) (Pls. 1E-F, 2A-B). Occasionally, one can see snails, such as *Littorina* (Pl. 2D), feeding on the mats, but grazers do not commonly survive in this hyper-

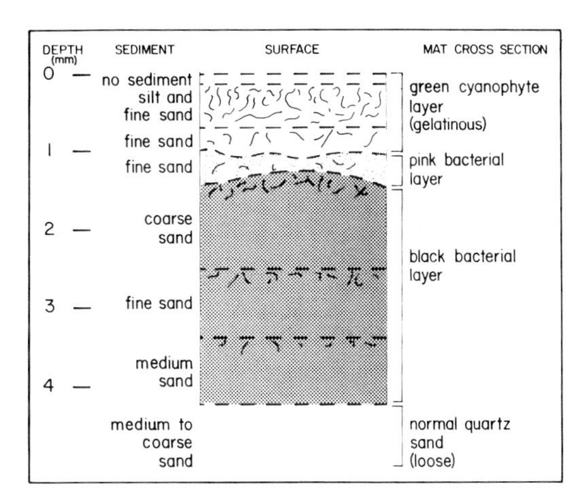

Fig. 2.—Generalized vertical cross section of algal mats in the spit area of Plum Island during June, 1976. Note typical green cyanophyte, pink bacterial, and black bacterial layers. Wiggly lines represent algal filament concentrations.

Explanation of Plate 1

Algal mats in the spit area of Plum Island (see Fig. 1).

Fig. A.— Western end of metahaline pond at high tide with marginal algal mats covered in October, 1977.

B.— Same area at mid-tide with small tidal channelways and mats (darker areas) uncovered in June, 1976 (trenching shovel at right for scale).

C.— Seaward end of meandering tidal channel (October, 1977) that connects to pond.

D.— Close-up of algal mats of Figure B showing mat-covered older channelways.

E.— Non-desiccated and continuous high intertidal mats associated with salt tolerant grasses (*Spartina*) near dune field in June, 1976.

F.— Desiccated thin, low supratidal mats in October, 1978 (11 cm × 30 cm card at right for scale).

G.— Cohesive wet mats cut from flood tide-covered area in October, 1977.

H.— Gas dome at high tide being punctured in October, 1977.

I.— Partially dried, high intertidal gas dome cut for examination at low tide in October, 1978 (6 cm long knife for scale).

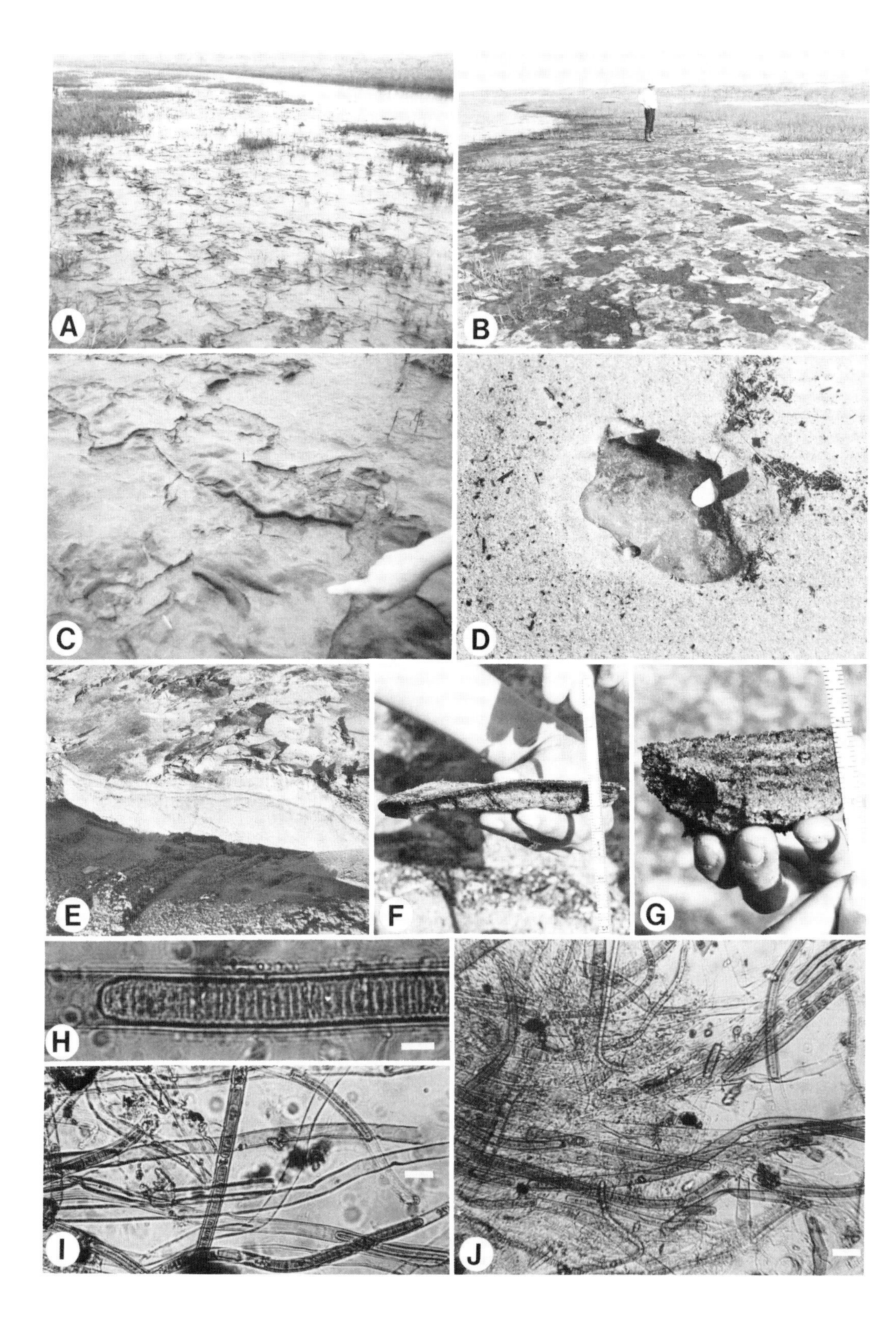
A
B
C
D
E
F
G
H
I
J

saline environment. Gelatinous algal filaments of the mats stabilize the sediment surface.

Sediment is trapped and bound together into a mat by the gelatinous (mucilaginous), intertwined algal filaments to form a cohesive "rubbery" (Pl. 1G), dark brownish-green surface layer that is green in cross section (Fig. 2; Pl. 2F-G). Filamentous blue-green algae dominate (Pl. 2H-J) the aerated upper layer of the mats, while anaerobic bacteria dominate the lower layers. Texturally, the sediment is silt and fine-grained sand, presumably of predominantly aeolian origin. However, in this area medium to coarse sand layers also are found that are due to infrequent overwash. Mineralogically, the sand is almost entirely quartz. No gypsum was found under these mats, and they are uncemented by carbonates.

Internal structure.—The algal mat is really a vertically structured community (Golubic, 1973, 1976a). At Plum Island it is composed of three major zones or layers (Fig. 2) that can be easily recognized by their different colors in cross section in the field (Pl. 2F-G) and under the dissecting microscope. Thicknesses of these layers are variable. The greenish surface layer ranges from less than 1 mm to about 2 mm thick. The salmon pink second layer may be discontinuous or continuous laterally and range up to 0.3 mm thick. A black third layer that is usually present ranges from 0.5 cm to 10 cm thick (Pls. 2F-G, 3H) and contains decaying buried older mats. Yellowish white, cleanly washed quartz sand underlies the black layer (Pls. 2E-G, 3F).

Green surface layer.—In most of the algal mats of the spit area of Plum Island, the photosynthetic and aerobic upper algal layer is gelatinous and composed of up to five sublayers each of which is a tenth to a few tenths of a millimeter thick (Fig. 2). At the surface there is a thin, slippery, gelatinous sublayer which is underlain by a dark, brownish-green sublayer dominated by the large filamentous cyanophyte *Lyngbya* (Pl. 2H-I) with some bundles of the filamentous cyanophyte *Microcoleous*. Silt and sand may or may not be present in this second sublayer, and at least 50 percent of the *Lyngbya* filaments between the sediment grains are vertical in orientation. The next two sublayers occur in either order: one is a *Microcoleous*-dominated, light or kelly green gelatinous silt and sand sublayer, while the other is a fine-grained sand sublayer with a few empty *Lyngbya* sheaths. The underlying fifth sublayer, which is not always present, is clear, gelatinous, filament-free and sand-free. This last sublayer resembles the uppermost surface sublayer and is possibly the top of a previous algal mat surface that was covered by a significant amount of sediment. Where the channel enters the metahaline pond (Pl. 1A) the algal mats are well-developed but dominated by *Microcoleous* instead of *Lyngbya*, possibly due to lower salinity.

Within this green algal layer, there are living coccoid cyanophytes, possibly *Entophysalis*, and an unidentified smaller oscillatorian filamentous cyanophyte (Pl. 2J). Live *Euglena*, diatoms, and nematode worms also are present throughout this layer.

Pink layer.—Underneath the algal layer, there is a pinkish or salmon-colored layer which may be horizontally discontinuous or continuous and which contains clusters of photosynthetic, anaerobic purple sulphur bacteria (Golubic, 1973, 1976a; Tropper, 1976). This layer may be vertically repeated, as well as discontinuous, because buried, partly decayed, prior mat surfaces at 2 and 3 mm depths may contain thin, pink layers beneath dense, dark, thin laminae of organic matter. Living cyanophytes are uncommon in this layer, but their empty sheaths are abundant. The upper and lower boundaries are gradational.

Black layer.—The black lower layer comprises the bulk of the mat community, varying from the usual few

EXPLANATION OF PLATE 2

Algal mats and associated organisms in the spit area of Plum Island (see Fig. 1).

FIG. A.— Previously desiccated mats being flooded by high tide in October, 1977.

B.— Same area as in Figure A in June, 1976, at low tide showing where partially floated and shifted mats leave dark bare sand areas (black layer as in Fig. 2) after resettling.

C.— Close-up of Figure A showing wrinkled mats that result from disturbance due to partial flotation by flood tide and incomplete resettlement during ebb tide.

D.— Herbivorous snails (*Littorina*) feeding on incompletely covered mats in May, 1978, after storm overwash sedimentation.

E.— Trench showing several generations of older overwash sediment-covered mats and desiccated surface mats in October, 1978 (6 cm long knife at center for scale).

F.— Cut piece of mat, June, 1976, illustrating vertical layering (scale in inches; see Fig. 2).

G.— Cut piece of mat in June, 1976, illustrating at least four major episodes of burial sedimentation and renewed surface mat formation (scale in inches; see Fig. 2).

H.— Filament of the blue-green alga *Lyngbya* sp. from surface layer of the mat illustrating its thick sheath and trichome (scale bar = 10 μm).

I.— Empty and filled sheaths of *Lyngbya* sp. (scale bar = 40 μm).

J.— Smaller, blue-green algal filaments associated with *Lyngbya* sp. (scale bar = 40 μm).

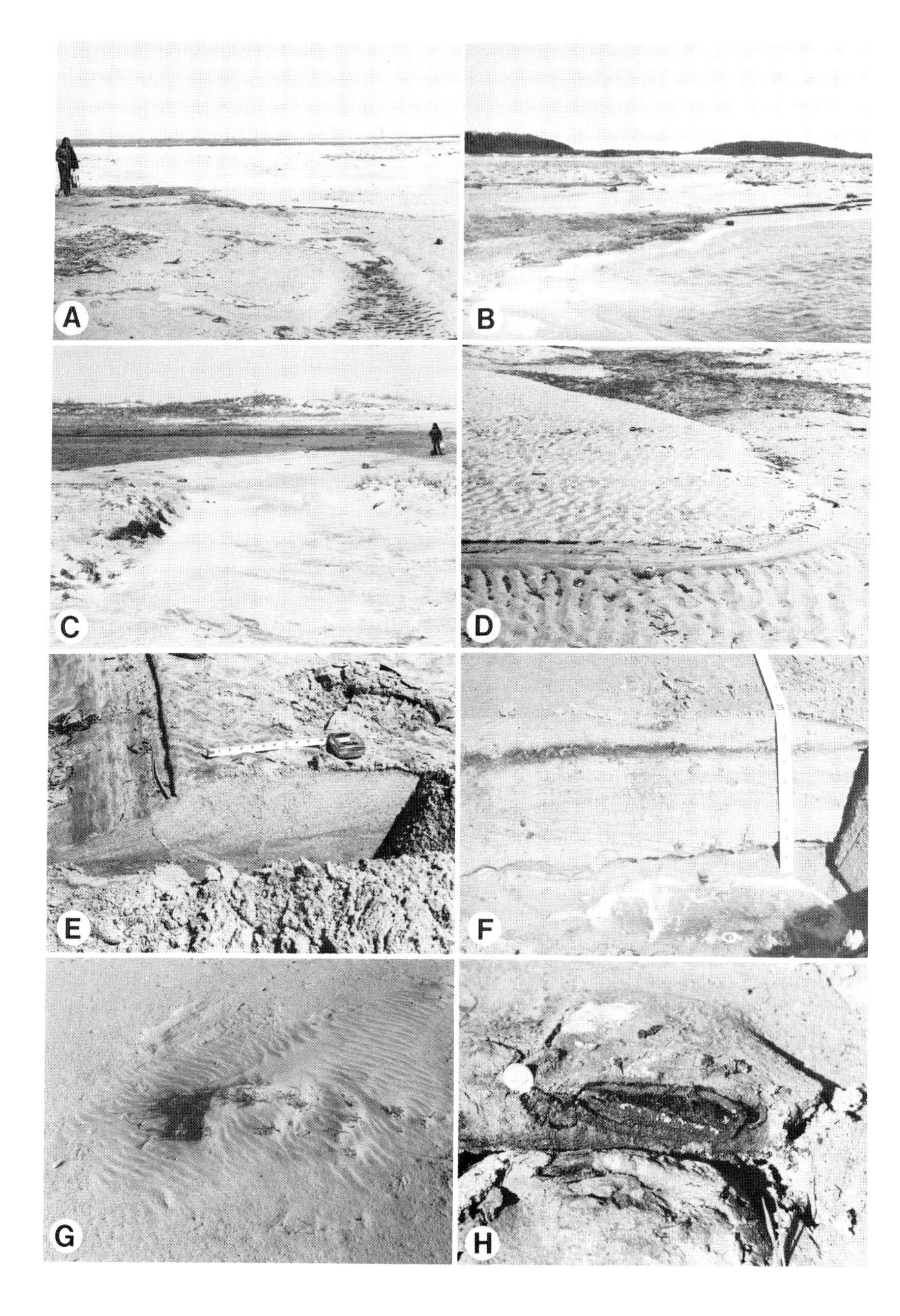
A
B
C
D
E
F
G
H

millimeters to nearly 10 cm in thickness (Pls. 2F-G, 3H). This black zone is characterized by anoxic conditions associated with bacterial decay of organic matter of buried surface mats (see Golubic, 1976a), which can be recognized by silt and sand laminae containing concentrations of *Lyngbya* sheaths in mostly horizontal positions (Fig. 2). Where this zone was found to be very thick, i.e., 10 cm, the mat surface had been rolled over onto itself up to several times, possibly by a storm or vigorous currents at high tide, to form one of several elongated mounds at the high tide entrance to the pond (Pls. 2B, 3H). Here, the mats are thickest and illustrate several generations of development separated by white quartz sand layers. For example, at one sampling station four mat horizons occur in about 4 cm (Pl. 2G). In upper supratidal areas where desiccation cracks (Pl. 1F) aerate the underside of the mat, an orangish light brown color may characterize all or part of this layer, probably due to oxidation or due to the development of other bacteria(?).

Sediments.—The mat sediments are composed predominantly of quartz with subordinate amounts of feldspar and trace amounts of heavy minerals. Texturally, the sediment is coarse silt to fine sand. Mean grain size varies from 1.5 to 2 phi in the spit area, but silt (4 phi and finer) and fine-grained sand dominate sediment within the algal mat itself. Sediment laminations are distinguishable by these textural changes and the presence of organic matter. All mat samples examined microscopically exhibited fine-grained silt or sandy silt laminae with abundant gelatinous organic matter, including filamentous and coccoid cyanophytes and their empty sheaths. Beneath the green algal and pinkish bacterial layers, the black bacterial layer contains fine sand laminae alternating with empty sheath-bearing, sandy silt. One to two recognizable laminae per millimeter were counted microscopically. Thirty to 60 cm deep trenches reveal buried thin black layers of partially decomposed prior mats covered by up to 30 cm of quartz sand (Pls. 2E, 3F). Probably, these prior surfaces were covered by storm-generated washovers and/or windblown sands.

Large structures.—The large structures associated with these algal mats include gas domes, desiccation polygons, elongated ridges, rolled mats, mounds, and resettled floated mats. Numerous gas domes (Schwarz et al., 1975) up to 10-15 cm across can be seen at high tide when they are nearly covered by water (Pl. 1H). They occur between the mid-tide and high tide levels. Such domes may remain intact at low tide and even dry out to produce large air-filled cavities that can be cut with a knife (Pl. 1I). These structures are probably produced by entrapped air and decomposition gases lifting parts of the mat at flood tide where it is not dissected by desiccation cracks. They are fragile and ephemeral.

During the summer and fall, desiccation polygons are extensively developed in the supratidal mats (Pls. 1F, 2A). They become increasingly smaller away from the waterline and towards the adjacent grass and low dunes where the mats are thinner and less well developed. The edges of the polygons are irregular and curled up. Dry polygons have the consistency of a thin, dry "pizza crust." Salt tolerant plants, such as *Spartina* and *Salicornia*, sparsely colonize these areas.

As the tidal channel enters the pond area, elongate mounds form between narrow channelways for the flooding and ebbing tidal waters (Pl. 1B, D). Although bare sand floors the large channelways, thin mats may be present in the narrower and shallower channels (Pl. 1D). Mat development is thicker on the elongated mounds. Occasionally, such mounds may be composed of a mat that was rolled over on itself several times, as can be seen by its internal structure (Pl. 3H). Enrolled mats along the channel area and near the adjacent low dune field occasionally form mounds that are colonized by *Spartina* and other salt tolerant plants (Pl. 3G). These local areas must be rich enough in organic matter, stabilized and high enough to support the grass which does not, in general, colonize the open flat mat area.

Occasionally, the in-coming high tide actually lifts and partially floats the upper layer of previously dried mats in the high intertidal areas (Pl. 2A, C). With the out-going tide, the upper mat layer of large mat areas resettles imperfectly to form a wrinkled surface (Pl. 2C). But in areas with small mats and small desiccation polygons, the upper layer may resettle over other mats,

EXPLANATION OF PLATE 3

Algal mats and sedimentation in the spit area of Plum Island (see Fig. 1).

FIG. A.— Large overwash fan (white sand) from oceanward side (east) of pond at low tide in May, 1978.
B.— Smaller overwash fan from open ocean encroaching on pond at high tide in May, 1978.
C.— Same as Figure B but looking towards pond showing opening (throat) through low dune field.
D.— Close-up of toe of overwash fan in Figures B and C at low tide (6.4 cm square tape case in center foreground).
E.— Trench cut into toe of overwash fan of Figure D showing cross-stratification (scale in inches).
F.— Trench into another overwash fan (May, 1978) showing an older overwash, a buried older algal mat (black horizon), and younger overlying overwash sand (scale in inches).
G.— Wind-blown sand covering algal mat and trapped around a vegetated mound in October, 1978.
H.— Similar mound in an earlier stage of development in May, 1978, showing rolled algal mats in vertical cross-section.

leaving the original area of development bare (Pl. 2B) and exposing the black layer underneath (Fig. 2).

Storms or high winds may cause extensive aeolian sedimentation or large overwash fans to cover the algal mats. When such events occur, the new surface will be recolonized by the upwardly motile algae if the new surface is still within the algal mat environment. Plate 2E illustrates a buried older mat with a desiccated new surface mat 10 to 15 cm above it, while Plate 3F shows a deeper mat buried by overwash sand with no new mat development above it.

Physical Sedimentation Effects

Algal mat growth affects physical sedimentation in part of the interdune area of the Plum Island spit by stabilizing the sand in several ways. The algae bind silt and sand with mucous and filaments to form a cohesive sediment surface that inhibits erosion. Silt seems to be preferentially incorporated in the mats, although sand is also present. After storms, which can cause burial of mats by washovers and/or aeolian sediments, the motile algae move upwards to recolonize and bind the new surface. This aids in the filling of ponded low areas of the spit. During high tides, partially floated and even overturned and rolled mats and/or desiccated mat polygons tend to form mounds that are more easily colonized by root-bearing vascular plants, such as grasses, which further stabilize the sediment. These vegetated mounds further trap windblown and water transported sediment between plant leaves and stems to form incipient dunes that gradually coalesce to develop still higher topography (Pl. 3G).

PRESERVATION POTENTIAL

It will probably not be easy to identify ancient quartzose algal laminites and stromatolites, even though modern examples are becoming increasingly well-recognized at many latitudes (Conover, 1962; Schwarz et al., 1975; Gunatilaka, 1975; Jones and Cameron, 1978a). The preservation potential of the organic matter in these quartzose algal mats is very low probably because of the lack of precipitation of mineral matter (Awramik et al., 1976). Lack of cementation by carbonate mineral precipitation also minimizes the preservation potential of quartzose algal mats and stromatolites as distinct biogenic sedimentary structures (Davis, 1968). Calcium carbonate precipitation and cementation, in addition to trapping and binding, in stromatolites in carbonate environments has probably lead to a preservational bias towards limestones. Therefore, criteria are needed for recognizing the remains of ancient siliciclastic algal laminites which may have been extensively developed along non-carbonate coasts in the geologic past.

Laminations and sedimentary structures associated with preserved siliceous diatom frustules (see Tropper, 1976) and possible pyrite formation along organic-rich layers are potential criteria for recognizing quartzose algal laminites from ancient tidal flats formed at high paleolatitudes. Of these criteria, laminations and sedimentary structures will probably prove most useful in the field.

The Plum Island algal mats can be characterized as differentially concentrating silt and fine-grained sand and as being composed of alternating silt and fine sand laminae (≤1 um). The associated upper and lower sedimentation units are thicker, coarser, usually storm generated, and windblown or water-laid.

Desiccation polygons of varying sizes, including small ones with crinkled and curled edges, should be fairly recognizable in ancient coastal quartz arenites. Locally transported small polygons may appear as hazy or ghost-like flat pebble clasts in flat pebble conglomerates if the surrounding sands are coarser.

Slightly undercut and elongated ridges between small channelways and ellipsoidal mounds that have thicker and more numerous laminae than the intervening lows may be recognizable in cross section. Overturned mats forming thickened mounds and ridges, as illustrated by Gill (1977, Fig. 9D, p. 992) for the Late Silurian Salina Group of Michigan, would also be quite diagnostic. Large resettled floated mats, unless their wrinkled surface is apparent, and uncemented gas domes would most probably be either unrecognizable or poorly preserved in siliciclastic rocks.

CONCLUSIONS

Modern coastal quartzose algal mats at high latitudes in many regions are similar structurally and florally to those in carbonate sediments at lower latitudes today. Although quartzose algal mats have been reported rarely from the stratigraphic record, they are probably widespread but difficult to recognize. Quartzose algal mats (algal laminites) could prove to be a paleoclimatic indicator once criteria for their recognition are developed. Preservational potential is low probably due to the absence of carbonate cementation. Possible criteria for recognition include fine sand and silt laminae, desiccation cracks, elongated ridges, mounds, and overturned mats.

The quartzose algal mats in the spit area of Plum Island, Massachusetts, are vertically stratified into a thin, upper, greenish layer dominated by blue-green algae, a thinner, middle, pinkish, anaerobic, photosynthetic bacterial layer, and a lower, black, anaerobic bacterial layer containing decomposing organic matter. Larger structures include desiccation polygons, gas domes, elongated ridges, mounds, disrupted mats due to partial flotation, and rolled mats.

These algal mats affect physical sedimentation by stabilizing silt and sand after each episode of deposition. The gelatinous materials of the algae bind sediment grains to form a leathery surface layer that inhibits erosion. With continued aggradation, overturned and rolled mats produce higher mounds that become colonized by vascular plants. The latter trap windblown sands and initiate dune development on low, flat areas above the high tide line.

ACKNOWLEDGMENTS

We thank Stanley M. Awramik of the University of California, Gerald M. Friedman of Rensselaer Polytechnic Institute, H. Allen Curran of Smith College, and Emily J. Hoffman and Duncan M. FitzGerald of Boston University for critically reading various drafts of this paper. Hans J. Hofmann of the University of Montreal provided helpful discussion and several references. The officials of the Parker River National Wildlife Refuge are gratefully acknowledged for their cooperation and permission to conduct field work on Plum Island. We also thank Ms. Hoffman for helping with the drafting of Figures 1 and 2 and the preparation of the plates, Lillian Paralikis for typing the manuscript and John Stewart for general assistance at several stages. Cameron acknowledges initial support from the United States National Science Foundation through grants EAR 76-84233 and EAR 79-11200 to S. Golubic and himself. Financial assistance for completion of research and manuscript came from the Natural Sciences and Engineering Research Council of Canada grant A8426 to Cameron. Jones acknowledges partial support from a Petroleum Research Grant from the American Chemical Society.

REFERENCES

AWRAMIK, S.M., 1971, Precambrian columnar stromatolite diversity: reflection of metazoan appearance: Science, v. 174, p. 825–827.

_____, GEBELEIN, C.D., AND CLOUD, P., 1978, Biogeologic relationships of ancient stromatolites and modern analogs, in Krumbein, W.E. (ed.), Environmental Biogeochemistry and Geomicrobiology: Ann Arbor, Michigan, Ann Arbor Science, v. 1, p. 165–178.

_____, HAUPT, A., HOFMANN, H.J., AND WALTER, M.R., 1979, Stromatolite bibliography 2: Precambrian Research, v. 9, p. 105–166.

_____, HOFMANN, H.J., AND RAABEN, M.E., 1976, Bibliography, in Walter, M.R. (ed.), Stromatolites: Amsterdam, Elsevier, p. 705–771.

BATHURST, R.G.C., 1967, Sub-tidal gelatinous mat, sand stabilizer and food, Great Bahama Bank: Jour. Geology, v. 75, p. 736–738.

_____, 1975, Carbonate Sediments and Their Diagenesis: Amsterdam, Elsevier, 658 p.

BASAN, P.B., AND FREY, R.W., 1977, Actual-paleontology and neoichnology of salt marshes near Sapelo Island, Georgia, *in* Crimes, T.P., and Harper, J.C. (eds.), Trace Fossils 2: Geol. Jour. Spec. Issue No. 9, p. 41–70.

BLACK, M., 1933, The algal sediments of Andros Island, Bahamas: Philos. Trans., Ser. B, CCXXII, p. 165–192.

CAMERON, B., 1979, Physical effects of quartzose algal mats and dune development—Plum Island spit, Massachusetts, *in* Aubrey, D.G., Proc. of a Workshop on Coastal Zone Research in Massachusetts (Nov. 27–28, 1978), W.H.O.I., Tech. Report 79–40, p. 33–34.

_____, AND JONES, J.R., 1977, New evidence for barrier island migration: New England-St. Lawrence Valley Geographical Society, American Association of Geographers, Proceedings, v. 6, p. 94–97.

_____, AND _____, 1979, Landward migration of barrier island environments: discriminant function analysis of sand size frequency distributions: Geol. Soc. America Ann. Mtg. Abs. with Programs, v. 11, no. 1, p. 6.

COASTAL RESEARCH GROUP, 1969, Coastal environments: northeastern Massachusetts and New Hampshire: Soc. of Econ. Paleontologists and Mineralogists Field Trip Guidebook, Contr. No. 1—Coastal Research Group, Univ. of Massachusetts, Dept. of Geology Publ. Ser., 462 p.

CONOVER, J.T., 1962, Algal crusts and the formation of lagoon sediments, *in* Marshall, N. (ed.), The Environmental Chemistry of Marine Sediments: Proceed. of Symp., Kingston, R.I., Univ. of Rhode Island Graduate Sch. of Oceanogr., Occasional Publ. No. 1, p. 69–76.

CUSSEY, R., AND FRIEDMAN, G.M., 1976, Antipathetic relationships among algal structures, burrowers, and grazers in Dogger (Jurassic) carbonate rocks, southeast of Paris, France: Am. Assoc. Petroleum Geologists Bull., v. 60, p. 612–616.

DAVIS, R.A., 1966, Willow River Dolomite: Ordovician analogue of modern algal stromatolite environments: Jour. Geology, v. 74, p. 908–923.

_____, 1968, Algal stromatolites composed of quartz sandstone: Jour. Sed. Petrology, v. 38, p. 953–955.

DONALDSON, J.A., 1963, Stromatolites in the Denault Formation, Marion Lake, coast of Labrador, Newfoundland: Geol. Surv. Can. Bull. 102, p. 1–33.

EDWARDS, J.M., AND FREY, R.W., 1977, Substrate characteristics within a Holocene salt marsh, Sapelo Island, Georgia: Senckenbergiana Marit., v. 9, p. 215–259.

EVANS, G., 1965, Intertidal flat sediments and their environments of deposition in the Wash: Q.J. Geol. Soc. London, v. 121, p. 209–245.

FARRELL, S.C., 1969, Growth cycle of a small recurved spit, Plum Island, Massachusetts, *in* Coastal Environments: Northeastern Massachusetts and New Hampshire: Soc. Econ. Paleontologists and Mineralogists Field Trip Guidebook, Cont. No. 1—Coastal Research Group, Univ. of Massachusetts, Dept. of Geology Publ. Ser., p. 316–337.

FREY, R.W., AND BASAN, P.B., 1978, Coastal salt marshes, *in* Davis, R.A. (ed.), Coastal Sedimentary Environments: New York, Springer-Verlag, p. 101–169.

Friedman, G.M., 1980, Reefs and evaporites at Ras Muhammad, Sinai Peninsula: A modern analog for one kind of stratigraphic trap: Israel Jour. Earth Sciences, v. 29, p. 166–170.

_____, G.M., Amiel, A.J., Braun, M., and Miller, D.C., 1973, Generation of carbonate particles and laminites in algal mats—Example from sea-marginal hypersaline pool, Gulf of Aqaba, Red Sea: Am. Assoc. Petroleum Geologists Bull., v. 57, p. 541–557.

Garrett, P., 1970, Phanerozoic stromatolites: noncompetitive ecologic restriction by grazing and burrowing animals: Science, v. 169, p. 171–173.

Gebelein, C.D., 1969, Distribution, morphology, and accretion rate of recent subtidal algal stromatolites, Bermuda: Jour. Sed. Petrology, v. 39, p. 49–69.

_____, 1976, The effects of the physical, chemical and biological evolution of the earth, *in* Walter, M.R. (ed.), Stromatolites: Amsterdam, Elsevier, p. 499–515.

Gill, D., 1977, Salina A-1 sabkha cycles and late Silurian paleogeography of the Michigan Basin: Jour. Sed. Petrology, v. 47, p. 979–1017.

Ginsburg, R.N., Rezak, R., and Wray, J.L., 1971, Geology of calcareous algae: Comp. Sediment. Lab., Univ. Miami, Notes for a Short Course, 61 p.

Golubic, S., 1973, The relationship between blue-green algae and carbonate deposits, *in* Carr, N.G., and Whitton, B.A. (eds.), The Biology of Blue-Green Algae: Oxford, Blackwell Scientific Publications, p. 434–472.

_____, 1976a, Organisms that build stromatolites, *in* Walter, M.R. (ed.), Stromatolites: Amsterdam, Elsevier, p. 113–126.

_____, 1976b, Taxonomy of extant stromatolite-building cyanophytes, *in* Walter, M.R. (ed.), Stromatolites: Amsterdam, Elsevier, p. 127–140.

_____, and Awramik, S.M., 1974, Microbial comparison of stromatolite environments: Shark Bay, Persian Gulf and the Bahamas: Geol. Soc. America Ann. Mtg. Abs. with Programs, v. 6, no. 7, p. 759–760.

_____, and Hofmann, H.J., 1976, Comparison of Holocene and Mid-Precambrian Entophysalidaceae (Cyanophyta) in stromatolitic algal mats: Cell division and degradation: Jour. Sed. Petrology, v. 50, p. 1074–1082.

Gunatilaka, H.A., 1972, A Survey of the Geochemistry and Diagenesis of Recent Carbonate Sediments from Connemara, Western Eire [unpub. thesis]: U.K., Reading University.

_____, 1975, Some aspects of the biology and sedimentology of laminated algal mats from Mannar Lagoon, Northwest Ceylon: Sed. Geology, v. 14, p. 275–300.

_____, 1977, Environmental significance of upper Proterozoic algal stromatolites from Zambia, *in* Flugel, E. (ed.), Fossil Algae: New York, Springer-Verlag, p. 74–79.

Heckel, P.H., 1972, Recognition of ancient shallow marine environments, *in* Rigby, J.K., and Hamblin, W.K. (eds.), Recognition of Ancient Sedimentary Environments: Soc. Econ. Paleontologists Mineralogists Spec. Pub. 16, p. 226–296.

Hoffman, P., 1973, Recent and ancient algal stromatolites: Seventy years of pedagogic cross-pollination, *in* Ginsburg, R.N. (ed.), Evolving Concepts in Sedimentology: Baltimore, Maryland, Johns Hopkins Univ. Press, p. 178–191.

Horodyski, R.J., and Vonder Haar, S.P., 1975, Recent calcareous stromatolites from Laguna Mormona (Baja California) Mexico: Jour. Sed. Petrology, v. 45, p. 894–906.

Jones, J.R., 1977, An alternative hypothesis for barrier island migration [unpub. Ph.D. thesis]: Boston, Boston Univ., 177 p.

_____, and Cameron, B., 1976, Sedimentary and geomorphic origin and development of Plum Island, Massachusetts: An example of a barrier island system, *in* Cameron, B. (ed.), Geology of Southeastern New England: Boston University, NEIGC Field Trip Guidebook: Princeton, N.J., Science Press, p. 188–204.

_____, and _____, 1977, Landward migration of barrier island sands under stable sea level conditions: Jour. Sed. Petrology, v. 47, no. 4, p. 1475–1483.

_____, and _____, 1978a, Algal mats in coastal quartz sands: Massachusetts and Nova Scotia: Geol. Soc. America Ann. Mtg. Abs. with Programs, v. 10, no. 2, p. 49.

_____, and _____, 1978b, Identification of potential storm hazard areas, *in* Jones, J.R. (ed.), Proceedings Volume, Blizzard of '78 Conference, Boston State College, p. 7–13.

_____, and _____, 1979a, Reply: Landward migration of barrier island sands under stable sea level conditions: Jour. Sed. Petrology, v. 49, no. 1, p. 325–330.

_____, and _____, 1979b, Reply: Landward migration of barrier island sands under stable sea level conditions: Jour. Sed. Petrology, v. 49, no. 1, p. 332–333.

_____, and _____, 1979c, On the identification of former coastal storm deposits, *in* Psuty, N.P. (ed.), Proceedings, Comm. on Mar. Geog., Ann. Mtg., Am. Assoc. Geog., New Orleans, p. 13–23.

Kellerhals, P., and Murray, J.W., 1969, Tidal flats at Boundary Bay, Fraser River Delta, British Columbia: Bull. Canadian Petroleum Geology, v. 17, p. 67–91.

Kepper, J.C., 1974, Antipathetic relation [sic] between Cambrian trilobites and stromatolites: Am. Assoc. Petroleum Geologists Bull., v. 58, p. 141–143.

Leeder, M.R., 1982, Sedimentology, process and product: London, George Allen and Unwin, xvi and 344 p.

Le Gall, J., and Larsonneur, C., 1972, Séquences et environnements sédimentaires dans la Baie des Veys (Manche): Rev. Geogr. Geol. Dyn., v. 14, p. 189–204.

Leonard, J.E., Cameron, B., Pilkey, O.H., and Friedman, G.M., 1981, Evaluation of cold-water carbonates as a paleoclimatic indicator: Sed. Geology, v. 28, p. 1–28.

Logan, B.W., 1961, *Cryptozoan* and associated stromatolites from the Recent of Shark Bay, Western Australia: Jour. Geology, v. 69, p. 517–533.

McIntire, W.G., and Morgan, J.P., 1963, Recent geomorphic history of Plum Island, Massachusetts, and adjacent coasts: Louisiana State Univ. Coastal Studies Ser., No. 8, 44 p.
Monty, C., 1977, Evolving concepts on the nature and the ecological significance of stromatolites, *in* Flugel, E. (ed.), Fossil Algae: New York, Springer-Verlag, p. 15–35.
Neumann, A.C., Gebelein, C.D., and Scoffin, T.P., 1970, The composition, structure, and erodability of subtidal mats, Abaco, Bahamas: Jour. Sed. Petrology, v. 40, p. 274–297.
Park, R.K., 1976, A note on the significance of lamination in stromatolites: Sedimentology, v. 23, p. 379–393.
Playford, P.E., and Cockbain, A.E., 1969, Algal stromatolites: deepwater forms in the Devonian of Western Australia: Science, v. 165, p. 1008–1010.
______, and ______, 1979, Modern algal stromatolites at Hamelin Pool, a hypersaline barred basin in Shark Bay, Western Australia, *in* Walter, M.R. (ed.), Stromatolites: Amsterdam, Elsevier, p. 389–411.
______, ______, Druce, E.C., and Wray, J.L., 1976, Devonian stromatolites from the Canning Basin, Western Australia, *in* Walter, M.R. (ed.), Stromatolites: Amsterdam, Elsevier, p. 543–563.
Pomeroy, L.R., 1959, Algal productivity in salt marshes of Georgia: Limnol. and Oceanogr., v. 4, p. 386–397.
Radwanski, A., and Szulczewski, M., 1965, Jurassic stromatolites of the Villany Mountains (southern Hungary): Ann. Univ. Scient. Budapest, Sec. Geol., v. 9, p. 87–107.
Reineck, H.-E., 1972, Tidal flats, *in* Rigby, J.K., and Hamblin, W.K. (eds.), Recognition of Ancient Sedimentary Environments: Soc. Econ. Paleontologists and Mineralogists Spec. Pub. 16, p. 146–159.
Schwarz, H.U., Einsele, G., and Herm, D., 1975, Quartz-sandy, grazing-contoured stromatolites from coastal embayments of Mauritania, West Africa: Sedimentology, v. 22, p. 539–561.
Semikhatov, M.A., 1976, Experience in stromatolite studies in the U.S.S.R., *in* Walter, M.R. (ed.), Stromatolites: Amsterdam, Elsevier, p. 337–358.
______, Gebelein, C.D., Cloud, P., Awramik, S.M., and Benmore, W.C., 1979, Stromatolite morphogenesis-progress and problems: Canadian Jour. Earth Sciences, v. 16, p. 992–1015.
Thompson, R.W., 1968, Tidal flat [sic] sedimentation on the Colorado River delta, northwestern Gulf of California: Geol. Soc. America Memoir, 107, 133p.
Tropper, C.B., 1976, Algal mat diatoms of a Massachusetts salt marsh [unpub. Ph.D. thesis]: Boston, Boston University, 109 p.
Walter, M.R., 1976, Stromatolites: Amsterdam, Elsevier, 790 p.
______, 1977, Interpreting stromatolites: American Scientist, v. 65, p. 563–571.
Williams, L.G., 1959, The role of algae in stabilizing beach sand: Bull. of Furman Univ., v. 33, p. 61–63.

TUBES OF THE MODERN POLYCHAETE *DIOPATRA CUPREA* AS CURRENT VELOCITY INDICATORS AND AS ANALOGS FOR *SKOLITHOS - MONOCRATERION*

JOHN H. BARWIS
Shell Oil Company, P.O. Box 481, Houston, Texas 77001

ABSTRACT

Tidal channel hydrodynamics exert a pronounced influence on the distribution and geometry of dwelling burrows of the modern polychaete *Diopatra cuprea* and the sediments in which they occur. These burrows share strong distributional similarities with the trace fossils *Skolithos* and *Monocraterion*, a relationship which permits more accurate and detailed interpretations of ancient tidal deposits.

Diopatra cuprea builds lined, vertical tubes in a wide range of restricted and open-marine environments. The positions and orientations of these tubes on estuarine point bars in South Carolina reflect several aspects of back-barrier tidal processes and their resultant deposits. Tubes are concentrated on the upstream ends of point bars, which are exposed primarily to ebb currents. On the intertidal bar crest, where population density is highest, tubes are arranged in rows normal to flow. Tube apertures are oriented at 45 degrees to the row axis and flow direction. On the ebb-dominated bar apex, strongly ebb-oriented scour pits form around the detritus-reinforced tube caps.

Several characteristics of *D. cuprea* tubes resemble features of the trace fossils *Monocraterion* and *Skolithos*. Like *Skolithos*, tubes are subcylindrical, vertical, straight, and unbranched. The upper portions of high intertidal tubes are surrounded by *Monocraterion*-like funnels; these occur in thin-bedded muddy sands of the bar crest. Tubes without funnel-tops occur in clean, well-sorted sands of the bar flank. This stratigraphic position is analogous to reported Cambrian *Skolithos - Monocraterion* assemblages.

INTRODUCTION

Facies analysis has benefited on two different levels by the widespread acceptance and use of trace fossils as true sedimentary structures. On a broad scale, spatial distributions of trace fossil assemblages have been correlated to their enclosing litho- and biofacies and thus, by inference, to the physico-chemical characteristics that define any depositional environment. Most of these results have come from studies of ancient depositional sequences, not only because rocks lend themselves more easily to three-dimensional observation, but also because fewer geologists study modern depositional environments. On a smaller scale, traces have been shown to provide information about sedimentation rate, current direction, and competition between organisms. Ichnology as a sedimentologic tool on this detailed level will probably develop its maximum potential only to the extent that it combines physical process studies with direct observation of trace production in the field or in aquaria. Only then can a trace maker and its lebensspuren be definitely related to the range of processes that shape a bed.

This paper examines the relationship between the physical processes of an estuarine depositional setting, their resultant deposits, and the burrowing habits of the onuphid polychaete *Diopatra cuprea*. The general distribution of *D. cuprea* within an estuary is evaluated in terms of large-scale hydrographic and geomorphic controls. Characteristics of single burrows and burrow groups are described in terms of their response to current velocity, water depth, and sediment transport rate. Finally, *D. cuprea* tubes are analyzed in terms of their usefulness in interpreting the *Skolithos* ichnofacies.

DEPOSITIONAL ENVIRONMENT

The *Diopatra cuprea* population described in this paper was observed on a tidal creek point bar in the salt marsh system of Kiawah Island, South Carolina (Fig. 1). Supplemental observations of *D. cuprea* throughout the South Carolina salt marshes also were made.

Kiawah Island occupies a fifteen kilometer stretch on the north flank of the Georgia Embayment in the beach-ridge barrier complex described by Brown (1977). Geomorphology and facies distribution of these barriers are typical of those on mesotidal mixed-energy coasts (Hayes, 1975; Hayes and Kana, 1976; Nummedal et al., 1977; Hubbard et al., 1979). Kiawah itself is a short, inlet-bounded complex of Pleistocene and Holocene quartzose beach ridges. These ridges are backed by marsh and tidal channel deposits (Fig. 1), with which they are intercalated (Barwis, 1978; Moslow, 1980; Wojtal and Moslow, 1980; Wojtal, 1981).

Hydrography

Bass Creek drains a small salt marsh that occupies a swale between bifurcated beach ridges at the eastern end of Kiawah Island (Fig. 1). Mean tidal range is 1.8 meters; spring and neap extremes are 2.4 and 0.8 meters, respectively (Ward, 1981). Bankfull conditions are exceeded only on the higher spring tides. The creek is protected from swell, so the only wave influences are wind-driven surface chop and boat wakes. Normal marine salinities prevail, since the only fresh water entering the system is a minor contribution from rain runoff. The two hydrographic characteristics of Bass Creek that most affect point bar deposition are ebb-dominance and time-velocity asymmetry.

Tidal wave storage characteristics of a mesotidal marsh complex may generate shallow-water tidal components (overtides) that lengthen flood stage duration (Byrne et al., 1975). Closed-system net discharges are identical for both tides, so mean ebb velocities must therefore be higher than mean flood velocities to accomodate the relatively shorter duration of the ebb stage. This regular occurrence of higher ebb velocities is termed *ebb dominance*, a phenomenon predicted theoretically by Mota Oliviera (1970). Ebb dominance controls the position and shape of point bars in Bass Creek in that below certain depths a net seaward sediment-transport direction is assured. The geomorphic result is that point bars occur seaward of their associated meander bend apexes (Barwis, 1978). This is a common feature of tidal point bars (Van Veen, 1950; Land and Hoyt, 1966) and of fluvial point bars as well (Fisk, 1947). Ebb dominance affects the distribution of point bar infaunas because it influences the lateral distribution of current velocities and resultant substrate textures. Upstream ends of points bars are exposed to higher bed shear and therefore tend to be composed of relatively coarser and better sorted sediments.

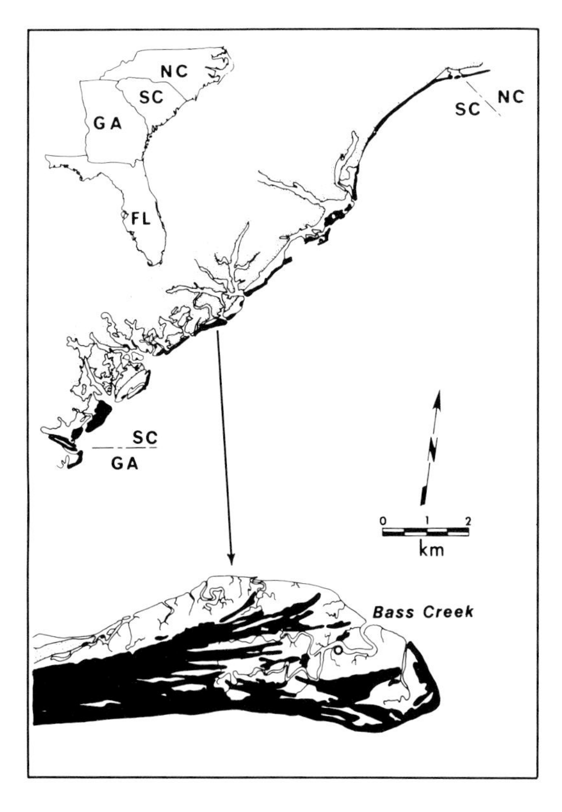

Fig. 1.—Location of Kiawah Island, South Carolina, and geomorphology of the northeastern end of the island. Point bar shown in Figure 2 is at the meander bend located by an open circle.

Maximum tidal current velocities seldom occur at mid-tide, when the rate of change of water surface elevation is at a maximum. Rather, velocity maxima occur later in the tidal cycle, temporally shifted from mid-tide toward times of high and low water. This delay is caused by inertial and frictional effects, and by changes in the effective flood basin area as the marsh or bay floods or drains (Keulegan, 1967; King, 1974). The occurrence of velocity maxima at different tidal stages means that shallow areas are often emergent prior to the onset of maximum ebb velocities (Van Veen, 1950; Price, 1963). The most important sedimentologic result of this time-velocity asymmetry is the development of mutually evasive flood and ebb channels, a situation which profoundly influences sand-body morphology and bedform distribution (Van Veen, 1950; Oomkens and Terwindt, 1960; Price, 1963; Hayes et al., 1973; Barwis, 1978).

Point Bar Sedimentology

A large polychaete population inhabits the upstream end (apex) of the point bar shown in Figure 2a. Trailing downstream (i.e., the ebb direction) from the apex are two lobes that act as margins to the three partially segregated channels. Current velocity distributions (Fig. 2b) are such that the outer channel is ebb-dominated and the two inner channels are flood-dominated. Despite its partially protected position behind the inner bank of the meander bend, the apical end is most influenced by ebb flows for two reasons. First, maximum flood current speed occurs while the bar crest and much of the apex are still subaerially exposed. In addition, the topographically higher bar crest, being downstream from the apex, shields the upstream end from flood currents.

Sediment texture and bedform distribution were described by Barwis (1978) and are briefly summarized here. The downstream extremity of the inner lobe is composed of very soft to thixotropic clay, and supports no *Diopatra cuprea*. The outer lobe consists of fine to very fine, well-sorted quartz sand mixed with widely varying amounts of carbonate bioclastic debris. *D. cuprea* is sparse in this area. The crest and apical areas are composed of thinly interbedded sand, mud, and shell hash. Bedforms are restricted to current ripples that are usually washed out during the falling tide. During spring tide ebb flows, upstream portions of the bar consist of flat beds.

The entire bar comprises an upward-fining sequence which is generally similar to those associated with nonmarine point bars, except for conspicuous differences

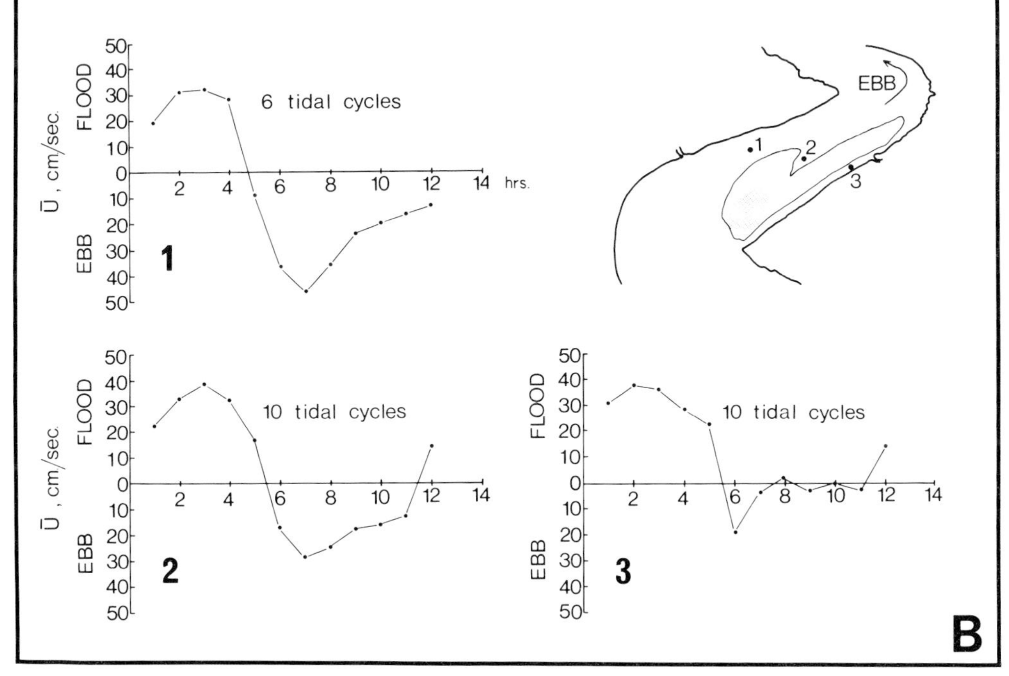

Fig. 2.—Geomorphology and hydrography of the point bar hosting the polychaete field discussed in this paper. *A*, Low-oblique aerial photograph, with ebb direction toward the top. Boat on outer flank is 5 m long. Photo taken approximately 30 minutes before spring low tide. *B*, Time-averaged current velocity distributions on the bar shown in A. Each point is the vector mean of 16 instantaneous velocities, evenly spaced throughout a lunar hour (e.g., station 3, hour 2 is the vector mean of 160 velocities). Note ebb-dominant main channel and well-developed current segregation. Shaded area is the *Diopatra cuprea* field.

in biogenic sedimentary structures (Pryor, 1967; Chamberlain, 1975). Animals are far more abundant in estuarine point bar deposits (Mayou and Howard, 1969; Howard, et al., 1975), and are responsible for extensive bioturbation.

Tidal Creek Biofacies

Bass Creek point bars support a rich faunal assemblage, the infaunas of which comprise mostly annelids, bivalves, and crustaceans. As in other southeastern estuaries, polychaete feeding and dwelling structures are the most abundant lebensspuren (Howard and Frey, 1980). In general, species diversity increases from bar to bar in a seaward direction and is highest on banks and bar crests, as described by Land and Hoyt (1966) and by Howard (1971). A large proportion of the epifaunal remains (primarily molluscan shells) represents an allochthonous open-marine death assemblage (Barwis, 1978).

Faunal zonations in Bass Creek and their associated biogenic structures are identical to those in the Georgia tidal creeks described by Frey and Howard (1969), Howard (1971), Howard and Frey (1973), and Howard et al. (1973). Bio- and ichnofacies details of these faunal assemblages are beyond the scope of this paper. The interested reader should refer to Howard and Frey (1975), Frey and Basan (1978), Frey and Howard (1980), and Howard and Frey (1980) for extensive summaries.

POLYCHAETES AS CURRENT INDICATORS

Polychaete burrows can be good indicators of current direction in several ways. Exposed, mucilagenous burrow linings of *Streplospio benedicti* often lie recumbent on the sediment surface, aligned parallel to flow (Frey, 1970). More rigid burrow linings, like those produced by *Onuphis microcephala* and *Diopatra cuprea*, provide obstructions to flow. As a result, aligned scour pits and current lineations form behind them. Although these particular depositional effects are not produced by active polychaete response to current regime, the burrow linings themselves have several unique features that do reflect active tube-building responses to the surrounding flow.

Diopatra Burrowing Habits

An onuphid polychaete, *Diopatra cuprea* is a predaceous carnivore than generally inhabits sandy beds in estuarine, open-bay, and shelf environments (Frey and Howard, 1969; Hertweck, 1972; Howard and Dörjes, 1972). In intertidal environments, the worm dwells in a vertical, subcylindrical, lined burrow having an inside diameter of 6 to 8 mm. In subtidal environments the tubes tend to be sinuous and oriented at various angles (Howard and Frey, 1975).

Burrow linings are secreted as mucous by the worm and later stiffen to a thin, parchment-like material. Tube tops are semi-rigid and protrude from 1 to 6 cm above the bed. During tube construction, tube exteriors are reinforced with detritus picked up by the worm from the adjacent bed (Fig. 3). On Bass Creek bars this material consists almost entirely of *Mulinia* valves, sand grains, and pine-needle fragments. Reinforcement not only increases tube strength, a condition necessary for protection and adequate flushing (Myers, 1972), but also aids in predator detection by increasing the effective tube diameter (Brenchley, 1976). Reinforcement probably also increases feeding efficiency by aiding in the trapping of edible detritus (Mangum et al., 1968). The animal controls tube geometry by se-

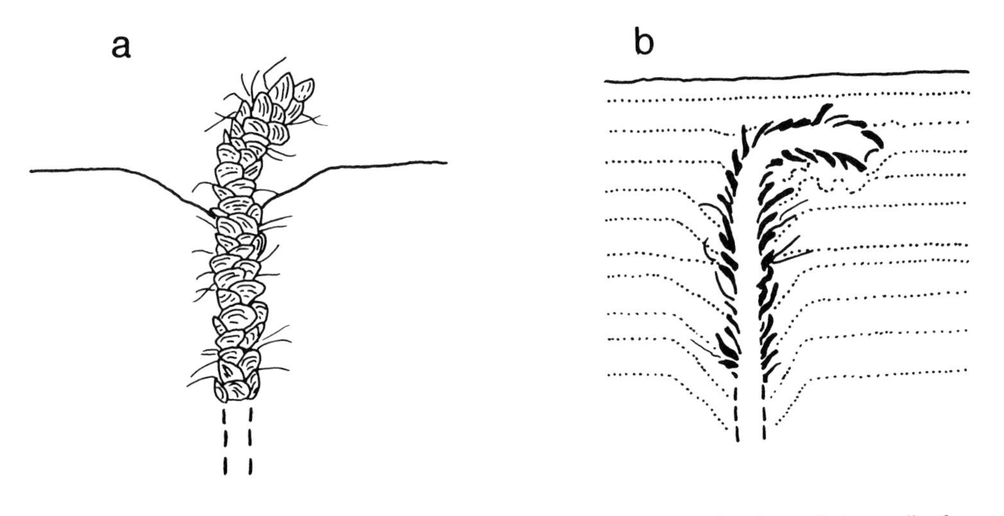

Fig. 3.—Sketch of reinforced *Diopatra cuprea* tubes. *A*, Valves of *Mulinia* sp., sand grains, and pine needles form armor material. Tube aperture, obscured from view, is normal to plane of page. *B*, Longitudinal section through tube after burial and compaction. Modified after Myers (1970).

lective armoring depending on sedimentation rate, and by trimming if bed erosion causes tube length to be excessive (Myers, 1972).

Tube tops have the shape of a periscope or inverted "J" (Fig. 3). This configuration helps improve feeding efficiency (Mangum et al., 1968). Myers (1972) found that the plane of the tube aperture at the end of this "J" is most often parallel to flow, an orientation which was presumed to aid in flushing, inasmuch as it allows the maximum development of laminar flow past the opening. Laminar flow conditions probably also aid in feeding, as the animal detects food by chemo-reception (Mangum and Cox, 1971).

Tube - Current Relationships

Almost all of the *Diopatra cuprea* tubes on the point bar shown in Figure 2 are located on the bar apex, where population density is between 150 and 200 individuals per square meter. Densities on the downstream end of the bar and in the subtidal channel range from 0 to 10 tubes per square meter. The bar apex location probably supports the densest population for three reasons. First, biological productivity is higher in the salt marsh than in the open marine environment. Consequently, Bass Creek ebb currents carry greater concentrations of organic debris than do flood currents (Ward, 1981). Upstream ends of bars offer more exposure to these food-laden currents and are therefore better dwelling sites for suspension feeders like *D. cuprea*.

Fig. 4.—*Diopatra cuprea* field on bar apex shown in Figure 2. View from inner bank, looking nearly normal to flow. Note tube rows and washed-out ripples. Current-parallel sand wisps form in the ebb shadows of tube rows (right center). Scale is one meter.

Second, feeding efficiency is in part a function of the total volume of water passing any given tube (Mangum et al., 1968). Even if food concentrations in ebb and flood currents were identical, the combination of ebb dominance and time-velocity asymmetry results in locally higher net ebb discharges at bar apex tube sites, the equality of flood and ebb discharges for the total channel cross-section notwithstanding.

Finally, intertidal depth preference may be the result of predation by fish, a possibility suggested by Myers (personal communication), who has observed fish feeding on *D. cuprea* heads. Subtidal tubes would be more vulnerable because they are available to predators throughout the tidal cycle. However, subtidal tubes are locally abundant in Georgia estuaries and the inner shelf (Frey, personal communication). Despite this fact, and because very little information is available that relates *D. cuprea* population dynamics to depositional environment, it seems reasonable to assume that the influence of predation is a significant factor in tube distribution.

Sedimentation patterns may inhibit *Diopatra* establishment in other areas on the bar. The intertidal outer flank is inhospitable because the few tubes built there become buried by megaripple migration. Cutbank instability caused by erosion and slumping limits *D.*

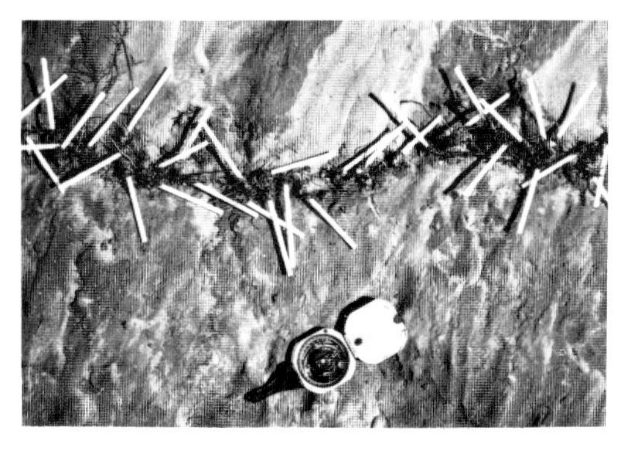

Fig. 5.—Orientations of tube tops shown by soda straws inserted into tube cap apertures. Note tendency of apertures in alternate tubes to face obliquely up or down current. No two adjacent apertures face each other directly.

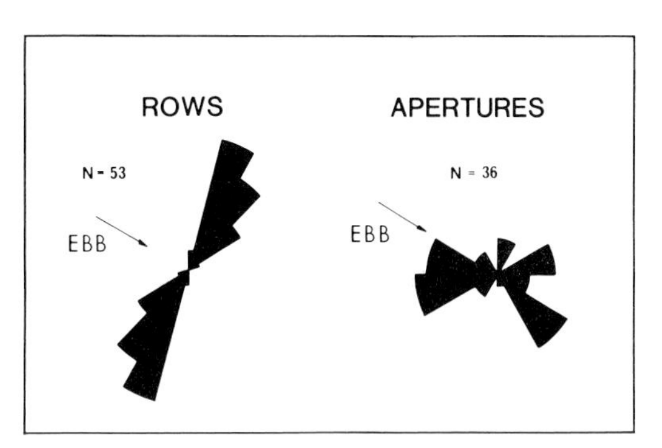

Fig. 6.—Summary of row and tube aperture orientations. Ebb direction is ESE (120 degrees). Mean row alignment axis is 217 degrees. Mean values for aperture orientation modes are 122 and 260 degrees.

cuprea to stable or aggrading channel banks. The absence of *D. cuprea* on the muddy inner lobe may be attributed to substrate instability, prohibitively low current velocities (i.e., inadequate food supply), or lack of material for tube reinforcement. The relative importance of these factors is unknown, inasmuch as tubes were occasionally seen without ornamentation, and sparsely distributed individual tubes were observed in soft mud substrates.

Tubes in the *Diopatra cuprea* field are arranged in scores of rows ranging in length from 15 to 150 centimeters (avg. 47 cm), as shown in Figure 4. These rows are linear to gently undulatory and are aligned perpendicular to flow (Figs. 5, 6), a pattern first described by Frey and Howard (1969), and later by Myers (1972) and Lawrence (1974). Average tube spacing within a row is 1.3 cm. Rows are arranged *en echelon* so that only a part of any row lies immediately downcurrent from any adjacent row. The densest spacing between rows is 20 cm at the bar apex and spacing increases downstream to a maximum of 2.3 m. Average overall spacing is 83 cm. Lawrence (1974) proposed that these current-normal rows and their *en echelon* arrangement provide the optimum distribution pattern for feeding efficiency. Although the most highly organized *D. cuprea* populations are on point bar apexes, tube rows are also established on nonbarred creek banks (Frey and Basan, 1978; Frey and Howard, 1980). These rows are also aligned normal to flow.

In Bass Creek, most tube apertures are oriented at 45 degrees to flow (Figs. 5, 6). Very few tube apertures are parallel to flow, which is contrary to observations by Myers (1972) for tubes occurring singly and in small groups. This arrangement may be a compromise between the maintenance of optimum flushing and feeding orientation (aperture parallel to flow) and the avoidance of directly facing another tube. Such an avoidance could provide two advantages: (1) *Diopatra cuprea* apparently competes aggressively with other individuals of the species whose tubes are too close; occasionally tube tops are clipped off by neighboring worms (Brenchley, personal communication). (2) In addition to catching suspended food particles, *D. cuprea* browses on the bed near its tube. Facing an adjacent tube would therefore force individuals to compete for browsing area. Facing 45 degrees to both current direction and row orientation would maximize the feeding area and minimize the hydraulic disadvantage of facing directly up- or down-current.

Effects on Sedimentation

Tube-building polychaetes generally enhance substrate stability (Fager, 1964). During ebb flows, sand is deposited in turbulent wakes behind the *Diopatra cuprea* tubes, resulting in wispy, current-parallel sand ribbons that often exceed a meter in length (Fig. 4). Scour pits as much as a centimeter deep and four centimeters in diameter form around the base of the tube. On the higher, muddier portions of the bar crest, where current shielding effects are absent, the scour pits are nearly round (Fig. 7a). On the bar apex, where shielding is at a maximum and ebb-dominance is pronounced, scour pits are elongate in the ebb direction (Fig. 7b). Scour pits become partially backfilled during neap tides, when current velocities are relatively lower.

Scour pits are poorly developed or absent on low intertidal and subtidal portions of the bar flank. There are two probable reasons for this paucity. First, this area is exposed to continuous winnowing by wave chop, which quickly smooths the cohesionless, clean sand bed. In contrast, mud laminae on the bar crest provide coherence and thus aid in scour pit preservation. Second, scour pits form more readily on the bar crest, where flows are commonly very shallow and upper flow regime conditions are temporarily established, even at low velocities (for example, during emergence). Near the low tide level, on the other hand,

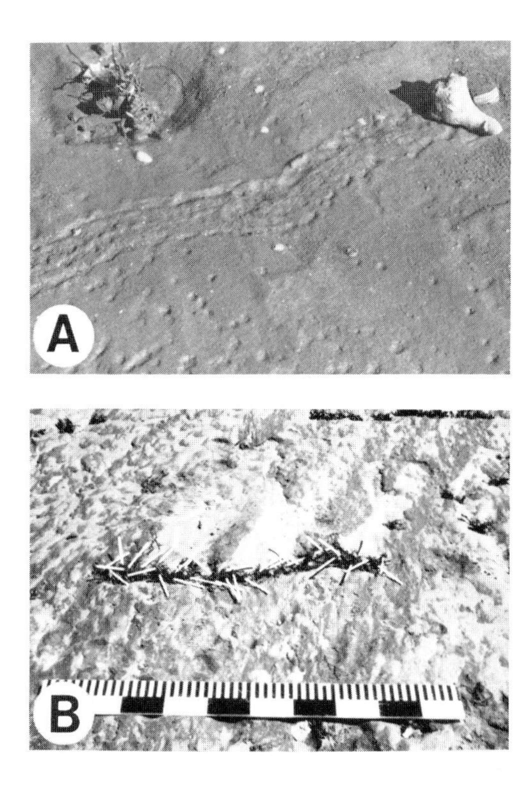

Fig. 7.—Scour pit shapes of *Diopatra cuprea*. *A*, Slightly oval, clay-draped pit around isolated tube on bar crest. Ebb direction is from right to left, roughly parallel to the hermit crab trail. Crab domicile is shell of juvenile knobbed whelk, *Busycon carica*. *B*, Asymmetrical, sand-filled scour pits behind tube row on bar apex. Ebb direction is toward top. Scale is one meter.

Froude number is never very high, because nearly slack-water conditions temporally bracket low tide. Regardless of cause, this zonation is especially significant when interpreting *Skolithos - Monocraterion* assemblages.

ANALOGOUS ANCIENT TRACES

The rendering of any modern-ancient trace analogy involves two related but slightly different perspectives. One approach involves evaluation of morphologic similarities between traces, to establish possible congeneric or conspecific origins for biogenic structures in deposits of widely varying age. The other approach involves examination of ichnocoenose distributional similarities, individual morphologic differences notwithstanding, that reflect the physical processes which control depositional environment and tracemaker behavior. The salient features of *Diopatra cuprea* burrows provide a good example of why these two perspectives are not mutually exclusive; although *D. cuprea* tubes are similar to one trace fossil, they are more similar to other fossil dwelling tubes in their relation to enclosing lithofacies.

Burrow Morphology

In intertidal areas, where *D. cuprea* burrows are vertical, the mucoid tubes have *Skolithos*-like affinities. Subtidal tubes are sinuous and oriented at various angles (Howard and Frey, 1975), and would probably be best compared to the trace fossil *Sabellarifex*, as described by Häntzschel (1975). Major differences between *Skolithos* and *D. cuprea* lie in the structure of the tube lining. *Skolithos* tubes are lined with sand only, and have never been described with shell-hash reinforcement or ornamented tube tops.

Chiefly because of its mucoid lining, *D. cuprea* was suggested as a possible modern analog for makers of Pleistocene *Skolithos* in North Carolina (Curran and Frey, 1977). However, these authors showed *Onuphis microcephala* to be a more plausible analog, based on tube size, morphology, and orientation. In addition, *O. microcephala* tubes and the Pleistocene *Skolithos* are continuous through relatively undeformed bedding surfaces. This is in marked contrast to the funnel tops associated with *D. cuprea* tubes, as shown in Figure 3 (see also Frey and Howard, 1972, their Fig. 8).

Comparison of Diopatra cuprea *Tubes to* Diopatrichnus roederensis

Diopatrichnus roederensis is the name given by Kern (1978) to a cylindrical trace fossil from a sandy mudstone unit of the Eocene Mission Valley Formation in southern California. Although named for the modern genus *Diopatra, D. roederensis* has a tube diameter two to three times greater than that of either *Diopatra cuprea* or *D. obliqua*. Aside from the size difference, the Eocene trace is morphologically identical to *D. cuprea*. Mollusc valve ornamentation is attached in the same imbricate fashion, and tubes are essentially straight and perpendicular to bedding. Although not as well-developed as *Diopatra cuprea* tube apertures, *Diopatrichnus roederensis* displays a slightly canted tube top (Kern, 1978, his Fig. 3). Kern also describes what could be a section of unreinforced tube that is similar to the unreinforced tube *D. cuprea* constructs following periods of rapid sedimentation (Myers, 1972).

Detailed facies relationships within the Mission Valley Formation are unknown. Kennedy and Moore (1971) describe the sequence as "friable, light olive gray fine grained sandstone containing several concretionary layers 0.5 m thick that contain molluscan fossils." These authors characterize the formation as "mostly marine," a description which could place the unit nearly anywhere below the intertidal zone. A detailed comparison of Bass Creek *Diopatra cuprea* environments with those of *Diopatrichnus roederensis* is therefore impractical.

Comparison of Diopatra cuprea *Tubes to* Skolithos-Monocraterion

The two tubular burrow forms *Monocraterion* and *Skolithos* were originally considered as products of separate organisms (Westergard, 1931). Hallam and Swett (1966) later concluded that the two traces were created by one burrower exposed to variable sedimentary conditions. They suggested that *Monocraterion* represents an escape structure associated with environments subject to higher sedimentation rates. In their analysis of the Cambrian Chickies Quartzite, Goodwin and Anderson (1974) presented compelling evidence supporting the one-organism hypothesis. Their observations revealed that *Monocraterion* was always connected to a subjacent *Skolithos* tube. In addition, the apparently membranous *Skolithos* tube walls were observed to be vertically continuous through their funnel-shaped tops. This arrangement negates escape origin for the funnel tops, which were interpreted by Goodwin and Anderson as resulting from polychaete feeding activity. *Skolithos* tubes without *Monocraterion* tops were interpreted as having formed in areas subject to higher velocity flows, which removed the funnel tops by erosion. This interpretation was supported by parallel distributions of sedimentary facies and burrow types, and by the truncation of *Skolithos* by the foresets of large-scale cross-strata. These Cambrian *Monocraterion* trace fossils were later interpreted as backfilled scour pits around *Skolithos* tubes (Barwis and Goodwin, 1979).

Goodwin and Anderson's (1974) paleoenvironmental model is remarkably similar to facies and burrowing patterns on the Bass Creek point bar. Chickies Quartzite *Skolithos - Monocraterion* traces are similar to Bass Creek *Diopatra cuprea* tubes in the following ways:

1) Burrow length, diameter, and orientation are nearly identical.
2) Burrows are arranged in rows, although to a lesser degree in the Chickies Quartzite.
3) Burrow walls are outlined by grains coarser than the mean size of the surrounding bed.
4) Tube density is lowest in subtidal environments.
5) Tubes with funnel tops are most common in the high intertidal environment.
6) Depth, diameter, and plan view shape of funnel tops are nearly identical.
7) Tubes are continuous through funnel-filling sediment. Individual laminae are preserved within the funnels.

Diopatra burrowing patterns suggest, however, a slightly different interpretation of the relationship between bedforms and *Skolithos - Monocraterion* burrow types.

If a Bass Creek channel profile is compared to Goodwin and Anderson's (1974) environmental model (Fig. 8), the most striking similarity is the restriction of *Monocraterion* tube types to poorly-sorted, thin-bedded sediments displaying small-scale cross-stratification. Goodwin and Anderson interpreted this correlation to be the result of lower flow-regime conditions on bar crests, which promoted *Monocraterion* preservation. The presence of flat, graded beds was invoked as evidence of these conditions.

Flow conditions on the crests of modern tidal point bars span a wide range of depths and velocities. During late-stage ebb flows, Froude number can approach unity for short periods, especially during spring tides. Scour pits and flat beds or washed-out ripples are developed during these periods. The bed as a whole, however, remains muddy and poorly sorted because of the ephemeral nature of the rapid flow events and because of the relatively high bed shear required to erode mud deposited during slack-water intervals (Postma, 1967). During the low-velocity flows surrounding high slack-water, lower flow-regime flat bed and ripple geometries characterize the bed. Thus, a given point on the bar crest can be subject to both lower and upper flow-regime flat bed conditions. Scour pits may therefore be ephemerally generated features on a substrate otherwise subjected to much lower bed shear. Nevertheless, upper and lower flow-regime flat beds have yet to be reliably discerned on the basis of grading, so the presence of flat beds alone cannot be used to indicated lower flow-regime conditions.

Myers (1970, 1972) offers another explanation for the funnel tops which is related neither to feeding, as suggested by Goodwin and Anderson (1974), nor to scouring, as suggested by Barwis and Goodwin (1979). Myers found that, following sudden burial, the animal burrows up to the new sediment-water interface and builds an unreinforced tube in the process. An armored tube top is then constructed, which the animal pulls down 4 to 6 cm into the substrate, thereby creating a depression around the tube (Myers, 1972, p. 354). Thus, although funnel tops are not directly created by an escape act itself, they may be an inevitable result. The depressions created by pulling-down may therefore significantly enhance scour-pit formation.

Myers (1970, 1972) notes that this pulling-down occurs only when sudden bed accretion exceeds 5 centimeters. During three years of visits to the Bass Creek bar, no bedform with a height exceeding 5 centimeters was ever observed on the densely populated bar crests, so the pits in those areas are wholly attributable to scour. Even though cross-bed set thicknesses in the Chickies Quartzite *Monocraterion* fields are medium-scale, which is thick enough to involve Myers's pulling-down mechanism, no tube tops were ever observed by Goodwin and Anderson (1974), nor were sections of reinforced versus unreinforced tube. Funnel tops are oval-shaped, however, so some degree of scour is probable, notwithstanding the fact that funnel-top long axes are parallel to the tube rows (Goodwin and Anderson, 1974, p. 782).

The highest population densities of *Skolithos* in the Chickies Quartzite are in the largest-scale cross-beds,

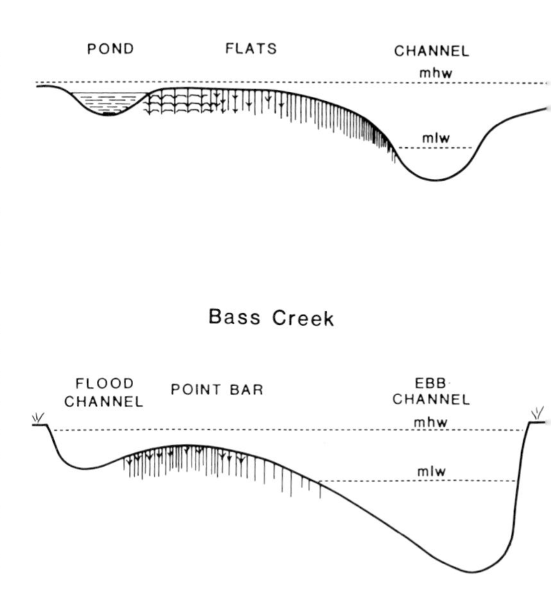

Fig. 8.—Comparison of Bass Creek point bar environments to the Cambrian Chickies Quartzite paleoenvironmental model of Goodwin and Anderson (1974). *Skolithos* and the modern *Skolithos*-like tubes are straight, vertical lines; *Monocraterion* and the *Monocraterion*-like tubes are lined with v-shaped tops. Note topographic coincidence of the position of *Monocraterion* forms, and the disparity in relative population densities. Bass Creek profile is at bar apex, upstream from lobe bifurcation.

where tangential sets range to 50 centimeters in thickness (Fig. 8). This is in marked contrast to the Bass Creek analog, where *Diopatra cuprea* is sparse to absent near megaripples. Nothing observed on modern bars helps to explain this disparity. The difference could be attributed to niche differences between the Cambrian trace-formers and modern polychaetes; the Chickies Quartzite *Skolithos*-producing worm may have been capable of tolerating higher sedimentation rates. In any case, Goodwin and Anderson present a convincing case for erosion of *Monocraterion* tops on a bar flank, an erosion event which would occur regardless of funnel origin.

CONCLUSIONS

1) In South Carolina tidal creeks, *Diopatra cuprea* populations normally are densest on upstream bar extremities, as a result of marsh productivity, ebb dominance, and time-velocity asymmetry.
2) Burrows are arranged in rows aligned normal to flow, a pattern which provides the most efficient use of finite food resources.
3) Tube apertures are oriented at 45 degrees to flow. The reasons for this orientation are uncertain but are probably related to competition and hydraulic efficiency.
4) Groups of *Diopatra cuprea* tubes in tidal-channel environments probably owe their degree of organization to the relatively consistent direction vector associated with channelized flows.
5) Tubes of the modern polychaete *Diopatra cuprea* resemble both *Skolithos* and *Monocraterion* traces. The degree of similarity to either trace depends on where a tube is located with respect to bar topography. *Monocraterion*-like forms are restricted to higher areas.
6) The distribution of *Diopatra cuprea* on mesotidal point bars represents a close modern analog to burrow distribution patterns described from certain Cambrian tidal flats.
7) *Diopatra cuprea* tube linings have a fabric similar to that described for the Eocene trace fossil *Diopatrichnus roederensis*.

ACKNOWLEDGMENTS

I wish to thank Mike Wadell and Larry Ward for their field support. Gail Brenchley, James D. Howard, David R. Lawrence, and Alan C. Myers provided informative discussions on the habits and habitats of *Diopatra cuprea*, and I gratefully acknowledge their advice. Peter Goodwin's ideas on animal-sediment relationships were invaluable, as was his companionship on Cambrian outcrops. Maria Balzarini and Rita Monahan made suggestions which improved an earlier version of the manuscript. I appreciate the thorough, constructive reviews of Robert Frey and Peter Goodwin. Typing and drafting were provided by Shell Oil Company.

REFERENCES

BARWIS, J.H.,1978, Sedimentology of some South Carolina tidal creek point bars, and a comparison with their fluvial counterparts, *in* Miall, A.D. (ed.), Fluvial Sedimentology: Canadian Soc. Petroleum Geologists, Mem. No. 5, p. 129–160.

______, AND GOODWIN, P.W., 1979, Modern and ancient examples of annelid burrows as current direction indicators (abs.): Am. Assoc. Petroleum Geologists Bull., v. 63, p. 415.

BRENCHLEY, G.A., 1976, Predator detection and avoidance: ornamentation of tube-caps of *Diopatra* sp. (Polychaeta - Onuphidae): Marine Biology, v. 38, p. 179–188.

BROWN, P.J., 1977, Variations in South Carolina coastal morphology: Southeastern Geology, v. 18, p. 249–264.

BYRNE, R.J., BULLOCK, P., AND TYLER, D.G., 1975, Response characteristics of a tidal inlet: A case study, *in* Cronin, L.E. (ed.), Estuarine Research: New York, Academic Press, V. II, p. 201–216.

CHAMBERLAIN, C.K., 1975, Recent lebensspuren in nonmarine aquatic environments, *in* Frey, R.W. (ed.), The Study of Trace Fossils: New York, Springer-Verlag, p. 431–458.

CURRAN, H.A., AND FREY, R.W., 1977, Pleistocene trace fossils from North Carolina (U.S.A.), and their Holocene analogues, *in* Crimes, T. P., and Harper, J.C. (eds.), Trace Fossils 2: Liverpool, Seel House Press, p. 139–162.

FAGER, E.W., 1964, Marine sediments: Effects of a tube-building polychaete: Science, v. 143, p. 356–359.

FISK, H.N., 1947, Fine-grained alluvial deposits and their effects on Mississippi River activity: Vicksburg, Mississippi, U. S. Waterways Exp. Sta., 2 Vols., 82 p.

FREY, R.W., 1970, Environmental significance of Recent marine lebensspuren near Beaufort, North Carolina: Jour. Paleontology, v. 44, p. 507–519.

______, (ed.), 1980, Excursions in Southeastern Geology, Volume II: Atlanta, Geol. Soc. America, Abs. with Prog., Ann. Mtg., 310 p.

______, AND BASAN, P.B., 1978, Coastal salt marshes, *in* Davis, R.A., Jr. (ed.), Coastal Sedimentary Environments: New York, Springer-Verlag, p. 101–169.

______, AND HOWARD, J.D., 1969, A profile of biogenic sedimentary structures in a Holocene barrier island—salt marsh complex, Georgia: Gulf Coast Assoc. Geol. Soc. Trans., v. 19, p. 427–444.

______, AND ______, 1972, Georgia coastal region, Sapelo Island, U.S.A.: Sedimentology and biology. IV. Radiographic study of sedimentary structures made by beach and offshore animals in aquaria: Senckenbergiana Maritima, v. 4, p. 169–182.

______, AND ______, 1980, Physical and biogenic processes in Georgia estuaries. II. Intertidal facies, *in* McCann, S.B. (ed.), Sedimentary Processes and Animal Sediment Relationships in Tidal Environments: Halifax, Nova Scotia, Geol. Assoc. Canada, p. 183–220.
GOODWIN, P.W., AND ANDERSON, E.J., 1974, Associated physical and biogenic structures in environmental subdivision of a Cambrian tidal sand body: Jour. Geology, v. 82, p. 779–794.
HALLAM, A., AND SWEET, K., 1966, Trace fossils from the Lower Cambrian Pipe Rock of the northwest Highlands: Scottish Jour. Geol., v. 2, p. 101–106.
HÄNTZSCHEL, W., 1975, Trace fossils and problematica, *in* Teichert, C. (ed.), Treatise on Invertebrate Paleontology Part W, Supplement 1: New York, Geol. Soc. America, and Lawrence, Univ. Kansas Press.
HAYES, M.O., 1975, Morphology of sand accumulation in estuaries: An introduction to the symposium, *in* Cronin, L. E. (ed.), Estuarine Research: New York, Academic Press, v. II, p. 3–22.
HAYES, M.O., AND KANA, T.W., eds., 1976, Terrigenous Clastic Depositional Environments: Columbia, Univ. South Carolina, Tech. Rept. No. 11-CRD, 302 p.
______, OWENS, E.H., HUBBARD, D.K., AND ABELE, R.W., 1973, The investigation of form and process in the coastal zone, *in* Coates, D.R. (ed.), Coastal Geomorphology: Binghamton, Proc. Third Ann. Geomorphology Sym., p. 11–41.
HERTWECK, G., 1972, Georgia coastal region, Sapelo Island, U.S.A.: Sedimentology and biology. V. Distribution and environmental significance of lebensspuren and in-situ skeletal remains: Senckenbergiana Maritima, v. 4, p. 125–167.
HOWARD, J.D., 1971, Trace fossils as paleoecological tools, *in* Howard, J.D., Valentine, J.W., and Warme, J.E. (eds.), Recent Advances in Paleoecology and Ichnology: Washington D.C., Am. Geol. Inst., p. 184–212.
______, AND DÖRJES, J., 1972, Animal-sediment relationships in two beach-related tidal flats; Sapelo Island, Georgia: Jour. Sed. Petrology, v. 42, p. 608–623.
______, ELDERS, C.A., AND HEINBOKEL, J.F., 1975, Estuaries of the Georgia coast, U.S.A.: Sedimentology and biology. V. Animal-sediment relationship in estuarine point bar deposits, Ogeechee River-Ossabaw Sound, Georgia: Senckenbergiana Maritima, v. 7, p. 181–203.
______, AND FREY, R.W., 1973, Characteristic physical and biogenic sedimentary structures in Georgia estuaries: Am. Assoc. Petroleum Geologists Bull., v. 57, p. 1169–1184.
______, AND______, 1975, Estuaries of the Georgia coast, U.S.A.: Sedimentology and biology. II. Regional animal-sediment characteristics of Georgia estuaries: Senckenbergiana Maritima, v. 7, p. 33–103.
______, AND ______, 1980, Holocene depositional environments of the Georgia coast and continental shelf, *in* Howard, J. D., DePratter, C.B., and Frey, R.W. (eds.), Excursions in Southeastern Geology: The Archaeology-Geology of the Georgia Coast: Georgia Geol. Surv., Guidebook 20, p. 66–134.
______, ______, AND REINECK, H.E., 1973, Holocene sediments of the Georgia coastal area, *in* Frey, R.W., ed., The Neogene of the Georgia Coast: Athens, Georgia Geol. Soc., 8th Ann. Field Trip, Dept. Geol., Univ. Georgia, p. 1–58.
HUBBARD, D.K., OERTEL, G., AND NUMMEDAL, D., 1979, The role of waves and tidal currents in the development of tidal inlet sedimentary structures and sand body geometry; examples from North Carolina, South Carolina, and Georgia: Jour. Sed. Petrology, v. 49, p. 1073–1091.
KENNEDY, M.P., AND MOORE, G.W., 1971, Stratigraphic relations of Upper Cretaceous and Eocene formations, San Diego coastal area, California: Am. Assoc. Petroleum Geologists Bull., v. 55, p. 709–722.
KEULEGAN, G.H., 1967, Tidal flow in entrances; water-level fluctuations of basins in communications with seas: U.S. Army Corps of Engineers, Comm. on Tidal Hydrology, Tech. Bull. No. 14, 111 p.
KERN J.P., 1978, Paleoenvironment of new trace fossils from the Eocene Mission Valley Formation, California: Jour. Paleontology, v. 52, p. 186–194.
KING, D.B., 1974, The Dynamics of Inlets and Bays: Gainesville, Florida, Coastal and Oceanographic Eng. Lab., Univ. Florida, Tech. Rept. 22, 82 p.
LAND, L.S., AND HOYT, J.H., 1966, Sedimentation in a meandering estuary: Sedimentology, v. 6, p. 191–207.
LAWRENCE, D.R., 1974, Models for pattern diversity in the fossil record (abs.): Geol. Soc. America, Abs. with Prog., Ann Mtg., v. 6, p. 840–841.
MANGUM, C.P., AND COX, C.D, 1971, Analysis of the feeding response in the onuphid polychaete *Diopatra cuprea* (Bosc.): Biology Bull., v. 140, p. 215–229.
______, SANTOS, S.L., AND THODES, W.R., 1968, Distribution and feeding in the onuphid polychaete, *Diopatra cuprea* (Bosc.): Marine Biology, v. 2, p. 33–40.
MAYOU, T.V., AND HOWARD, J.D., 1969, Recognizing estuarine and tidal-creek sand bars by biogenic sedimentary structures (abs.): Am. Assoc. Petroleum Geologists Bull., v. 53, p. 731.
MOSLOW, T.F., 1980, Stratigraphy of mesotidal barrier islands [unpublished Ph.D. thesis]: Columbia, Univ. South Carolina, 186 p.
MOTA OLIVIERA, I.B., 1970, Natural flushing ability of tidal inlets: New York, Am. Soc. Civil Engineers, Proc. 12th Conf. Coastal Eng., p. 1827–1845.
MYERS, A.C., 1970, Some paleoichnological observations on the tube of *Diopatra cuprea* (Bosc.): Polychaeta, Onuphidae, *in* Crimes, T.P., and Harper, J.C. (eds.), Trace Fossils: Liverpool, Seel House Press, p. 331–334.
______, 1972, Tube-worm sediment relationships of *Diopatra cuprea* (Polychaete: Onuphidae): Marine Biology, v. 17, p. 350–356.

Nummedal, D., Oertel, G., Hubbard, D.K., and Hine, A.C., 1977, Tidal inlet variability— Cape Hatteras to Cape Canaveral: Coastal Sediments '77, Am. Soc. Civil Engineers, p. 543–562.

Oomkens, E., and Terwindt, H.J.H., 1960, Inshore estuarine sediments in the Haringvliet (Netherlands): Geol. en Mijnbouw, v. 39, p. 701–710.

Postma, F.J., 1967, Sediment transport and sedimentation in the estuarine environment, *in* Lauff, G. H. (ed.), Estuaries: Washington, Am. Assoc. Advancement Science, p. 158–179.

Price, W.A., 1963, Patterns of flow and channeling in tidal inlets: Jour. Sed. Petrology, v. 33, p. 279–290.

Pryor, W.A., 1967, Biogenic directional features on several recent point bars: Sedimentary Geology, v. 1, p. 235–245.

Van Veen, J., 1950, Ebb- and flood-channel systems in the Dutch tidal waters: Tijdschr. Koninkl. Ned. Aardrijkskundig Genootschap (2), v. 67, p. 303–325.

Ward, L.G., 1981, Suspended material transport in marsh tidal channels, Kiawah Island, South Carolina: Marine Geology, v. 40, p. 139–154.

Westergard, A.H., 1931, *Diplocraterion*, *Monocraterion* and *Scolithus* from the Lower Cambrian of Sweden: Sver. Geol. Undersokn, Ser. C., Arh. Uppa. No. 372 (Arsbok 25), p. 1–25.

Wojtal, A.M., 1981, Back-barrier stratigraphy of Kiawah Island, South Carolina [unpublished Ph.D. Thesis]: Columbia, Univ. South Carolina, 179 p.

______, and Moslow, T.F., 1980, Stratigraphy of barrier and back-barrier facies, Kiawah Island, South Carolina, *in* Frey, R.W., (ed.), Excursions in Southeastern Geology, V. II: Atlanta, Georgia, Geol. Soc. America, Abs. with Prog., Ann. Mtg., p. 289–303.

THE *GLOSSIFUNGITES* ICHNOFACIES: MODERN EXAMPLES FROM THE GEORGIA COAST, U.S.A.

S. GEORGE PEMBERTON[1] AND ROBERT W. FREY
Alberta Research Council, Terrace Plaza, 4445 Calgary Trail South, Edmonton, Alberta T6H 5R7; and
Department of Geology, University of Georgia, Athens, Georgia 30602

ABSTRACT

The *Glossifungites* ichnofacies, albeit areally restricted, constitutes a distinct intermediary between freely shifting, particulate substrates of the *Skolithos* ichnofacies and fully consolidated substrates of the *Trypanites* ichnofacies. All three typically occur within, but are not restricted to, littoral to shallow sublittoral, relatively high-energy zones. Substrates of the *Glossifungites* ichnofacies ordinarily are palimpsest—altered by contemporary physical and biogenic processes—and may exhibit relict features indicative of conditions prior to semiconsolidation of sediments. Characteristic lebensspuren consist of unlined domiciles of suspension-feeding or foraging animals, such as certain bivalves, crustaceans, and polychaetes, and these crosscut relict physical and biogenic sedimentary structures.

On the Georgia coast, the ichnofacies occurs most commonly among coherent, partially dewatered, Holocene salt marsh muds formerly buried underneath transgressive beach ridges but now being exhumed by beach or tidal stream erosion. Relict in-situ biotic features include roots of the grass *Spartina*, burrows of the crab *Uca*, and valves of the snail *Littorina*, the mussel *Geukensia*, and the oyster *Crassostrea*. On St. Catherines and Petit Chou Islands, the substrates now are occupied by three intergradational ichnocoenoses: (1) a petricolid assemblage, (2) a petricolid-pholad-crustacean assemblage, or (3) a petricolid-crustacean-polydoran assemblage. Distributions of these traces seem to be governed mainly by incidence of wave versus current energy, duration of subaerial exposure, density of relict root mats, and variations in sand content, cohesiveness, and related geotechnical properties of the substrate. In the rock record, diagnostic biogenic structures among the assemblages would include: firm- (as opposed to hard-) substrate *Gastrochaenolites* and (or) closely related ichnogenera, made by petricolid and pholad bivalves; small *Diplocraterion*, made by the polychaete *Polydora*; essentially unlined *Palaeophycus* and *Thalassinoides*, made by polychaetes and shrimp, respectively; and *Psilonichnus*, or unlined crab burrows. Of the last three, some forms occur in relict (pre-semiconsolidation) assemblages as well.

INTRODUCTION

During the past three decades, perhaps the single most significant concept to evolve from ichnology concerns the grouping of ichnofossils in recurring assemblages or ichnofacies. This scheme originated with and evolved from careful examination of the distribution and association of numerous ichnofossils from units of varying age and environmental implication (Seilacher, 1954, 1955, 1958) and culminated in the formal designation of six distinct assemblages (Seilacher, 1963, 1964, 1967). Seilacher (1967) noted that, in marine environments, many parameters that govern the abundance and distribution of tracemakers (such as temperature, food supply, and intensity of wave or current agitation) tend to change progressively with water depth. Hence, some of the named ichnofacies constitute the basis for a relative scale of bathymetry. Numerous authors have stressed, however, that the distribution of ichnofossils is ultimately linked with the complex interactions of environmental parameters and as such should not be regarded as unique depth indicators (Frey, 1971; Hayward, 1976; Crimes, 1977; Frey and Seilacher, 1980).

Originally, Seilacher (1963, 1964, 1967) recognized two basic littoral assemblages: the *Glossifungites* ichnofacies for stable, coherent substrates subject to water turbulence, and the *Skolithos* ichnofacies for shifting sediments subject to rapid deposition-erosion cycles. Of the two, the *Glossifungites* ichnofacies has been a source of considerable confusion; some authors restricted it to semiconsolidated, coherent substrates (e.g., Frey, 1975, Table 2.1), others viewed it as representing fully lithified units (e.g., Kennedy, 1975, p. 377), while still others regarded it as indicative of both semiconsolidated and consolidated substrates (e.g., Rhoads, 1975, p. 149, 157). In an attempt to clarify the situation, Frey and Seilacher (1980) redefined the *Glossifungites* ichnofacies concept, restricting it to firm but unlithified marine surfaces, and introduced the new *Trypanites* ichnofacies for fully lithified marine substrates or organic substrates such as bone, shell, and wood (Table 1).

Distinctions between the *Glossifungites* and *Trypanites* ichnofacies are well justified by numerous modern occurrences of the two, but these differences remain poorly elaborated. The disparity stems at least partly from insufficient literature on firm-substrate traces in general. Thus, the main objectives of this paper are (1) to review the concept of the *Glossifungites* ichnofacies, (2) to report the characteristics of selected modern examples from the Georgia coast, and (3) to evaluate the paleoecological significance of these deposits.

[1]Present address: Department of Geology, University of Alberta, Edmonton, Alberta T6G 2E3

THE *GLOSSIFUNGITES* ICHNOFACIES: GENERAL FEATURES

The *Glossifungites* ichnofacies is typical of firm but unlithified substrates, whether intertidal to subtidal omission surfaces or stable, cohesive sediments either in protected, low-energy settings or in places of somewhat higher energy where the semiconsolidated substrate offers resistance to erosion. Most such substrates consist of dewatered muds (Fig. 1).

Because dewatering and consequent firm-substrate burrowing occur commonly during prolonged subaerial exposure, the requisite substrates and ichnofacies tend to develop intertidally or supratidally. Dewatering also occurs with increasing depth of burial of sediments, of course, but such deposits remain unavailable to contemporary tracemakers unless they are subsequently exhumed by erosion. Exhumation occurs most commonly in shallow-water or intertidal conditions, as with major storms or slight tectonic or eustatic changes in sea level and consequent transgression or regression. On the present Georgia coast, for example, beaches are gradually retreating as sea level rises, and the accompanying erosion commonly unearths old marsh muds (Fig. 1B) formerly buried beneath beach sands (Frey and Howard, 1969; Howard et al., 1972, 1973; Morris and Rollins, 1977; Frey and Basan, 1981). A somewhat analogous situation was reported by Schäfer (1972) for the southern part of the North Sea, where extensive intertidal and subtidal peat deposits have been exhumed. The end result, bathymetrically, is essentially the same as that of muds dewatered intertidally or supratidally.

Desiccation processes may impart mud cracks, syneresis structures, or even large-scale blocky jointing to the deposit, and these, in turn, may influence the distribution and configuration of burrows (Schroder, 1982). Conversely, the distribution and configuration of burrows also may influence the pattern of desiccation polygons (cf. Baldwin, 1974; Kues and Siemers, 1977; Metz, 1980).

Whatever their origin, such substrates are considerably less common than are those of most other coastal and nearshore environments. The *Glossifungites* ichnofacies therefore tends to have a more limited geographic extent.

In addition, because they are intermediate between unconsolidated sediments of the *Skolithos* ichnofacies

TABLE 1.—RECURRING, PREDOMINANTLY LITTORAL, TRACE FOSSIL ASSEMBLAGES AND THEIR ENVIRONMENTAL IMPLICATIONS. (AFTER FREY AND SEILACHER, 1980)

Characteristic trace fossils	Typical benthic environment
Trypanites ichnofacies (hard substrates)	
Cylindrical to vase-, tear-, or U-shaped to irregular domiciles of endolithos, oriented normal to substrate surfaces, or shallow anastomosing systems of borings (sponges, bryozoans); excavated mainly by suspension feeders or "passive" carnivores. Raspings and gnawings of algal grazers, as well as other organisms (chitons, limpets, echinoids). Diversity generally low, although borings or scrapings of given kinds may be abundant.	Consolidated marine littoral and sublittoral omission surfaces (rocky coasts, beachrock, hardgrounds) or organic substrates (wood, shell, bone). Bioerosion is as important as, and indeed accelerates, physical erosion of the substrate (Warme 1975; Bromley 1978). Intergradational with the *Glossifungites* ichnofacies.
Glossifungites ichnofacies (firm substrates)	
Vertical cylindrical, U- or tear-shaped borings or boring-like structures, or sparsely to densely ramified dwelling burrows; protrusive spreiten in some, developed mostly through growth of animals. Fan-shaped *Rhizocorallium Diplocraterion*. Many intertidal species (e.g., crabs) leave the burrows to feed; others are mainly suspension feeders. Diversity typically low, but given kinds of structures may be abundant.	Firm but unlithified marine littoral and sublittoral omission surfaces, or stable, coherent substrates either in protected, low-energy settings (salt marshes; Letzsch and Frey, 1980, Fig. 5) or in areas of somewhat higher energy where semiconsolidated substrates offer resistance to erosion (Frey and Basan, 1981).
Skolithos ichnofacies (shifting substrates)	
Vertical cylindrical or U-shaped dwelling burrows; protrusive and retrusive spreiten in some, developed mainly in response to substrate aggradation or degradation (escape structures); forms of *Ophiomorpha* consisting predominantly of vertical or steeply inclined shafts. Animals chiefly suspension feeders. Diversity is low, yet given burrows may be adundant.	Littoral to infralittoral, relatively high-energy conditions; clean, well-sorted, shifting sediments, subject to abrupt erosion or deposition. (Higher energy increases physical reworking and obliterates biogenic sedimentary structures, leaving a preserved record of physical stratification; Rhoads, 1975, Fig. 9.8; Howard 1975.)

and fully consolidated ones of the *Trypanites* ichnofacies, these semiconsolidated substrates are occupied by a diverse ethologic mixture of traces; most are clearly burrows, but others closely simulate borings. Whereas the distinction between burrowers and borers ordinarily is clear (Frey, 1973), here the boundary—or our perception of it—may be largely semantic. Domiciles excavated by the burrowing polychaete *Nereis* may be extremely similar in watery modern salt marsh muds (Frey et al., 1973, Fig. 1B) and partially

FIG. 1.—Modern example of the *Glossifungites* ichnofacies, St. Catherines Island beach, Georgia. Coherent muds or sandy muds crop out in the intertidal zone and adjoin shifting sands of the *Skolithos* ichnofacies. *A*, Hummocky surface of a veneer of densely rooted palmetto (*Serenoa repens*) paleosol overlying salt marsh mud overlying, in turn, a wave-cut platform of Pleistocene beach sand (Station 1, Morris and Rollins, 1977). *B*, Salt marsh mud (Station 8, Morris and Rollins, 1977).

dewatered relict marsh muds (Fig. 2D), for example, and the boring polychaete *Polydora*, noted for its excavations in mollusk shells and other consolidated carbonate substrates (Warme, 1975), may appear here as well (see Figs. 10D, 11A). On the other hand, some organisms, such as the bivalve *Petricola pholadiformis* (Fig. 2B), are adapted almost exclusively to the excavation of semiconsolidated sediments (Schäfer, 1972).

One can argue whether the latter are true borings—i.e., whether the excavation cuts through, rather than around, individual sediments grains, or whether the organism possesses the requisite adaptations for boring (e.g., Purchon, 1955). But the fact remains that pouch-shaped traces virtually identical to pholad borings in hard substrates are common among *Glossifungites* ichnofaunas. Thus, in terms of trace fossil morphology—which, in turn, is the basis for trace fossil taxonomy (Pemberton and Frey, 1982)—certain ichnogenera ordinarily considered to be indicative of the Trypanites ichnofacies do indeed occur in the *Glossifungites* ichnofacies.

Animals that exploit these substrates tend to construct permanent domiciles. Because of sediment cohesiveness, reinforced burrow linings generally are unnecessary and the walls are sharply defined. Pronounced firmness of the substrate also may preclude intrastratal deposit feeding, even though the sediments may contain appreciable organic matter. Most animals thus are suspension feeders and construct vertical or steeply inclined dwelling structures. Among U-shaped burrows, protrusive spreiten develop more through the growth of the animal and its increased depth of burrowing than through substrate aggradation of degradation; for the same reason, the lower part of the "U" tends to be broader than the upper part. Ramified burrow systems also may be present (Fig. 2C); some of these are occupied by communal suspension feeders whereas others are occupied by carnivores or deposit feeders that leave the burrows to forage.

Trace fossils most characteristic of the *Glossifungites* ichnofacies include certain forms of *Rhizocorallium* (formerly assigned to the synonymous ichnogenus *Glossifungites* Lominicki, 1886), *Diplocraterion*, *Gastrochaenolites* and (or) related ichnogenera, *Thalassinoides*, and *Psilonichnus*. The arms of this form of *Rhizocorallium* are divergent or fan-shaped rather than being parallel, as in soft-sediment forms of the ichnogenus developed through feeding activities. Fürsich (1974a) placed this form in the ichnospecies *R. jenense*, which he interpreted as the dwelling burrow of a suspension feeder. Even though *Glossifungites* has been placed in synonymy with *Rhizocorallium* (Fürsich, 1974a; Häntzschel, 1975), we favor retention of the name for this ichnofacies, only, because the name is well entrenched in the literature. To rename it the "*Rhizocorallium* ichnofacies" would cause unnecessary confusion, especially considering the common occurrence of other forms of that ichnogenus in the *Cruziana* ichnofacies. *Diplocraterion* is distinguished from *Rhizocorallium* primarily by its vertical orientation (Fürsich, 1974b); it too is interpreted as the dwelling burrow of a suspension feeder.

At present, the ichnogenus *Trypanites* is too broadly conceived; plans include its forthcoming redefinition, together with that of various *Gastrochaenolites* and morphologically related forms (R. G. Bromley, 1981, personal communication). In the *Glossifungites* ichnofacies, *Gastrochaenolites*-like forms are limited chiefly to the sac-like or pouched excavations typical of pholad clams. The *Thalassinoides* structure is essentially unlined, commonly exhibits well-developed chelaped sculptings, and tends to be excavated in the form of a three-dimensional boxwork. *Psilonichnus* (Fürsich, 1981) consists of irregularly developed shafts and tunnels such as those excavated by some modern crabs; like most of the above ichnogenera, it is by no means restricted to the *Glossifungites* ichnofacies.

Not all the above structures need to be present in any given occurrence of the *Glossifungites* ichnofacies. But the overall assemblage is dominated by such firm-substrate excavations (Fig. 2A).

Substrate conditions and requisite behavior are the chief criteria distinguishing this assemblage from the *Trypanites* ichnofacies developed in hard, consolidated substrates (Table 1). There, ichnospecies of *Trypanites* and related ichnogenera are much more diverse, as are borings excavated by numerous kinds of tracemakers other than pholads. Among pholads, specific morphological variations within domiciles may be correlated with substrate consistency (Röder, 1977), whether firm substrates of the *Glossifungites* ichnofacies or fully consolidated ones of the *Trypanites* ichnofacies. Furthermore, true borings of the *Trypanites* ichnofacies penetrate tests and shells within the substrate just as readily as the lithified sediment (Bromley, 1975, Figs. 18.6, 18.16B); in all other ichnofacies, tests and shells or rock clasts are skirted by deeply penetrating excavations.

Another potential point of confusion or ambiguity in documenting the *Glossifungites* ichnofacies is that most such substrates are palimpsest. For example, normally soft muds of certain tidal flat or estuarine deposits might exhibit a *Cruziana* ichnofacies prior to dewatering and semiconsolidation of sediments. This relict assemblage of soft-sediment trace fossils would then be crosscut by the new assemblage of firm-sediment trace fossils. The situation is analogous to the pre- through post-ommission suites of trace fossils discussed by Bromley (1975, p. 400-402). In spite of the possible confusion, however, such overprinted suites, once understood, are themselves good evidence for *Glossifungites*-type conditions.

MODERN EXAMPLES OF THE *GLOSSIFUNGITES* ICHNOFACIES FROM THE GEORGIA COAST

As noted previously, the literature reveals a dearth of detailed information on modern examples of the *Glossifungites* ichnofacies. Notable exceptions include the "boggy layers" along the Massachusetts coast (Gould, 1870), peat deposits of the southern part of the North Sea (Schäfer, 1939, 1972), and relict salt marsh muds from both Cabretta Island (Hoyt et al., 1966; Frey and Howard, 1969; Howard et al., 1973; Howard and Frey, 1980; Frey and Basan, 1981) and St. Catherines Island, Georgia (Morris and Rollins, 1977). During the course of this project, three additional deposits were examined on the barrier island coast of Georgia: Petit Chou, Wassaw, and Sea Islands (Fig. 3).

In general, the origin and post-depositional history of Georgia deposits follows a similar pattern. A salt-marsh origin was postualted for St. Catherines Island muds by Morris and Rollins (1977), based on the presence of abundant, relict salt-marsh features, especially remnants of faunas and floras. Similar observations were made among exposures on Cabretta Island (Frey and Howard, 1969; Howard et al., 1973; Howard and Frey, 1980; Frey and Basan, 1981; present study), and Petit Chou, Wassaw, and Sea Islands (present study).

Therefore, the overall abundance and environmental position of these relict muds suggest that extensive coastal marshes and their protective barriers originally extended farther seaward than their present positions.

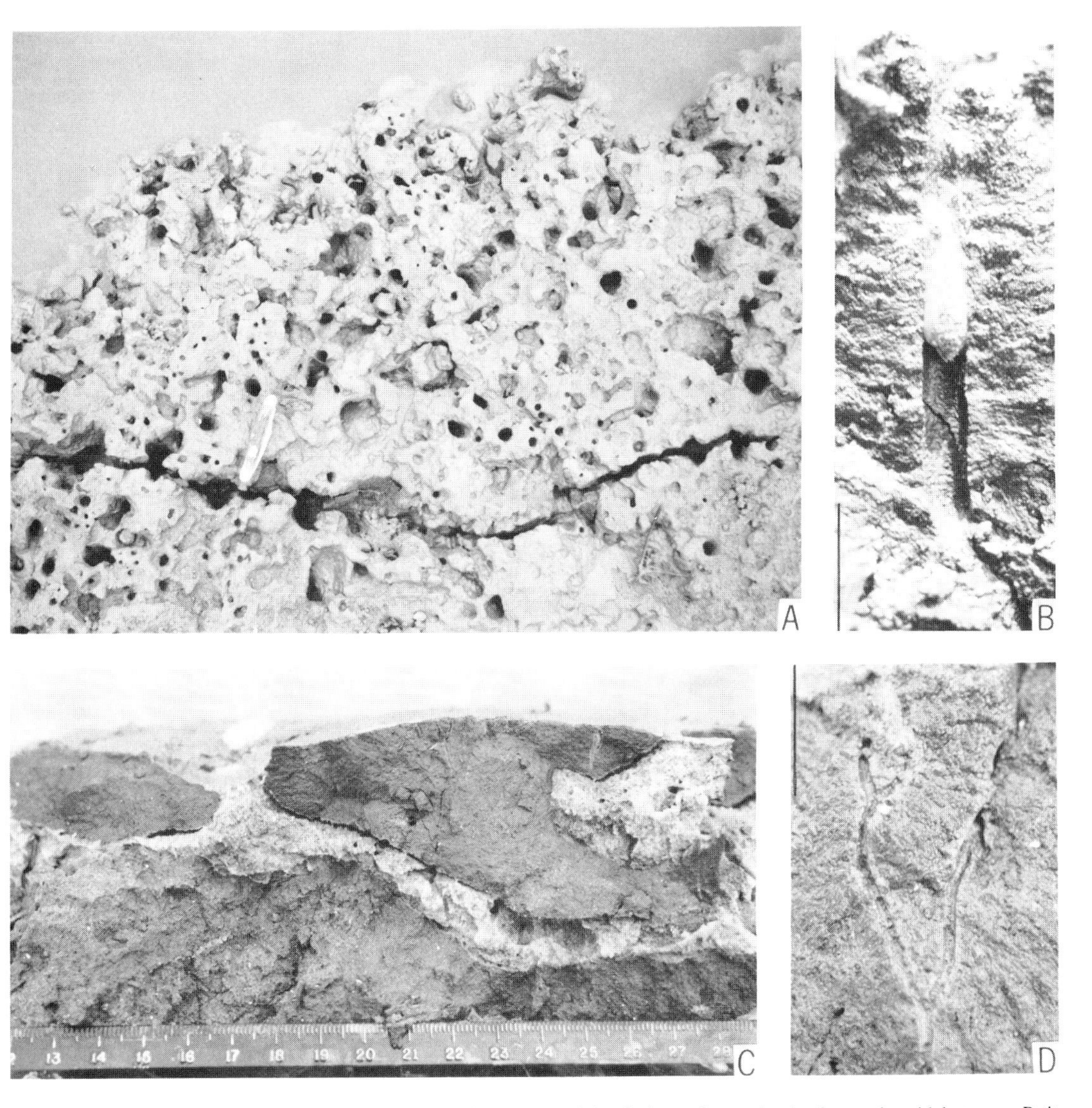

FIG. 2.—Characteristic lebensspuren of modern *Glossifungites* ichnofacies; coherent bank of estuarine tidal stream, Petit Chou Island, Georgia. *A*, Bioeroded, current-scoured bedding surface with tensional crack. *B*, Excavation of in-situ dead bivalve, *Petricola pholadiformis*. *C*, Abandoned, sand-filled burrow of shrimp *Upogebia affinis*. *D*, Occupied burrow of polychaete *Nereis succinea*. Bar scales = 1 cm.

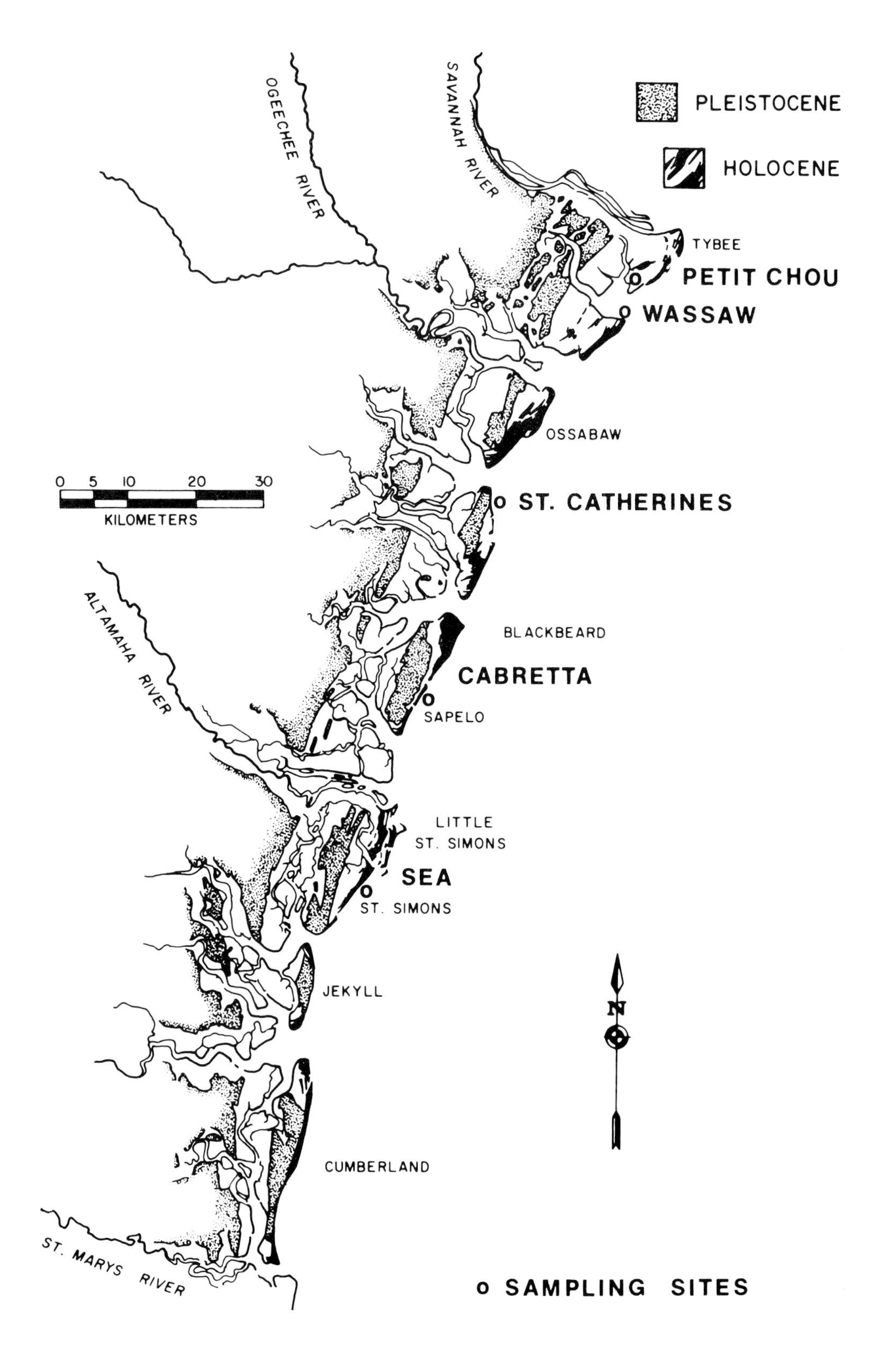

PLEISTOCENE
HOLOCENE
OGEECHEE RIVER
SAVANNAH RIVER
TYBEE
PETIT CHOU
WASSAW
OSSABAW
0 5 10 20 30
KILOMETERS
ST. CATHERINES
BLACKBEARD
CABRETTA
SAPELO
ALTAMAHA RIVER
LITTLE
ST. SIMONS
SEA
ST. SIMONS
JEKYLL
N
CUMBERLAND
ST. MARYS RIVER
o SAMPLING SITES

Radiometric dating indicates ages of 500 to 1,000 years B.P. for Cabretta examples (Howard et al., 1972, p. 7). These and most other relict salt marsh muds now exposed on erosional beaches in Georgia lie at essentially the same elevation as modern marshes, into which the old muds commonly grade (e.g., Frey and Basan, 1981). Subsequent beach sands and marginal marine sediments were transported and deposited over these original marsh sediments, rendering them unavailable for biotic exploitation. Active erosion, brought about by storm incidence and a general rise in late Holocene sea level (DePratter and Howard, 1981), then exhumed the old muds (Howard and Frey, 1980; Frey and Basan, 1981). This prolonged period of burial resulted in some degree of dewatering and thus semiconsolidation of the substrate.

Not only salt marsh deposits are involved, however. The "brownish peat" overlying dark gray muds mentioned by Morris and Rollins (1977, p. 100) is in fact a youthful terrestrial paleosol densely rooted by the saw palmetto *Serenoa repens*; detrital leaf fronds and in-situ large lateral roots and small vertical rootlets are common in the deposit, as are scattered fragments of the soft-rush Bermuda grass *Cynodon dactylon*. A similar association of live plants occurs on closely adjacent island soils. Present bathymetric position of the paleosol—upper to mid-intertidal zone—indicates that sea level was slightly lower at the time of its formation. It perhaps is related to sea-level rise following the late Holocene lowstand of sea level, 4,500 to 2,400 years B.P. (DePratter and Howard, 1981).

The paleosol, where eroded into peculiar hummocks (Fig. 1A), remains more than 40 cm thick. It overlies relict low-marsh deposits (cf. Frey and Basan, 1978)—the "dark gray muds" of Morris and Rollins (1977)—densely vegetated by roots and stem-stubble of the cordgrass *Spartina alterniflora*. The upper part of the marsh deposit yielded a radiocarbon date of 475 years B.P.; a sample approximately 0.5 m deeper yielded a date of 1325 years B.P. (J. Groce, 1982, personal communication). High-marsh sands, which normally overlie low-marsh muds, evidently were stripped away prior to development of the paleosol.

Both the paleosol and relict mud are now occupied by modern organisms of the *Glossifungites* ichnofacies. The muds, in turn, constitute a thin veneer overlying a wave-cut bench in Pleistocene foreshore sands replete with *Ophiomorpha nodosa*, identical to modern burrows of the shrimp *Callianassa major* (Wunderlich, 1972, Pl. 1, fig. 3; Frey et al., 1978). A *Skolithos* ichnofacies thus disconformably underlies the *Glossifungites* ichnofacies here, and relict substrates of the latter include both marine and nonmarine biofacies.

In order to document such biotic and ichnologic characteristics of the modern *Glossifungites* ichnofacies, including both relict and palimpsest aspects, detailed examinations were made on St. Catherines and Petit Chou Islands (Fig. 4). Numerous supplementary observations were made on Wassaw, Cabretta, and Sea Islands (Fig. 3). The paleoecologic and taphonomic aspects of Cabretta deposits were enumerated by Frey and Basan (1981). All of these exposures change from time to time, depending mainly upon the frequency and intensity of storms. With beach erosion during high-energy conditions, for example, the outcrops increase in extent; with sand accumulation during low-energy conditions, the exposures decrease in size. Our present data were gathered mainly in 1980, at times of low to moderate energy levels. Most areas of exposure were of intermediate size. A visit to the sampling sites during 1981, a time of higher energy conditions, revealed outcrops of substantially larger size. These and other differences will be recounted in a separate report.

St. Catherines Island

Extensive relict marsh muds cropping out along the north beach of St. Catherines Island (Fig. 3) were described recently by Morris and Rollins (1977). Bathymetrically, these deposits are positioned in the lower to upper foreshore and are separated laterally by extensive areas of more typical sandy substrates (Fig. 1). Examination of numerous outcrops revealed that two similar yet distinct assemblages of infaunal organisms are present.

The most diverse assemblage was recorded from the deposit adjacent to Seaside Inlet (station 8 of Morris and Rollins, 1977), dominated by petricolid and pholad bivalves and crustaceans (Table 2). Numerically, the dominant species was *Petricola pholadiformis* (Fig. 5), represented by profuse live and dead individuals. Interspersed among these were numerous pholad clams, including live and dead specimens of *Barnea truncata* (Fig. 6D), *Cyrtopleura costata* (Fig. 6A-C), and *Pholas campechiensis* (Fig. 6E). Specimens of in-situ dead, articulated shells of *Tagelus plebeius* and *Mercenaria mercenaria* also were recorded. Crustaceans present included the crabs *Uca pugnax* and *Pan-*

←

FIG. 3.—Index map of study areas along Georgia coast. The present study was concerned chiefly with mud deposits of Petit Chou and St. Catherines Islands. Petit Chou muds are exposed primarily to estuarine tidal currents whereas St. Catherines muds are exposed mainly to beach surf. St. Catherines deposits are confined to a narrow, discontinuous zone at the northern tip of the island (Morris and Rollins, 1977, Fig. 2). In symbol for Holocene, black is sand and white is salt marsh/tidal stream deposits. (Base map courtesy of J. D. Howard.)

FIG. 4.—Examples of relict salt marsh mud deposits studied. *A*, Irregularly eroded surface penetrated by biogenic structures and littered with detrital bivalve shells, St. Catherines Island (Station 8, Morris and Rollins, 1977). *B*, Coherent mud outcrops along tidal channel, Petit Chou Island; mud deposit is continuous but usualy is covered by thin veneer of sand.

opeus herbsti; although not observed in this study, common individuals of the thalassinidean shrimp *Upogebia affinis* and rare specimens of the caridean shrimp *Alpheus* sp. were reported by Morris and Rollins (1977).

Additional infaunal associates included the polychaetes *Nereis succinea* (Fig. 7A), *Drilonereis longa*, *Diopatra cuprea* (Fig. 7B), and *Onuphis microcephala*. A single specimen of the nemertean *Cerebratulus lacteus* was observed.

As mentioned previously, a relict biota indicative of a salt marsh also was present and included: dense root mats of the cordgrass *Spartina alterniflora* (Fig. 8D), articulated, in-situ shells of the oyster *Crassostrea virginica* (Fig. 8A) and the mussel *Geukensia demissa* (Fig. 8C), as well as infilled burrows of the fiddler crab *Uca* sp. (Fig. 8B). This relict fauna, especially the *Uca* burrows, has resulted in contemporary differential erosion and an extremely irregular surface topography consisting of numerous pits and protuberances (Figs. 4A, 8B), many of which eventually become filled with sand and shell debris deposited by incoming tides. Examination of this shell drift revealed a diverse suite of biotic components, dominated by bivalves (Table 3); with burial, some of these are incorporated into the omission surface of the deposit.

Farther to the north of the Seaside Inlet deposit are several laterally disjointed outcrops (Station 2, Morris and Rollins, 1977), which display a somewhat different infaunal assemblage (Table 4). The dominant species is *Petricola pholadiformis*, which in places forms dense "pavements" or pocked surfaces (Fig. 5). Associated organisms include the bivalves *Barnea truncata* and *Tagelus plebeius*, and the polychaetes *Nereis succinea* and *Drilonereis longa*. In contrast to the Seaside Inlet deposit, no live crustaceans were observed; however, a similar suite of relict salt marsh organisms was recorded. Surficial shell debris, although present, was mostly in the form of hash, reflecting the somewhat greater wave energies affecting these outcrops.

Polydorans were not observed at Station 2 during our main study in 1980, but a few small, local populations were seen during a brief visit in 1981. Thus, unlike the large, persistent populations on Petit Chou Island (see Figs. 10D, 11A), *Polydora ?websteri* is considered to be an ephemeral component here.

Between Stations 2 and 1 of Morris and Rollins (1977), in the position of the lower backshore, we discovered tiny outcrops of relict high-marsh sand containing stems and roots of the saltwart *Salicornia virginica* and the rush *Juncus roemerianus*. No animal traces were observed, evidently due to insufficient exposures. On Wassaw, Cabretta, and Sea Islands, larger high-marsh exposures of this type yielded both relict and modern burrows of the fiddler crab *Uca pugilator* (e.g., Frey and Basan, 1981).

Our census of organisms at Station 1 (Fig. 1A) agrees generally with that of Morris and Rollins (1977, p. 101–107), including the predominance of *Petricola pholadiformis*, except that 1) the surficial "*Brachidontes recurvus* pavement" (= *?Ischadium recurvum*) no longer exists on the palmetto paleosol, and 2) we commonly found modern burrows—some occupied and others abandoned and filled with modern beach sand—of the polychaete *Nereis succinea* and the sand fiddler *Uca pugilator*. Occurrence of this crab in the paleosol is not unusual, considering the present texture and sand content, but its occurrence in relict low-marsh muds is exceptional; the mud fiddler *U. pugnax* ordinarily exploits such substrates (see Fig. 11B), in places even reexcavating the fills from relict burrows of the same species. Also unusual, in our experience, was the occurrence of *Nereis succinea* inside hollow (partially decayed) roots of *Spartina alterniflora*; the worm seemed to be using them as temporary dwelling tubes.

We visited Station 7 of Morris and Rollins (1977) only briefly. No attempt was made to document conditions there. However, as at Station 1, no modern mussel clusters were observed.

The surfaces of muds at all stations are irregular, having numerous small pits, potholes, protuberances, and local surge channels (Fig. 5A). Scattered clumps of the green alga *Ulva lactea* and patches of an unidentified microfilamentous blue-green alga locally colonize substrates associated with relict *Uca* burrows, rendering them somewhat more resistant to erosion. In places, contemporary desiccation produces extensive mud cracking (Fig. 9A) and eventual large-scale blocky jointing (Fig. 9B) which, when subjected to wave action, results in the generation of numerous mud clasts. These are quickly rounded by surf; most are destroyed but some lodge in potholes and are preserved with relict shells (Table 3). Some outcrops erode into tiered small scarps; surf action on these also produces large numbers of mud clasts (Fig. 9C).

TABLE 2.—LIST OF MODERN INFAUNAL ORGANISMS IN RELICT MUD DEPOSIT ADJACENT TO SEASIDE INLET, ST. CATHERINES ISLAND, GEORGIA (STATION 8, MORRIS AND ROLLINS, 1977)

ORGANISM	DENSITY[1]
BIVALVES	
Petricola pholadiformis	A
Barnea truncata	VC
Pholas campechiensis	R-C
Cyrtopleura costata	R-C
Tagelus plebeius	C
Mercenaria mercenaria	R-C
POLYCHAETES	
Nereis succinea	A
Drilonereis longa	VC
Diopatra cuprea	R-C
Onuphis microcephala	R
NEMERTEAN	
Cerebratulus lacteus	R
CRUSTACEANS	
Upogebia affinis[2]	C
Alpheus sp.[2]	R
Uca pugnax	R
Panopeus herbsti	R

1. Relative density: VA = very abundant, A = abundant, VC = very common, C = common, R = rare.
2. *Upogebia affinis* and *Alpheus* sp. reported by Morris and Rollins (1977) from this deposit but were not observed in present study.

Most relict sediments are thoroughly bioturbated. Internal bedding, where present, is somewhat like that of other salt marsh-tidal creek environments (Edwards and Frey, 1977, Pl. 8, fig. 37; Frey and Basan, 1981) and consists of thin, discontinuous laminae penetrated by burrows and roots.

The biogenic structures present characterize the relict (Suite A) and palimpsest (Suite B) aspects of these deposits. Suite A, which in most instances is indicative of original salt marsh substrates, consists predominantly of abundant *Spartina alterniflora* stem and root systems (Fig. 8D) or root-penetration structures and vertical, irregular, unlined shafts constructed by the fiddler crab *Uca* sp. cf. *U. pugnax* (Fig. 8B). In places, the relict substrates represent high-marsh sand and, at Station 1, the palmetto paleosol. Botanical characteristics of these two deposits were outlined previously; no relict animal traces were observed there, although some have been seen in high-marsh sands elsewhere. Suite A thus is indicative of the pre-semiconsolidated character of the substrate. Superimposed upon it is a new assemblage (Suite B) constructed by burrowing or

FIG. 5.—Bioerosion by bivalve *Petricola pholadiformis*, St. Catherines Island. *A*, Intensely excavated surface broken up in background by wave-surge channels, somewhat like a spur-and-groove system. *B*, Abundant in-situ, mostly dead specimens. (Station 2, Morris and Rollins, 1977; *P. pholadiformis* is similarly abundant at Station 8.)

semiboring animals now occupying the deposits, including the paleosol. Suite B is dominated by petricolids, pholads, and crustaceans (Figs. 5, 6).

The petricolid and pholad bivalves construct single-entrance, pouch-shaped structures, which are non-isodiametric and terminate in a rounded to elongated, cupulate chamber (Morris and Rollins, 1977, Figs, 10, 11). In the rock record, such structures would be classified as *Trypanites*, *Gastrochaenolites* or related ichnogenera (Bromley, 1970, 1972; Pemberton et al., 1980); the respective ichnogenera presently await taxonomic revision, as mentioned previously. Distinct morphological and size variations are a function of the particular species involved and the cohesiveness of the substrate (Röder, 1977). For example, at Station 8, four species were observed (*Petricola pholadiformis*, *Barnea truncata*, *Cyrtopleura costata*, and *Pholas campechiensis*), which tend to occupy distinct horizons within the deposit. However, not all of these species were ubiquitous; although *P. pholadiformis* and *B. truncata* were recovered from all outcrops, *C. costata* and *P. campechiensis* were much more restricted. This distribution seems to be a function of the degree of water turbulence affecting the outcrop, the density of root matter within the substrate, and its coherence. *C. costata* and *P. campechiensis* were found in more protected zones adjacent to the tidal channel at Seaside Inlet (Station 8), in cohesive substrates mostly devoid of *Spartina* roots.

Schäfer (1972) and Röder (1977) indicated that *Petricola* and *Barnea* tend to be mechanical borers that are well suited to firm substrates, whereas *Cyrtopleura* and *Pholas* tend to be rock borers. In fact, Purchon (1955) suggested that *Petricola pholadiformis* is not a

FIG. 6.—In-situ pholad bivalves. *A*, Dead *Cyrtopleura costata* being exhumed by surf erosion. *B*, *C*, Siphonal passage and bifurcated siphon of live *C. costata*, at low tide. *D*, Live *Barnea truncata*, with constricted, displaced siphon. *E*, Dead, fragmentary *Pholas campechiensis*. Bar scales = 2 cm. (A, Cabretta Island, Frey and Howard, 1969; B-E, St. Catherines Island, Station 8, Morris and Rollins, 1977.)

true borer at all because it lacks the specialized boring devices of the pholads. Nevertheless, all these *Gastrochaenolites* or *Trypanites*-like excavations were observed to truncate stems and roots of relict salt marsh grasses, and such cross-cutting relationships commonly are taken as criteria for true boring behavior. Whether individual clay particles were truncated (boring) or were removed "grain by grain" (burrowing) remains difficult to determine; from the cohesiveness of the substrate, we suspect some clay particles were at least distorted and that, from the organisms' point of view, the process of excavation differed relatively little from that of true boring in poorly cemented rock (e.g., chalk).

The crustaceans present also construct several easily recognizable dwelling structures, most of which are similar to those of their counterparts in present-day unconsolidated muds. There, burrows of the shrimp *Upogebia affinis* consist of an irregular, ramifying system, 1 to 2 cm in diameter and up to 50 cm deep, which enlarge at points of bifurcation (Frey and Howard, 1969, Pl. 4, fig. 5: Howard and Frey, 1975, Pl. 5, fig. 19, Pl. 9, fig. 44). The walls tend to be smooth, and no distinct lining is present. *Alpheus heterochaelis* constructs a similar type of structure, but it differs from that of *Upogebia affinis* in having a more dense, closely integrated network of interconnected branches in the upper part, anastomosing downward to form a central trunk (Howard and Frey, 1975, Pl. 10; Basan and Frey, 1977, Table 3). In fossil form both structures would be called *Thalassinoides* (Frey and Howard, 1975; Howard and Frey, 1975). Crab burrows

TABLE 3.—LIST OF SURFICIAL NONINDIGENOUS MOLLUSK SHELLS ON RELICT MUD DEPOSIT ADJACENT TO SEASIDE INLET, ST. CATHERINES ISLAND, GEORGIA[1]

RELICT SHELLS FROM MUD DEPOSIT[2]			
BIVALVES		GASTROPODS	
		Ilyanassa obsoleta	C
Crassostrea virginica	A	*Littorina irrorata*	VC
Geukensia demissa	A	*Urosalpinx cinerea*	R
EXOTIC DETRITAL SHELLS[2]			
BIVALVES		*Spisula solidissima*	R
Abra aequalis	VC	*Strigilla mirabilis*	R
Anadara brasiliana	C	*Tellina versicolor*	A
Anadara ovalis	C	GASTROPODS	
Corbula swiftiana	C	*Busycon canaliculatum*	R
Dinocardium robustum	R	*Busycon carica*	R
Donax variabilis	A	*Crepidula plana*	C
Dosinia discus	C	*Ilyanassa obsoleta*	C
Ensis directus	R	*Littorina irrorata*	VC
Macoma balthica	R	*Nassarius trivittatus*	C
Macoma constricta	R	*Nassarius vibex*	R
Mulinia lateralis	A	*Oliva sayana*	R
Noetia ponderosa	R	*Olivella mutica*	R
Raeta plicatella	R	*Polinices duplicatus*	C
Solen viridis	VC	*Terebra dislocata*	C

1. Shell identifications by S.W. Henderson, Unversity of Georgia.
2. Relative abudance based on 20 randomly selected square meter quadrats, where A = more than 25; VC = 11 to 25; C = 3 to 10; and R = 2 or less valves.

present include those constructed by *Uca pugnax* and *Panopeus herbsti*. In uncrowded conditions, such as those typified by these deposits, *U. pugnax* burrows tend to be vertical, straight to slightly sinuous, unbranched shafts having no discernible lining (Frey, 1970; Allen and Curran, 1974; Basan and Frey, 1977). *Panopeus herbsti*, on the other hand, constructs highly irregular burrow complexes, generally 2 to 4 cm in diameter and 10 cm deep, which are horizontally branched and multi-entranced (Basan and Frey, 1977, Pl. 3d). The burrow walls are generally irregular and unlined but exhibit prolific claw sculpturings (see Fig. 10A). Burrows such as those constructed by *Uca* and *Panopeus* are referable to the ichnogenus *Psilonichnus* (Fürsich, 1981).

Additional biogenic structures present include those attributable to the polychaetes *Nereis succinea*, *Drilonereis longa*, *Onuphis microcephala*, and *Diopatra cuprea*. *Nereis succinea* creates highly variable systems ranging from simple, irregularly U-shaped, sparsely branched structures to complex systems with prominent vertical and horizontal components (Frey et al., 1973, Fig. 1B; Howard and Frey, 1975). Burrows of *Drilonereis longa* display similar morphological variations, ranging from predominantly horizontal systems (Howard and Frey, 1975) to vertically inclined, multi-branched burrows 1 to 1.5 mm in diameter (Hertweck, 1972, Fig. 3B). Because of the omnivorous to predaceous foraging of *N. succinea* and *D. longa*, their semipermanent burrows tend to remain open and fill passively; therefore, in fossil form both might be referable to the ichnogenus *Palaeophycus* (Pemberton and Frey, 1982).

TABLE 4.—LIST OF MODERN INFAUNAL ORGANISMS IN RELICT MUD DEPOSIT NORTH OF SEASIDE INLET, ST. CATHERINES ISLAND, GEORGIA (STATION 2, MORRIS AND ROLLINS, 1977).

ORGANISM	DENSITY*
BIVALVES	
Petricola pholadiformis	VA
Barnea truncata	C
Tagelus plebeius	R
POLYCHAETES	
Nereis succinea	A
Drilonereis longa	C
Polydora ? websteri	R

*Relative density: VA = very abundant, A = abundant, C = common, R = rare.

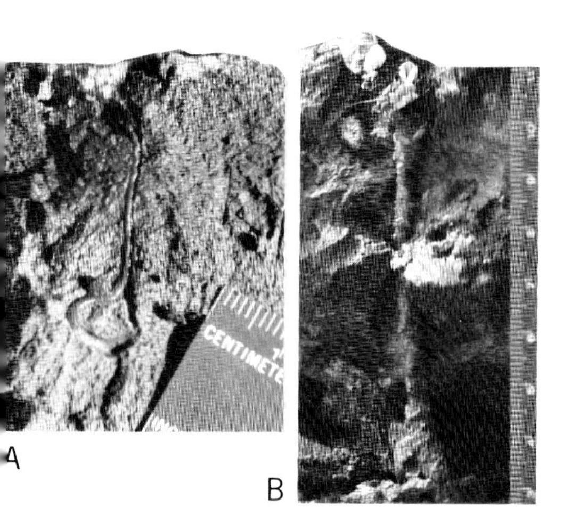

FIG. 7.—Polychaete lebensspuren, St. Catherines Island. A, In-situ, tiny, live *Nereis succinea*. *B*, In-situ, occupied dwelling tube of *Diopatra cuprea*, ornamented at top by small shells of bivalve *Mulinia lateralis*. (Station 8, Morris and Rollins, 1977.)

Onuphis microcephala constructs thin (4-8 mm in diameter), vertical, chitinous tubes, which extend downward as much as 45 cm into the substrate (Frey and Howard, 1969, Pl. 2, fig. 2; Howard and Dörjes, 1972, Figs. 3c, 8, 13, 14; Hertweck, 1972, Fig. 11b; Howard and Frey, 1975, Pl. 4, figs. 9, 13, Pl. 8, fig. 36). The tubes are good analogs for the ichnogenus *Skolithos* (Curran and Frey, 1977). *Diopatra cuprea* constructs vertical, subcylindrical dwelling burrows approximately 6 to 10 mm in diameter and 30 cm in length (Frey and Howard, 1969, Pl. 3, fig. 3; Hertweck, 1972, Fig. 10b; Howard and Frey, 1975, Pl. 8, fig. 34, Pl. 9, fig. 41, Pl. 14, fig. 60). The walls are lined with a thin parchment and the upper one-third of the tube is reinforced with debris commonly consisting of small disarticulated bivalve shells (Myers, 1972). This unique behavior may result in downwarping of sediment laminae surrounding the upper part of the tube (Frey and Howard, 1972, Fig. 8), suggesting that in fossil form such structures might be referable to the ichnogenus *Monocraterion*. However, such downwarping would be less likely in semiconsolidated than in unconsolidated muds. Kern (1978) erected the ichnogenus *Diopatrichnus* for the distinct morphology of the dwelling tube itself. The burrow of *Polydora ?websteri*, equivalent to the ichnogenus *Diplocraterion*, is described subsequently.

The potential ichnocoenose of these deposits therefore would consist of the following elements: a relict suite of *Spartina* root systems and *Psilonichnus* (or in rare cases, *Serenoa* root systems) crosscut by abundant *Trypanites*, *Gastrochaenolites*, and (or) related forms, *Thalassinoides*, and *Palaeophycus*; common *Psilonichnus*; rare *Skolithos*, *Diopatrichnus*, and—possibly—*Monocraterion*; and very rare *Diplocraterion*.

Petit Chou Island

In contrast to the seaside deposits of St. Catherines Island, relict marsh muds crop out along back-barrier tidal streams adjacent to Petit Chou Island (Fig. 4B). A salt marsh origin for these muds is indicated by: 1) in-situ articulated shells of the mussel *Geukensia demissa* and the marsh gastropod *Littorina irrorata*, 2) numer-

FIG. 8.—Relict salt marsh features of mud deposits, St. Catherines Island. *A*, In-situ beds of oyster *Crassostrea virginica* along former tidal creek-mud flat complex adjacent to marsh. *B*, Exhumed burrow (reopened aperture) and bioturbate textures of fiddler crab *Uca* sp. cf. *U. pugnax*, and detrital valves of relict *C. virginica*. *C*, In-situ valves of mussel *Geukensia demissa*, truncated by surf abrasion; valve of modern *Anadara ovalis* is lodged in mussel clump (right center); mud mostly covered by thin veneer of sand. *D*, In-situ, well-preserved roots and stems of cordgrass *Spartina alterniflora*, representing a former low-marsh environment.

FIG. 9.—Present-day physical alteration of mud deposits. *A*, Desiccation polygons, partly filled with modern beach sand. *B*, Blocky jointing, the result of prolonged desiccation and development, by erosion, of an adjacent, small-scale scarp. *C*, Mud clasts plucked from outcrop and rounded by surf; rippled beach sand and relict, detrital shells of *Crassostrea virginica* lie adjacent to mud deposit. (A, C, St. Catherines Island; B, Petit Chou Island.)

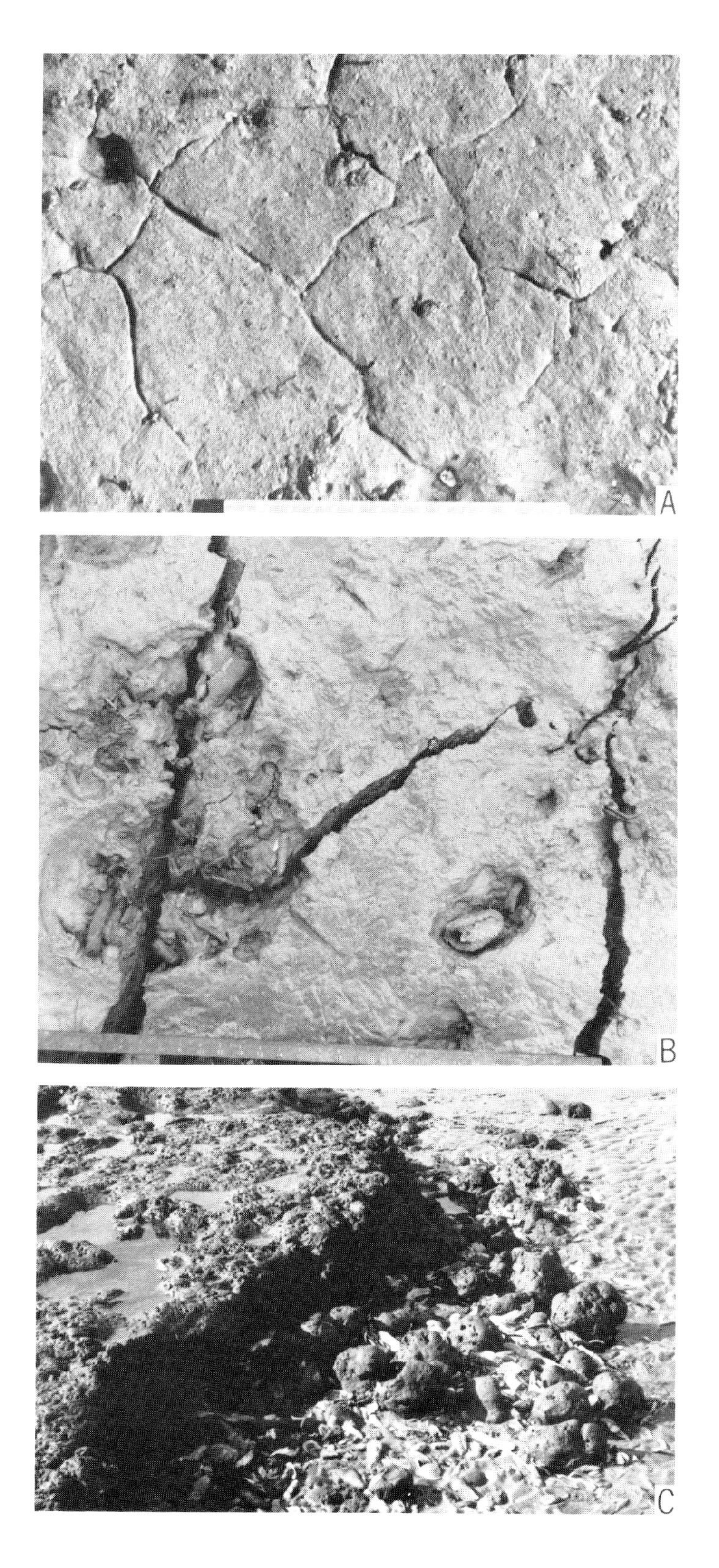
A
B
C

ous *Spartina alterniflora* roots and stems, and 3) common unlined burrows of the fiddler crab *Uca pugnax* (Fig. 10C). These deposits have been exhumed by lateral erosion of the tidal channel. The infaunal biotic components of these muds closely resemble the suite found at Station 8 (adjacent to Seaside Inlet, St. Catherines Island) and are characterized by a crustacean-petricolid-polydoran association (Table 5; Figs. 2, 10, 11).

Numerically, the Petit Chou deposit is dominated by the crustaceans *Panopeus herbsti* (Fig. 10A), *Upogebia affinis* (Fig. 10B), and *Uca pugnax* (Fig. 11B). Bivalves include abundant *Petriocola pholadiformis* (Fig. 2B) and rare *Barnea truncata* and *Tagelus*

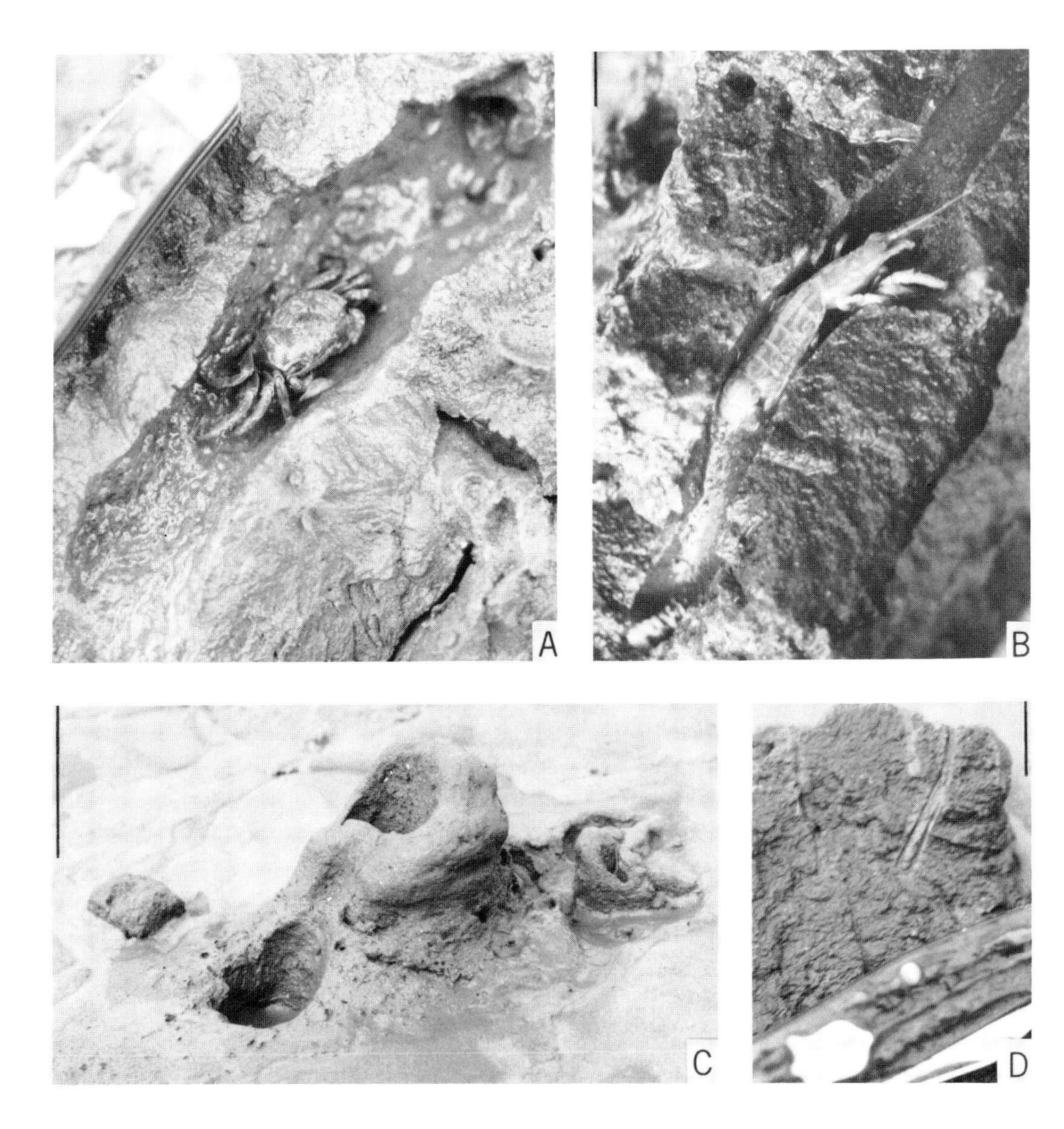

FIG. 10.—Characteristic lebensspuren, Petit Chou Island. Compare with Fig. 2. *A*, Live in-situ crab *Panopeus herbsti;* unlined burrow walls are claw-sculpted. *B*, Live in-situ shrimp *Upogebia affinis:* burrow walls are smooth but essentially unlined. *C*, Relict chimneyed aperture of fiddler crab burrow, *Uca* sp. cf. *U. pugnax*, exhumed by tidal current erosion of substrate surface. *D*, Fresh burrow of polychaete *Polydora* ?*websteri*. Bar scale for C = 3 cm; all others = 1 cm.

plebeius. Associated polychaetes include *Nereis succinea* (Fig. 2D), *Drilonereis longa*, and *Polydora ?websteri* (Figs. 10D, 11A). One specimen of an eel, *Ophichthus gomesi*, also was recovered from a distinct dwelling structure.

Aside from *Polydora ?websteri* and *Ophichthus gomesi*, biogenic structures produced by these organisms were described in the preceding section. The sinuous, vermiform burrows of *O. gomesi* (Howard and Frey, 1975, Fig. 2A) are strikingly like those made by certain invertebrates; the burrow is smooth-walled and lined with thin coats of mucus. In the rock record, it would constitute an errant form of *Palaeophycus*.

Although well known as a borer in skeletal substrates (Bromley, 1970; Warme, 1975), *Polydora* also is found in semiconsolidated muds and peats (Schäfer, 1972). Burrows of *Polydora* tend to be vertical, U-shaped dwelling structures with slightly divergent, rather than parallel, arms. The walls tend to be lined with sand and ornamented by numerous fine sculptings, possibly created by external bristles. Protrusive spreiten may be present, which seem to be related more to the growth of the organism and its increased depth of burrowing than to sediment aggradation or degradation. In fossil form, spreiten burrows of *P. ?websteri* would be named *Diplocraterion* (Fürsich, 1974b); those lacking spreiten would be assigned to *Arenicolites* (Häntzschel, 1975).

The virtual restriction of *P. ?websteri* to Petit Chou muds seemed puzzling at first; however, Schäfer (1972, p. 273) indicated that its distribution may be controlled somewhat by the sand content of the substrate. Edwards and Frey (1977, Figs. 3, 4) showed that the mean percent of sand varies considerably in salt marsh environments at Sapelo Island, Georgia, ranging from 2 percent in creek banks, 1 percent to 19 percent in low-marsh environments, to 82 percent in high-marsh environments. Therefore, the presence or absence of *P. ?websteri* may simply be a function of which marsh environment the relict deposit represents. The worm is abundant locally in similar Sea Island deposits.

TABLE 5.—LIST OF MODERN INFAUNAL ORGANISMS IN MUD DEPOSIT ADJACENT TO TIDAL CHANNEL, PETIT CHOU ISLAND, GEORGIA

ORGANISM	DENSITY*
BIVALVES	
Petricola pholadiformis	A
Barnea truncata	R-C
Tagelus plebeius	R
POLYCHAETES	
Nereis succinea	C-A
Drilonereis longa	C
Polydora ? websteri	A
CRUSTACEANS	
Panopeus herbsti	VA
Upogebia affinis	A
Uca pugnax	C-A
VERTEBRATES	
Ophichthus gomesi	R

*Relative density: VA = very abundant, A = abundant, C = common, R = rare.

As with the St. Catherines deposit, the ichnocoenose of Petit Chou muds would consist of two distinct suites, a relict assemblage of unlined crab burrows and root systems (Suite A) crosscut by a firm-sediment assemblage consisting of *Thalassinoides*, *Psilonichnus*, *Palaeophycus*, *Diplocraterion* and (or) *Arenicolites*, and one or two morphotypes of *Gastrochaenolites* or related ichnogenera (Suite B).

Aside from the presence of *Polydora ?websteri*, other differences in biotic components of the Petit Chou and St. Catherines examples relate to the density and complexity of crustacean and bivalve structures. Petit Chou, being in a more environmentally tranquil setting, is dominated by the infaunal crustaceans *Panopeus herbsti*, *Upogebia affinis*, and locally, *Uca pugnax*; the burrowing activities of these have resulted in irregular, ropy surfaces (Fig. 2A), which are accentuated even further by tidal current erosion. The St. Catherines deposits, on the other hand, are subjected to direct wave action, which inhibits the activity of mobile organisms such as crustaceans and favors the suspension-feeding petricolid and pholad bivalves.

PALEOECOLOGICAL-ICHNOLOGICAL SIGNIFICANCE

The concept of recurring assemblages of ichnofossils has long been used as a powerful tool in the interpretation of ancient depositional systems. The *Skolithos*, *Trypanites*, *Cruziana*, *Zoophycos*, and *Nereites* ichnofacies are well documented in both modern and ancient environments. The *Glossifungites* ichnofacies, on the other hand, has received considerably less attention and, with the exception of the *Scoyenia* ichnofacies (Frey et al., 1984), remains the least understood.

Examination of relict deposits on the Georgia coast, especially Cabretta, St. Catherines, and Petit Chou Islands, revealed that the ichnocoenose consists of two major elements: Suite A—a relict component, indicative of the original genesis of the deposit, and Suite B a modern component, indicative of prevailing environmental conditions.

Suite A is potentially more variable and reflects more diverse origins than Suite B. No traces of terrestrial animals were observed in the St. Catherines paleosol, but dense relict root systems of *Serenoa repens* bear clear testament to the original environment and its striking contrast with subjacent low-marsh muds. The latter are characterized mainly by dense relict root systems of *Spartina alterniflora* and burrows of *Uca pugnax*. On Cabretta Island, different ecophenotypes of *S. alterniflora* also have been documented, together with relict burrows of the crabs *Sesarma reticulatum* and possibly *Panopeus herbsti* and *Eurytium limosum* (Frey and Basan, 1981). The paleoecological and ichnological significance of these and associated marsh organisms was indicated by Basan and Frey (1977) and Edwards and Frey (1977). Relict high-marsh sands and their characteristic faunas and floras also are well known on the Georgia coast, but these deposits are rarely sufficiently firm to entice a diverse modern community of *Glossifungites*-type organisms. Furthermore, most such deposits crop out in the position of the backshore and thus are exposed subaerially for most of the tidal cycle. Hence, relict and palimpsest burrows of *Uca pugilator* are the dominant traces present.

Although the relict assemblage is similar in St. Catherines and Petit Chou low-marsh deposits, at least three distinct assemblages of Suite B organisms were observed: a petricolid assemblage, a petricolid-pholad-crustacean assemblage, and a petricolid-crustacean-polydoran assemblage (Table 6). Like conspicuous variations well known in other ichnofacies (Table 1), these variations in the modern *Glossifungites* ichnofacies merely reflect local environmental conditions and are to be expected in the fossil record. Major differences noted among these assemblages are attributable to numerous interrelated factors:

1) Notable differences exist in the relative importance of crustaceans and infaunal bivalves (Table 6). Petit Chou, situated in the more tranquil back-barrier environment, is influenced predominantly by tidal currents, and its assemblage is dominated by the crustaceans *Panopeus herbsti*, *Uca pugnax*, and *Upogebia affinis*. Petricolid and pholad bivalves, on the other hand, dominate the more foreshore-like, wave influenced deposits of St. Catherines Island. In these outcrops, crustaceans are present mainly adjacent to Seaside Inlet, which is not subjected to the full brunt of incoming waves; tidal currents are also important here. Therefore, crustaceans—especially the crabs, being more mobile surface deposit feeders or scavengers—would tend to be favored by low energy levels, and their characteristic, extensive burrow systems would seem to exclude the larger pholad bivalves. Higher energy levels, in contrast, would favor infaunal, suspension-feeding bivalves and would tend to preclude mobile surface-feeding organisms. The only major exception observed is the common occurrence of *Uca*

TABLE 6.—BIOTIC ASSEMBLAGES AND POTENTIAL ICHNOCOENOSES FROM RELICT MUD DEPOSITS ON ST. CATHERINES AND PETIT CHOU ISLANDS

TRACE-MAKING ORGANISM	ASSEMBLAGE* 1	2	3	PROBABLE ICHNOGENUS
		RELICT LOW-MARSH ORGANISMS		
PLANTS				
Spartina alterniflora	X	X	X	(root-penetration patterns)
CRUSTACEANS				
Uca pugnax	X	X	X	*Psilonichnus* type A
		MODERN COHESIVE-SUBSTRATE ORGANISMS		
BIVALVES				
Petricola pholadiformis	X	X	X	*Gastrochaenolites* type A**
Barnea truncata	X	X	X	*Gastrochaenolites* type B**
Pholas campechiensis	X			*Gastrochaenolites* type C**
Cyrtopleura costata	X			*Gastrochaenolites* type D**
Tagelus plebeius	X		X	(elongate burrow)
Mercenaria mercenaria	X			(body-size burrow)
POLYCHAETES				
Nereis succinea	X	X	X	*Palaeophycus* type A
Drilonereis longa	X	X	X	*Palaeophycus* type B
Diopatra cuprea	X			*Diopatrichnus*/*Monocraterion*
Onuphis microcephala	X			*Skolithos*
Polydora ?websteri		X	X	*Diplocraterion*/*Arenicolites*
CRUSTACEANS				
Upogebia affinis	X		X	*Thalassinoides* type A
Alpheus sp.	X		X	*Thalassinoides* type B
Uca pugnax	X		X	*Psilonichnus* type A
Panopeus herbsti	X		X	*Psilonichnus* type B

*Assemblage 1: petricolid-pholad-crustacean suite, Station 8, St.Catherines Island.
Assemblage 2: petricolid suite, Station 2, St. Catherines Island.
Assemblage 3: petricolid-crustacean-polydoran suite, Petit Chou Island.
**or related, pholad-type ichnogenera, as explained in the text.

pugilator in waveswept relict deposits at Station 1; there, extremely dense root mats of *Serenoa* and *Spartina* so stabilize the substrate that the crabs are able to retreat safely into their burrows during high tide.

2) The relative density of the relict *Spartina* root mat also seems to influence the distribution of some infaunal bivalves. Smaller species such as *Petricola pholadiformis* and *Barnea truncata* are not obviously affected, although *B. truncata* was not observed in the most dense mats. The larger species, *Pholas campechiensis* and *Cyrtopleura costata,* in contrast, are restricted to areas relatively free of roots. From a paleoecological standpoint this observation is significant because it may aid in delineating relict marsh zones in the rock record, where roots or root-penetration structures are not likely to be so well preserved.

3) Relative cohesion and other geotechnical properties of the substrate seem to influence the distribution of certain infaunal bivalves. *C. costata* and *P. campechiensis* are more nearly restricted to the firmer substrates, whereas the distribution of *P. pholadiformis* and *B. truncata* is more nearly uniform. The relatively root-free areas in which *C. costata* and *P. campechiensis* were observed may have been subject to more inten-

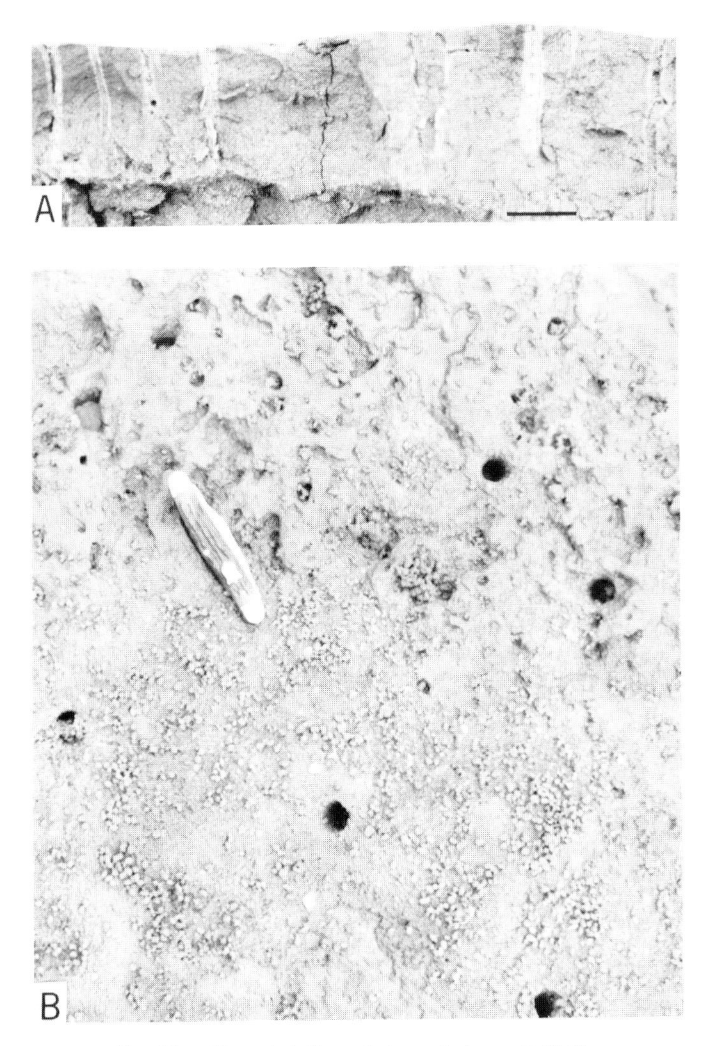

FIG. 11.—Characterisitic omission-suite traces, Petit Chou Island. *A*, Concentration of domiciles of polychaetes *Polydora ?websteri* and *Nereis succinea* along discontinuity surface. Bar scale = 1 cm. *B*, Reoccupation of substrate by fiddler crab *Uca pugnax;* fresh burrows and feeding pellets dot the foreground, becomming admixed with relict burrows of the same species (Figs. 2A, 10C).

sive dewatering, thus resulting in a more cohesive substrate, although not sufficiently firm to exclude *P. pholadiformis* and *B. truncata*, which excavate predominantly by mechanical means (Schäfer, 1972). In contrast, animals such as the polychaetes *Nereis succinea* and *Drilonereis longa* and the shrimp *Upogebia affinis* seem to range through the entire spectrum, from watery, unconsolidated modern muds (Howard and Frey, 1975; Basan and Frey, 1977), to the slightly to markedly dewatered, semiconsolidated relict muds.

4) Textural composition of sediments evidently influences certain animals. Although the polychaetes *Nereis succinea* and *Drilonereis longa* were found in similar densities in all three assemblages, *Polydora ?websteri* essentially was recovered only from the back-barrier Petit Chou deposit. As mentioned previously, its distribution may be influenced by the initial sand content of the marsh mud. Similarly, the mud fiddler *Uca pugnax* prefers relict low-marsh muds and the sand fiddler *U. pugilator* ordinarily prefers relict high-marsh sands (Frey and Basan, 1981). Occurrence of the latter at St. Catherines Station 1 is perhaps explained by the peculiar sandy plant-mud fabric of the contemporary substrate surface.

5) The duration of subaerial exposure also plays a significant role in distribution of most biotic components found in the three assemblages. Because of the laterally extensive occurrence of St. Catherines outcrops, these bathymetric effects are somewhat more pronounced than at the Petit Chou deposit. Among suspension-feeding bivalves, *C. costata* and *P. campechiensis* (where present) are situated along the most seaward margin of the deposit, in areas not subject to appreciable periods of subaerial exposure. *P. pholadiformis* and *B. truncata*, although interspersed through most of the outcrop, are concentrated in the lower to middle zones. Crustaceans show a similar pattern; *Uca pugnax* is situated in the more landward reaches of the deposit and *Panopeus herbsti* and *Upogebia affinis* are situated in more intermediate to seaward positions. In all outcrops examined, no pattern emerged with regard to the burrowing polychaete distribution; both *Nereis succinea* and *Drilonereis longa* are somewhat ubiquitous. Dense infestation was observed even in the higher, landward margins of the deposits, which are subject to extensive periods of subaerial exposure. Ichnologically, these areas are dominated by relict components of the ichnocoenose. The tube dwellers, *Diopatra cuprea* and *Onuphis microcephala*, are more characteristic of seaward parts, however. The overall effect, therefore, is a somewhat crude bathymetric zonation, discernible by utilizing the distribution and interrelationships of both palimpsest and relict biotic elements.

6) Sand derived from the adjacent, contemporaneous *Skolithos* ichnofacies influences the preservation of all biotic components as well as the distribution of certain palimpsest components. Sandblasting by surf may scour or truncate in-situ shells (Fig. 8C), and accumulations of sand in potholes and surge channels (Figs. 5A, 6A) may bury detrital shells (Table 3). Incorporation of beach sand within densely vegetated relict substrate surfaces (Fig. 1A) evidently facilitates feeding there by the sand fiddler *Uca pugilator*. More importantly, migrations of sand waves across mud surfaces (Fig. 4B) tend to fill excavations (Fig. 2C) and to stifle contemporary bivalves; thus, any given substrate is apt to contain more dead bivalves (Fig. 5) than live ones. Accumulations of large quantities of sand, such as that currently involved in beach progradation (e.g., Frey and Basan, 1981), may rebury the entire outcrop and emplace a new *Skolithos* ichnofacies over it.

Finally, as is obvious from the foregoing discussion, several types of Suite A traces are recurrent in Suite B. Indeed, the potential diversity of recurrent tracemakers is substantial (Table 7). Those listed in the table are grouped according to habitat fidelity, e.g., although the bivalves *Mercenaria mercenaria* and *Tagelus plebeius* are known from palimpsest low-marsh muds, they do not occur in modern (or therefore relict) marsh deposits. All species listed have been found in both present-day marshes/creeks and their palimpsest equivalents. Some have not yet been documented unequivocally in relict condition (e.g., *Alpheus heterochaelis*) but undoubtedly occur; their traces are equally as preservable as well-documented relict burrows from equivalent marsh zones or creek banks (e.g., *Uca pugnax*). Several other tracemakers probably could be added to the list. The polychaete *Heteromastus filiformis* and the crab *Sesarma reticulatum* are known from both modern and relict marshes, for example, and may yet be found live in palimpsest ones.

Equally important, none of these tracemakers (Table 7) ranges into the *Trypanites* ichnofacies, and none of the "borers" (Table 6) occurs in modern or relict marshes and tidal creeks. Recurrent burrowers (Suites A + B), together with contemporaneous mixtures of burrows and borings or boring-like structures (Suite B), are exclusive traits of the *Glossifungites* ichnofacies as developed on the Georgia coast.

ANCIENT EXAMPLES OF THE *GLOSSIFUNGITES* ICHNOFACIES

Ancient examples of the *Glossifungites* ichnofacies, like most modern occurrences, are not well documented. Seilacher (1967) noted an example from the Eocene of Maryland but did not elaborate. Bromley (1975) reviewed the literature on omission surfaces and indicated that distinct preomission, omission, and postomission ichnofossil suites could be discerned; such omission suites, if generated in semiconsolidated substrates, would represent the *Glossifungites* ichnofacies. Kennedy (1967) described such a situation

TABLE 7.—ORGANISMS POTENTIALLY OCCURRING IN PRESENT-DAY, RELICT, AND PALIMPSEST SALT MARSH/TIDAL CREEK DEPOSITS, GEORGIA COAST*

Taxa	High-marsh sand	Low-marsh mud	Creek banks
POLYCHAETES			
Diopatra cuprea			x
Onuphis microcephala			x**
Nereis succinea	x	x	x
CRUSTACEANS			
Alpheus heterochaelis		x	x
Panopeus herbsti		x	x
Uca pugilator	x		x**
Uca pugnax		x	x
Upogebia affinis			x
BIVALVES			
Mercenaria mercenaria			x
Tagelus plebeius			x

*Includes data from present study; Basan and Frey, 1977; Morris and Rollins, 1977; Frey and Basan, 1981.
**Most common in sandy creek banks; other organisms listed here prefer muddy creek banks.

from Cenomanian chalks in southern England, as did Sellwood (1970) from the lower Pliensbachian Belemnite Marl of England. Schroder (1982) recognized a distinct omission horizon in Upper Cretaceous-Lower Tertiary rocks of east-central Georgia that consisted of an irregular upper surface of slightly sandy mud penetrated by *Rhizocorallium*-like burrows and pholad-type *Gastrochaenolites*, representing an exposed semiconsolidated substrate. Similarly, the "firmground fauna" reported by Fürsich et al. (1981, Table 3) represents the *Glossifungites* ichnofacies.

SUMMARY AND CONCLUSIONS

Environmentally, the chief distinction between the three predominantly marine, littoral to shallow sublittoral ichnofacies (*Trypanites*, *Glossifungites*, and *Skolithos*) is based on substrate consistency. The *Trypanites* ichnofacies is indicative of fully consolidated omission surfaces (rocky coasts, beachrock, hardgrounds) or organic substrates (wood, shell, bone), and is intergradational with the *Glossifungites* ichnofacies, which develops in firm but unlithified omission surfaces. The gradation between these two ichnofacies includes the mutual occurrence of pholad-type traces that are termed "borings" in consolidated substrates but which might be called "burrows" in semiconsolidated substrates. The difference—at the boundary between ichnofacies—is perhaps more a matter of semantics than of fundamental alterations in animal behavior. The *Skolithos* ichnofacies, in contrast, indicates unconsolidated, shifting sediments subject to abrupt erosion or deposition.

At many places on the Georgia coast, exhumation of relict salt marsh deposits (or at one site on St. Catherines Island, a palmetto paleosol) has placed semiconsolidated substrates in a geographic and facies position ordinarily occupied only by sand beaches containing a suite of biogenic structures indicative of the *Skolithos* ichnofacies. A suite of opportunistic organisms, many of them indigenous to neither beaches nor salt marshes, has subsequently occupied the relict deposits. Similarly, lateral migration of an estuarine tidal stream on Petit Chou Island has exposed old marsh deposits now occupied by opportunistic estuarine animals. The cumulative record of all such deposits thus consists of a relict soft-substrate assemblage of diverse origins (i.e., various original habitats), overprinted by a firm-substrate suite constituting the *Glossifungites* ichnofacies.

Environmental gradients associated with these palimpsest substrates, together with the varied adaptations of individual organisms, yield variations within the ichnofacies comparable to those observed elsewhere in the *Skolithos* and *Trypanites* ichnofacies; not all characteristic ichnogenera are present in every occurrence of these ichnofacies, recent or ancient. Among the more novel of these "facies crossers" is the reoccupation of semiconsolidated substrates by certain organisms—such as foraging polychaetes—that ordinarily dwell in unconsolidated substrates. In addition to these traces, the prevalent mixture of contemporary burrows and borings or boring-like structures is uniquely indicative of the *Glossifungites* ichnofacies on the Georgia coast. Diagnostic criteria from these deposits, outlined herein, should aid in the recognition of additional ancient examples of the ichnofacies.

Most modern processes acting upon semiconsolidated Georgia deposits involve erosion, and hence destruction, of these palimpsest substrates. However, subsequent reburial of such muds on Cabretta Island (Frey and Basan, 1981) shows that preservation indeed occurs. Comparable deposits should occur in the rock record and would include a complex mixture of relict and palimpsest features overlain by beach sands or related deposits (*Skolithos* ichnofacies).

ACKNOWLEDGMENTS

For their valuable reviews of the manuscript, we thank H. A. Curran and R. G. Bromley. In field sampling, valuable assistance was rendered by S. W. Henderson and G. S. Duncan. Most photographs were printed by S. C. Murphy and M. C. Standridge. The research was supported by personnel and facilities at the Skidaway Institute of Oceanography, for work on Petit Chou and Wassaw Islands, and by a grant from the St. Catherines Island Research Program, which is administered by the American Museum of Natural History and supported by the Edward J. Noble Foundation.

REFERENCES

ALLEN, E.A., AND CURRAN, H.A., 1974, Biogenic sedimentary structures produced by crabs in lagoon margin and salt marsh environments near Beaufort, North Carolina: Jour. Sed. Petrology, v. 44, p. 538–548.

BALDWIN, C.T., 1974, The control of mud crack patterns by small gastropod trails: Jour. Sed. Petrology, v. 44, p. 695–697.

BASAN, P.B., AND FREY, R.W., 1977, Actual-palaeontology and neoichnology of salt marshes near Sapelo Island, Georgia, *in* Crimes, T.P., and Harper, J.C. (eds.), Trace Fossils 2: Geol. Jour., Spec. Issue 9: Liverpool, Seel House Press, p. 41–70.

BROMLEY, R.G., 1970, Borings as trace fossils and *Entobia cretacea* Portlock, as an example, *in* Crimes, T.P., and Harper, J.C. (eds.), Trace Fossils: Geol. Jour., Spec. Issue 3: Liverpool, Seel House Press, p. 49–90.

______, 1972, On some ichnotaxa in hard substrates, with a redefinition of *Trypanites* Mägdefrau: Paläontol. Zeitschr., v. 46, p. 93–98.

______, 1975, Trace fossils at omission surfaces, *in* Frey, R.W. (ed.), The Study of Trace Fossils: New York, Springer-Verlag, p. 399–428.

CRIMES, T.P., 1977, Trace fossils of an Eocene deep-sea sand fan, northern Spain, *in* Crimes, T.P., and Harper, J.C. (eds.), Trace Fossils 2: Geol. Jour., Spec. Issue 9: Liverpool, Seel House Press, p. 71–90.

CURRAN, H.A., AND FREY, R.W., 1977, Pleistocene trace fossils from North Carolina (U.S.A.) and their Holocene analogues, *in* Crimes, T.P., and Harper, J.C. (eds.), Trace Fossils 2: Geol. Jour., Spec. Issue 9: Liverpool, Seel House Press, p. 139–162.

DEPRATTER, C. B., AND HOWARD, J.D., 1981, Evidence for a sea level lowstand between 4500 and 2400 years B.P. on the southeast coast of the United States: Jour. Sed. Petrology, v. 51, p. 1287–1295.

DOUVILLE, H., 1908, Perforations d'annelides: Bull. Soc. Geol. France, Ser. 4, v. 7, p. 361–370.

EDWARDS, J.M., AND FREY, R.W., 1977, Substrate characteristics within a Holocene salt marsh, Sapelo Island, Georgia: Senckenberg. Marit., v. 9, p. 215–259.

FREY, R.W., 1970, Environmental significance of recent marine lebensspuren near Beaufort, North Carolina. Jour. Paleontology, v. 44, p. 507–519.

______, 1971, Ichnology—The study of fossil and recent lebensspuren, *in* Perkins, B.F. (ed.), Trace Fossils, A Field Guide: Louisiana State Univ., School Geosci., Misc. Publ. 71-1, p. 91–125.

______, 1973, Concepts in the study of biogenic sedimentary structures: Jour. Sed. Petrology, v. 43, p. 6–19.

______, 1975, The realm of ichnology, its strengths and limitations, *in* Frey, R.W. (ed.), The Study of Trace Fossils: New York, Springer-Verlag, p. 13–38.

______, AND BASAN, P.B., 1978, Coastal salt marshes, *in* Davis, R.A., Jr., ed., Coastal Sedimentary Environments: New York, Springer-Verlag, p. 101–169.

______, AND ______, 1981, Taphonomy of relict Holocene salt marsh deposits, Cabretta Island, Georgia: Senckenberg. Marit., v. 13, p. 111–155.

______, ______, AND SCOTT, R.M., 1973, Techniques for sampling salt marsh benthos and burrows: Amer. Midland Naturalist, v. 89, p. 228–234.

______, AND HOWARD, J.D., 1969, A profile of biogenic sedimentary structures in a Holocene barrier island-salt marsh complex, Georgia: Gulf Coast Assoc. Geol. Socs., Trans., v. 19, p. 427–444.

______, AND ______, 1972, Georgia coastal region, Sapelo Island, U.S.A.: Sedimentology and biology. VI. Radiographic study of sedimentary structures made by beach and offshore animals in aquaria: Senckenberg. Marit., v. 4, p. 169–182.

______, AND ______, 1975, Endobenthic adaptations of juvenile thalassinidean shrimp: Geol. Soc. Denmark, Bull., v. 24, p. 283–297.

______, ______, AND PRYOR, W.A., 1978, *Ophiomorpha:* Its morphologic, taxonomic, and environmental significance: Palaeogeogr., Palaeoclimatol., Palaeoecol., v. 23, p. 199–229.

______, AND MAYOU, T.V., 1971, Decapod burrows in Holocene barrier island beaches and washover fans, Georgia: Senckenberg. Marit., v. 3, p. 53–77.

______, PEMBERTON, S.G., AND FAGERSTROM, J.A., 1984, Morphological, ethological, and environmental significance of the ichnogenera *Scoyenia* and *Ancorichnus:* Jour. Paleontology, v. 58, p. 511–528.

______, AND SEILACHER, A., 1980, Uniformity in marine invertebrate ichnology: Lethaia, v. 13, p. 183–207.

FÜRSICH, F.T., 1974a, Ichnogenus *Rhizocorallium:* Paläontol. Zeitschr., v. 48, p. 16–28.

______, 1974b, On *Diplocraterion* Torell 1870 and the significance of morphological features in vertical, spreiten-bearing, U-shaped trace fossils: Jour. Paleontology, v. 48, p. 952–962.

______, 1981, Invertebrate trace fossils from the Upper Jurassic of Portugal: Comun. Serv. Geol. Portugal, v. 67, p. 153–168.

______, KENNEDY, W.J., AND PALMER, T.J., 1981, Trace fossils at a regional discontinuity surface: the Austin/Taylor (Upper Cretaceous) contact in central Texas: Jour. Paleontology, v. 55, p. 537–551.

GOULD, A., 1870, Report on the invertebrates of Massachusetts: Boston, W.B. Binney.

HÄNTZSCHEL, W., 1975, Trace fossils and problematica, *in* Teichert, C. (ed.), Treatise on invertebrate paleontology, Pt. W, Misc., Supp. 1: Lawrence, Kansas, Geol. Soc. America and Univ. Kansas Press, p. W1–W269.

HAYWARD, B.W., 1976, Lower Miocene bathyal and submarine canyon ichnocoenoses from Northland, New Zealand: Lethaia, v. 9, p. 149–162.

HERTWECK, G., 1972, Georgia coastal region, Sapelo Island, U.S.A.: Sedimentology and biology. V. Distribution and environmental significance of lebensspuren and in-situ skeletal remains: Senckenberg. Marit., v. 4, p. 125–167.

Howard, J.D., and Dörjes, J., 1972, Animal-sediment relationships in two beach-related tidal flats; Sapelo Island, Georgia: Jour. Sed. Petrology, v. 42, p. 608–623.
______, and Frey, R.W., 1975, Estuaries of the Georgia coast, U.S.A.: Sedimentology and biology. II. Regional animal-sediment characteristics of Georgia estuaries: Senckenberg. Marit., v. 7, p. 33–103.
______, and ______, 1980, Holocene depositional environments of the Georgia coast and continental shelf, *in* Howard, J. D., DePratter, C.B., and Frey, R.W. (eds.), Excursions in Southeastern Geology: the Archaeology-Geology of the Georgia Coast: Georgia Geol. Surv., Guidebook 20, p. 66–134.
______, ______, and Reineck, H.-E., 1972, Georgia coastal region, Sapelo Island, U.S.A.: Sedimentology and biology. I. Introduction: Senckenberg. Marit., v. 4, p. 3–14.
______, ______, and ______, 1973, Holocene sediments of the Georgia coastal area, *in* Frey, R.W., ed., The Neogene of the Georgia Coast: Athens, Georgia, Georgia Geol. Soc., 8th Ann. Field Trip, Guidebook, p. 1–58.
Hoyt, J.H., Henry, V.J., Jr., and Howard, J.D., 1966, Pleistocene and Holocene sediments, Sapelo Island, Georgia and vicinity: Guidebook, Field Trip 1, Southeast. Sect., Geol. Soc. America, p. 6–27.
Kennedy, W.J., 1967, Burrows and surface traces from the Lower Chalk of southern England: British Mus. (Nat. Hist.), Geol., Bull., v. 15, p. 127–167.
______, 1975, Trace fossils in carbonate rocks, *in* Frey, R.W. (ed.), The Study of Trace Fossils: New York, Springer-Verlag, p. 377–398.
Kern, J.P., 1978, Paleoenvironment of new trace fossils from the Eocene Mission Valley Formation, California: Jour. Paleontology, v. 52, p. 186–194.
Kues, B.S., and Siemers, C.T., 1977, Control of mudcrack patterns by the infaunal bivalve *Pseudocyrena:* Jour. Sed. Petrology, v. 47, p. 844–848.
Letzsch, W.S., and Frey, R.W., 1980, Erosion of salt marsh tidal creek banks, Sapelo Island, Georgia: Senckenberg. Marit., v. 12, p. 201–212.
Lominicki, A.M., 1886, Slodkowodny utwar trzece orzedny na Podulu galicyjskiem: Akad. Umiejet Krakow, Kom. Fizyogr., v. 20, p. 48–119.
Metz, R., 1980, Control of mudcrack patterns by beetle larvae traces: Jour. Sed. Petrology, v. 50, p. 841–842.
Morris, R.W., and Rollins, H.B., 1977, Observations on intertidal organism associations of St. Catherines Island, Georgia. I. General description and paleoecological implications: Bull. Amer. Mus. Nat. Hist., v. 159, p. 89–128.
Myers, A.C., 1972, Tube-worm-sediment relationships of *Diopatra cuprea* (Polychaeta: Onuphidae): Marine Biology, v. 17, p. 350–356.
Pemberton, S.G., and Frey, R.W., 1982, Trace fossil nomenclature and the *Planolites-Palaeophycus* dilemma: Jour. Paleontology, v. 56, p. 843–881.
______, Kobluk, D.R., Yeo, R.K., and Risk, M.J., 1980, The boring *Trypanites* at the Silurian-Denovian disconformity in southern Ontario: Jour. Paleontology, v. 54, p. 1258–1266.
Purchon, R.D., 1955, The functional morphology of the rock-boring lamellibranch *Petricola pholadiformis* Lamarck: Jour. Mar. Biol. Assoc. U. K., v. 34, p. 257–258.
Rhoads, D.C., 1975, The paleoecological and environmental significance of trace fossils, *in* Frey, R.W. (ed.), The Study of Trace Fossils: New York, Springer-Verlag, p. 147–160.
Röder, H., 1977, Zur Beziehung zwischen Konstruktion und Substrat bei mechanisch bohrenden Bohrmuscheln (Pholadidae, Teredinidae): Senckenberg. Marit., v. 9, p. 105–213.
Schäfer, W., 1939, Fossile und rezente Bohrmuschel-Besiedlung des Jadegebietes. Senckenbergiana, v. 21, p. 227–254.
______, 1972, Ecology and Palaeoecology of Marine Environments: Edinburgh and Chicago, Oliver and Boyd and Univ. Chicago Press, 568 p.
Schroder, C.H., 1982, Trace fossils of the Oconee Group and basal Barnwell Group of east-central Georgia: Georgia Geol. Survey, Bull. 88, 125 p.
Seilacher, A., 1954, Die geologische Bedeutung fossiler Lebensspuren: Deutsche Geol. Gesell. Zeitschr., v. 105, p. 214–227.
______, 1955, Spuren und fazies im Unterkambrium, *in* Schindewolf, O. H., and Seilacher, A., Beiträge zur Kenntnis des Kambriums in der Salt Range (Pakistan): Akad. Wiss. Lit. Mainz, Math.-Nat. Kl., Abhandl., v. 10, p. 372–399.
______, 1958, Zur ökologischen Charakteristik von Flysch und Molasse: Eclog. Geol. Helvetiae, v. 51, p. 1062–1078.
______, 1963, Lebensspuren und Salinitätsfazies: Fortschr. Geol. Rheinld. u. Westfal., v. 10, p. 81–94.
______, 1964, Biogenic sedimentary structures, *in* Imbrie, J., and Newell, N.D. (eds.), Approaches to Paleoecology: New York, J. Wiley and Sons, p. 296–316.
______, 1967, Bathymetry of trace fossils: Marine Geology, v. 5, p. 413–428.
Sellwood, B.W., 1970, The relation of trace fossils to small sedimentary cycles in the British Lias, *in* Crimes, T.P., and Harper, J.C. (eds.), Trace Fossils: Geol. Jour., Spec. Issue 3: Liverpool, Seel House Press, p. 489–504.
Warme, J.E., 1975, Borings as trace fossils and the processes of marine bioerosion, *in* Frey, R.W. (ed.), The Study of Trace Fossils: New York, Springer-Verlag, p. 181–227.
Wunderlich, F., 1972, Georgia coastal region, Sapelo Island, U.S.A.: sedimentology and biology. III. Beach dynamics and beach development: Senckenberg. Marit., v. 4, p. 47–79.

THE TRACE FOSSIL ASSEMBLAGE OF A CRETACEOUS NEARSHORE ENVIRONMENT: ENGLISHTOWN FORMATION OF DELAWARE, U.S.A.

H. ALLEN CURRAN
Department of Geology, Smith College, Northampton, Massachusetts 01063

ABSTRACT

Trace fossils are abundant and extremely well-preserved in the unconsolidated, siliciclastic sediments of the Late Cretaceous Englishtown Formation exposed along the Chesapeake and Delaware Canal in Delaware. The trace fossil assemblage is characteristic of a nearshore environment and can be divided into two subassemblages. The upper part of the formation is dominated by densely packed *Ophiomorpha nodosa* shafts and tunnels, *Skolithos linearis*, and a delicate branching burrow that forms vertical polygonal nets. The subassemblage that characterizes the middle and lower parts of the sequence consists of *Macaronichnus segregatis* and a *Skolithos*-like form, both of which created mottled horizons, isolated *O. nodosa* shafts and tunnels, *Schaubcylindrichnus coronus*, and probable echinoid burrows. Each trace fossil taxon is described, and modern tracemaker analogs are discussed to enable further interpretation of the trace fossils and their paleoenvironmental significance. The Englishtown Formation is interpreted to have been deposited in a prograding nearshore environment with a low energy, *O. nodosa*-dominated, upper shoreface—lower foreshore zone similar to parts of the modern shallow-water Sea Isles coast of Georgia.

INTRODUCTION

Trace fossils are reasonably well known from Late Cretaceous siliciclastic rocks of the Western Interior of North America. Several studies in the region (e.g. Howard, 1966; Frey and Howard, 1970; Campbell, 1971; Maberry, 1971; Cotter, 1975; and McLane, 1982) have demonstrated the successful use of trace fossil assemblages to define more concisely the shallow-water paleodepositional environments of Cretaceous epeiric sea margins. The potential criteria for use of Cretaceous ichnofacies models in the Western Interior have been well summarized by Frey and Howard (1982).

By contrast, the trace fossils of Late Cretaceous rocks of the Atlantic Coastal Plain are poorly known and generally have not been used in paleodepositional environment interpretations. Outcrops of the Englishtown Formation in Delaware reveal the presence of abundant and well-preserved trace fossils. The goals of this paper are: 1) to describe the trace fossils of the Englishtown Formation; and 2) to demonstrate how the trace fossil assemblages can be used as an aid to a more refined interpretation of the paleodepositional environments represented by the Englishtown Formation.

THE GEOLOGIC SETTING

Outcrop Locality

The largest and best exposure of Cretaceous strata in Delaware and particularly of the Englishtown Formation occurs along the north bank of the Chesapeake and Delaware Canal about 1.5 km east of the New Summit Bridge where Rt. 896 crosses the canal (Fig. 1). Here the formations of the Matawan Group crop out in a steep cliff of 15 to 20 m height above canal water level (Fig. 2). Traveling south from Newark on Rt. 896, the cliff exposure can best be reached by turning east on Delaware Rt. 71 at its intersection with Rt. 896, about 1 km north of the New Summit Bridge. Several unimproved access roads to the canal banks join Rt. 71 on its right (south) side about 1.4 km beyond the Rt. 896

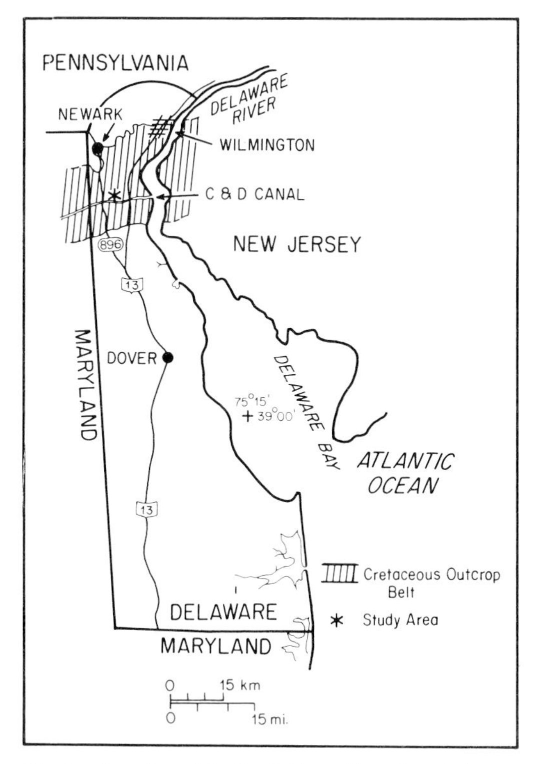

Fig. 1.—Location of the Englishtown Formation study area on the north bank of the Chesapeake and Delaware Canal, Delaware.

intersection. Once at the canal's edge, one should follow the canal bank road east to its end at the cliff exposure. In 1982 the character of the outcrop was modified by the Corps of Engineers through extension of the canal bank dike along the front of the outcrop face. This will prevent undercutting of the outcrop by canal waters and subsequent slumping, but it may have the long-term effect of causing the outcrop face to become more vegetated, possibly steeper, and thus less accessible.

Stratigraphy

The Englishtown Formation is the middle unit in the exposures of the Matawan Group along the north bank of the Chesapeake and Delaware Canal (Figure 2). These exposures were mapped by Pickett (1970), and the stratigraphy of the Matawan Group in Delaware, Maryland, and New Jersey was described in detail by Owens et al. (1970, 1977). Earlier literature (e.g. Groot et al., 1954; Jordan, 1962; and other papers) often used the name Wenonah Formation for the quartz sand unit along the canal, but it is now known that the Wenonah sands do not extend from New Jersey into Delaware as previously thought. Rather, it is the Englishtown beds that are continuous from New Jersey to Delaware (Owens et al., 1970, Fig. 4). Based on the planktonic foraminiferal and dinoflagellate assemblages of samples from test boreholes located immediately to the south of the canal, the Englishtown Formation has been determined to be of Late Campanian age (Houlik et al., 1983; Pickett, 1983).

The Englishtown Formation is generally thought to represent a regressive phase of sedimentation on the Late Cretaceous Mid-Atlantic Coastal Plain and to have been deposited in a nearshore, shallow-water environment (Pickett, 1970; Owens et al., 1977, Fig. 79; Houlik, 1983, Fig. 2). The formation normally has sharp contacts with the underlying Merchantville and overlying Marshalltown Formations, although the Merchantville-Englishtown contact is somewhat gradational at the study site.

Along the Chesapeake and Delaware Canal, the Englishtown Formation consists of light gray to white to buff, highly micaceous, well-sorted, very fine to fine quartz sand. Beds in the uppermost part of the formation show some textural coarsening, with a slight increase in medium to coarse quartz grains. Glauconite grains, an abundant constituent of the Merchantville and Marshalltown Formations, are sparse in the Englishtown sands except near the top of the formation. Here glauconite is common where, for the most part, the glauconite grains appear to have been "piped in" as burrow fill material from the overlying Marshalltown beds. Staining by iron oxide is prevalent throughout the Englishtown sands and can be particularly concentrated in the upper and lowermost beds of

Fig. 2.—The Upper Cretaceous Matawan Group beds form a steep cliff at the study area site. Mv = Merchantville Formation, E = Englishtown Formation, Mt = Marshalltown Formation.

the formation. In some cases, this staining has enhanced the preservation and recognition potential of individual trace fossils, but, where concentrated, the staining tends to obscure the trace fossils. Small chunks of dark, woody material occur sparsely throughout all but the uppermost beds of the formation.

Physical sedimentary structures are not prominent; rather, the Englishtown Formation for the most part has a massive, burrow-mottled to heavily bioturbated appearance. However, some weak tabular cross bedding occurs in the middle and lower parts of the formation. Where discernable, individual beds are thin; and light-colored, clean sands often are interbedded with darker, clayey, silty sand layers. Although bioturbation has destroyed most of the original physical structures, burrowing activity was not so complete that individual burrow boundaries and forms cannot be recognized. The degree of bioturbation is not uniform throughout the formation; certain horizons are particularly heavily bioturbated as indicated in Figure 3.

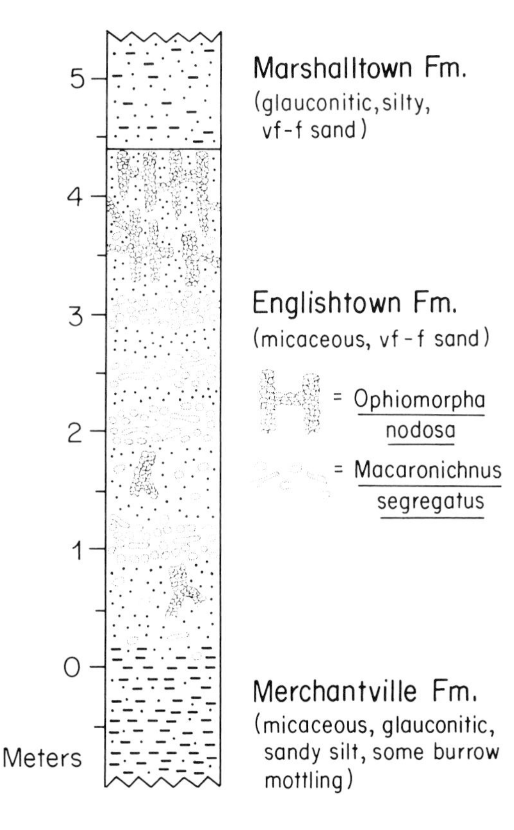

Fig. 3.—Generalized stratigraphic column for the Matawan Group at the study area site. The column shows the maximum thickness for the Englishtown Formation and depicts the distribution pattern of the two most conspicuous trace fossils within the formation.

Mega- and microfossils occur only sporadically in Englishtown outcrops (Owens et al., 1970, 1977), and no body fossils were found at the study site by the author. Thus traces are the dominant fossils preserved in the Englishtown Formation along the Chesapeake and Delaware Canal. The compacted but unconsolidated nature of the Englishtown sediments makes the formation ideal for the study of trace fossils because outcrop faces easily can be scraped down and individual forms studied in their full three-dimensional aspect. Furthermore, the fine texture and "fluffy" nature of the sandy matrix has permitted wind and rain to etch out in full relief extremely delicate trace fossils that either would not be recognized or could not be studied well in most settings. Thus the Englishtown Formation presents a situation of truly exceptional trace fossil preservation and provides an excellent example of a Late Cretaceous nearshore trace fossil fauna.

SYSTEMATIC ICHNOLOGY

In this section the Cretaceous trace fossils from the Englishtown Formation at the Chesapeake and Delaware Canal locality are described. These trace fossils were formed by a variety of shallow-water marine invertebrates, and they represent primarily dwelling, dwelling/feeding, feeding, and locomotion activities. To the fullest extent possible, each trace fossil is discussed and interpreted in terms of similar fossil forms, potential modern analogs, and paleoenvironmental significance.

Ichnogenus MACARONICHNUS Clifton and Thompson, 1978
Macaronichnus segregatis Clifton and Thompson, 1978
Pl. 1B, C, D

Description.—Intrastratal, irregularly meandering burrows, horizontal or gently inclined to bedding, and circular in cross section. Burrow diameters range from 0.7 to 1.5 cm, with an average diameter of 1.0 cm. Burrow walls lined with a concentration of micaceous material that forms a distinct rim which is darker than the burrow fill material and the surrounding matrix. By contrast, burrow fill material is light-colored quartz sand, texturally similar to but cleaner than the sand of the host sediments. Burrows tend to be densely packed (Pl. 1B) and normally are unbranched, but branching can occur (Pl. 1D); burrow crossovers and interpenetrations are common (Pl. 1C).

Discussion and interpretation.—The specimens of this ichnospecies are highly abundant in the Englishtown Formation, particularly in the lower two-thirds of the sequence (Fig. 3). The Englishtown specimens compare closely with specimens of *Macaronichnus segregatis* originally described by Clifton and Thompson (1978). The primary difference is that the Englishtown burrows are larger in diameter (average diameter = 1.0 cm) than the specimens described by Clifton and Thompson which had a diameter range of 0.3 to 0.5 cm. In the Englishtown specimens, the dark rim of material lining the burrows is composed primarily of iron oxide-stained muscovite flakes with minor biotite and heavy mineral grains. This type of lining is consistent with the findings of Clifton and Thompson for specimens of *M. segregatis* from other areas. On surfaces cut parallel to bedding in the Englishtown sands, individual *M. segregatis* specimens often could be traced for distances of 20 cm or more before the burrow would move above or below the cut plane.

Clifton and Thompson (1978) interpreted *Macaronichnus segregatis* as the burrow of a deposit-feeding organism, very probably a marine polychaete. My observations of the Englishtown specimens conform to this interpretation of a burrow constructed by a deposit feeder selectively ingesting sand and segregating out the mica flakes that form the burrow lining. Thus the burrow fill sediment has been modified by the organism's feeding activity.

This has important implications for the ichnogeneric assignment of these specimens because *Macaronichnus* clearly is close in form to the well-known ichnogenera *Planolites* and *Palaeophycus*, and, in fact, this relationship was discussed by Clifton and Thompson (p. 1293–1295). However, further comment is made here in light of the recent extensive review by Pemberton and Frey (1982) of the *Planolites-Palaeophycus* "dilemma". Pemberton and Frey redefined *Planolites* as an unlined burrow infilled with sediments having a textural character unlike that of the host rock. The infill represents sediment processed by the tracemaker, most often by deposit-feeding activity (Pemberton and Frey, p. 865). *Palaeophycus* is redefined as a lined burrow filled with sediment typically identical to that of the surrounding matrix. Here the infill represents passive, gravity-induced sedimentation into the open, lined burrows (Pemberton and Frey, p. 852). In a sense, *Macaronichnus* is the third point of an ichnogeneric triangle—a *lined* burrow with *modified* infill material. However, *Macaronichnus* remains in need of further definition because no formal type specimens were designated by Clifton and Thompson and no synonymy was prepared.

Clifton and Thompson (1978, Table 1) recognized several occurrences of *Macaronichnus segregatis* in rocks ranging from Jurassic to Pleistocene age, all thought to be of shallow marine origin. The marine polychaete, *Ophelia limacina*, was suggested by Clifton and Thompson as a modern tracemaker analog for *M. segregatis* burrows. In experiments conducted in aquaria by Clifton and Thompson, specimens of *Ophelia limacina* produced burrows closely resembling *M. segregatis*. Species of the Family Opheliidae have a cosmopolitan distribution and are known to be errant, deposit-feeding burrowers in shallow marine waters (Gosner, 1971; Barnes, 1974). It seems quite probable that the *M. segregatis* burrows of the Englishtown Formation were formed by similar deposit-feeding polychaetes.

Ichnogenus OPHIOMORPHA Lundgren, 1891
Ophiomorpha nodosa Lundgren, 1891
Pl. 2A, B, C; Pl. 3C; Pl. 5A

Description.—Branched burrow systems of straight to gently curved shafts and horizontal to gently inclined tunnels with clayey walls 2–4 mm thick. Walls smooth on the interior surface and distinctly mammilated on the exterior surface, most often with a single-pellet form of construction. Outside burrow diameter ranges from 0.5 to 3 cm with an average of about 1.5 cm for well developed tubes. Individual shaft or tunnel segments sometimes measurable for lengths of 35 to 40 cm. No apertural necks observed. Burrows filled with sediment like that of the surroundings matrix except near the top of the Englishtown Formation where infill material often is rich in glauconite grains from the overlying Marshalltown Formation.

Discussion and interpretation.—The *Ophiomorpha* burrows described here are assigned to the ichnospecies *O. nodosa* based on the predominant single-pellet mode of wall formation, following the diagnosis of Frey et al. (1978). However, pellet form can be somewhat variable, ranging from discoid (Pl. 5A) to oval and/or blocky (Pl. 2A, C).

This is the only trace fossil previously described from the Englishtown Formation, and it certainly is the most obvious trace fossil in outcrop. Jordan (1962), Groot et al. (1954), and earlier papers used the name *Halymenites major* Lesquereux for the form, which was listed as "Insertae Sedis" by Groot et al. (p. 54) and noted to have been identified by various previous authors as worm tubes, mollusk borings, or plant remains.

Ophiomorpha nodosa is most abundant at and near the top of the Englishtown Formation where shafts are

EXPLANATION OF PLATE 1

FIG. A.— *Skolithos linearis*, pen top = 6 cm.
B.— *Macaronichnus segregatis*, vertical suface of zone mottled by dense occurrence of this ichnospecies; bar scale = 2 cm.
C.— *Macaronichnus segregatis*, horizontal surface; burrow diameter = 1.0–1.2 cm.
D.— *Macaronichnus segregatis*, vertical surface; burrow diameter = 1.2–1.5 cm.

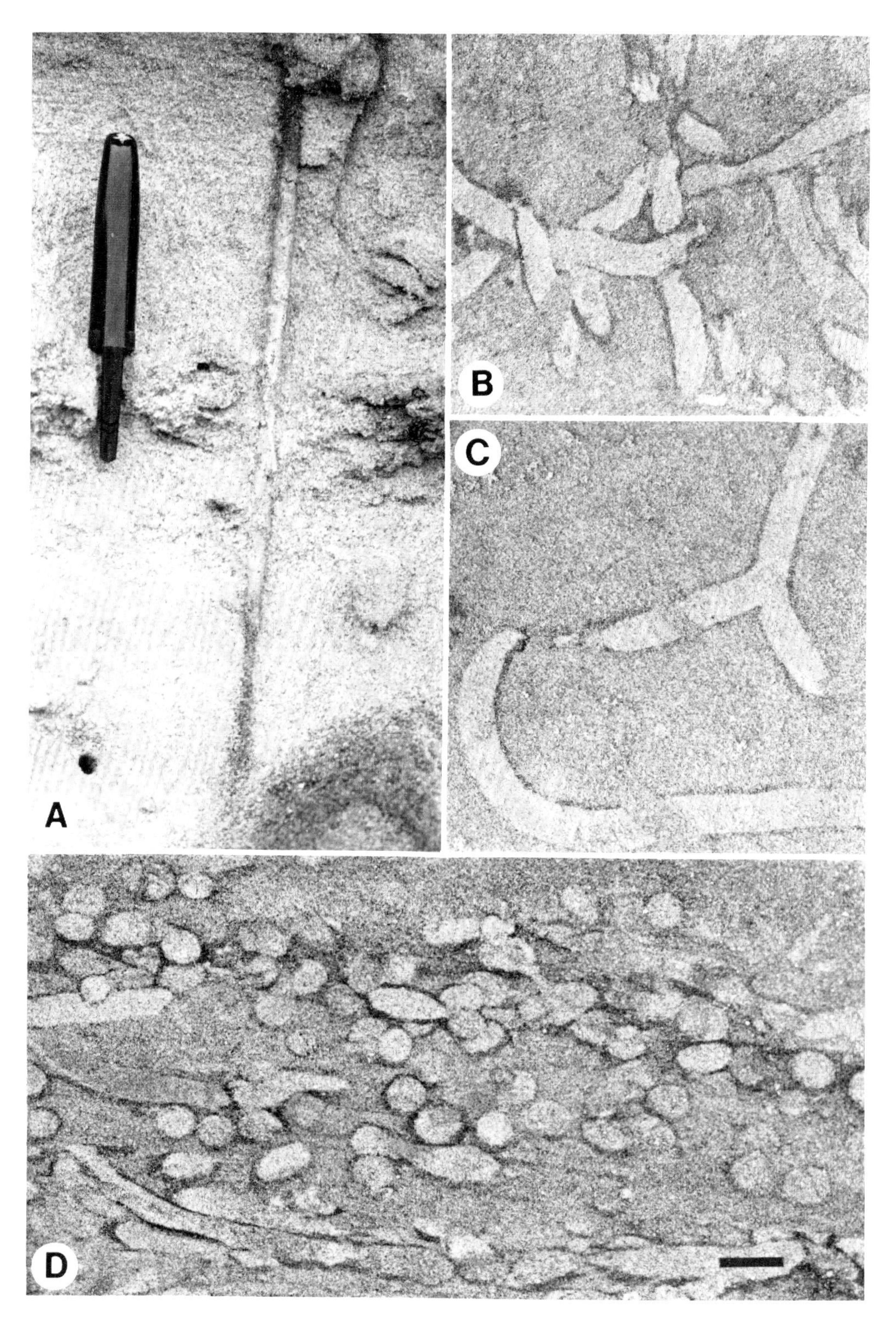
A
B
C
D

very closely spaced (i.e. "wall-to-wall") in a stack configuration (Pl. 2C), similar to that illustrated by Chamberlin and Baer (1973, Text-Fig. 4a,f) and Frey et al. (1978, Fig. 2A). In the downward direction the shafts often give way to an irregular maze structure (e.g. Frey et al., Fig. 2F). However, fully developed and preserved mazes as described by Pickett et al. (1971, Pl. 28, Fig. 7) from an Englishtown outcrop on the Chesapeake and Delaware Canal at St. Georges, a short distance east of study site, cannot be clearly discerned in the cliff exposures. Horizontal exposures, as once were present at St. Georges, are needed to reveal a full maze structure like those diagramed by Chamberlin and Baer (1973, Text-Fig. 4a) and Frey et al. (1978, Fig. 11). Below the top third of the Englishtown exposure, specimens of *Ophiomorpha nodosa* occur mostly as isolated shaft and tunnel segments (Fig. 3). Here the segments frequently are surrounded by dense concentrations of *Macaronichnus segregatis* (Pl. 2A.)

Small diameter shafts (0.5–0.75 cm, Pl. 2B) are common in the Englishtown sediments, and these shafts have a mammilated exterior surface identical in every respect to larger shafts except that individual pellets are smaller. It seems quite likely that these small-diameter shafts were formed by juvenile individuals of the organism that constructed *Ophiomorpha nodosa*; furthermore, the common occurrence of these small shafts suggests that juveniles made up a considerable portion of the tracemaker population. Full-sized shafts of *O. nodosa*, particularly near the top of the Englishtown exposure, often serve as an attachment anchor line for a delicate, branching burrow that runs along the shaft (Pl. 3C). These burrows will be discussed further in a subsequent section.

The analog relationship of the modern burrows of callianassid shrimp, particularly burrows formed by the ghost shrimp *Callianassa major* Say, to *Ophiomorpha nodosa* has been well documented by Weimer and Hoyt (1964) and Frey et al. (1978). The author fully concurs with the interpretation of Pickett et al. (1971) that the *O. nodosa* burrows of the Englishtown Formation were formed by a callianassid shrimp species with habits very similar to *C. major*.

Ophiomorpha nodosa with Burrowed Walls
Pl. 3A, B

Description.—Walls of apparent *Ophiomorpha nodosa* shafts intensely infested with small-diameter, sinuous, predominantly vertical, sometimes irregularly branching burrows. Original shaft wall material largely eroded so that the weakly consolidated, small-diameter burrow walls and/or infill materials are preserved in full relief. Wall burrows of two diameter size-classes, one of about 0.5 mm (Pl. 3A) and the other of about 2–3 mm (Pl. 3B).

Discussion and interpretation.—Gentle erosion by wind and rain of the Englishtown sands often reveals the presence of small and extremely delicate trace fossils, and these shaft wall burrows are a prime example. Although wall material has been removed, the form of the shafts and their occurrence with typical *Ophiomorpha nodosa* suggests that these shafts also are *O. nodosa* specimens. The obviously burrowed *O. nodosa* shafts were found near the top of the Englishtown Formation, but they are never common, probably because special conditions of slow erosion with cliff overhang protection are needed to reveal the wall burrows. However, the phenomenon may be very common in *O. nodosa* walls from this locality but simply has not been revealed in most cases. Bromley and Frey (1974, Fig. 5 and Pl. 1) illustrated specimens of *Gyrolithes davreuxi* Saporta that have burrowed walls similar to those described here; occurrence is very common (80%) in the *G. davreuxi* specimens studied by Bromley and Frey. These wall burrows originally were identified as *Chondrites* by Saporta, but, as Bromley and Frey (p. 316) noted, the small wall burrows do not show the constant branching angle or straightness of course found in typical *Chondrites*. Nonetheless, the specimens illustrated by Bromley and Frey show more consistent branching than those figured here, so I am reluctant to assign the Englishtown specimens to *Chondrites*.

The burrows in *Ophiomorpha nodosa* shafts undoubtedly were constructed to support dwelling/feeding activity; possibly the shaft wall material was used by the secondary burrower as a food source. Whether the secondary burrowing occurred during the time of occupation of the primary burrow by its tracemaker or soon after burrow abandonment cannot be determined. However, here one may well have an example of fossil commensalism. A precise taxonomic assignment for the wall burrower also is not possible, but small polychaetes or similar small vermiform animals would seem to be the mostly likely tracemakers. The two diameter size-classes of wall burrows in the Eng-

EXPLANATION OF PLATE 2

FIG. A.— Isolated shaft of *Ophiomorpha nodosa* in lower part of Englishtown Formation; circular cross-sections of *Macaronichnus segregatis* can be seen in the surrounding sediment; scale = 15 cm.

B.— Shafts of *Ophiomorpha nodosa*; arrow indicates small diameter shaft (= 0.7 cm) presumably formed by a juvenile animal of the tracemaker species; bar scale = 2 cm.

C.— Closely spaced *Ophiomorpha nodosa* shafts and tunnels at top of the Englishtown Formation; shaft diameters = 2.0–2.5 cm.

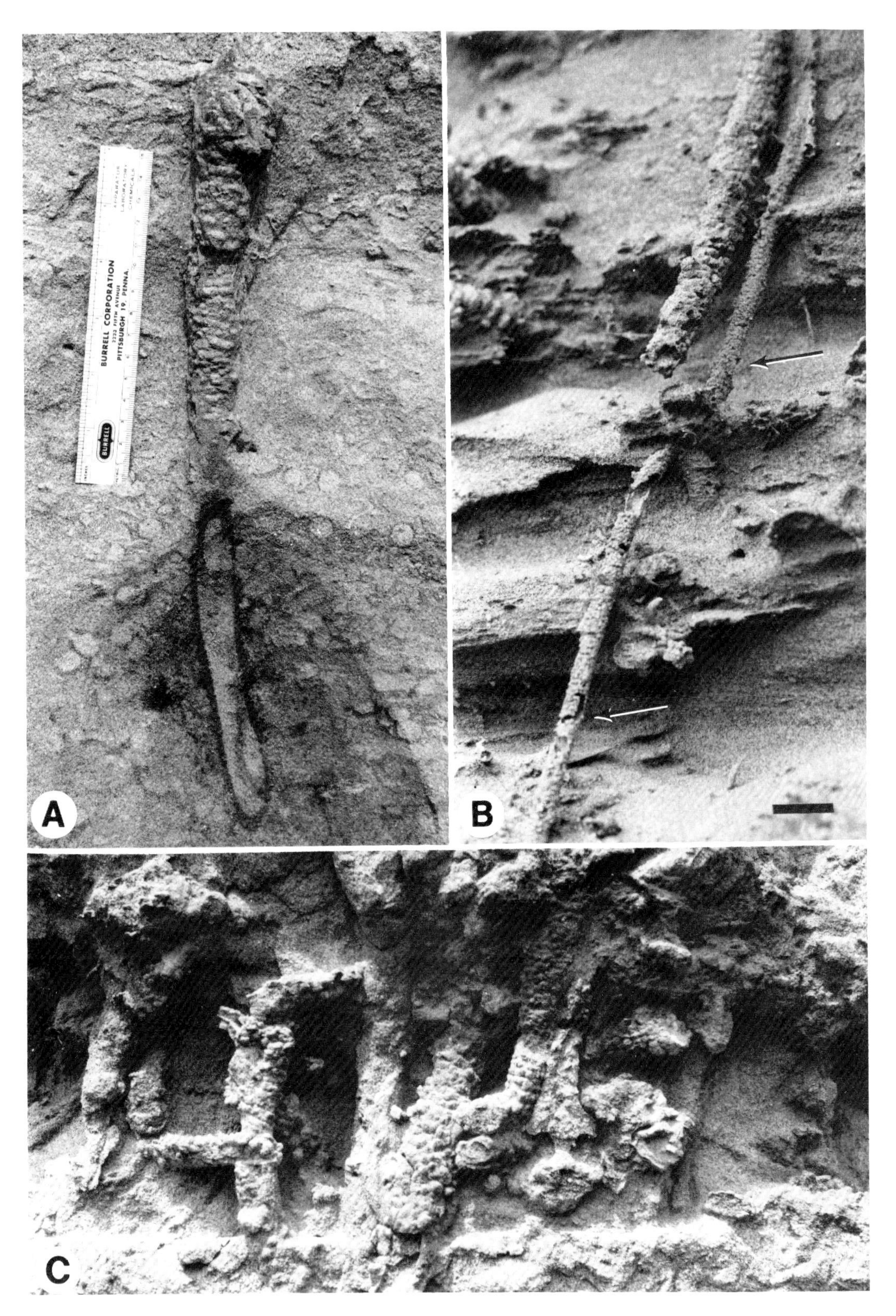
A
B
C
BURRELL CORPORATION
PITTSBURGH 19, PENNA.
BURRELL

lishtown examples suggest that at least two different tracemaker species were involved.

Ichnogenus SCHAUBCYLINDRICHNUS
Frey and Howard, 1981
Schaubcylindrichnus coronus
Frey and Howard, 1981
Pl. 4C, D; ? Pl. 5B, C

Description.—Bundles of closely spaced, vertical to gently curved, clean quartz sand-lined tubes; number of tubes in a sheaf usually 3 to 8. Upper part of tubes more nearly vertical than the lower extremities which commonly approach the horizontal. Individual tubes typically 5–7 mm in diameter with wall linings 1–2 mm thick. Interconnections between tubes not observed. Lower extremities of tubes apparently merge with unlined, clean quartz sand burrow infillings about 5–6 mm in diameter that form a radiate pattern on horizontal planes (Pl. 5B) and distinctively mottled horizons in vertical cuts (Pl. 5C).

Discussion and interpretation.—The Englishtown specimens are morphologically identical to specimens of this ichnospecies described by Frey and Howard (1981) from Late Cretaceous strata in Utah. The Delaware occurrence appears to be the first recorded from a Mesozoic formation beyond the United States western interior.

Radiate patterns (Pl. 5B) and mottled horizons in the vertical (Pl. 5C) are burrow infillings or areas of tracemaker animal-sorted clean quartz sand that apparently occur at the extremities of the tubes. No single cut made by the author in the unconsolidated Englishtown sediments completely and unequivocally confirmed this relationship, but well-defined tubes and radiate patterns have been traced through serial cuts to very close juxtaposition, leading to this interpretation that the radiate patterns are a continuum from the tubes.

Frey and Howard (1981) suggested that *Schaubcylindrichnus coronus* represents the tracemaking activity of an obligatorily gregarious suspension or deposit feeding animal. The Englishtown specimens clearly support the interpretation that *S. coronus* was formed by a deposit feeding, vermiform animal foraging from the base of its dwelling tube. The radiate pattern of clean, well-sorted quartz grains was the result of the foraging activity, with some of the quartz grains being used by the tracemaker to form its tube lining. In the Englishtown, the radiate patterns often were found in horizons rich in organic matter, further supporting the foraging activity interpretation. Frey and Howard (p. 802) compared this pattern of activity to that of the modern polychaete *Clymenella torquata* (see also Howard and Frey, 1975, p. 52), but this polychaete does not normally form clusters of tubes, so the analogy is not complete.

Ichnogenus SKOLITHOS Haldeman, 1840
Skolithos linearis Haldeman, 1840
Pl. 1A

Description.—Vertical to steeply inclined, straight to slightly curved, unbranched, subcylindrical burrows, 2.5–5 mm in diameter and traceable for lengths up to 20 cm. Burrow walls distinct, formed of agglutinated sand grains, and often stained with a dark-colored, apparently organic residue.

Discussion and interpretation.—The Englishtown specimens clearly fall within the range for this ichnospecies as defined by Alpert (1974). Well-defined specimens of *Skolithos linearis* are never abundant in the Englishtown exposure; specimens occur most commonly in the upper third of the formation in close association with *Ophiomorpha nodosa* shafts. The association is very similar to that shown by Curran (1976, Text-fig. 5), and the specimens are morphologically like those described by Curran and Frey (1977) from Pleistocene siliciclastic beds in North Carolina.

The Englishtown *Skolithos linearis* specimens were dwelling tubes most probably formed by polychaetes. Curran and Frey (1977, p. 148–150) reviewed the characteristics of a variety of modern polychaetes that form dwelling tubes similar to *S. linearis*. Tubes formed by the polychaete *Onuphis microcephala* are particularly similar in form to the Englishtown *S. linearis* specimens. Along the low-energy Georgia Sea Isles coast, *O. microcephala* tubes often occur in association with *Callianassa major* burrows (Hertweck, 1972; Curran and Frey, Pl. 5a) in a manner much like that found in the Englishtown Formation.

? *Skolithos* sp.
Pl. 4A, B

Description.—Delicate, vertical to steeply inclined, straight to gently curving to somewhat meandering, subcylindrical burrows, 1–1.5 mm in diameter and sometimes traceable for lengths of up to 6 cm. Branching not common but does occur (Pl. 4A), although with

EXPLANATION OF PLATE 3

FIG. A.— Intensely burrowed wall of an apparent *Ophiomorpha nodosa* shaft; shaft is 1.2 cm in diameter.
B.— Burrowed walls of *Ophiomorpha nodosa* shafts near top of the Englishtown Formation; large shaft in center is 3.0 cm in diameter.
C.— *Ophiomorpha nodosa* shafts and tunnels near top of the Englishtown Formation, 2.0–2.5 cm in diameter. Note delicate branching burrow that used a shaft as an anchor line; inset photo shows close-up of the branching burrow, which is 1.5 mm in diameter.

A
B
C

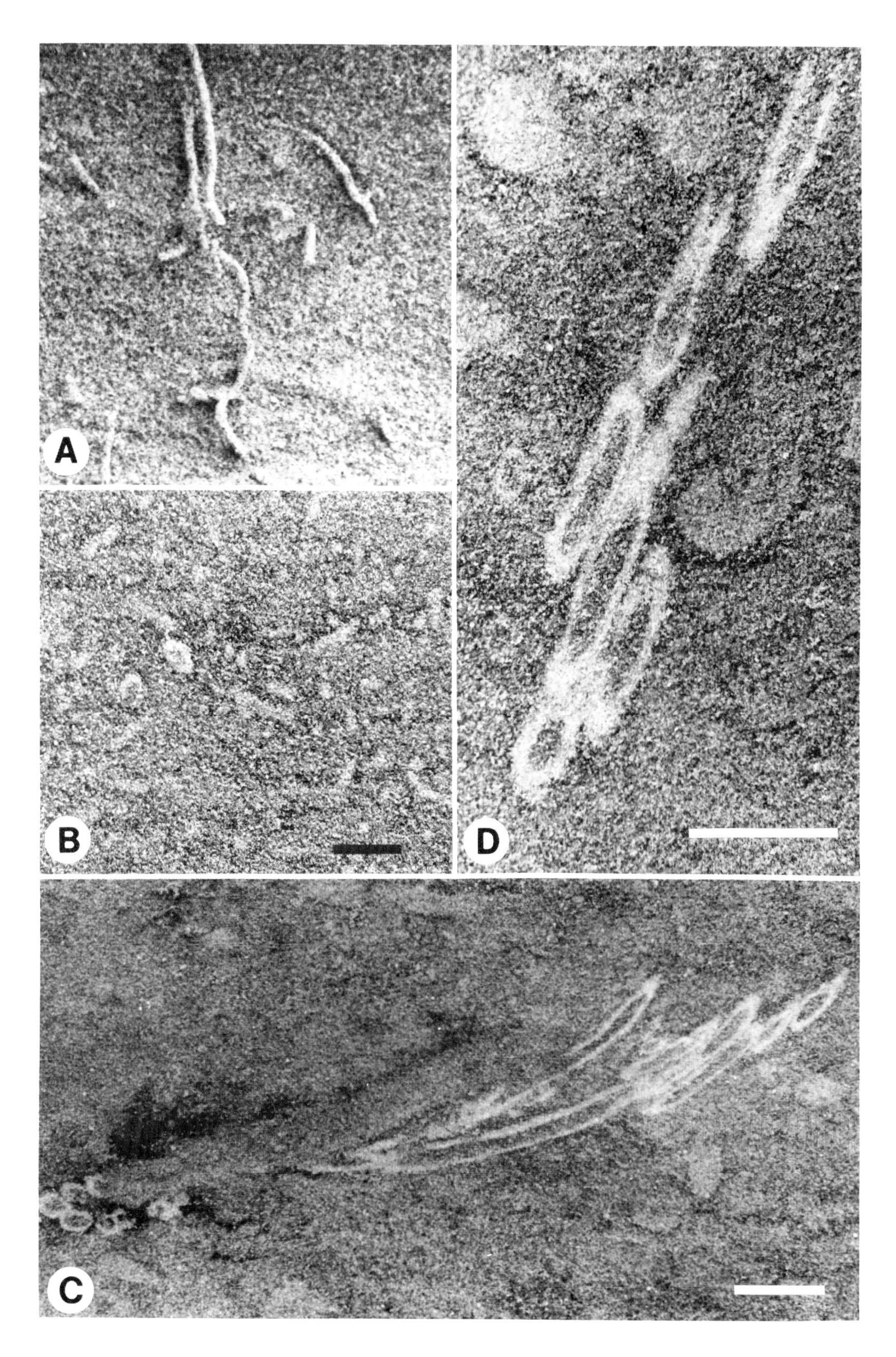
A
B
D
C

EXPLANATION OF PLATE 4

FIG. A.— *Skolithos* sp., burrows are about 1.0 mm in diameter.

B.— Sediment mottled by the dense occurrence of ? *Skolithos* sp. burrows; vertical surface, lower part of the Englishtown Formation; bar scale = 1.0 cm.

C.— *Schaubcylindrichnus coronus*, vertical surface; bar scale = 2.0 cm.

D.— Close-up of *Schaubcylindrichnus coronus* tubes, vertical surface; bar scale = 2.0 cm.

EXPLANATION OF PLATE 5

FIG. A.— *Ophiomorpha nodosa* shaft and tunnel; tunnel wall shows well-developed single pellet morphology; shaft diameter = 1.5 cm.

B.— Radiate pattern apparently formed at distal end of *Schaubcylindrichnus coronus* tubes by foraging activity of the tracemaker animal; horizontal surface.

C.— Vertical surface showing mottled horizon formed by the presumed foraging activity at the lower extremities of *Schaubcylindrichnus coronus* tubes; bar scale = 3.0 cm.

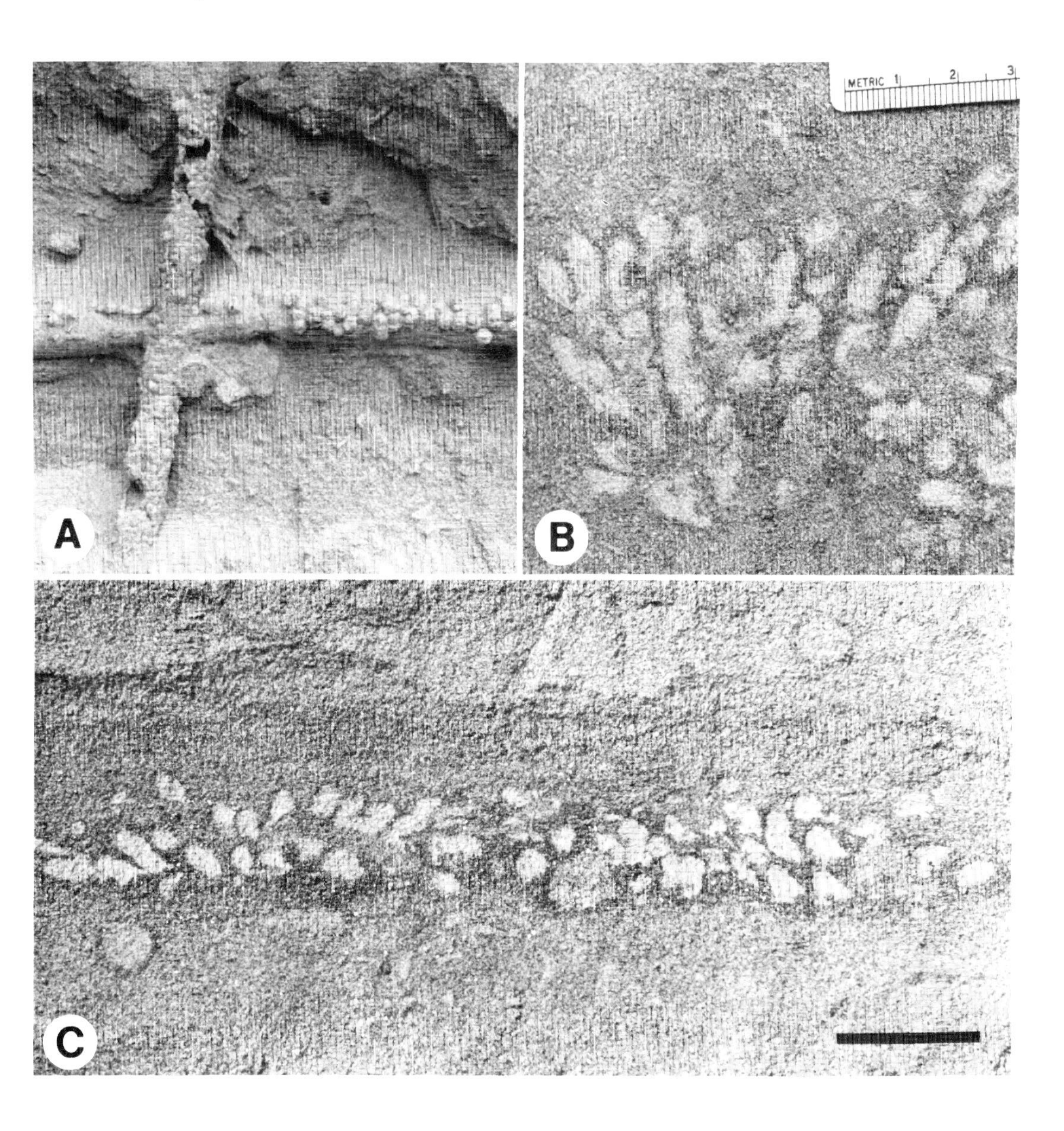

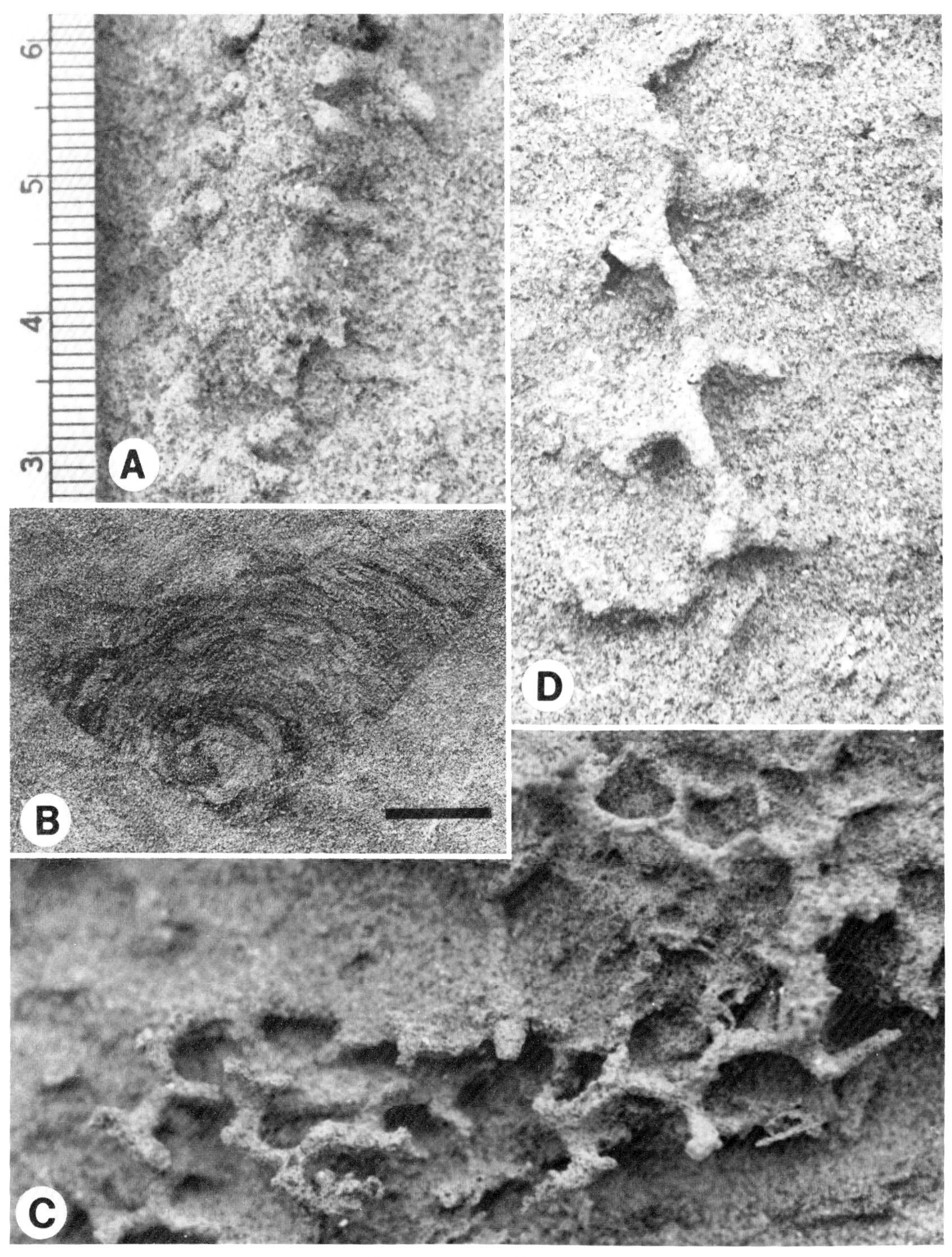

EXPLANATION OF PLATE 6

FIG. A.— Delicate branching burrow wrapped around a presumed *Ophiomorpha nodosa* shaft; scale in cm.

B.— Probable echinoid burrow, vertical surface, lower part of the Englishtown Formation; bar scale = 3.0 cm.

C.— Polygonal-patterned net of delicate branching burrows in a vertical exposure, upper part of the Englishtown Formation; burrow diameter = 1.5–2.0 mm.

D.— Close-up of delicate branching burrow; burrow diameter = 1.5 mm.

no regular pattern. Burrow walls formed of agglutinated fine sand and silt grains; burrow infill material usually lighter colored than the surrounding matrix (Pl. 4B).

Discussion and interpretation.—These burrows differ from the traditional morphological concept of *Skolithos* in that some branching does occur, but the pattern of branching is not nearly as ordered as with *Chondrites* and no second order branching was observed. The Englishtown specimens most closely resemble *Skolithos pusillus* as figured by Frey and Howard (1982, Fig. 4K), but the match is not complete because, although *S. pusillus* is shown to branch occasionally, it tends to be a straighter form than those of the Englishtown Formation. Consequently, only a tentative assignment to the ichnogenus *Skolithos* is made here.

The walls of these small diameter, very delicate burrows probably were originally mucous-lined, with sand and silt grains adhered to the sticky cylindrical surfaces. In the Englishtown, these small burrows are highly abundant, so much so that the burrows cause mottled sediment horizons (Pl. 4B). Such horizons are not nearly as prominent as those formed by *Macaronichnus segregatis* burrows, but they are present, particularly in the lower part of the Englishtown exposure, below the zone of concentrated *Ophiomorpha nodosa* occurrence.

These burrows probably were the dwelling tubes for small polychaetes, with the occasional branching representing feeding probes. In sandy horizons, branching is not common, but it may increase considerably in horizons with more organic-rich sediment (see Pl. 4B), and thus the burrow may begin to resemble *Chondrites*. Potentially close modern analogs would be burrows of the polychaete *Heteromastus filiformis*, particularly as figured by Howard and Frey (1975, Figs. 4A, 5A–D, 6D, G) from estuarine sediments of the Georgia coast, and/or the polychaete *Capitomastus* cf. *aciculatus* as figured by Hertweck (1972, Fig. 3a) from the Georgia shallow-water offshore zone. Both authors noted that in sandy sediments branching of these modern burrows decreased and the burrows tended to be more nearly vertical than in muddy sediments. These observations also seem to apply in the Cretaceous example.

Polygonal-patterned and Wrap-around Burrows
Pl. 3C; Pl. 6A, C, D

Description.—Delicate burrows that branch in an ordered manner to form polygonal patterns in the vertical plane (Pl. 6C). Burrows 1.5–2 mm in diameter; vertical segments sinuous with branches spaced at an even interval of about 5–6 mm (Pl. 6D). Some burrows observed to wrap around and branch away from probable *Ophiomorpha nodosa* shafts (Pl. 6A); other burrows may use an *O. nodosa* shaft as an anchor point and extend along the length of the shaft (Pl. 3C); polygonal net pattern may form away from the shaft.

Discussion and interpretation.—These very delicate burrows and burrow networks are a further example of the excellent preservation of trace fossils found in the fine sands of the Englishtown Formation. Their occurrence is very common in the upper part of the Englishtown exposure, particularly in the zone dominated by *Ophiomorpha nodosa*. The walls of the burrows probably originally were mucous-lined, with fine grains adhered to the surfaces. The forms figured here show different morphologic aspects, but close inspection in outcrop revealed that both the wrap-around and anchor-line forms give rise laterally to the polygonal burrow nets.

Again, these burrows probably were formed to support combined dwelling/feeding activity; polychaetes are the likely tracemaker group, but the author is not aware of any modern polychaete that forms vertically oriented polygonal patterns of this sort. The Englishtown forms somewhat resemble *Paleodictyon*, but that trace fossil is always oriented horizontal to bedding. Further study may indicate that this form requires a new trace fossil name.

Probable Echinoid Burrows
Pl. 6B

Description.—Structures in the vertical plane consisting of somewhat concentric, alternating light- and dark-colored sediment layers; overall pattern oval to sub-triangular in cross-sectional outline, with heights of 4–8 cm and lengths of 10–14 cm. Pattern of layers in the lower center part of the structure nearly circular.

Discussion and interpretation.—These structures occur rarely, as isolated examples in the lower part of the Englishtown exposure. They were best studied in vertical cross-sectional cuts perpendicular to the long axis of the structure (see Pl. 6B). Good longitudinal cuts were difficult to obtain, consequently the complete morphology of the form at this locality is not known.

The light-colored layers of the structure are composed of fine quartz sand, and the dark layers are concentrations of iron oxide-stained mica flakes and organic matter. These structures are interpreted as having been formed by an animal burrowing through the substrate. They most closely resemble the burrows of spatangoid echinoids, particularly as figured by Hertweck (1972, Fig. 9b; Pl. 4), Howard et al. (1974, Pl. 1, Fig. 2), and Bromley and Asgaard (1975, Figs. 8, 10, 28). The circular zone in the lower center part of the Englishtown structures appears to correspond well in form and position to the echinoid burrow canal structures figured by Bromley and Asgaard. However, in the present case, more specimens, and particularly longitudinal cuts, are needed for study before further interpretations and a positive identification can be made.

INTERPRETATION OF THE ENGLISHTOWN PALEODEPOSITIONAL ENVIRONMENT

The marine sediments of the Englishtown Formation have been shown to have been deposited in a nearshore

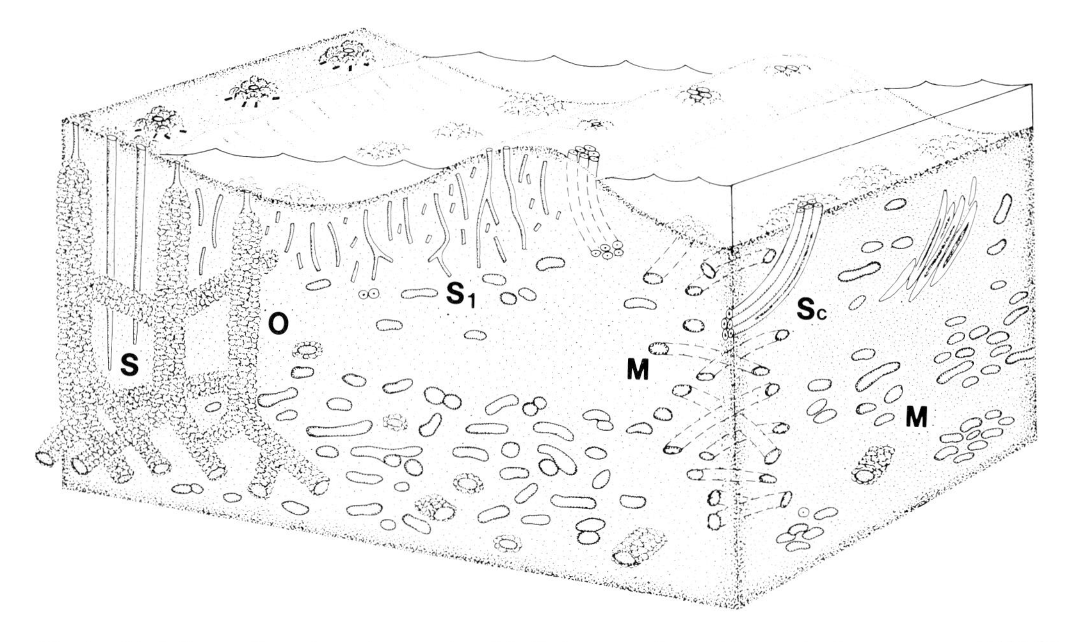

Fig. 4.—Block diagram illustrating an interpretation of the Cretaceous seafloor and its organism traces as represented by the Englishtown Formation. Upper part of the formation represents a lower foreshore-upper shoreface environment; middle and lower parts of the formation represent a shoaling, middle-lower shoreface environment. Longshore bars like the one shown here would not necessarily have been emergent at low tide. M = *Macaronichnus segregatis*; O = *Ophiomorpha nodosa*; S = *Skolithos linearis*; S_1 = ? *Skolithos* sp.; Sc = *Schaubcylindrichnus coronus*.

environment undergoing regression based on stratigraphic relationships and sedimentary characteristics (Pickett, 1970; Owens et al., 1970, 1977), body fossils (Owens et al., 1970), and microfossils (Houlik et al., 1983). The dominant presence of the trace fossils *Ophiomorpha nodosa* and *Macaronichnus segregatis* in the Englishtown Formation confirms this interpretation. *O. nodosa* is well established as a nearshore indicator; although, as Frey et al. (1978, p. 217) correctly pointed out, the modern analog burrows formed by *Callianassa major* and other callianassid species can have a wide distribution in nearshore subenvironments. The previously reported occurrences of *M. segregatis* (Clifton and Thompson, 1978) also suggest that this trace fossil is an excellent indicator of nearshore deposits. The environmental significance of *Schaubcylindrichnus coronus*, also a common trace fossil in the Englishtown beds, is less well known, but its previously reported occurrences all are from nearshore zones (Frey and Howard, 1981).

Given this general environmental setting, the trace fossil assemblage of the Englishtown Formation can be used to define in greater detail the depositional environments of these sediments. In vertical section, the Englishtown trace fossils are distributed between two somewhat overlapping subassemblages as suggested in Figure 3. The upper part of the formation is characterized by a trace fossil subassemblage consisting of densely packed *Ophiomorpha nodosa* shafts and tunnels, *Skolithos linearis*, and delicate branching burrows that form polygonal patterns. The trace fossil subassemblage that characterizes the middle and lower parts of the formation consists of *Macaronichnus segregatis* and ? *Skolithos* sp., both of which form distinctly mottled horizons, isolated *O. nodosa* shafts and tunnels, *Schaubcylindrichnus coronus*, and probable echinoid burrows.

A reconstruction of the Englishtown seafloor environment with its traces is given in Figure 4. The dense occurrence of *Ophiomorpha nodosa* with very closely spaced shafts (Pl. 2C) and a resulting lack of physical sedimentary structures dominates the upper part of the formation and suggest a lower foreshore-upper shoreface environment. The matrix of well-sorted, very fine, micaceous quartz sands indicates generally low wave energy conditions like those found today along the Georgia Sea Isles coast (Howard and Reineck, 1972). Howard and Scott (1983, p. 180) found the same sort of "wall-to-wall" *Ophiomorpha nodosa* occurrence in the sandy, burrowed and laminated facies of a Pleistocene prograding siliciclastic sequence in Florida. They theorized that as offlap progresses, long *O. nodosa* shafts formed in the lower foreshore penetrated through to the underlying upper shoreface, also *Ophiomorpha*-bearing beds. The result is a strong "*Ophiomorpha* overprint" in the underlying upper shoreface beds. This same process appears to have been operative during the development of the upper part of the Englishtown sequence. The association of *Skolithos linearis* with *O. nodosa* in these uppermost Englishtown beds also supports a lower foreshore-upper shoreface environment by analogy with the occurrence of *Onuphis microcephala* and *Callianassa major* burrows in the lower foreshore zone off Sapelo Island, Georgia (Hertweck, 1972).

The middle and lower parts of the Englishtown sequence probably were deposited in the middle to lower shoreface zone, possibly in a shoaling environment with nearshore bars. Again, the beds are well bioturbated, often with zones mottled by *Macaronichnus segregatis* and ? *Skolithos* sp., but some poorly defined planar cross bedding does occur. This interpretation is consistent with the generally nearshore, shoaling environment of occurrence for *M. segregatis* as interpreted by Clifton and Thompson (1978). The deposit feeder that made these burrows probably worked well below the substrate surface (10's of centimeters), as illustrated by Clifton and Thompson (Fig. 4). The more isolated occurrence of *Ophiomorpha nodosa* shafts and tunnels in this part of the sequence is consistent with shoaling substrate conditions which would not be conducive to the formation of well defined, shaft-dominated *Ophiomorpha* systems as found in the lower foreshore-upper shoreface zone (Frey et al., 1978).

In summary, the Englishtown Formation is interpreted as having been deposited in a prograding, low energy, shoaling, nearshore environment with an *Ophiomorpha nodosa*-dominated upper shoreface-lower foreshore and a *Macaronichnus segregatis*-dominated middle to lower shoreface. The Englishtown sequence resembles closely the nearshore, sandy part of a Pleistocene barrier island beach-to-offshore sequence in Florida described by Howard and Scott (1983).

ACKNOWLEDGMENTS

I thank Thomas E. Pickett, Delaware Geological Survey, for introducing me to the Englishtown localities along the Chesapeake and Delaware Canal and for helpful discussions regarding the Cretaceous stratigraphy of Delaware. Ronald L. Martino, Marshall University, and Thomas Pickett read the manuscript and offered constructive criticism. Marshall Schalk, Smith College, aided in the preparation of photographs. The field work for this study was supported by a grant from the Smith College Committee on Aid to Faculty Scholarship.

REFERENCES

Alpert, S.P., 1974, Systematic review of the Genus *Skolithos*: Jour. Paleontology, v. 48, p. 661–669.

Barnes, R.D., 1974, Invertebrate Zoology, 3rd edition: Philadelphia, W.B. Saunders Co., 870 p.

Bromley, R.G., and Asgaard, U., 1975, Sediment structures produced by a spatangoid echinoid: A problem of preservation: Bull. Geol. Soc. Denmark, v. 25, p. 261–281.

______, and Frey, R.W., 1974, Redescription of the trace fossil *Gyrolithes* and taxonomic evaluation of *Thalassinoides*, *Ophiomorpha* and *Spongeliomorpha*: Bull. Geol. Soc. Denmark, v. 23, p. 311–335.

Campbell, C.V., 1971, Depositional model—Upper Cretaceous Gallup Beach shoreline, Ship Rock Area, Northwestern New Mexico: Jour. Sed. Petrology, v. 41, p. 395–409.

Chamberlain, C.K., and Baer, J.L., 1973, *Ophiomorpha* and a new thalassinid burrow from the Permian of Utah: Brigham Young Univ., Geol. Studies, v. 20, p. 79–94.

Clifton, H.E., and Thompson, J.K., 1978, *Macaronichnus segregatis*: A feeding structure of shallow marine polychaetes: Jour. Sed. Petrology, v. 48, p. 1293–1302.

Cotter, E., 1975, Late Cretaceous sedimentation in a low-energy coastal zone: The Ferron Sandstone of Utah: Jour. Sed. Petrology, v. 45, p. 669–685.

Curran, H.A., 1976, A trace fossil brood structure of probable callianassid origin: Jour. Paleontology, v. 50, p. 249–259.

______, and Frey, R.W., 1977, Pleistocene trace fossils from North Carolina (U.S.A.), and their Holocene analogues, *in* T.P. Crimes and J.C. Harper (eds.), Trace Fossils 2: Geological Journal Special Issue No. 9, Liverpool, Seel House Press, p. 139–162.

Frey, R.W., and Howard, J.D., 1970, Comparison of Upper Cretaceous ichnofaunas from siliceous sandstones and chalk, Western Interior region, U.S.A., *in* T.P. Crimes and J.C. Harper (eds.), Trace Fossils: Geological Journal Special Issue No. 3, Liverpool, Seel House Press, p. 141–166.

______, and ______, 1981, *Conichnus* and *Schaubcylindrichnus*: Redefined trace fossils from the Upper Cretaceous of the Western Interior: Jour. Paleontology, v. 55, p. 800–804.

______, and ______, 1982, Trace fossils from the Upper Cretaceous of the Western Interior: Potential criteria for facies models: The Mountain Geologist, v. 19, p. 1–10.

______, ______, and Pryor, W.A., 1978, *Ophiomorpha*: Its morphologic, taxonomic, and environmental significance: Palaeogeogr., Palaeoclimatol., Palaeoecol., v. 23, p. 199–229.

Gosner, K.L., 1971, Guide to Identification of Marine and Estuarine Invertebrates: New York, Wiley-Interscience, 693 p.

Groot, J.J., Organist, D.M., and Richards, H.G., 1954, Marine Upper Cretaceous formations of the Chesapeake and Delaware Canal: Delaware Geological Survey, Bull. 3, p. 64.

Hertweck, G., 1972, Georgia coastal region, Sapelo Island, U.S.A.: Sedimentology and biology V. Distribution and environmental significance of lebensspuren and in-situ skeletal remains: Senckenberg. Marit., v. 4, p. 125–167.

HOULIK, C.W., Jr., OLSSON, R.K., AND AURISANO, R.W., 1983, Upper Cretaceous (Campanian-Maestrichtian) marine strata in the subsurface of northern Delaware: Southeastern Geology, v. 24, p. 57–65.

HOWARD, J.D., 1966, Characteristic trace fossils in Upper Cretaceous sandstones of the Book Cliffs and Wasatch Plateau: Utah Geological and Mineralogical Survey, Bull. 80, p. 35–53.

______, AND FREY, R.W., 1975, Estuaries of the Georgia coast, U.S.A.: Sedimentology and biology, II: Regional animal-sediment characteristics of Georgia estuaries: Senckenberg. Marit. v. 7, p. 33–103.

______, AND REINECK, H.-E., 1972, Georgia coastal region, Sapelo Island, U.S.A.: Sedimentary and biology, IV: Physical and biogenic structures of the nearshore shelf: Senckenberg. Marit., v. 4, p. 81–123.

______, ______, AND RIETSCHEL, S., 1974, Biogenic sedimentary structures formed by heart urchins: Senckenberg. Marit., v. 6, p. 185–201.

______, AND SCOTT, R.M., 1983, Comparison of Pleistocene and Holocene barrier island beach-to-offshore sequences, Georgia and Northeast Florida coasts, U.S.A.: Sedimentary Geology, v. 34, p. 167–183.

JORDAN, R.R., 1962, Stratigraphy of the sedimentary rocks of Delaware: Delaware Geological Survey, Bull. 9, p. 51.

MABERRY, J.O., 1971, Sedimentary features of the Blackhawk Formation (Cretaceous) in the Sunnyside District, Carbon County, Utah: U.S. Geological Survey Professional Paper 668, 44 p.

MCLANE, M., 1982, Upper Cretaceous deposits in south-central Colorado—Codell and Juana Lopez Members of Carlile Shale: Am. Assoc. Petroleum Geologists Bull., v. 66, p. 71–90.

OWENS, J.P., MINARD, J.P., SOHL, N.F., AND MELLO, J.F., 1970, Stratigraphy of the outcropping Post-Magothy Upper Cretaceous formations in southern New Jersey and northern Delmarva Peninsula, Delaware and Maryland: U.S. Geological Survey Professional Paper 674, 60 p.

______, SOHL, N.F., AND MINARD, J.P., 1977, A field guide to Cretaceous and Lower Tertiary beds of the Raritan and Salisbury Embayments, New Jersey, Delaware, and Maryland: Guidebook for AAPG/SEPM Annual Convention, Washington, D.C., 113 p.

PEMBERTON, S.G., AND FREY, R.W., 1982, Trace fossil nomenclature and the *Planolites-Palaeophycus* dilemma: Jour. Paleontology, p. 843–881.

PICKETT, T.E., 1970, Geology of the Chesapeake and Delaware Canal area: Delaware Geological Survey, Geologic Map Series, No. 1.

______, 1983, Correlation chart for Delaware, Column 9—Chesapeake and Delaware Canal, *in* Jordan, R.R. and Smith, R.V., coordinators, Atlantic Coastal Plain Correlation Chart: Am. Assoc. Petroleum Geologists, COSUNA Project, Tulsa.

______, KRAFT, J.C., AND SMITH, K., 1971, Cretaceous burrows—Chesapeake and Delaware Canal, Delaware: Jour. Paleontology, v. 45, p. 209–211.

WEIMER, R.J., AND HOYT, J.H., 1964, Burrows of *Callianassa major* Say, geologic indicators of littoral and shallow neritic environments: Jour. Paleontology, v. 38, p. 761–767.

PART III
ASSEMBLAGES OF BIOGENIC STRUCTURES IN MARINE SHALE-FORMING ENVIRONMENTS AND DEEP WATER BASINS

INTRODUCTION.—This final section of the volume concerns trace fossil assemblages from deeper water environments that often are oxygen deficient and the sites for significant accumulation of fine-grained sediment. Douglas Jordan's paper describes trace fossil assemblages from Devonian black shales in Kentucky and then illustrates how these assemblages can be used to interpret paleodepositional environments in shaly basins. The different assemblages are shown to vary with sediment type and color and are thought to be controlled by oxygen availability at the time of deposition. Table 1 presents a synthesis of the occurrence of trace fossils in black shales and the interpreted paleoenvironments and O_2 content, and Figure 12 illustrated Jordan's proposed model for the occurrence of trace fossil assemblages in a marine, deep-water basin.

The enigmatic depositional origin of the Ordovician "trilobite shales" of Ohio is the topic of the paper by Danita Brandt Velbel. Her study combines ichnologic, taphonomic, and sedimentologic data to show that these shales were deposited rapidly, but not necessarily by turbidity currents. Trace fossils, particularly various types of *Chondrites*, are shown to be useful in determining the relative timing of events of deposition.

The paper by Brian Edwards is a comprehensive study of modern bioturbation in sediments at oxygen-deficient, bathyal depths in the Santa Cruz Basin off the southern California coast. Although the basin floor sediments support a high density and diversity of benthic macrofauna, traces preserved in the sediment are low in density and diversity. Based on his findings from the Santa Cruz Basin, Edwards thinks that ancient sediments deposited in similar basins should have nondescript bioturbate textures with any distinct trace fossils biased toward those forms made by deeply burrowing infauna.

A.A. Ekdale has the final say in this volume (he also had the opening word with his article on eolianites, thus clearly demonstrating his breadth as an ichnologist) with his study of bioturbation in core sections from DSDP cores spanning Cretaceous anoxic episodes in the Atlantic Ocean. The study reveals a general decrease in intensity of bioturbation and diversity of trace fossils with increasing unoxidized carbon and total sulfur content in the sediment. *Chondrites* is most common in organic-rich layers, often occurring as the sole trace fossil. Ekdale concludes that the *Chondrites*-producing organism(s) actually has preferred oxygen deficient environments throughout its history.

TRACE FOSSILS AND DEPOSITIONAL ENVIRONMENTS OF UPPER DEVONIAN BLACK SHALES, EAST-CENTRAL KENTUCKY, U.S.A.

DOUGLAS W. JORDAN
Reservoirs Incorporated, 1151-C Brittmore Rd., Houston, Texas 77043

ABSTRACT

Trace fossil assemblages are recognized in "anoxic" black shales and these assemblages are valuable in interpreting paleodepositional environments in shaly basins. Trace fossils in the Upper Devonian black shales of east-central Kentucky (Ohio or New Albany Shale) occur within and below dolomites, gray shale, and in black shale. A diverse suite of epifaunal and infaunal deposit-feeding traces is preserved in interbedded dolomite and shale of the lower part of the Huron Member. *Planolites* is the most common trace fossil and is associated with *Chondrites* (Type A), *Zoophycos*, *Phycodes*, *Cruziana*, *Trichichnus*, *Teichichnus*, *Laevicyclus*, and larged ribbed burrows. In interbedded gray and black shale of the lower part of the Huron Member, *Planolites*, *Chondrites* (Type B), *Zoophycos*, *Teichichnus*, and *Rhizocorallium* are common on black shale bedding planes. In gray and black shale interbeds of the Three Lick Bed, a less diverse assemblage of traces occurs, including *Planolites*-like burrows, *Chondrites* (Type C and Type D), and *Zoophycos*. These ichnogenera are included in Seilacher's (1967) *Cruziana* and *Zoophycos* ichnofacies.

Devonian black shales were deposited in an environment of deposition that was transgressive over the axis of the present Cincinnati Arch in east-central Kentucky. Sediments representing the shallow-water carbonate environment of the Middle Devonian Boyle Dolomite interfinger with Upper Devonian shales. Upsection, carbonates and black shales are replaced by black and gray shales. Black shales represent total anoxic conditions while gray shale represents periodic oxygenation events (possibly movement of the dysaerobic zone) which allowed burrowing to occur for short time intervals. At the beginning of the Early Mississippian, the depositional regime changed, and deltaic sands and prodeltaic muds replaced black mud deposition.

INTRODUCTION

Trace fossils have been studied extensively in siliclastic rocks and less extensively in carbonate rocks. Occasionally they have been reported from shales, but usually only when the shales are associated with sandstones or limestones. The occurrence of trace fossils in black shales has received even less attention.

This study reports trace fossils from "anoxic" Upper Devonian black shales in east-central Kentucky along the western margin of the Appalachian Basin. The sedimentology, stratigraphy, geochemistry, paleontology, and structure of the Upper Devonian black shales in the Appalachian Basin have been investigated during the last 100 years (Orton, 1882; Grabau, 1906; Hard, 1931; Campbell, 1946; Conant and Swanson, 1961; Potter, et al., 1980). In east-central Kentucky, trace fossils in the black shales have been either ignored or misinterpreted. The purpose of this study is to document biogenic structures in the black shales and show how they aid in the reconstruction of depositional environments.

Black shales of Late Devonian age crop out in two areas in Kentucky. The first area lies in easternmost Kentucky along the folded Appalachians. The second outcrop belt parallels and lies on opposite sides of the Cincinnati Arch. To the south, this belt joins, and is continuous through, Tennessee.

The area of this study lies along the western margin of the Appalachian Basin on the eastern side of the Cincinnati Arch (Fig. 1). The outcrop belt is 20 miles wide and over 90 miles long and extends from Vanceburg (Section 10, Lewis County) on the Ohio River southwestward to Berea (Section 1, Madison County). The outer boundary of this area is known as the "knobs", which refers to limestone-capped hills overlying less resistant black shales of Devonian and Mississippian age. The Upper Devonian section thins from more than 250 feet at Vanceburg to 90 feet at Berea.

STRATIGRAPHY

Upper Devonian strata in east-central Kentucky (Ohio or New Albany Shale) lie both conformably and unconformably above either the Middle Devonian Boyle Dolomite or the Olentangy Shale (Middle to Upper Devonian), depending upon locality. The Upper Devonian black-shale sequence is divided into the Huron Member, the Three Lick Bed, and the Cleveland Member (Provo et al., 1978) and is shown in Figure 2.

The Huron Member is subdivided into a lower and an upper part (Jordan, 1979). The lower part consists of black shale interbedded with dolomite near its base and black shale interbedded with greenish-gray shale towards the top. The highest greenish-gray shale marks the top of the unit. The upper part of the Huron Member comprises the middle and upper part of Huron Member of Provo (1977) who recognized these units in the subsurface of east-central Kentucky. It consists of black shale locally interbedded with greenish-gray shale bearing the alga *Foerstia*. This fossil zone can be traced throughout most of the Appalachian Basin (Schopf and Schwietering, 1970).

The Three Lick Bed, described by Provo et al. (1978), consists of three greenish-gray shale beds interbedded with black shale. This unit can be traced throughout much of the western margin of the Appalachian Basin, although it thins to the south from Vanceburg to the Cincinnati Arch. The Cleveland Member consists of black shale containing phosphate nodules.

The black-shale sequence in Kentucky is referred to as the Ohio, New Albany, or Chattanooga Shale, depending on its location. The difference in nomenclature between the Ohio and New Albany Shale is due to the pinching out of beds southwestward from Ohio. The Ohio Shale is Upper Devonian and is overlain conformably by the Upper Devonian and Lower Mississippian greenish-gray Bedford Shale, the Lower Mississippian Berea Sandstone, and the still younger black Sunbury Shale in Ohio and northeast Kentucky. In east-central Kentucky, the Bedford and Berea disappear while the Sunbury continues. Here the black-shale sequence is referred to as the New Albany Shale. In Tennessee, the entire sequence, including equivalents of the Sunbury Shale, is called the Chattanooga Shale.

Figure 3 shows generalized Devonian stratigraphy through Ohio and Pennsylvania. Gradual filling of the Appalachian Basin and westward progradation of sediments began during the Middle Devonian, and by Late Devonian time successively younger beds were progressively deposited toward the west. Devonian sediments are thickest (up to 10,000 feet) in the geosynclinal trough comprising New York, Pennsylvania, West Virginia, Virginia, and easternmost Kentucky. These sediments thin toward the west and south in Ohio (over 500 feet), eastern Kentucky (100-300 feet), Tennessee (1-50 feet), and Alabama (1-30 feet).

Because of this thinning of the sequence, most workers suggest that the Cincinnati Arch was a shoal area in the Appalachian Basin (McFarlan and White, 1952; Ayrton, 1963; Oliver et al., 1967; Summerson and Swann, 1970; Schopf and Schwietering, 1970; Schwietering, 1970, 1979; Provo, 1977; Cluff, 1980).

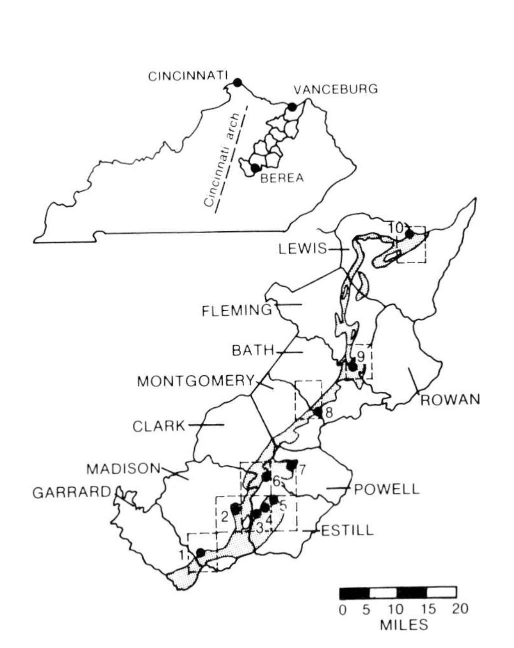

Fig. 1.—Outcrop area and county names in east-central Kentucky (Devonian outcrop belt shaded). Numbers refer to described sections and dashed rectangles are standard U.S.G.S. quadrangles.

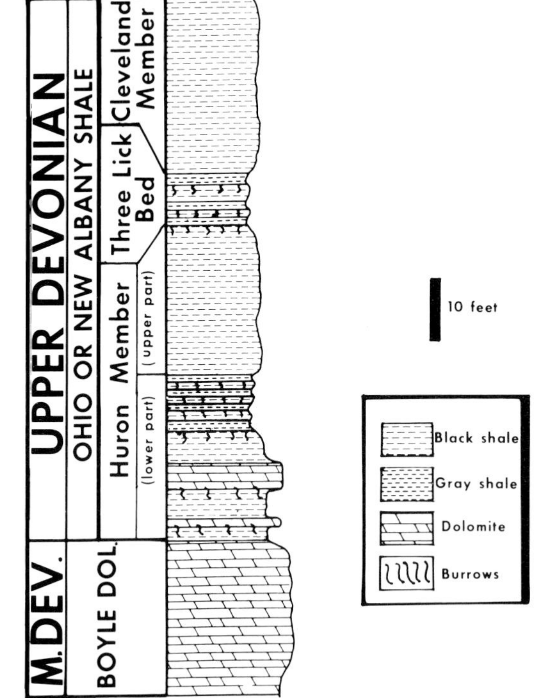

Fig. 2.—Stratigraphic column for Upper Devonian black shales in east-central Kentucky.

Others envision the Arch either as an emergent feature (Wells, 1944; Campbell, 1946; Berkheiser, 1971) or as a proto-Arch along the axis of the present Cincinnati Arch (Holbrook, 1964; Swager, 1978). Trace fossil and stratigraphic evidence described in this report support the concept of a shoaling environment immediately to the east of the Arch.

SEDIMENTOLOGY

Shale, the dominant lithology of the Upper Devonian sequence, changes lithologic character throughout east-central Kentucky. Most of the shale is brownish-black, but beds of greenish-gray shale and mudstone occur typically in the lower part of the Huron Member and the Three Lick Bed, and locally in the lower portion of the upper part of the Huron Member. Dolomite and dolomitic shale occur in the lower part of the Huron Member. Other minor constituents include phosphatic bone beds, pyritic laminae, phosphate nodules, carbonate concretions, and limestone with cone-in-cone structure.

Typical weathered sections reveal the black shales to be very fissile and cut by joints which impart a splintery, hackly character. In fresh samples, the shale is brownish-black, massive, and fractures subconchoidally. Shale outcrops exhibit a variety of colors from reddish-rust to yellowish burnt-orange to grayish-black due to chemical weathering, mineralization, and organic content.

The dark shales derive their color from abundant macerated organic matter. Other constituents include clay-sized quartz, pyrite, mica, illite, and chlorite. These shales commonly exhibit a ribbed weathering pattern on outcrops due to alternating hard quartz, mica, and clay-rich layers, and soft organic-rich layers. Discontinuous lamination is characteristic of these shales. Laminae are the result of quartz, carbonate, and organic material segregation.

Most of the flora in the black shale is in the form of macerated organic material. Remains of identifiable plant fossils are abundant and consist of the sporelike *Tasmanites*, the alga *Foerstia*, and *Callixylon* tree fragments. Remains of benthic organisms are rare in the black shales and consist predominantly of phosphatic remains of linguloid brachiopods. Fish fragments and conodonts are locally common.

Greenish-gray shales (hereafter referred to as "gray" shales) contain less organic matter and lack fissility and lamination. Gray shales appear massive on fresh outcrops, break into small prisms upon weathering, and become slippery and muddy when wetted. The gray shale forms "re-entrants" in the black shale sequence because it is softer and less resistant to weathering.

Dolomite in the lower part of the Huron Member could also be classed as dolomicstone because it is extremely fine-grained. Gradational beds from dolomicstone upwards to argillaceous dolomicstone to dolomudstone to dolomitic shale are common. Grading and

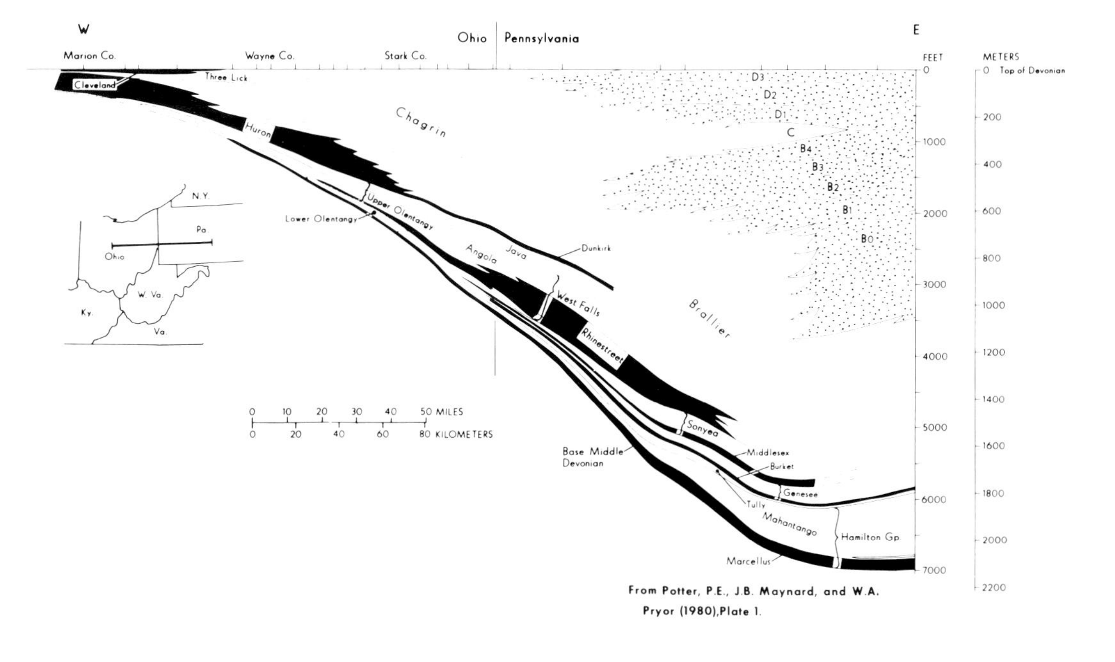

Fig. 3.—Schematic cross-section of Appalachian Basin fill from Pennsylvania to Ohio (from Potter, Maynard, and Fryor, 1980). A similar fill sequence occurs in east-central Kentucky.

upwards-fining sequences are present in the dolomite. Dolomite laminae in dolomitic shale locally are several grains thick and contain a mixture of dolomite rhombs, quartz, glauconite, fish fragments, pyrite, chert, and spores.

The abundance of dolomite and dolomitic shale in the lower part of the section indicates a transition between the Middle Devonian Boyle Dolomite and Late Devonian black-shale deposition. Bedding is typically horizontal and undeformed (Fig. 4A), but at several

Fig. 4.—Upper Devonian black shales. *A*, Nearly complete section of black shale, including the Middle Devonian Boyle Dolomite (1), interbedded dolomite and black shale (2.1), and interbedded gray and black shales (2.2) of the lower part of the Huron member, the upper part of the Huron Member (3), the Three Lick Bed (4), and the Cleveland Member (5). Section 7, entrance ramp to northbound Mountain Parkway, Clay City, Powell County, Kentucky. *B*, Soft-sediment slumping involving interbedded dolomite and black shale, lower part of the Huron Member, Section 2, State Route 52 near the village of Waco, Madison County, Kentucky.

localities (Waco, Section 2; Berea, Section 1) slumping occurs due to deformation after lithification of the Boyle Dolomite but prior to deposition of the overlying black shales (Fig. 4B).

TRACE FOSSILS

Energy conditions at the depositional interface often determine the behavior of benthic animals and consequently, the character of resultant biogenic sedimentary structures. Substrate type, rate of deposition, and availability of food and oxygen also influence the behavior of benthic organisms and define the area where a potential trace-maker will live. Unlike body fossils, trace fossils are rarely transported. Seilacher (1967) recognized trace fossil assemblages that were characteristic of shelf, slope, and basinal environments. These ichnofacies result from the fact that benthic animals in these different environments tend to possess mechanisms of food acquisition suited to their environment and are distributed accordingly. The ichnofacies are named according to the most common trace fossils. Generally, as water depth increases, sediment becomes finer (wave energy decreases) and the amount of food particles incorporated in the sediment increases. The major food source is organic matter which is held in suspension (high wave energy conditions) or deposited at the sediment-water interface. Suspension feeders typically inhabit nearshore waters in vertical burrows, while deposit-feeders efficiently exploit the nutrient-rich substrate in offshore waters and leave horizontal burrows (Seilacher, 1967).

In the Upper Devonian black shales in this study, the presence of trace fossils usually indicates an oxygenated, nontoxic environment. Most trace fossils in the black shales are horizontal burrows which were left by deposit-feeding animals. Changes in oxygen levels within the Appalachian Basin are charted by noting the size, shape, and association of these traces (Jordan, 1979).

Body fossils of benthic organisms are rare in the black shales. The occasional occurrence of predominantly pelagic fossils provides few clues for interpreting depositional environments and physical sedimentary structures usually are lacking. Thus, trace fossils can be the major tool for interpreting the sedimentary processes and depositional environments of shale sequences.

Trace fossils have been rarely reported in Devonian black shales outside of the Appalachian Basin. An exception to this is the trace fossils described from the Upper Devonian Three Forks Formation of Montana and Utah (Rodriguez and Gutschick, 1970).

From the Appalachian Basin, Kindle (1912), Grabau (1919), and Grossman (1944), reported impressions of "seaweeds" in the Upper Devonian of Ohio and New York, as well as marks attributed to worm burrows and animal tracks. Byers (1972, 1973) presented one of the first analyses of biogenic structures in a black-shale sequence. He was able to trace a time-parallel black shale unit basinward in the Upper Devonian of New York. Corbo (1977) reported a sequence of trace fossils in the Upper Devonian shale of New York which demonstrated transition from a deep water, turbidite environment to a shelf environment. Griffith (1977) presented a study of trace fossils in the Upper Devonian black shales of north-central Kentucky and noted an upwards-deepening depositional environment for strata.

Prior to this study, very little was known about trace fossils in the Devonian black shales of east-central Kentucky. Savage (1930) noted the occurrence of *Taonurus* (= *Zoophycos*) in brownish dolomite below black shale. Campbell (1946) reported similar marks in gray and black shale and referred to them as "*Spirophyton*-like" traces. Berkheiser (1971) found bifurcating and horizontal burrows in the bottom part of the Upper Devonian. Provo (1977) and Provo et al. (1978) noted traces beneath greenish-gray shales.

However, no systematic description or analyses of the biogenic sedimentary structures has been presented. This study concentrates on the interpretation of the biogenic structures in east-central Kentucky black shales.

Trace fossils in the Devonian black shales occur within and below the dolomite and dolomudstone which overlie black shale in the lower part of the Huron Member, and within and below gray shale beds in the lower part of the Huron Member and the Three Lick Bed (Table 1). Trace fossils are recognized where there is a color contrast between shales and/or dolomite, due to a change in organic content, chemical or mineralogic composition. The association of trace fossils and different lithologies is important in recognizing that the sediment emplacement was intimately associated with burrowing.

SYSTEMATIC ICHNOLOGY

PLANOLITES Nicholoson, 1873
Fig. 5A-B

Description.—*Planolites* consists of horizontal to subhorizontal meandering, unbranched tubes, which are circular in cross section where uncompacted. *Planolites* has been interpreted as the burrow of deposit-feeding worms, and is the most common trace fossil in the Upper Devonian sequence. In interbedded dolomite and shale, it is preserved on the bottom surface of the dolomite bed (Fig. 5A). Dolomite filling the burrow is locally graded. The length of *Planolites* ranges from 1.25–17.5 cm and the diameter of burrows ranges between 0.25–1.25 cm.

In black and gray shales of the lower part of the Huron Member, *Planolites* commonly exceeds 25 cm in length (Fig. 5B). In all cases it has smooth and unornamented walls. The burrow was made in black mud and was passively filled with gray mud.

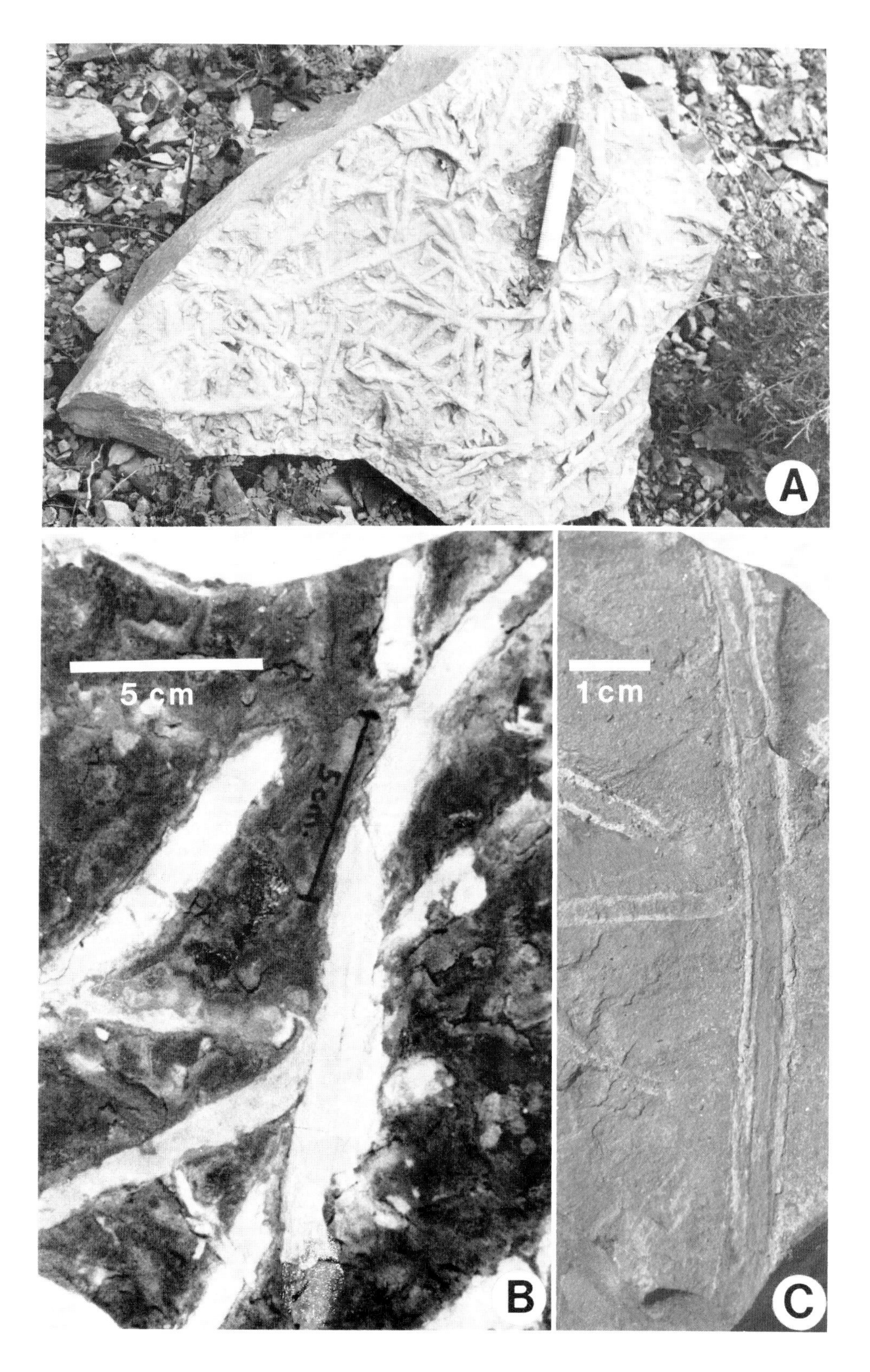
A
5 cm
5cm.
B
1 cm
C

PLANOLITES-like burrows
Fig. 5C

Description.—In the Three Lick Bed, traces resembling *Planolites* have thin rims that surround the burrow (Fig. 5C). This "*Planolites-like*" burrow is typically 12 cm long, 0.7 cm in diameter and does not branch. The burrow rims, in most cases, occupy one-fourth of the diameter of the burrow. The rims are composed of quartz, mica, organics, and clay, whereas the fill is composed of slightly coarser quartz, mica, organics, clay, and pyrite.

CHONDRITES Sternberg, 1833
Fig. 6A-D; 7A-B

Description.—*Chondrites* consists of a horizontal to inclined burrow system exhibiting branching (up to several orders) from a main stem. It has been interpreted as the trace of a deposit-feeding sipunculid worm having a retractable proboscis which allowed the animal to work from a fixed point to mine efficiently the substrate (Simpson, 1956). In all cases *Chondrites* represents simple, shallow feeding burrows.

Four types of *Chondrites* are recognized in this study. Type A (Fig. 6A) occurs in the lower part of the Huron Member below a dolomite bed in black shale. The specimens are very small (0.5–3.0 mm long, 0.05–0.5 mm wide) and show only two orders of branching at about 45°. Specimens extend 1–2 cm below the dolomite-shale interface and are filled with dolomite (see Fig. 11A). *Chondrites* Type A is cross-cut by other burrows.

Chondrites Type B occurs in black shale of the lower part of the Huron Member just below gray shale. Gray shale fills the burrows. Most commonly, *Chondrites* Type B is 1–4 cm long, from the base of a stem to the tip of the longest branch, and 1–5 mm wide. One to three orders of branching are common. *Chondrites* Type B has many forms: very straight, branching forms (Fig. 6B); short, stubby forms (Fig. 6C); and "crinoid-calyx" types (Fig. 6D).

TABLE 1.—TRACE FOSSILS FROM UPPER DEVONIAN BLACK SHALES IN EAST-CENTRAL KENTUCKY AND THEIR PALEOENVIRONMENTAL INTERPRETATIONS

STANDARD STRATIGRAPHIC SECTION			THIS STUDY		TRACE FOSSILS	ENVIRONMENTAL INTERPRETATION (AND ICHNOFACIES)	REMARKS
OHIO OR NEW ALBANY SHALE		CLEVELAND MEMBER		CLEVELAND MEMBER	NONE	OUTER SHELF TO SLOPE	NO OXYGEN
		THREE LICK BED		THREE LICK BED	PLANOLITES-LIKE*, CHONDRITES C&D, AND ZOOPHYCOS	OUTER SHELF TO SLOPE (LOWER CRUZIANA TO UPPER ZOOPHYCOS)	VARIABLE OXYGENATED ZONES (0.1-1.0 ml/l dissolved O_2) INTERMIXED WITH POORLY OXYGENATED ZONES
	HURON MEMBER	UPPER PART	HURON MEMBER	UPPER PART	NONE	MIDDLE TO OUTER SHELF?	NO OXYGEN
		MIDDLE PART			PLANOLITES*, CHONDRITES B*, RHIZOCORALLIUM, ZOOPHYCOS, AND TEICHICHNUS IN INTERBEDDED GRAY AND BLACK SHALE	MIDDLE TO OUTER SHELF (MIDDLE TO LOWER CRUZIANA)	VARIABLE OXYGENATED ZONES (0.1-1.0 ml/l dissolved O_2) INTERMIXED WITH POORLY OXYGENATED ZONES
		LOWER PART		LOWER PART	PLANOLITES*, CHONDRITES A*, ZOOPHYCOS*, TEICHICHNUS, TRICHICHNUS, RIBBED BURROW, CRUZIANA, PHYCODES, AND LAEVICYCLUS IN INTERBEDDED DOLOMITE AND SHALE	INNER TO MIDDLE SHELF (UPPER CRUZIANA)	
		BOYLE DOLOMITE		BOYLE DOLOMITE	PLANOLITES*, CRUZIANA, AND RUSOPHYCUS	PLATFORM OR INNER SHELF (UPPER CRUZIANA)	FULL MARINE WITH OXYGEN EXCEEDING 1.0 ml/l dissolved O_2.

*ABUNDANT

Fig. 5.—*Planolites* - *A*, Underside of dolomite in lower part of Huron Member. *Planolites* formed in black mud and was filled with carbonate (convex hyporelief). Section 3, State Route 52 near town of Winston, Estill County, Kentucky. *B*, *Planolites* on black shale bedding plane, filled with gray shale. Lower part of Huron Member, Section 7, entrance ramp to northbound Mountain Parkway, Clay City, Powell County, Kentucky. *C*, *Planolites*-like burrows on black shale bedding plane, filled with gray shale in the Three Lick Bed. Note thin rims which represent a size selecion of quartz, mica, and clays. Section 10, State Route 10, 3.9 miles west of Vanceburg, Lewis County, Kentucky.

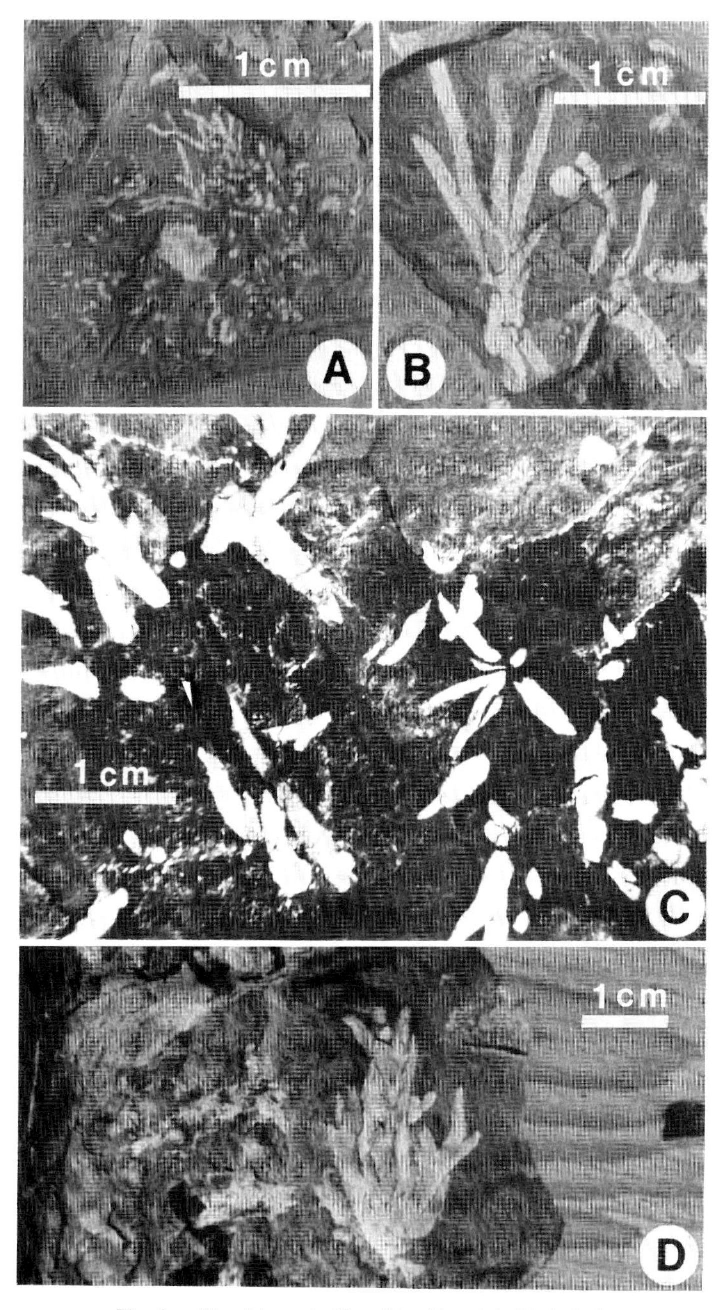

Fig. 6.—*Chondrites* - *A*, *Chondrites* Type A in black shale and filled with dolomite. Section 6, State Route 89, 1 mile north of Hargett, Powell County, Kentucky. *B*, Straight branching *Chondrites* Type B in black shale filled with gray shale. Section 9, Interstate 64 west of Morehead, Rowan County, Kentucky. *C*, Short, stubby *Chondrites* Type B. Section 9, Interstate 64 west of Morehead, Rowan County, Kentucky. *D*, "Crinoid-calyx" *Chondrites* Type B in the lower part of the Huron Member. Section 9, Interstate 64 west of Morehead, Rowan County, Kentucky.

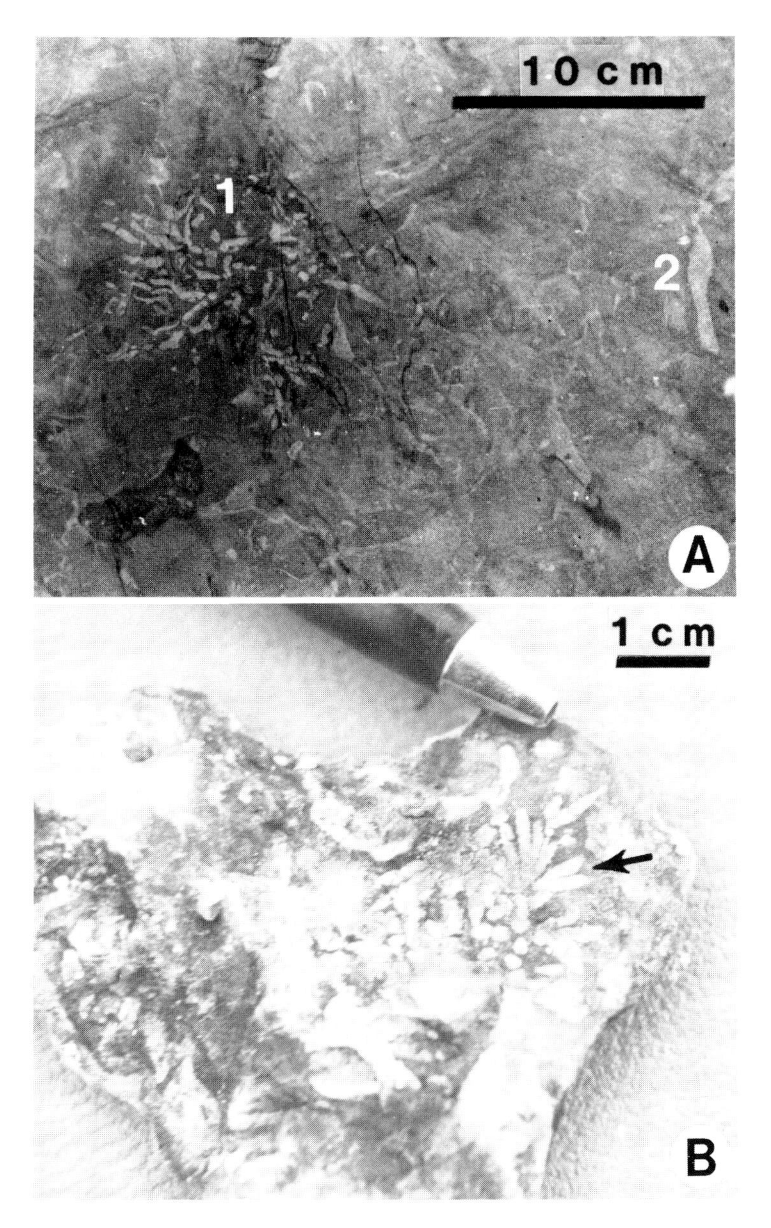

Fig. 7.—*A*, *Chondrites* Type C (1) in a creek bed, Section 8, Highway 965, 0.45 mile north of Hope, Kentucky. *Zoophycos* (2) also associated on bedding planes of the Three Lick Bed with *Chondrites*. *B*, *Chondrites*, Type D (arrow). Section 1, pit adjacent to exit ramp, Interstate 75 north, Berea, Madison County, Kentucky. Note that both types of *Chondrites* have many orders of branching, reflecting stringent conditions for highly specialized burrowing activity.-

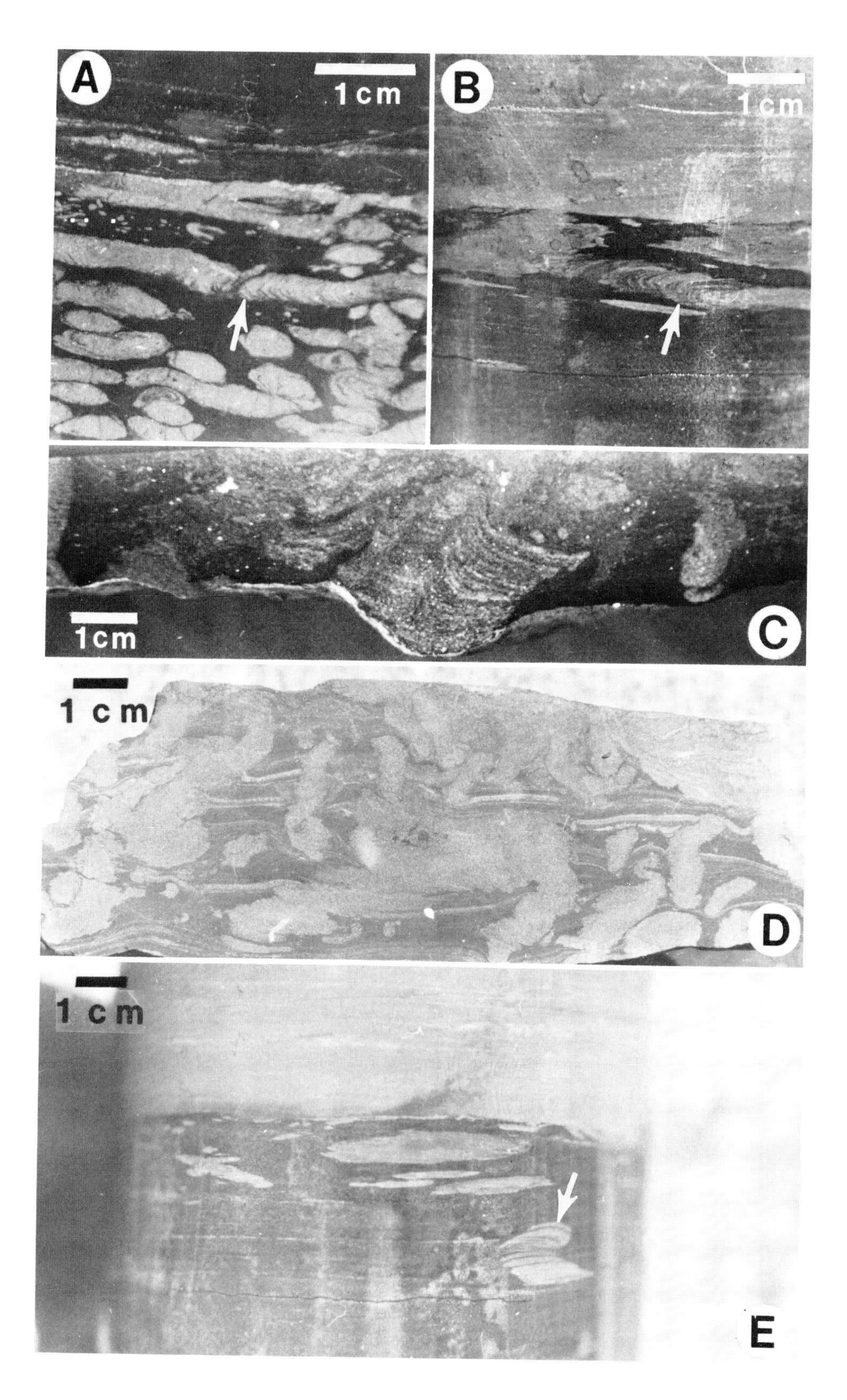
A
1 cm
B
1 cm
1 cm
C
1 cm
D
1 cm
E

Chondrites Type C (Fig. 7A) is found in black shale just below gray shale in the Three Lick Bed. It differs from other *Chondrites* by having four to five orders of branching. It is larger, with typical lengths of branches reaching 5 cm.

Chondrites Type D (Fig. 7B) is found in the Three Lick Bed. This form resembles Type A but is slightly larger. The stem is 5–10 mm long, branches are 1–2 cm long, and the width is 0.25–0.5 mm. Many orders of complex branching occur, as in Type C. Type D occurs in black shale and is filled with gray shale.

Simpson (1956) discussed two types of preservation associated with *Chondrites*. The first, burial preservation, occurs after burrowing where overlying sediment passively fills voids in the burrowed substrate. The second is concealed bed-junction preservation. Here again, overlying sediment passively fills the burrow. Later, however, the overlying sediment is removed by erosion. Further deposition of a different sediment type overlies the burrowed horizon.

This latter type of preservation occurs at many dolomite-shale interbeds (for example, see Fig. 11A). Many *Chondrites* (and other burrows) are filled with dolomite although no dolomite is present above. Instead, dolomitic shale containing isolated ripple marks formed of dolomite overlies the burrow zone. These cases demonstrate that currents may have been operating directly after the burrows were formed and filled, truncating the upper part of the burrow zone and removing carbonate (now dolomite).

ZOOPHYCOS Massalongo, 1855
Figs. 8A-B; 7A

Description.—*Zoophycos* has spreiten produced by reworking of sediment into a series of closely-spaced, J-shaped patterns. The spreiten consist of mined and unmined sediments. Each arc or "swoop" of the burrow represents the action of a vermiform organism as it moved in the substrate and ingested sediment. A new burrow segment was constructed alongside after the animal retracted and proceeded on a new course. This type of burrow represents the extreme type of efficient mining of a nutrient-rich substrate.

In interbedded dolomite and shale, the best specimens of *Zoophycos* are preserved in vertical section as tunnels lying roughly parallel to bedding (Fig. 8A). These tubes are up to 10 cm long and 0.5 cm wide. Spreiten are represented as alternating dark (unburrowed) and light (burrowed and filled) lunae.

In the gray and black shale of the Huron Member, *Zoophycos* shows similar size and form in sections parallel to bedding (Fig. 8B). *Zoophycos* occurs just below gray and black shale interfaces as well as in the gray shale where slightly lighter bands cut across a dark gray background.

In the Three Lick Bed, *Zoophycos* is broad and contains concentric laminae which are up to 3 cm thick (Fig. 7A). These laminae contain a few black interior wisps which represent spreiten.

TEICHICHNUS Seilacher, 1955
Fig. 8C-E

Description.—*Teichichnus* often first appears to be a vertical burrow, but closer inspection reveals that it represents the progressive vertical movement of a horizontal burrower. *Teichichnus* has been interpreted as a feeding burrow of a fixed worm which mined horizontally the substrate essentially from a fixed position. *Teichichnus* in interbedded dolomite and shale is commonly 1–5 cm high and 2 cm wide (Fig. 8C). On bedding planes, *Teichichnus* is filled with dolomite, and chevron organic layers (spreiten) represent tube building. The length of the tube is 5–10 cm.

A problem exists in identifying burrows at one section (Waco, Section 2). Interbedded dolomite and shale exhibit penecontemporaneous slumping (Fig. 4B). Vertical, poorly defined burrows which may represent *Teichichnus* have crude organic layers (spreiten) and horizontal tube components (Fig. 8D). These burrows were constructed in soupy mud because their outlines are neither sharp nor distinct. Dolomitic laminae are deflected upward and downward demonstrating horizontal mining beneath the muddy surface. Other burrows are lacking, probably due to the inhospitable conditions of the substrate.

Teichichnus also is recognized within black shale beneath the gray shale of the lower part of the Huron Member (Fig. 8E). It is smaller (1.0 cm high, 0.5 cm wide) than *Teichichnus* in the dolomite and shale. *Teichichnus* is not found in the Three Lick Bed.

RHIZOCORALLIUM Zenker, 1836
Fig. 9A

Description.—*Rhizocorallium* is an oblique to horizontal, unbranched U-shaped tube, containing spreiten. It is preserved as flattened tubes on bedding planes of black shale and filled with gray shale in the lower part of the Huron Member. *Rhizocorallium* has

Fig. 8.—*Zoophycos* and *Teichichnus* is the lower part of the Huron Member. *A, Zoophycos* in interbedded dolomite and shale (arrow). View is perpendicular to bedding. Section 4, 3.7 miles west of Irvine, Estill County, Kentucky. *B, Zoophycos* (arrow) in interbedded gray and black shales. View is perpendicular to bedding. Section 4, State Route 52, 3.7 miles west of Irvine, Estill County, Kentucky. *C, Teichichnus*, with spreiten, sculptured in black shale and filled with dolomite. View perpendicular to bedding. Section 2, State Route 52, near village of Waco, Madison County, Kentucky. *D*, Polished section cut perpendicular to bedding of *Teichichnus* showing mottled burrow shapes in interbedded dolomite and shale. Burrows were made in soupy mud and later filled with carbonate (now dolomite). See Figure 4B. Section 2, State Route 52, near village of Waco, Madison County, Kentucky. *E, Teichichnus* (arrow) in black shale filled with gray shale. View perpendicular to bedding. Core from Fleming County, Kentucky.

1 cm
A
5 cm
2
1
B
15 cm
C
1 cm
1
2
D

generally been attributed to a deposit-feeding crustacean (Seilacher, 1967; Crimes, 1975).

Rhizocorallium is easily recognized because of its U-shaped tubes lying on bedding planes (Fig. 9A). The tubes are filled with gray shale and usually each "arm" of the tube is 5 cm long and 0.25 cm wide. The distance between the "arms" is about 1.0 cm. Spreiten commonly are 0.25 mm thick and the intervening gray shale layers are 0.75 mm thick.

Rhizocorallium tubes probably extended subhorizontally from the gray shale until the lower end of the tube became horizontal, parallel to bedding. After compaction only the bottom part of the burrow was preserved.

PHYCODES Richter, 1850
Fig. 9B

Description.—*Phycodes* is a thick, tightly branching, bundled burrow which emanates from a common source and lacks spreiten. *Phycodes* represents the systematic mining by a worm having a fixed base moving downward and outward through black shale, with subsequent filling by overlying dolomite.

Phycodes is found in the interbedded dolomite and shale only. It occurs on the basal dolomudstone bed above the bottommost black shale (Fig. 9B). It usually has seven dolomite-filled branches from a main stem, with each branch being about 5.0 cm long. Apparently, the *Phycodes*-producing organism burrowed into black mud from a common, central point much like *Chondrites*. *Chondrites*, however, is dendritic and has thinner branches of several orders. In polished vertical sections, *Phycodes* is associated with *Chondrites* Type A (Fig. 11A).

LARGE, RIBBED BURROWS
Fig. 9C-D

Description.—Large, ribbed, gently curving burrow extending obliquely in shale occurs at one horizon only: below the first dolomite bed in the lower part of the Huron Member (Fig. 9C). It is commonly 10-100 cm long and 10 cm in diameter. It protrudes down from the base of the dolomite bed at an angle of 15° and is locally flattened to an ellipsoid due to compaction (Fig. 9D). Bifurcation is common. The burrow has lateral striations ("ribs") lying perpendicular to the long axis of the burrow. The burrow is attributed to some type of arthropod, based upon its size and digging marks.

Striations on the burrow are aligned perpendicular to the axis. Individual marks are 0.8 mm apart and are convex away from where the burrow protrudes from the dolomite bed. Knobs (1 mm in diameter) are found on the sides of the burrow where "ribs" become absent.

The "ribs" on these burrows represent claw marks of a digging animal, possibly a large arthropod. Knobs represent debris pushed aside by the animal which dug and piled sediment to the side of the burrow.

The only previous account of this burrow is by Berkheiser (1971) who stated that there is a remote possibility that it represents an aestivation tube of dipnoan fish. However, the striations and horizontal orientation of the burrow rule out this interpretation.

Edward Landing of the University of Toronto studied similar traces from the Middle Devonian Hungry Hollow Formation in Ontario (personal communication, 1979). He found that the burrows had a similar size and striations, but exhibited a bilobed habit with a median furrow. On this basis, he named the burrow *Cruziana*. He attributed the burrow to the trilobite *Dipleura*, whose thoracic segments were found in the rock.

The large, ribbed burrow found in the Upper Devonian sections is not bilobed and, hence, it is differentiated from the type of burrow described by Landing as *Cruziana*.

CRUZIANA d'Orbigny, 1842
Fig. 10A

Description.—*Cruziana* is an unbranched, flattened, bilobed burrow, generally attributed to the activity of trilobites, with a median furrow and striations perpendicular to the burrow axis. *Cruziana* is preserved on the basal dolomite of the lower part of the Huron member containing large, ribbed burrows (Fig. 10A). It differs from the latter in size (smaller by one half) and it is bilobed. The length of *Cruziana* ranges from 5-10 cm.

The form of *Cruziana* in the interbedded dolomite and shale indicates a head-down, plowing movement into black mud, as postulated by Seilacher (1970). The burrow begins in the dolomite bed, runs along the bottom of the dolomite at the black shale interface, and deflects back upward. Delicate scratch marks are preserved by dolomite infilling the burrow cast made in firm mud.

TRICHICHNUS Frey, 1970
Fig. 10B

Description.—*Trichichnus* is an unbranched to locally branched, subvertical, cylindrical tube which generally does not exceed 2 cm in length and 0.5 mm

Fig. 9.—*A*, *Rhizocorallium* on black shale bedding plane filled with gray shale, lower part of Huron Member. Note spreiten. Section 7, entrance ramp to northbound Mountain Parkway, Clay City, Powell County, Kentucky. *B*, *Phycodes* (1) on the bottom of a dolomite bed. Burrows were made in black shale and filled by carbonate. Note also abundant *Planolites* (2). Section 3, State Route 52, near town of Winston, Estill County, Kentucky. *C*, Large, ribbed burrows occur below the first dolomite bed above black shale in the lower part of the Huron Member. Section 3, Route 52, near town of Winston, Estill County, Kentucky. *D*, Large, ribbed burrow with "ribs" (1) representing digging marks of an arthropod, and "knobs" which represent the diggings pushed to the sides of the burrow. Section 3, Route 52, near town of Winston, Estill County, Kentucky.

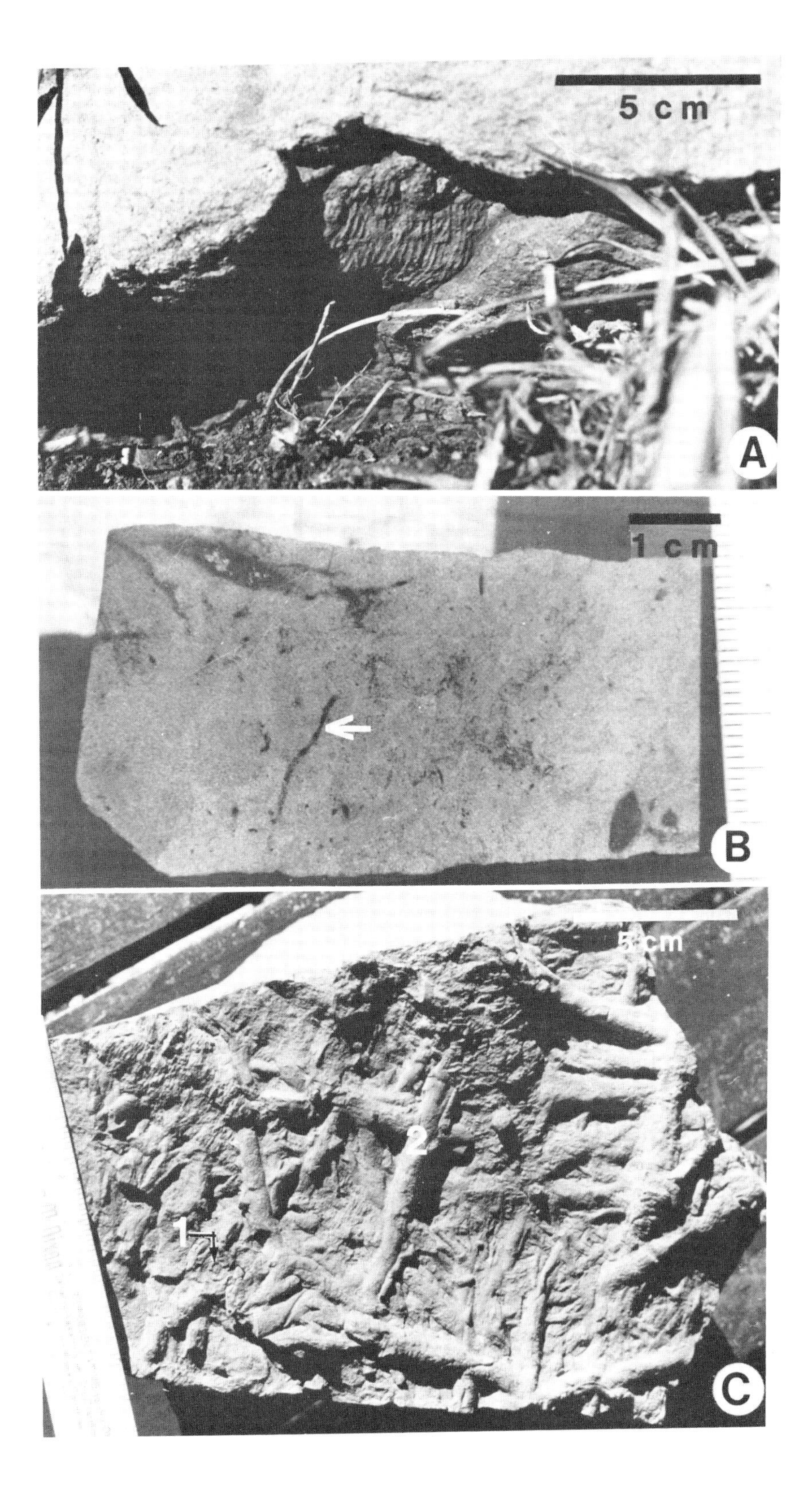
5 cm
A
1 cm
B
5 cm
2
1
C

in width. The origin of the trace fossil is still in doubt (Frey, 1970).

Trichichnus is found in dolomites of the lower part of the Huron Member. Pyrite fills the burrow and lines the burrow walls. The burrow is straight, inclined to bedding, and is dark brown against the gray dolomite background. The pyritic nature of the burrow probably represents decay and sulphide enrichment of organic slime along the burrow linings (Frey, 1970). It is commonly reported in marly or argillaceous sediments (Frey and Howard, 1970) and is widespread in chalks of the shelf environment (Kennedy, 1975).

LAEVICYCLUS Quenstedt, 1879
Fig. 10C

Description.—*Laevicyclus* is a button-like, circular, vertically spiral burrow typically 1 cm in diameter. *Laevicyclus* is found on the bottom of dolomite beds above black shales. Only the spiral upper part of the burrow is preserved as the burrow was made in black mud and filled with carbonate. Seilacher (1953) interpreted the trace fossil as a feeding burrow and compared it with the dwelling tube and scraping circles of Scolecolepsis, a Recent annelid worm (Häntzschel, 1962). Boyd (1975) noted the confusion between Recent feeding burrows similar to *Laevicyclus* and gas expulsion marks.

Laevicyclus commonly has been reported from ancient shallow water carbonates (Frey and Howard, 1970), as well as deep sea cores (Kennedy, 1975). It is associated in the lower part of the Huron Member with marine sediments interpreted as having been deposited in shallow water.

STRATIGRAPHIC DISTRIBUTION OF TRACE FOSSILS IN UPPER DEVONIAN SHALES

The remarkable feature of the distribution of trace fossils in the Upper Devonian black-shale sequence is the distinctive association of trace fossils in each of the stratigraphic units. The assemblages are summarized in Table 1.

The typical zonation of trace fossils in interbedded dolomite and shale is shown in Figure 11A. *Chondrites* Type A occurs just below a rippled erosion zone. *Planolites* is common below *Chondrites* and cuts through the small burrows. *Zoophycos* cuts across the shale above *Phycodes*.

A different assemblage of burrows is found in the interbedded gray and black shale of the lower part of the Huron Member. Figure 11B illustrates the typical assemblage of these beds. *Chondrites* Type B is filled with gray shale and occurs just below the interface of gray and black shale. It is associated with *Planolites* and *Rhizocorallium*. Immobile worm or worm-like organisms account for the shallowness of *Chondrites* Type A and *Rhizocorallium* (as well as *Zoophycos* and *Phycodes*).

In the Three Lick Bed, the diversity of burrow types is low (*Chondrites, Planolites, Zoophycos*) and only two sizes are represented. The abundance of one type (*Planolites*-like burrows) is great. *Chondrites* has two forms here. Types C and D are more complex than *Chondrites* from elsewhere in the section, and they are restricted to upper parts of the burrowed zone. *Zoophycos* is more intricate on the bedding planes than in the Huron Member.

DISCUSSION OF THE TRACE FOSSIL DISTRIBUTION PATTERNS

The various assemblages of trace fossils in Upper Devonian black shales (Fig. 12) are due to the varying oxygen levels in the basin. Burrows in modern aerated shallow-water environments are typically those of arthropods and worm-like organisms. Deeper-water burrows in poorly oxygenated environments are derived predominantly from worm-like organisms.

Byers (1973) found that as water depth changes with the configuration of basins, so does the oxygen in the water. As water shallows toward the basin margin, anoxic conditions give rise to better oxygenation. Benthic life is established with the organisms best adapted to initially low dissolved oxygen in the water column (small, infaunal polychaetes). As the basin shallows, body size, abundance, and diversity of organisms increases with increasing oxygen supply.

Based on its cross-cutting relationships, *Chondrites* Type A is interpreted as representing the pioneer burrower in the black mud. Later, infestation by larger infaunal deposit-feeders is represented by *Planolites, Zoophycos*, and *Phycodes*, which truncate *Chondrites* Type A. *Planolites* is the trace of a mobile deposit feeder and so it could explore the furthest reaches of the burrow zone. *Planolites* locally is vertical at the bottom of the zone, representing the animal's apparent attempt to 'test' the limits of the burrow zone, with the bottom being defined by the lowest limit of available oxygen. *Zoophycos* and *Phycodes*, both traces of immobile worms, efficiently mined the substrate in patterns near the top of the burrow zone.

Immobile worms or worm-like organisms also account for the shallowness of *Chondrites* Type A and *Rhizocorallium* (as well as *Zoophycos* and

Fig. 10.—*A*, *Cruziana* on underside of dolomite in the lower part of the Huron Member. Note well-preserved striations ("ribs"). Section 5, Highway 80, 0.4 mile east of Calloway Creek, near Irvine, Kentucky. *B*, *Trichichnus* (arrow) in dolomite from the lower part of the Huron Member. View is perpendicular to bedding. Section 2, State Route 52, near village of Waco, Madison County, Kentucky. *C*, Bottom of dolomite containing *Laevicyclus* (1) and *Planolites* (2). *Laevicyclus* was made in black mud and is filled by carbonate. Only the circular top of the spiral burrow is preserved on the base of the dolomite. Section 3, State Route 52, near town of Winston, Estill County, Kentucky.

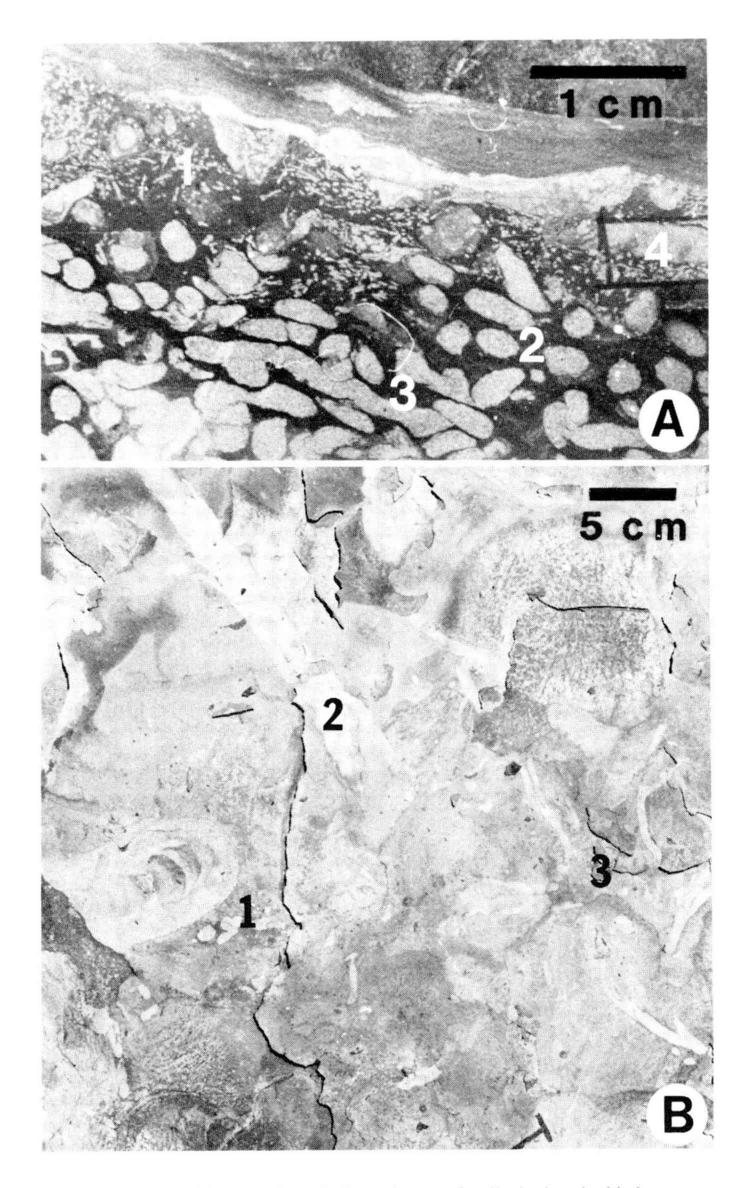

Fig. 11.—*A*, Association of trace fossils in interbedded dolomite and black shale, lower part of Huron Member. View is perpendicular to bedding. *Chondrites* Type A (1) was the first burrow in the black mud and followed by *Phycodes* (2), *Planolites* (3), and *Zoophycos* (4). Carbonate filled the burrows but was removed by erosion, resulting in concealed bed-junction preservation. Erosion is represented by a rippled zone. Black shale overlies the rippled zone. Section 4, State Route 52, 3.7 miles west of Irvine, Estill County, Kentucky. *B*, Association of trace fossils in interbedded gray and black shales, lower part of Huron Member. *Rhizocorallium* (1), *Planolites* (2) and *Chondrites* Type B (3) are common on a black shale bedding plane. Section 7, entrance ramp to north-bound Mountain Parkway, Clay City, Powell County, Kentucky.

Teichichnus) near the interface of black and gray shale in the interbedded black and gray shales in the lower part of the Huron Member. These animals needed to maintain a connection with the oxygenated water at the sediment/water interface. *Planolites*, on the other hand, represents the testing of the limits of oxygenation in the sediment, and, hence, is found below these burrows.

Trace fossils in the Three Lick Bed represent the exploitation of the substrate by a few deposit-feeding organisms. The intricate burrow systems may attest to a specialized and restricted type of burrowing habit.

Conditions were optimal for burrowers to live at certain horizons, for below these areas laminae are not disrupted. Where burrows extend down from gray shale and dolomite into black shale, vertical gradation occurs. Burrowing is abundant in the dolomite or gray shale bed and the entire bed is mottled and nonlaminated. Downwards, bioturbation is quite abundant at the interface of gray shale or dolomite and black shale, and burrows extend a few millimeters into the black shale. Below, burrowing decreases until, at the bottom of the zone of bioturbation, only a few isolated burrows are found. Laminated black shales are found below this. A gradient in the burrow zone is thus established.

The mud into which the burrows were made had to be firm enough to support the burrow walls, and allow overlying sediment to fill it in. Soft, water-laden mud would flow thixotropically and burrows would either not be preserved or would become mottled. The latter occurs at one section (see Fig. 4B). Animals burrowed into soupy mud, and burrows were filled with carbonate mud and preserved. Similar relationships were noted by Ghent and Henderson (1965) and Rhoads (1970).

Oxygen supplied to basinal sediments allowed organisms to inhabit the substrate and disrupt the sediment. Bioturbation and oxidation are reflected in the deposition of dolomite and gray shale while decreasing oxygen is reflected by the decrease in burrows vertically into black shale. An abrupt resumption of anoxic conditions is suggested by nonburrowed black shale which sharply overlies the bioturbated zones. This oxygenation event occurred periodically, given the abundance of black and gray shale interbeds (30-35 interbeds in many cases).

Griffith (1977) reported that the source of the oxygen which allowed burrowing animals to thrive in the sediments was possibly short-lived turbidity currents, as Sholkovitz and Soutar (1975) reported in the Santa Barbara Basin. While this is a possibility, no evidence of turbidity currents (graded bedding, scoured surfaces) is present. Fluctuations in the dysaerobic or oxygen-minimum zone (by whatever mechanism) contributed to periodic introduction of oxygen into the basin.

Byers (1973) noted that normal oxygenation in surface waters of today's oceans is about 7 ml/l dissolved oxygen (d.o.). Marine communities are little affected as this diminishes to 2 ml/l d.o. Further lowering excludes more bottom-dwelling fauna; animals become smaller in body size, less abundant, and less diverse.

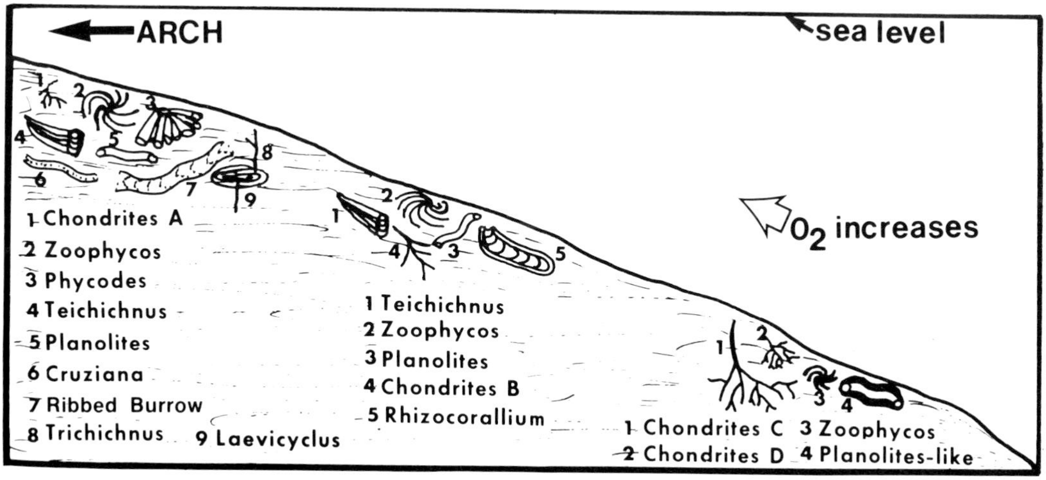

Fig. 12.—Composite diagram of burrow assemblages. Shallow-water burrows represent activities of arthropods and worm-like animals in oxygenated water, while deeper water trace fossils represent specialized deposit-feeding worms in progressively dysaerobic water. Periods of carbonate and gray shale deposition represent change to dysaerobic conditions at the sediment-water interface.

Small, vermiform, infaunal animals, which produce horizontal deposit-feeding burrows, dominate. Rhoads and Morse (1971) termed the full marine condition aerobic, from 0.1-1.0 ml/l d.o. dysaerobic, and below 0.1 ml/l anaerobic. They recognized that major changes in animal communities take place in the dysaerobic environment where calcareous-secreting animals are excluded. Today, this zone is dominated by protozoans, rotifers, gastrotrichs, tubellarians, nematodes, polychaetes, oligochaetes, lamellibranchs, gastropods, crustaceans, and tartigrades (Theede et al., 1969).

It is not difficult to imagine similar conditions when interpreting the burrows described in this study. For example, in the lower part of the Huron Member, very small infaunal burrows (*Chondrites* Type A) were established in black mud as oxygen became established on the anoxic sea bottom. Organic matter was consumed and oxidized as O_2 levels increased and gray shales formed. With further oxygenation, larger animals were established and their burrows cut the pioneering *Chondrites*.

Byers (1973) recognized burrows having circular halos which he interpreted as the result of a deposit-feeding vermiform animal exercising a size selection, ingesting clay and rejecting larger silt grains in its pathway. The "light" burrows of Byers (1973) which did not show a size selection of grains may correspond to the *Planolites* in the Huron member, whereas the "dark" burrows resemble *Planolites*-like burrows from the Three Lick Bed. The light burrows were found only in siltstones or claystones and were mechanically filled by them. Byers (1973) noted in the Appalachian Basin that these burrows occurred primarily in eastern, slightly coarser beds close to shore. Westward, in the deeper parts of the basin, in silty claystones, only dark burrows were found. This area was less favorable for organisms, but it was still able to support life. Burrow size, diversity, and abundance increased as siltier beds were encountered as the basin shallowed (Byers, 1973). The type of burrowing reflected the oxygen gradients, the dark burrows signifying less aerobic conditions in the distal parts of the basin.

PALEODEPOSITIONAL ENVIRONMENT AND OXYGEN LEVELS OF THE UPPER DEVONIAN BLACK SHALES IN EAST-CENTRAL KENTUCKY

Based upon the trace fossil assemblages, sedimentology, and stratigraphy, an upwards-deepening, transgressive setting is proposed for the Upper Devonian black-shale sequence in Kentucky. The Middle Devonian section records the initial infilling of the Appalachian Basin by deltaic sandstones, siltstones, and shales derived form eastern highlands (Fig. 13). By late Middle Devonian time, deposition of fine-grained material allowed gradual filling towards the western margin of the basin. The Boyle Dolomite was deposited on a regional unconformity along the flanks of a proto-Arch in Ohio and Kentucky. Although supratidal and intratidal environments have been reported in these rocks (Berkheiser, 1971), emergence of the Arch is questionable. Shallow subtidal environments containing rip-up clasts, grainstone mounds, cross-bedded dolograinstones, and a trace fossil assemblage of *Cruziana*, *Rusophycus*, and *Planolites*, suggest that a proto-Arch created a shoal area within the Devonian epicontinental sea.

Fine-grained clastic sediment was deposited on the carbonate platform during the Late Devonian. The interbedded dolomites and shale of the lower part of the Huron Member denote the shifting of paleoenvironments. The oxygenated environment represented by the deposition of dolomite (primary?) allowed small burrowers (*Chondrites* Type A) to burrow into the underlying substrate. Higher levels of dissolved oxygen in the shifting environments permitted large, more diverse infaunal deposit-feeding organisms to thrive. Shallow marine subtidal environments (inner to middle shelf) are recorded by traces of arthropods (large, ribbed burrows; *Cruziana*) and worm-like organisms (*Planolites*, *Zoophycos*, *Teichichnus*, *Laevicyclus*, *Trichichnus*, and *Phycodes*).

As transgressing seas further brought reducing conditions to the platform area, carbonate deposition ceased. While anoxic black muds accumulated on the bottom, planktonic and nektonic organisms abounded in the upper oxygenated waters of the basin and sank to the bottom after death. In the lower part of the Huron Member, shifts in the partially oxygenated (dysaerobic) zone are reflected by oxidized, totally bioturbated gray shale and partially burrowed organic-rich black shale.

Trace fossils indicative of a muddy, subtidal setting (middle to outer shelf) became established in the temporarily oxygenated environment. Infaunal deposit-feeding animals left essentially horizontal traces (*Planolites*, *Chondrites* Type B, *Teichichnus*, *Rhizocorallium*, *Zoophycos*), reflecting gradients from small to large benthos as oxygen levels increased. These traces indicate the middle to outer areas of the

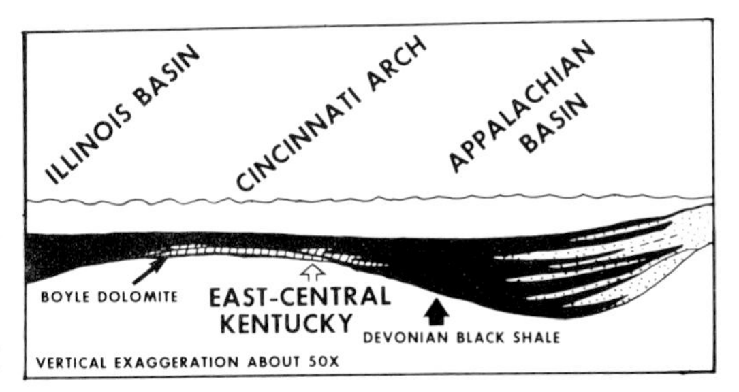

Fig. 13.—Model of Late Devonian deposition in the Appalachian Basin. Note interfingering of carbonate and black shale near the present Cincinnati Arch. Modified after Griffith.

Cruziana facies (predominance of deposit-feeding organisms).

Resumption of anoxic conditions is represented by deposition of black muds. Occasional oxygenation events (additional movement of the dysaerobic zone) represented by the omnipresent Three Lick Bed allowed burrowing to resume. Decreased diversity and increased abundance of trace fossils reflect deeper water conditions containing low levels of oxygen. *Planolites*-like burrows and complex feeding burrows (*Chondrites* Type C and D, *Zoophycos*) reveal outer shelf to slope environments in the lower *Cruziana* and upper *Zoophycos* ichnofacies. Deposit feeding appears to have become specialized and efficient, possibly to ensure maximum coverage of the substrate under less favorable conditions. Deposition of mud in an anoxic environment resumed, represented by the overlying Cleveland Member, and burrowing ceased.

Deltaic deposition finally reached the east-central Kentucky area at the end of the Late Devonian and beginning of Early Mississippian time. Prodeltaic mud (Bedford and Sunbury Shale Formations) and sands (Berea Sandstone Formation) in advance of the deltas replaced black mud as the dominant type of deposit as sedimentation exceeded basin subsidence.

CONCLUSIONS

Trace fossils in the Upper Devonian black shales of east-central Kentucky were formed during brief periods of oxygenation, possibly reflecting movements in the dysaerobic zone. The burrow assemblages in interbedded dolomite and shale suggest shallow water conditions, while the lower diversity burrow assemblages upsection in interbedded gray and black shale reflect progressively deepening water. Many of the burrowers in the black shale are characteristic of mud deposited in restricted environments: small, vermiform, infaunal deposit feeders. These trace fossils are commonly reported in clastic and carbonate environments as indicating shelf to slope water depths.

Trace fossil sequences can be recognized in shales. After all, the ichnofacies model proposed by Seilacher (1967) deals in part with burrows of deposit feeders which are recognized in shales interbedded with sandstones or limestones. One only has to remove the coarse clastic or carbonate input and establish an "Oxygenated Mud Gradient." Ekdale (this volume) also notes this "anaerobic ichnofacies model."

This study of Upper Devonian shales establishes a general tool for the refinement of an internal stratigraphy. Trace fossil assemblages in interbedded gray and black shales, for example, can help differentiate units (Three Lick Bed versus Huron Member) containing the interbeds. Future work in the study of trace fossils in black shale could possibly resolve basinal equivalents to the east-central Kentucky area and establish oxygen gradients in the Devonian black shales of easternmost Kentucky, southwestern Virginia, and West Virginia.

ACKNOWLEDGMENTS

This research was part of a master thesis at the University of Cincinnati and was partially supported by the U.S. Department of Energy under contract ERDA E-(40-1)-5201, the University of Cincinnati Sedimentology Fund, and the Fenneman Fund. I would like to thank W. A. Pryor for suggesting the project and guiding the course of the study, and P. E. Potter and D. L. Meyer for their helpful discussions of shale sedimentology. Roy C. Kepferle of Eastern Kentucky University provided many ideas on the stratigraphy of the area. J. Todd Stephenson of Amoco Production Company in Houston supplied information on Middle Devonian paleoenvironments in Kentucky. C. W. Byers and R. M. Slatt critically reviewed the manuscript. Maggie Draughon kindly typed the manuscript.

REFERENCES

Ayrton, W.G., 1963, Isopach and lithofacies map of the Upper Devonian of northeastern United States, *in* Symposium on Middle and Upper Devonian Stratigraphy of Pennsylvania and Adjacent States: Penn Geol. Survey Bull. G-39, p. 3–6.

Berkheiser, S.W., Jr., 1971, Petrographic analysis of the Boyle Dolomite (Devonian) of eastern Kentucky [unpub. M.A. thesis]: Richmond, Eastern Kentucky Univ., 100 p.

Boyd, D.W., 1975, False or misleading traces, *in* Frey, R.W. (ed.), The Study of Trace Fossils: New York, Springer-Verlag, p. 65–83.

Byers, C.W., 1972, Analysis of paleoenvironments in Devonian black shale by means of biogenic structures (abs.): Geol. Soc. America Ann. Mtg., Abs. with Programs, v. 7 (4), p. 464.

______, 1973, Biogenic structures of black shale paleoenvironments: [unpub. Ph.D. thesis]: New Haven, Yale Univ., 134 p.

Campbell, G., 1946, New Albany Shale: Geol. Soc. Amer. Bull., v. 57, p. 829–908.

Cluff, R.M., 1980, Paleoenvironment of the New Albany Shale group (Devonian-Mississippian) of Illinois: Jour. Sed. Petrology, v. 50, p. 767–780.

Conant, L.C., and V.E. Swanson, 1961, Chattanooga Shale and related rocks of central Tennessee and nearby areas: U.S. Geological Survey Prof. Paper 357, 91 p.

Corbo, S., 1977, Distribution and significance of trace fossils in Upper Devonian Glen Aubrey Formation (turbidite to shelf transition), New York: Amer. Assoc. Petroleum Geologists Bull., v. 61, no. 5, p. 774.

Crimes, T.P., 1975, The stratigraphic significance of trace fossils, *in* Frey, R.W. (ed.), The Study of Trace Fossils: New York, Springer-Verlag, p. 109–130.

FREY, R.W., 1970, Trace fossils of Fort Hays Limestone Member of Niobrara Chalk (Upper Cretaceous), west-central Kansas: Univ. Kansas Paleont. Contr. Art. 53, 41 p.

______, AND J.D. HOWARD, 1970, Comparison of Upper Cretaceous ichnofaunas from siliceous sandstones and chalk, Western Interior Region, U.S.A., in Crimes, T.P. and J.C. Harper (eds.), Trace Fossils: Geol. Jour., Spec. Issue No. 3, p. 141–166.

GHENT, E.D., AND R.A. HENDERSON, 1965, Significance of burrowing structures in the origin of convolute laminae: Nature, v. 207, p. 1286–87.

GRABAU, A.W., 1906, Types of sedimentary overlap: Geol. Soc. America Bull., v. 17, p. 567–636.

______, 1919, Significance of the Sherburne sandstone in Upper Devonian stratigraphy: Geol. Soc. America Bull., v. 30, p. 423–470.

GRIFFITH, C., 1977, Stratigraphy and paleoenvironment of the New Albany Shale (Upper Devonian) in north-central Kentucky [unpub. M.S. thesis]: Madison, Univ. of Wisconsin, 214 p.

GROSSMAN, W.L., 1944, Stratigraphy of the Genesee Group of New York: Geol. Soc. American Bull., v. 55, p. 41–76.

HÄNTZSCHEL, W., 1962, Trace fossils and problematica, *in* R. C. Moore (ed.), Treatise on Invertebrate Paleontology, pt. W, Miscellanea: Lawrence, Kansas, Geol. Soc. America and Univ. Kansas Press, p. W177–W245.

HARD, W.W., 1931, Black shale deposition in central New York: Amer. Assoc. Petroleum Geologists Bull., v. 15, p. 165–181.

HOLBROOK, C.E., 1964, Stratigraphic relationships of the Silurian and Devonian in Clark, Powell, Montgomery, and Bath counties [unpub. M.S. thesis]: Lexington, Univ. of Kentucky, 79 p.

JORDAN, D.W., 1979, Trace fossils and stratigraphy of the Devonian black shales in east-central Kentucky [unpub. M.S. thesis]: Cincinnati, Univ. of Cincinnati, 227 p.

KENNEDY, W.J., 1975, Trace fossils in carbonate rocks, *in* Frey, R.W. (ed.), The Study of Trace Fossils: New York, Springer-Verlag, p. 377–398.

KINDLE, E.M., 1912, The stratigraphic relations of the Devonian shales of northern Ohio: Amer. Jour. Science, v. 184, p. 187–213.

MCFARLAN, A.C., AND W.H. WHITE, 1952, Boyle-Duffin-Ohio Shale relationships: Kentucky Geol. Survey Bull., ser. 6, no. 10, p. 5–24.

OLIVER, W.A., JR., W. DEWITT, JR., J.A. DENNISON, D.M. HOSKINS, AND J.W. HUDDLE, 1967, Devonian of the Appalachian Basin, United States, *in* Oswald, D.H. (ed.), International Symposium on the Devonian System, Calgary, 1967: Calgary, Alberta Soc. Petroleum Geologists, v. 1, p. 1001–1040.

ORTON, E., 1882, A source of the bituminous matter of the black shales of Ohio: Proc. American Assoc. Adv. Science, v. 31, p. 373–384.

POTTER, P.E., J.B. MAYNARD, AND W.A. PRYOR, 1980, Final report of special geological, geochemical, and petrological studies of the Devonian shales in the Appalachian basin: Department of Energy, Eastern Gas Shales Project, Contract DE-AC21-76MC05201, 81 p.

PROVO, L.J., 1977, Stratigraphy and sedimentology of the radioactive Devonian-Mississippian shales of the central Appalachian basin [unpub. Ph.D. thesis]: Cincinnati, Univ. of Cincinnati, 177 p.

______, R. C. KEPFERLE, AND P.E. POTTER, 1978, Division of black Ohio Shale in eastern Kentucky: Amer. Assoc. Petroleum Geologists Bull., v. 62, no. 9, p. 1703–1713.

RHOADS, D.C., 1970, Mass properties, stability, and ecology of marine muds related to burrowing activity, *in* Crimes, T. P. and J.C. Harper (eds.), Trace Fossils: Geol. Jour. Spec. Issue No. 3, p. 391–406.

______, AND J.W. MORSE, 1971, Evolutionary and ecologic significance of oxygen-deficient marine basins: Lethaia, v. 4, p. 413–428.

RODRIGUEZ, J., AND R.C. GUTSCHICK, 1970, Late Devonian-Early Mississippian ichnofossils from western Montana and northern Utah, *in* Crimes, T.P. and J.C. Harper (eds.), Trace Fossils: Geol. Jour., Spec. Issue No. 3, p. 407–438.

SAVAGE, T.E., 1930, The Devonian rocks of Kentucky: Kentucky Geol. Survey, Ser. 6, 257 p.

SCHOPF, J.M., AND J.F. SCHWIETERING, 1970, The *Foerstia* Zone of the Ohio and Chattanooga Shales: U.S. Geol. Survey, Bull. 1294-H, 15 p.

SCHWIETERING, J.F., 1970, Devonian shales of Ohio and their eastern equivalents [unpub. Ph.D. thesis]: Columbus, Ohio State Univ., 100 p.

______, 1979, Devonian shales of Ohio and their eastern and southern equivalents: Morgantown Energy Research Center, MERCI CR-79-2, Energy Research and Development Administration, 68 p.

SEILACHER, A., 1953, Studien zur Palichnologie. I. Uber die Methoden der Palichnologie: Neues Jahrb. Geol. Paläont., Abhandl., v. 96, p. 421–452.

______, 1967, Bathymetry of trace fossils: Marine Geology, v. 5, p. 413–428.

______, 1970, *Cruziana* stratigraphy of "non-fossiliferous" Paleozoic sandstones, *in* Crimes, T.P. and J.C. Harper (eds.), Trace Fossils: Geol. Jour., Spec. Issue No. 3, p. 447–476.

SHOLKOVITZ, E., AND A. SOUTAR, 1975, Changes in the composition of the bottom water of the Santa Barbara Basin: effect on turbidity currents: Deep Sea Research, v. 22, p. 13–21.

SIMPSON, S., 1956, On the trace-fossil *Chondrites*: Quart. Jour. Geol. Soc. London, v. 112, p. 475–499.

SUMMERSON, C.H., AND D.H. SWANN, 1970, Patterns of Devonian sand on the North American craton and their interpretation: Geol. Soc. America Bull., v. 81, p. 469–490.

SWAGER, D.R., 1978, Stratigraphy of the Upper Devonian-Lower Mississippian shale sequence in the eastern Kentucky outcrop belt [unpub. M.S. thesis]: Lexington, Univ. of Kentucky, 116 p.

THEEDE, H., A. PONAT, K. HIROKI, AND C. SCHLIEPER, 1969, Studies on the resistance of marine bottom invertebrates to oxygen deficiency and hydrogen sulfide: Marine Biology, v. 2, p. 325–337.

WELLS, J.W., 1944, Middle Devonian bone beds of Ohio: Geol. Soc. America Bull., v. 55, 18 p.

ICHNOLOGIC, TAPHONOMIC, AND SEDIMENTOLOGIC CLUES TO THE DEPOSITION OF CINCINNATIAN SHALES (UPPER ORDOVICIAN), OHIO, U.S.A.

DANITA BRANDT VELBEL
Department of Geology, University of Cincinnati, Cincinnati, Ohio 45221

ABSTRACT

The depositional origin of Cincinnatian shales (Upper Ordovician) of Ohio has long been problematical. Some of these shales are characterized by a patchy distribution of *Chondrites* burrows, a lack of fissility, and an abundant, well-preserved trilobite fauna (mostly *Flexicalymene* and *Isotelus*). The spatial distribution and density of biogenic structures in these "trilobite shales" help clarify the relative timing of deposition and burrowing of the shales. The excellent preservation of the trilobite body fossils indicates that their burial by the mud was instantaneous. Petrographic thin-sections of these shales show decreasing density of fossil fragments upward within some shale horizons. These observations indicate rapid deposition of the "trilobite shales." The Cincinnatian shales compare favorably with other fossil localities, famous for excellent preservation of a fossil fauna in shale, which have been interpreted as turbidites. The full suite of sedimentary structures characteristic of classical turbidites is absent in the "trilobite shales." The ichnologic, taphonomic, and sedimentologic features of these shales, however, provide data that bear on new views of the rapid deposition of fine-grained sediment.

INTRODUCTION

After more than 72 years of debate, the origin of the alternating sequence of limestone and shale which characterizes the Cincinnatian Series is still unresolved. Theories include: deposition of the limestone as storm lags of skeletal fragments and of the shales by settlement of clay-sized particles from suspension (Bucher, cited in Rich, 1951); and characterizing the limestones as slow accumulations of indigenous fauna and the shales as higher-energy terrigenous influxes (Scotford, 1965). This study investigates the deposition of some Cincinnatian shales by examing their ichnologic, taphonomic, and sedimentologic characteristics.

Certain Upper Ordovician shales of the Cincinnatian Series near Cincinnati, Ohio are distinguished by an unusually abundant, well-preserved trilobite fauna. These "trilobite shales" occur at different stratigraphic horizons throughout the Cinncincinnatian Series. The thickness and lateral extent of these shales is difficult to determine because these sediments appear to be mineralogically and texturally identical to laterally and vertically adjacent shales that do not contain the abundant trilobite fauna. One such "trilobite shale" (Locality G, see Figs. 1 and 2) yielded several thousand specimens of *Flexicalymene*. These unusual trilobite occurrences raise questions that also bear on the deposition of these shales: Do the trilobites represent an autochthonous assemblage; i.e., did the trilobites die and slowly become buried *in situ* by the mud? Or do the trilobites represent an allochthonous assemblage, and does their preservation reflect rapid burial by mud?

The study area comprises the tri-state region of southwestern Ohio, southeastern Indiana, and north-central Kentucky (Fig. 1). Four "trilobite shales" were located during this investigation; their stratigraphic positions are indicated in Fig. 2 (see Appendix for precise location of the collecting sites). At these four localities trilobites were collected *in situ* (Fig. 3). These data are supplemented by a large (over 1200 specimens) collection from the University of Cincinnati Geology Museum. The museum collection is designated "G" on the locality map and stratigraphic column. Only the data from the four *in situ* collection localities (A, B, C, D) were used in the frequency histograms described in the "taphonomy" section.

SEDIMENTOLOGY

Macroscopically, the "trilobite shales" show no obvious bedding or lamination; they might be better termed "mudstones" in the sense of Lundegard and Samuels (1980). This homogeneous texture is disrupted at irregular intervals by thin (on the order of centimeters), discontinuous concentrations of fossils, commonly comprising small brachiopods and crinoid stems. The shales weather in a characteristic "blocky" pattern (Fig. 4), although fresh surfaces break conchoidally.

Mineralogically, the "trilobite shales" do not differ from other Cincinnatian shales. The primary minerals in the shale are quartz, illite, chlorite, and pyrite. The blue-gray color reflects a low organic content. Petrographic thin-sections of the shale show intervals of upward-fining fossil fragments (Fig. 5).

[1]Present address: Department of Geology, Yale University, New Haven, CT. 06511.

ICHNOLOGY

The "trilobite shales" are incompletely disrupted by biogenic structures. The sediment is not completely homogenized by bioturbation; rather, individual branches of *Chondrites* burrows can be distinguished in outcrop. Horizontal, oblique, and vertical sections of *Chondrites* burrows have a patchy distribution within the same shale horizon. The burrows are at least slightly compressed but are still recognizable in outcrop because they are filled with a lighter-colored mud. Figure 6 shows a longitudinal cross-section of a portion of a *Chondrites* burrow from these shales.

Osgood (1970) listed three different *Chondrites* burrow morphologies from the Cincinnatian of Ohio which he designated *Chondrites* types A, B, and C. His described specimens are from siltstones, but the *Chondrites* burrows in the "trilobite shales" most closely resemble Osgood's *Chondrites* type B (Fig. 7). The branches of these burrows may or may not interpenetrate; the tunnel systems range in position from oblique to parallel to bedding. The branching pattern appears to be random. The burrows have smooth walls and do not exceed a few millimeters in diameter; most are 1 mm or less in diameter. Osgood found that the sediment filling the *C.* type-B burrows was finer-grained than the sediment of the host rock. This difference is not observed in the *Chondrites* found in the "trilobite shales"; both burrow filling and host sediment are of similar size.

The vertical and lateral extent of individual *Chondrites* networks is difficult to determine. The burrows do not weather differentially, and the host "rock" is incompetent. The best approach to studying the extent of the burrow network is to impregnate the shale with epoxy, then section and X-ray the block of shale. This technique was not highly successful during the course of this study, and only fragments of *Chondrites* networks were recovered.

Limestones and siltstones above and below the "trilobite shales" show the same patchy distribution of biogenic structures but contain at least two additional ichnogenera: *Diplocraterion* and *Rusophycus*.

TAPHONOMY

The trilobites *Flexicalymene* and *Isotelus* are the most conspicuous fossils in these shales because of

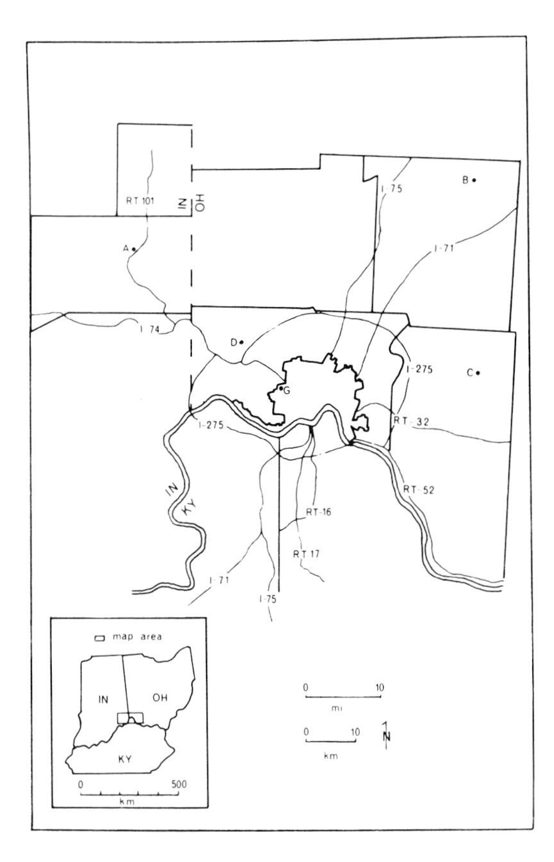

Fig. 1.— Locality map. City limits of Cincinnati in bold outline. For locality names see Appendix.

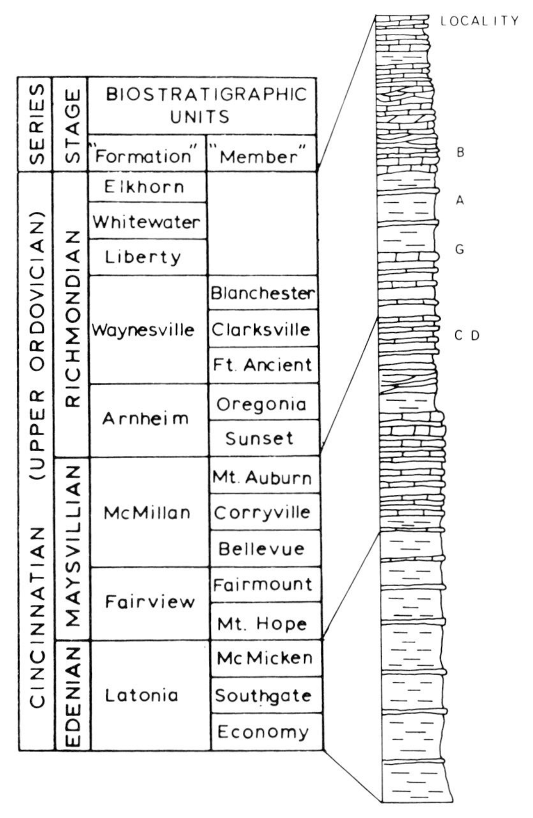

Fig. 2.—Relative stratigraphic positions of collecting localities.

their relatively large size and unusual abundance. Whole and fragmentary trilobite body fossils lie at different attitudes with respect to bedding; for example, dorsal or ventral surface up, or at some intermediate position. All orientations can occur in a single shale horizon, and all are present at each of the four localities (Fig. 8). *Flexicalymene* occurs enrolled (i.e., with the thorax curved so that the ventral surfaces of the cephalon (head shield) and pygidium (posteriormost segment) are in contact) and not enrolled. Nonenrolled trilobites greatly outnumber enrolled trilobites at the four *in situ* collecting localities (Fig. 9), but almost all the specimens from the museum collection (Locality G) are enrolled. The museum sample proba-

Fig. 3.—Trilobite *in situ*, ventral side up. Centimeter scale.

Fig. 4.—Weathered, vertical exposure of the "trilobite shale" showing its characteristic "blocky" weathering. Scale is 6 inches.

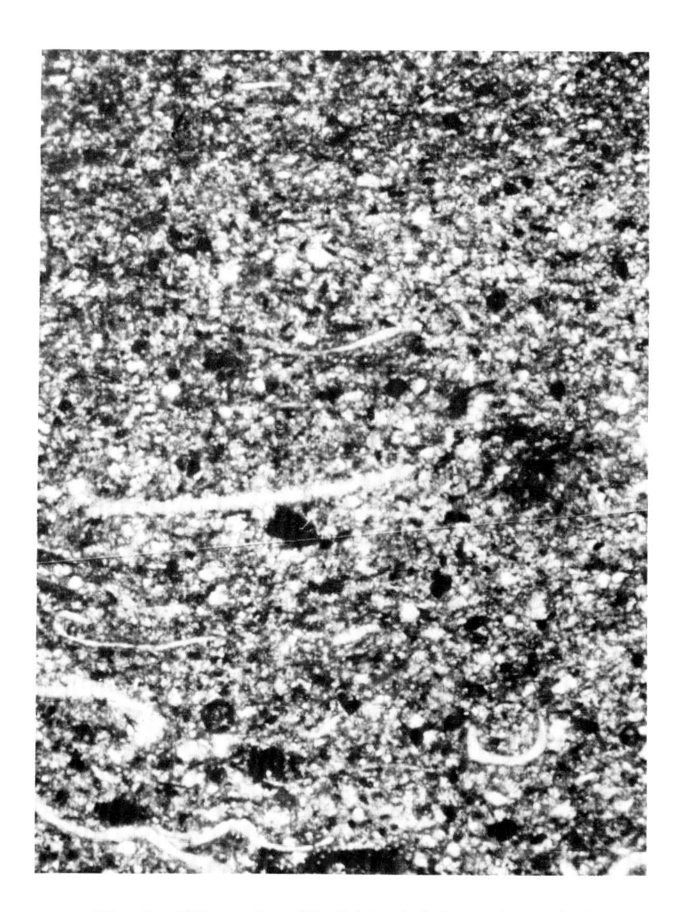

Fig. 5.—Thin-section of "trilobite shale" showing vertical grading of fossil fragments. Scale: field of view (long dimension) = 4.5 mm.

bly reflects weathering or collecting bias, because these trilobites were not collected *in situ*; therefore, these trilobites were omitted from Figure 9. The number of dissociated cranidia and pygidia do not differ significantly within localities (Fig. 10). A range of sizes of trilobites is present at each locality (Fig. 11) and there does not appear to be a consistent size sorting. Elongate trilobites and other fossil fragments show a roughly northwest-southeast orientation, without preference for anterior or posterior alignment (Fig. 12). Trilobites are well-preserved (i.e., whole body fossils with articulated thoracic segments are common) but most specimens are deformed, compressed in the plane of bedding. A survey of 464 trilobites from the museum collection (Locality G) revealed that these trilobites do not show a tendency to deform in a preferred direction, but that their orientation prior to compaction of the mud was random with respect to bedding (Fig. 13). This assumes that the deformation of the trilobites was due to compaction perpendicular to the bedding.

Trilobites are found throughout these shales, but they tend to occur in clusters of two or more, rather than singly. Other fossils are also common at each locality and are irregularly distributed within these shales. At Locality A, for example, the epifaunal pelecypod *Ambonychia* and shallow infaunal pelecypod *Modiolus* are preserved in two orientations: (1) with open, articulated valves lying convex-up in the plane of bedding, a post-mortem position; and (2) with closed valves oriented at an angle to bedding, possibly representing a "life position." The infaunal brachiopod *Lingula* is found oriented horizontally, with the commissure plane parallel to the bedding. In life, modern *Lingula* is oriented vertically. The small brachiopods *Onniella* and *Zygospira* appear to be randomly oriented and are generally associated with the trilobites. *Cornulites*, conularids, and crinoid fragments are also found in these shales, but the overall faunal diversity in the shales is low. Faunal abundance, with the exception of the trilobites, is also generally low.

INTERPRETATION

The density of biogenic structures is a function of turnover rate of the population of potential trace-

Fig. 6.—Longitudinal section through a horizontal *Chondrites* burrow. Scale: field of view (long dimension) = 4.5 mm.

Fig. 7.—*Chondrites* type-B. Bar scale = 1 cm. Redrawn after Osgood, 1970.

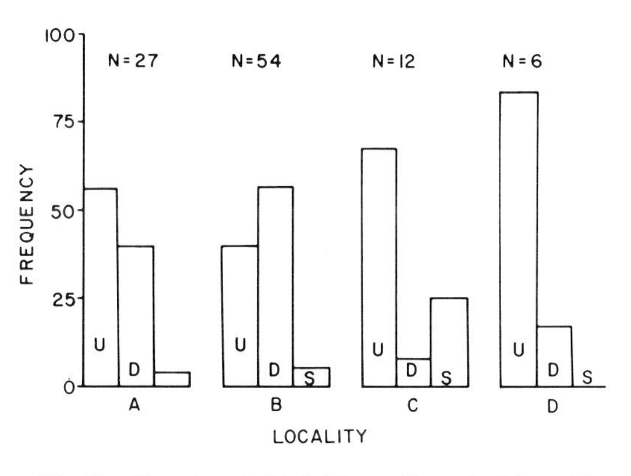

Fig. 8.—Dorsal-ventral trilobite positions. In this and the subsequent frequency diagrams the data are based on counts of *Flexicalymene* specimens. U = dorsal side up, D = dorsal side down, S = on side. Frequencies in percent of individuals.

makers, the mobility of the burrowing organisms, and the sedimentation rate (Rhoads, 1975, p. 149). The absence or relatively low density of biogenic structures in shale may be interpreted as the result of either an inhospitable environment for burrowing organisms during shale deposition (for example, low oxygen content; see Rhoads, 1975) or rapid deposition of the mud, which prevented disruption of the depositional fabric by the burrowing organisms (Howard, 1975). The relatively abundant epifaunal and shallow-infaunal pelecypod fauna in the "trilobite shales" and the low organic content of these shales suggest that the muds were deposited in an oxygenated rather than anaerobic environment. In such an aerobic environment, under conditions of slow sedimentation, one would predict

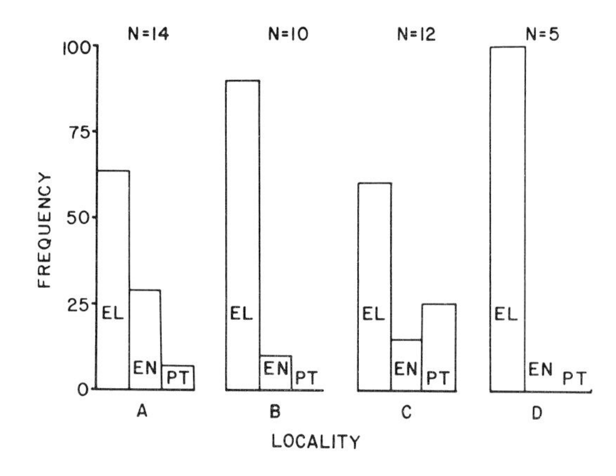

Fig. 9.—Trilobite enrollment frequencies: EL = elongate trilobites, EN = enrolled trilobites, PT = partially enrolled trilobites. Frequencies in percent of individuals.

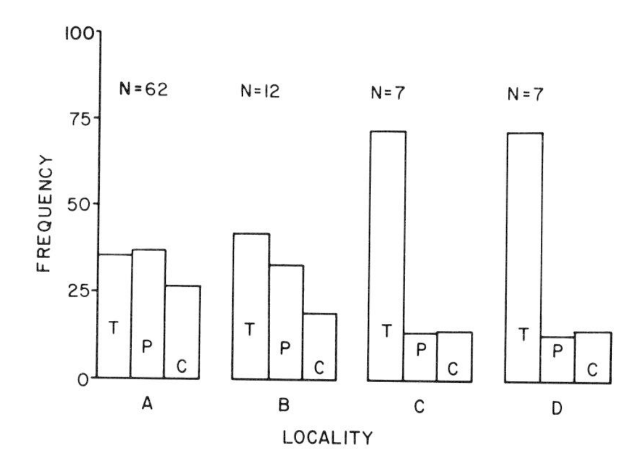

Fig. 10.—Occurrence of trilobite exoskeletal parts. T = thorax, P = pygidium, C = cephalon. Frequencies in percent of individuals.

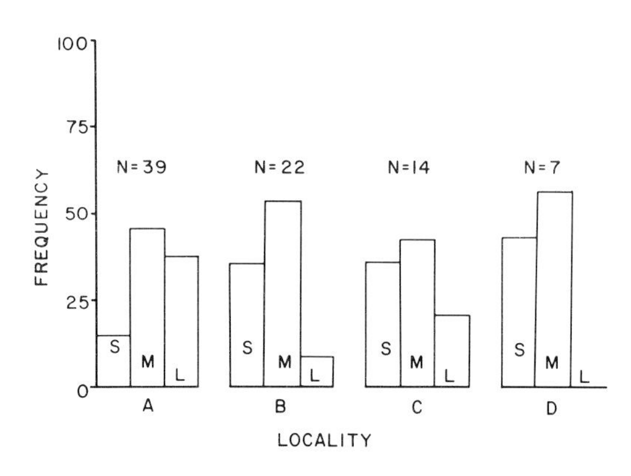

Fig. 11.—Size-distribution of trilobites. Size classes based on length of glabella: S = small (less than 5 mm), M = medium (5-10 mm), L = large (more than 10 mm). Frequencies in percent of specimens.

more extensive bioturbation than the patchy burrowing actually observed. The relatively low density and localized occurrence of biogenic structures in the "trilobite shales" imply rapid deposition of these shales.

The orientation data from trilobites recovered *in situ* from Richmondian (Upper-most Ordovician) shales show a tendency for a preferred direction, indicating current transport or at least alignment of the trilobites by currents. This direction is consistent with paleocurrent data reported by Hofmann (1966) which was based on ripple marks in Richmondian-age limestones. Elias (1979) recorded a northeast-southwest alignment of rugose corals at Locality A. If these cylindrical corallites were rolled along by the current (i.e., oriented perpendicular to the current direction as Elias suggests) the coral and trilobite orientation data are compatible. Varying size ranges in the trilobite collections, however, indicate a living population rather than a sorted, transported trilobite assemblage (Fortey, 1975, p. 333). This range of sizes and the excellent preservation of the trilobites indicate that transport of the trilobites was not extensive. Trilobite enrollment may have been a response to unfavorable environmental conditions (Emielity, 1963, p. 9). The presence of both whole enrolled and non-enrolled trilobites in the same shale horizon suggests that the trilobites were killed instantaneously–so quickly that some did not have time to enroll for protection. Complete preservation of the multisegmented trilobites is also a strong argument for rapid burial of the trilobites; for example, Schäfer (1972, p. 128) notes that arthropod carcasses are attacked immediately upon death of the animal, resulting in disarticulation of the exoskeleton. Immediate burial of the animal minimizes destruction by scavengers.

An alternative suggestion for this trilobite occurrence is that the trilobites may have been semi-infaunal suspension feeders which oriented themselves to exploit currents in feeding. Upon death, the trilobites were already partially imbedded in the substrate and remained more or less intact. Some trilobites were subsequently disturbed by burrowing organisms that overturned them resulting in the random dorsal-ventral attitude of the trilobite. Trilobites are generally regarded as vagrant benthic detritus feeders, and it is not known whether they may have aligned themselves preferentially in life with respect to currents to maximize feeding efficiency. Trilobites may have adapted a burrowing habit for several reasons, such as for protection from predators or for locating food (Bergström, 1973, p. 62). There is no evidence, however, that *Flexicalymene* did more than excavate shallow *Rusophycus pudicum* resting burrows: (1) the perfectly-interlocking enrollment ability of *Flexicalymene* presumably provided protection for its vulnerable ventral surface. Such trilobites with good enrollment capability would not have had to burrow to avoid predators (Osgood, 1975, p. 96); (2) several specimens of *Rusophycus pudicum* show a "worm" burrow terminating at the trilobite resting trace (see Bergström, 1973, p. 55). These structures have been interpreted as recording the capture of the "worm" by the trilobite. The implication here is that *Flexicalymene* did not have to burrow to obtain food. No biogenic structures within these "trilobite shales" are attributable to trilobites (e.g., *Rusophycus, Cruziana*). The burrows that are present are not large enough to have been produced by the trilobites. The sum of these observations indicates that the orientation of the trilobites in these shales was passive and attributable to limited transport by currents.

DEPOSITION OF THE "TRILOBITE SHALES"

The ichnologic, taphonomic, and sedimentologic evidence point to a relatively rapid deposition of these shales. The "trilobite shales" may represent events in which a turbid layer of silt and clay swept over the shelf, killing and rapidly burying the trilobites that lived nearby. Burrowing organisms, also disturbed during the event, re-established themselves and exploited the new substrate after it had stabilized. Rapid, recurrent influxes of fine sediment prevented extensive syndepositional bioturbation of the shale and precluded the establishment of an attached epifauna.

Graded bedding, scour or erosional features, and escape structures are characteristic of rapidly deposited sediments. The lack of these features in the "trilobite shales" requires explanation.

Graded Bedding and Erosional Features

Moore (1969) and Stanley (1969) both discussed situations in which sediment may be entrained and move as a "low-velocity, low-density" turbidity current, in which erosive potential is low, and graded bedding is absent because of the lack of coarse sedimentary components. The "trilobite shales" are composed of silt- and clay-sized particles, and grading of this material may not be obvious becuase of the very small grain sizes. Several thin sections of this shale show a basal layer of fossil fragments which grades upward into unfossiliferous shale, and may in fact represent a graded interval (Fig. 5).

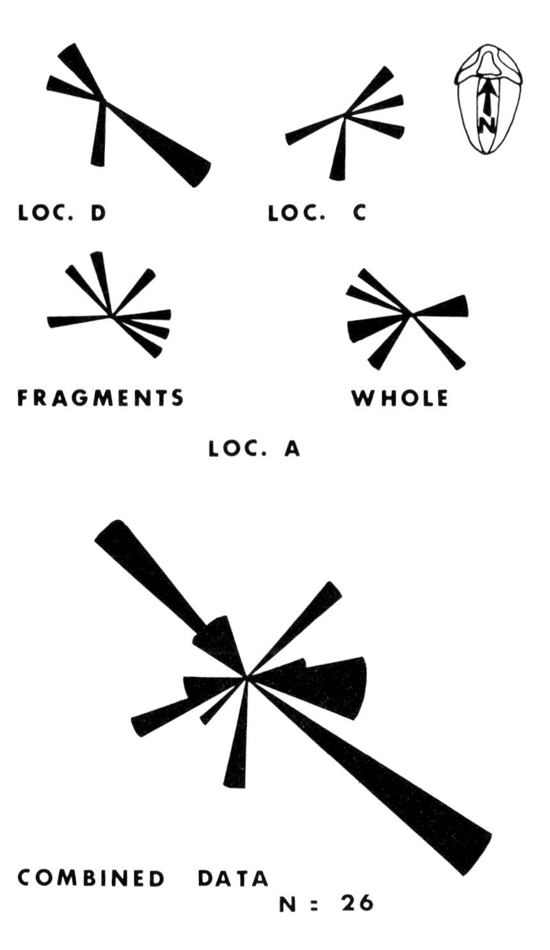

Fig. 12.—Trilobite orientations. Top: Orientation of trilobites and trilobite fragments at Localities D and C. Middle: Separate rose diagrams for fragments (N = 7) and whole trilobites (N = 8) at Locality A. Bottom: Combined data from the three Localities, N = 26.

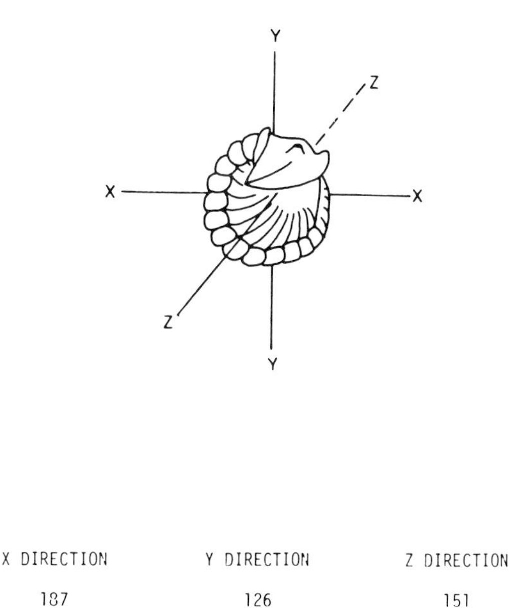

Fig. 13.—Deformation of trilobites from Locality G. Direction of deformation indicates the individual's orientation relative to bedding and prior to compaction of the mud.

Escape Structures

Assuming rapid deposition of the "trilobite shales," escape structures would not be present if: (a) there were no fauna present capable of escaping; (b) the fauna were killed during the influx of sediment; or (c) the thickness of the new sediment was sufficient to prevent escape.

Explanation (a) is rejected because the taphonomic data (articulation of the trilobites) indicate that the trilobites were not transported a great distance; rather, they lived in the area. Fortey (1975) suggests conditions under which explanation (b) might be true. He postulates that trilobites could not withstand a sudden temperature change as might accompany a rapid influx of terrigenous sediments; therefore, the trilobites would be killed immediately. Alternative (c) is difficult to document, since the boundaries between sedimentation events are not clear. Schäfer (1972, p. 111) maintained that a thin layer of rapidly deposited mud is sufficient to suffocate some benthic arthropods, so that thick accumulation of mud is not required. Another possibility is that the escape structures are there but have not been recognized. Until more is known about trilobite behavior, some combination of explanations (b) and (c) seems most likely to be the case.

COMPARISON WITH OTHER EXCEPTIONAL FOSSIL OCCURRENCES

Other localities famous for exceptional fossil preservation in shale share these attributes of the Cincinnatian "trilobite shales": excellent preservation of delicate forms; compressed fossil specimens; a preferred alignment of elongate fossils; randomly oriented (dorsal-ventral) fossils. Two examples are described briefly below.

The arthropod-dominated fauna of the Middle Cambrian Burgess Shale of British Columbia is "entombed in the sediment at all angles, many approximately parallel to bedding but others are on the side or vertical" (Whittington, 1975, p. 104). Whittington concluded that the Burgess Shale animals were trapped alive in a moving cloud of suspended sediment and buried as the mud settled. He argued that the animals did not have to be transported far, nor did the current have to be strong to disrupt and bury the fauna. Subsequent compaction of the shale distorted the individuals oriented obliquely to the bedding. This scenario was repeated at irregular intervals. Piper (1972) found graded bedding within the fossiliferous Burgess Shale horizons. He proposed a turbidite origin for the bed, reasoning that a turbidity current carrying only silt and clay could transport the animals a short distance with a minimum of disarticulation of fragile forms.

In his re-examination of "Beecher's trilobite bed" (Ordovician of New York), Cisne (1973) showed that specimens of *Triarthrus* are aligned, but with no preference for anterior/posterior or dorsal/ventral positions. Trilobite appendages are preserved and there is no size-sorting of the trilobites. Diversity in this bed is low, with relatively few taxa of brachiopods, trilobites, and graptolites comprising more than 90 percent of the total number of specimens. The shale shows some graded bedding and internal directional features. Cisne concluded that a "micro-turbidite" killed and buried the trilobites without transporting them far from their original habitat.

CONCLUSIONS

Although they are spatially and temporally separate, the "trilobite shales" of the Cincinnatian Series have common ichnologic, taphonomic, and sedimentologic characteristics. Presumably, comparable processes acted at each locality. Sedimentologic features typical of turbidites are absent, but the collective evidence points to rapid deposition of the Cincinnatian shales. Rapid non-turbiditic deposition of shale may be understood in light of emerging concepts of shale deposition. Wells, Prior, and Coleman (1980), for example, describe subaqueous mass movement of muds on slopes of as little as 0.03 to 0.08° at rates of 1-10 cm/min. The ichnologic, taphonomic, and sedimentologic characteristics of the "trilobite shales" indicate that mechanisms other than turbidity currents (in their classical sense) must be considered in order to account for the rapid deposition of these shales.

ACKNOWLEDGMENTS

This study was done as part of an M.S. thesis at the University of Cincinnati. I thank D. L. Meyer, W. A. Pryor (University of Cincinnati); D.B. Blake (University of Illinois); D.E. Schindel, D.W. Larson, M.A. Velbel (Yale); and J. Bergström (Sveriges geologiska underskning) for helpful discussions. M.F. Miller (Vanderbilt University), H.A. Curran (Smith College), and J. Cisne (Cornell University) reviewed the manuscript. I am also indebted to my colleagues at Cincinnati and Yale for their thought-provoking discussions. This project was supported by the Department of Geology, University of Cincinnati, and by a grant from Sigma Xi. S.M. Swapp (Yale) drafted the histograms.

APPENDIX

Locality Index

A—Bon Well Hill; roadcut, 1 mile north of Brookville, Indiana, on IN Rt. 101. Brookville quadrangle.

B—Caesar's Creek; dam spillway exposure, east of Waynesville, Ohio on OH Rt. 73. Warren County, Waynesville quadrangle.

C—Stonelick Creek; stream cut, off OH Rt. 131, 0.5 miles east of junction of 131 and OH 727. Clermont County, Goshen quadrangle.

D—Rumpke landfill; construction exposure, east of OH Rt. 27, 1 mile north of intersection of OH 27 and I-275. Colerain Twp. Hamilton County, Cincinnati West quadrangle.

G—Boudinot Avenue; construction exposure now covered, 2 blocks south of intersection of Boudinot Avenue and Westwood Northern Blvd., Cincinnati. Hamilton County, Cincinnati West quadrangle.

REFERENCES

BERGSTRÖM, J., 1973, Organization, life, and systematics of trilobites: Fossils and Strata, v. 2, p. 1–69.

CISNE, J.L., 1973, Beecher's trilobite bed revisited: Ecology of an Ordovician deepwater fauna: Peabody Museum Postilla, Yale University, v. 160, 25 p.

ELIAS, R.J., 1979, Late Upper Ordovician solitary rogose corals of eastern North American [unpub. Ph.D. dissertation]: Univ. Cincinnati, 514 p.

EMIELITY, J.G., 1963, Silurian trilobites of southeastern Wisconsin: Milwaukee Public Museum Publications in Geology, v. 1, 36 p.

FORTEY, R.A., 1975, Early Ordovician trilobite communities: Fossils and Strata, v. 4, p. 331–352.

HOFMANN, H.J., 1966, Ordovician paleocurrents near Cincinnati, Ohio: Jour. Geology, v. 74, p. 868–890.

HOWARD, J.D., 1975. The paleontological significance of trace fossils: *in* Frey, R.W. (ed.), The Study of Trace Fossils: New York, Springer-Verlag, p. 87–108.

LUNDEGARD, P.D., AND SAMUELS, N.D., 1980. Field classification of fine-grained sedimentary rocks, Jour. Sed. Petrology, v. 50, p. 781–786.

MOORE, D.G., 1969, Reflection profiling studies of the California continental borderland: Structure and quaternary turbidite basins: Geol. Soc. America Spec. Paper 107, 142 p.

OSGOOD, R.W., JR., 1970, Trace fossils of the Cincinnati Area: Palaeont. America, v. 41, p. 281–438.

______, 1975, The paleontological significance of trace fossils: *in* Frey, R.W. (ed.), The Study of Trace Fossils: New York, Springer-Verlag, p. 87–108.

PIPER, D.J.W., 1972, Sediments of the Middle Cambrian Burgess Shale, Canada: Lethaia, v. 5, p. 169–175.

RHOADS, D.C., 1975, The paleoecological and environmental significance of trace fossil: *in* Frey, R.W. (ed.), The Study of Trace Fossils: New York, Springer-Verlag, p. 87–108.

RICH, J., 1951, Three critical environments of deposition, and criteria for recognition of rocks deposited in each of them: Geol. Soc. America Bull., v. 62, p. 1–20.

SCHÄFER, W., 1972, Ecology and Palaeoecology of Marine Environments: Chicago, Univ. Chicago Press, 568 p.

SCOTFORD, D.M., 1965, Petrology of the Cincinnatian Series shales and environmental implications: Geol. Soc. America Bull., v. 76, p. 195–222.

STANLEY, D.J., 1969, Sedimentation in slope and base of slope environments: *in* Stanley, D.J. (ed.), The New Concept of Continental Margin Sedimentation. Amer. Geol. Inst., p. DJS-8 1–25.

WELLS, J.T., PRIOR, D.B., AND COLEMAN, J.M., 1980, Flowslides in muds on extremely low angle tidal flats, northeastern South America: Geology, v. 8, p. 272–275.

WHITTINGTON, H.B., 1975, Trilobites with appendages from the Middle Cambrian Burgess Shale, British Columbia: Fossils and Strata v. 4, p. 97–136.

BIOTURBATION IN A DYSAEROBIC, BATHYAL BASIN: CALIFORNIA BORDERLAND

BRIAN D. EDWARDS
U.S. Geological Survey, 345 Middlefield Road, Menlo Park, California 94025

ABSTRACT

Oxygen-deficient waters at the bathyal depths of the Santa Cruz Basin in the California Continental Borderland create harsh conditions for marine life. A feeding strategy approach is used for descriptions of the life habits (i.e., the organism's motility and its living position with respect to the substrate) of benthic fauna and is intended to provide insight into how these organisms disrupt slope and basin floor sediment. Washes of sediment recovered by box corers show that the basin supports a surprisingly high density and diversity of benthic macrofauna. In contrast, however, biogenic sedimentary structures preserved in the sediment are low in density and diversity. Furthermore, bottom photographs taken at 117 stations reveal a biogenically produced, microhummocky topography with few resolvable biogenic traces.

Recognizable biogenic traces are of three main classes: tracks and trails, depressions, and fecal matter. Echinoderms, the most abundant epifauna on the slope and adjacent basin floor, produce most of the large tracks and trails. Significantly, because of the soft, soupy nature of the surficial bathyal sediment, many tracks and trails there are less distinct than similar markings found at abyssal depths. Depressions made by asteroids, regular echinoids, and bottom dwelling fish are most common at moderate depths (<800 m). A characteristic circular depression made by a feeding, tubulous polychaete is restricted to the lowermost slope and the adjacent basin floor. Holothurian fecal strands dominate the feces types that can be seen in bottom photographs. These holothurian feces take the clothesline form common to the abyss.

Open burrows are common in bottom photographs but are scarce in box core slabs. Photography and X-ray radiography of box-core sediment slabs reveal relatively homogeneous, burrow-mottled sediment. The paucity of distinct biogenic structures results from a lack of sediment density contrast for radiography and from the thixotropic response of the sediment to biogenic disturbance. Traces at the sediment-water boundary and within the sediment have poor preservation potential. Sediment from silled bathyal environments similar to the Santa Cruz Basin will probably come to appear in the rock record as homogeneously mottled and bioturbated mudstones with few preserved biogenic structures.

INTRODUCTION

Benthic macrofauna of oxygen-deficient bathyal environments (200–2,000 m) are poorly represented in the rock record because of the development in those environments either of organisms with thin, fragile, calcareous skeletons or of organisms lacking hard parts (Rhoads and Morse, 1971). However, most macroinvertebrates are motile and produce traces (i.e., tracks, trails, burrows) of their activity. When preserved and used in concert with stratigraphic and lithologic data, these trace fossils are useful in defining depositional environments.

Literature on recent bathyal environments is sparse, and little is known about oxygen-deficient environments. However, the recognition of bathyal deposits in the geologic record (e.g., Crimes, 1975; Cook, 1979) attests to the importance of understanding the geologic and biologic processes that operate in this biologically stressful environment.

In this paper, I describe the occurrence of benthic macrofauna and the traces those organisms leave on the surface of and within clastic sediment on the slopes of the Santa Cruz Basin, a modern dysaerobic basin in the California Continental Borderland. A feeding strategy approach is used to describe the animal-sediment relationships because this method is useful in detailing the manner in which an animal disrupts sediment while performing its life functions (Rhoads and Young, 1970; Aller and Dodge, 1974; Jumars and Fauchald, 1977). Benthic fauna, identified to species level, have been classed in a community life structure (i.e., feeding strategies and life habits). "Feeding strategy" refers to an organism's general mode of feeding (i.e., as a suspension feeder, surface or subsurface deposit feeder, herbivore, grazer, scavenger, or parasite) and does not give information about its relative position in the food chain. "Life habit" is used to describe the organism's motility and its life position with respect to the sediment mass. Fauna are classed as (1) *sessile* if they are sedentary and immobile, (2) *discretely motile* if they are sessile except for periods of discrete motion, or (3) *motile* if their movement is frequent or continuous (Jumars and Fauchald, 1977). In any of these modes, an animal may be dominantly epifaunal or infaunal.

SETTING

Tectonics

Right-lateral shearing related to late Cenozoic transform faulting at the boundary of the American and Pacific plates (Atwater, 1970) and to multiple periods of folding has produced a series of basins that are separated by ridges and island platforms. This geologically complex physiographic province extends from Point Conception, California, to Isla Cedros, Baja California, Mexico (Crowell, 1974; Yeats et al., 1974; Moore, 1976) and is known as the California Continental Borderland (Shepard and Emery, 1941).

The Santa Cruz Basin, south of Santa Cruz Island, is the deepest basin in the borderland (Fig. 1). To the west lies the Santa Rosa-Cortes Ridge, an anticlinorium (Vedder et al., 1974) the eastern flank of which is a relatively featureless wall with an average declivity of 8° (Barnes, 1970). At its southern end, the Santa Cruz Basin is separated from the San Nicolas Basin by a gentle flexure in the sea floor that controls the flow of deep waters into and across the basin floor (Emery, 1960). The Santa Cruz-Catalina Ridge lies to the east and northeast and separates the basin from the shallower Santa Monica Basin. The Santa Cruz Basin encompasses an area of 1,740 km^2 and has a maximum floor depth of 1,966 m, with 885 m of closure below sill depth (Emery, 1960; Barnes, 1970). The basin floor proper lies within the 1,750 m isobath (Emery, 1960). Canyons in both the southern and northern ends of the basin have at their mouths fan and channel systems (Barnes, 1970), whereas the central and eastern parts of the floor remain exceptionally flat.

Hydrography

Water character in the Santa Cruz Basin is governed largely by a complex interplay of surface, intermediate, and deep water masses. The Southern California Eddy, a cyclonic gyre of the California Current (Sverdrup and Fleming, 1941), is permanently situated over or just west of the Santa Cruz Basin, and it dominates the surface circulation pattern of the northern borderland. During the spring and summer months, north to northwest winds intensify along the coast, force surface waters offshore, and thereby drive the flow of replacement water from below. These upwelled waters, which typically are limited to the upper 300 m of the water column (Skogsberg, 1936), are cold and rich in nutrients.

Intermediate (or Southern) waters stem from the equatorial pacific, flow northward, and mix with waters of the California Current system at depths of 200 to 500 m. As a result of thermal stratification, only minor vertical advection occurs (Emery, 1954), and so the warm, saline Southern Water maintains its characteristically low dissolved-oxygen levels and high nutrient content.

Drawing on comparative data about the temperature, salinity, and dissolved oxygen in adjacent basins, Emery (1954) has described a northward flow of deep water between the basins. Sills separating each basin impede these deep flows, thereby maintaining unique water characteristics in each basin which are similar to the water characteristics at the controlling sill.

Marine waters can be divided into three zones according to dissolved oxygen content: *aerobic*, where the oxygen content is $>1.0\ m\ell_{O_2}/\ell_{H_2O}$; *dysaerobic*, where the oxygen content is 0.1 to 1.0 $m\ell/\ell$, and *anaerobic*, where the oxygen content is $<0.1\ m\ell/\ell$

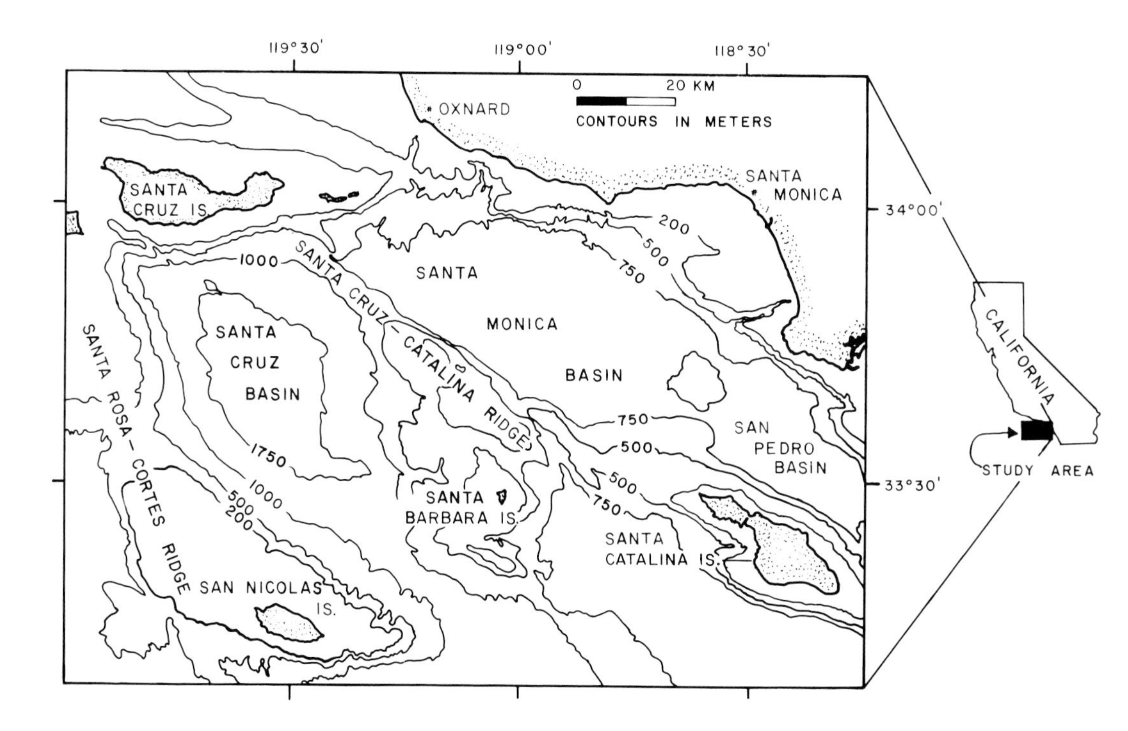

Fig. 1.—Location of Santa Cruz Basin in the California Continental Borderland.

(cf., Rhoads and Morse, 1971). Above 300 m, the waters of the central Santa Cruz Basin lie entirely in the aerobic zone and have a surface mixed layer to a depth of 40 m (Fig. 2). Below 300 m, the basin waters become dysaerobic, and everywhere below 500 m they contain less than 0.5 mℓ/ℓ of dissolved oxygen. Minimum dissolved-oxygen levels (0.19 mℓ/ℓ) occur at about 700 m but increase gradually to about 0.5 mℓ/ℓ at the central basin floor. Comparison of data on the quality of central basin water collected over a 20-year period shows that this zonal configuration has been stable for decades (Fig. 2; Bainbridge and Show, 1978).

Sediments

Surface sediment in the Santa Cruz Basin exhibits a general fining of grain size from the basin rim toward the central basin floor (Fig. 3). Typically, the rim has a veneer of poorly sorted, fine to very fine sand ($\bar{X}$ = 2.85 Φ = 13.9 μm) and the basin floor has a thick accumulation of moderately sorted, very fine silt to clay ($\bar{X}$ = 7.87 Φ = 4.3 μm). For descriptive purposes, textural data are grouped into four depth zones: *ridge*, at 200 to 500 m; *upper slope*, at 500 to 1,100 m; *lower slope*, at 1,100 to 1,700 m; and *basin floor*, at 1,700 to 1,966 m. Definition of the boundary between the ridge and the upper slope (500 m) is based on a change of mean grain size from sand to mud; the boundary between the upper slope and lower slope is drawn at sill depth (approximately 1,100 m), and the boundary between the lower slope and the basin floor is drawn at the base of slope (1,700 m).

Total sedimentation rates in the basin are on the order of 30 mg/cm²/yr for the central basin; they decrease to 9 mg/cm²/yr for the flanks (Barnes, 1970). Assuming an average water content of 110 percent by weight of dry sediment (Demars and Taylor, 1971), a salinity of 35 ‰, and a grain density of 2.7 g/cm³, these sedimentation rates are on the order of 0.4 mm/yr

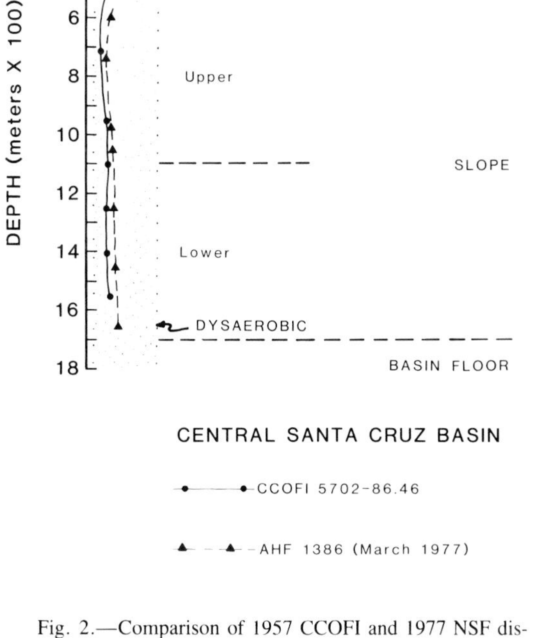

Fig. 2.—Comparison of 1957 CCOFI and 1977 NSF dissolved-oxygen data for central Santa Cruz Basin. Stippled pattern denotes dysaerobic zone.

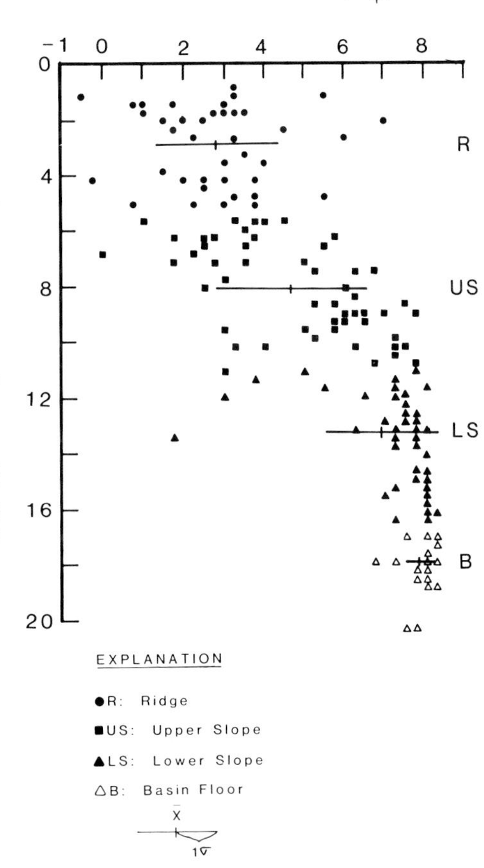

Fig. 3.—Bivariate plot of moment mean grain size ($\bar{X}_\phi$) vs. water depth, showing boundaries used in textural zonations. Bars represent 1 σ on either side of mean for each textural zone.

and 0.1 mm/yr respectively. These rates correspond well with values based on radiocarbon age dating techniques (e.g., 0.4 mm/yr, Emery and Bray, 1962; 32 mg/cm²/yr, Emery, 1960).

Much of the slope sediment is unstable and fails repeatedly by mass transport processes, as evidenced by box-core data (Edwards, 1979) and by acoustic profiling records (Nardin et al., 1979; Field and Edwards, 1980). This continued movement mixes the sediment and so provides an unstable substrate for the benthic macrofauna.

METHODS

Between 1976 and 1978, 209 stations were occupied in the Santa Cruz Basin by means of the University of Southern California vessel, R/V VELERO IV. Initial collections were made in 1975–76 as part of a comprehensive program investigating the benthic ecology, mi-

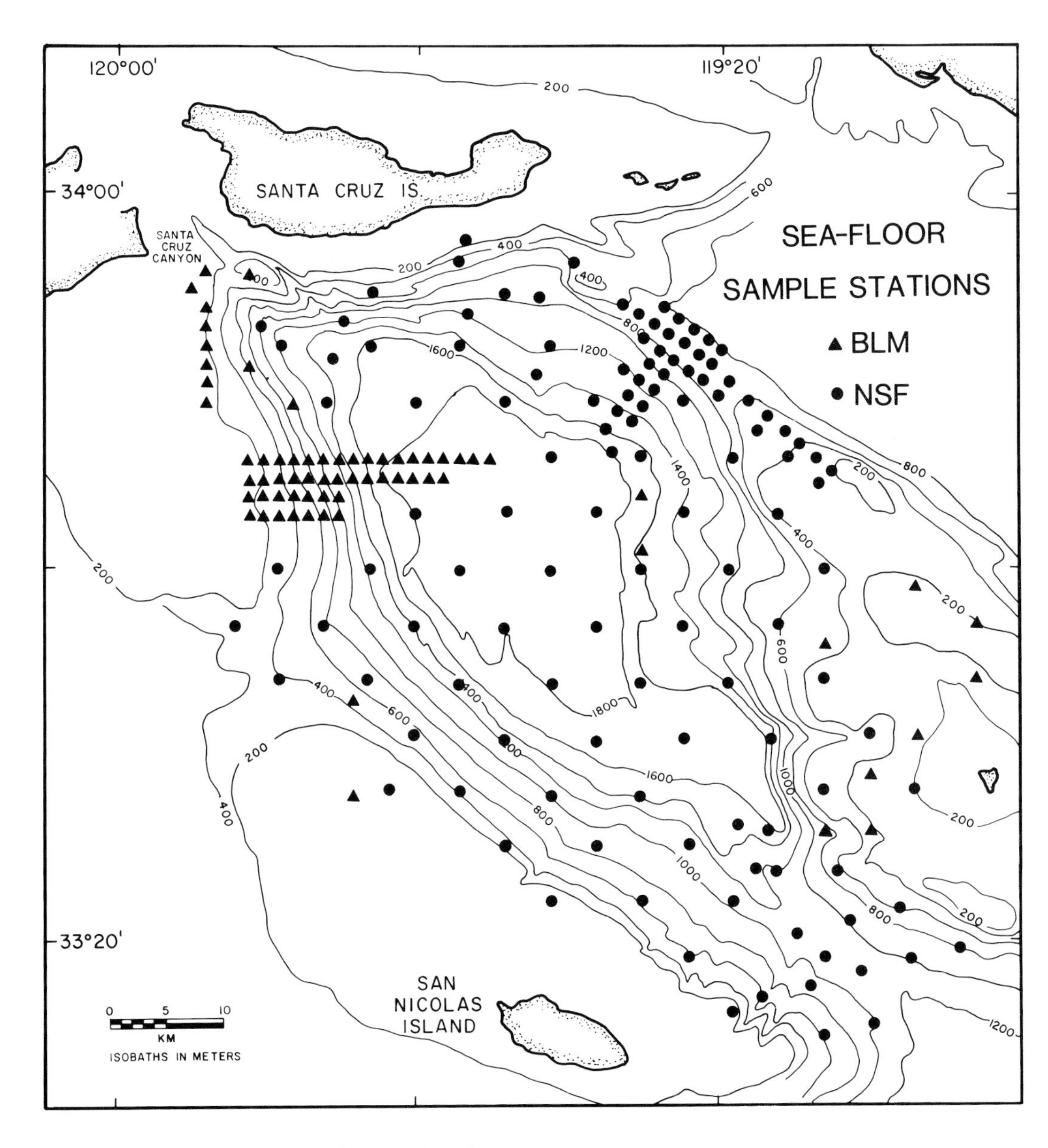

Fig. 4.—Location of occupied sea-floor sample stations.

cropaleontology, sedimentology, and geochemistry of the California Bight (Callahan and Shokes, 1978). Subsequent collections were made in 1976, 1977, and 1978. In all, 184 bottom sample sites, 3 hydrographic stations, and 22 E.G.&G. camera stations were occupied (Figs. 4 and 5).

The primary sampling device was a U.S. Naval Electronics Laboratory (USNEL)-style spade corer modified after the original Reineck design (Bouma and Marshall, 1964; Rosfelder and Marshall, 1967). The corer was equipped with a stainless-steel box and a Benthos deep-sea utility camera system designed to photograph the sea floor immediately prior to sampling. Subsamples of recovered cores were taken for sedimentological analysis, X-ray radiography, and micropaleontological analysis. The remainder of the core sample was gently washed on-board through nested 1.0-mm and 0.5-mm sieves using an overflow-barrel

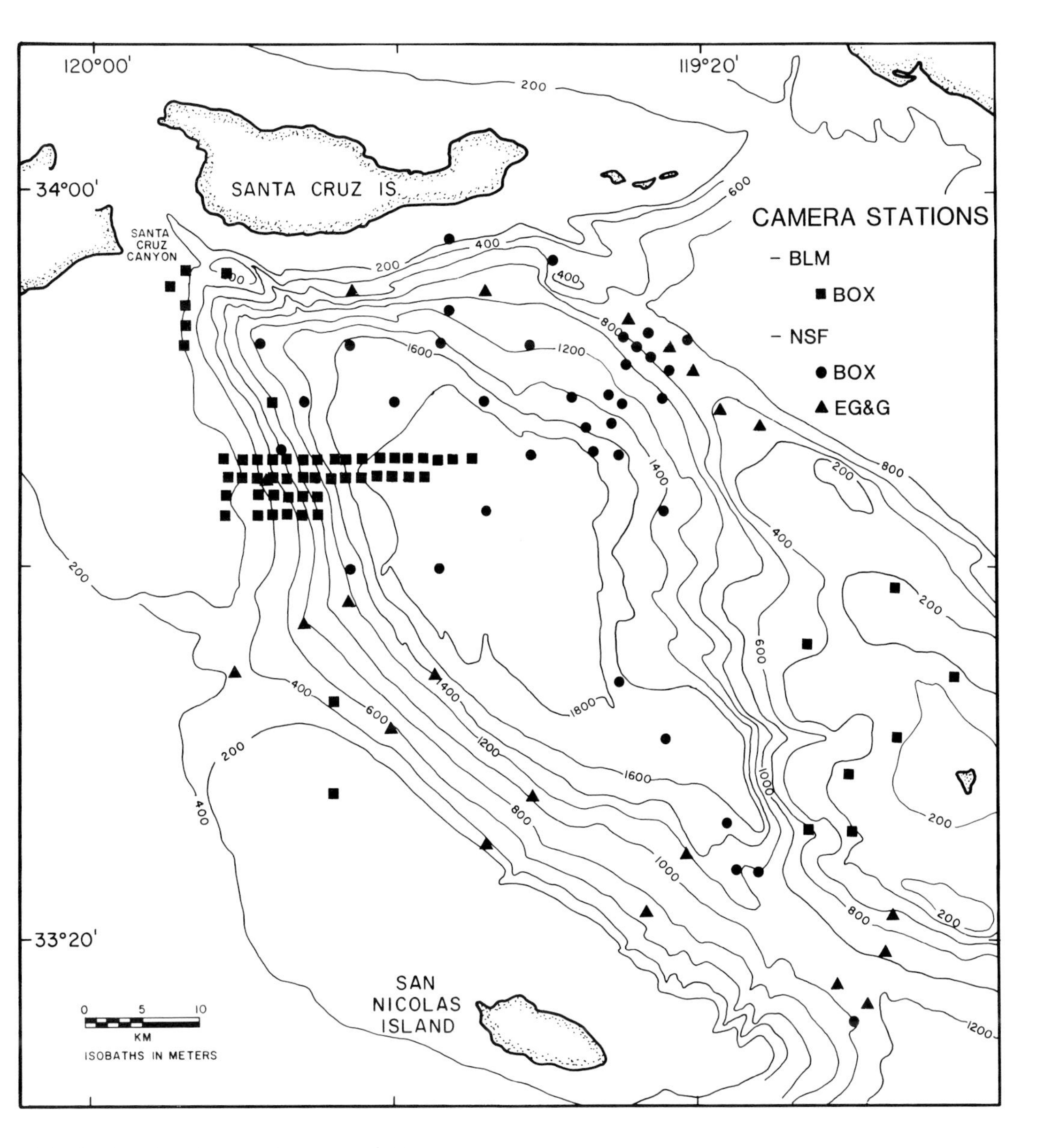

Fig. 5.—Location of occupied camera stations.

technique. The washed biological sample was placed in a 6% magnesium chloride solution in sea-water for 1 hour to relax the contained macrofauna prior to their preservation in a 10 percent buffered formalin solution. After 24 hours, the fauna were transferred to 70 percent ethanol for final storage. All faunal samples were analyzed by scientists of the Department of Biological Sciences, University of Southern California, Los Angeles, California, under the direction of Drs. K. Fauchald and G. F. Jones. Their biological identifications are summarized in Fauchald and Jones (1976).

BENTHIC MACROFAUNA

The benthic macrofauna (>1 mm) recovered from 10 box-core samples along a transect down the western flank of the basin (Fig. 6) were sorted, identified to species level, and classified according to feeding strategy and life habit (Fauchald and Jones, 1976; Edwards,

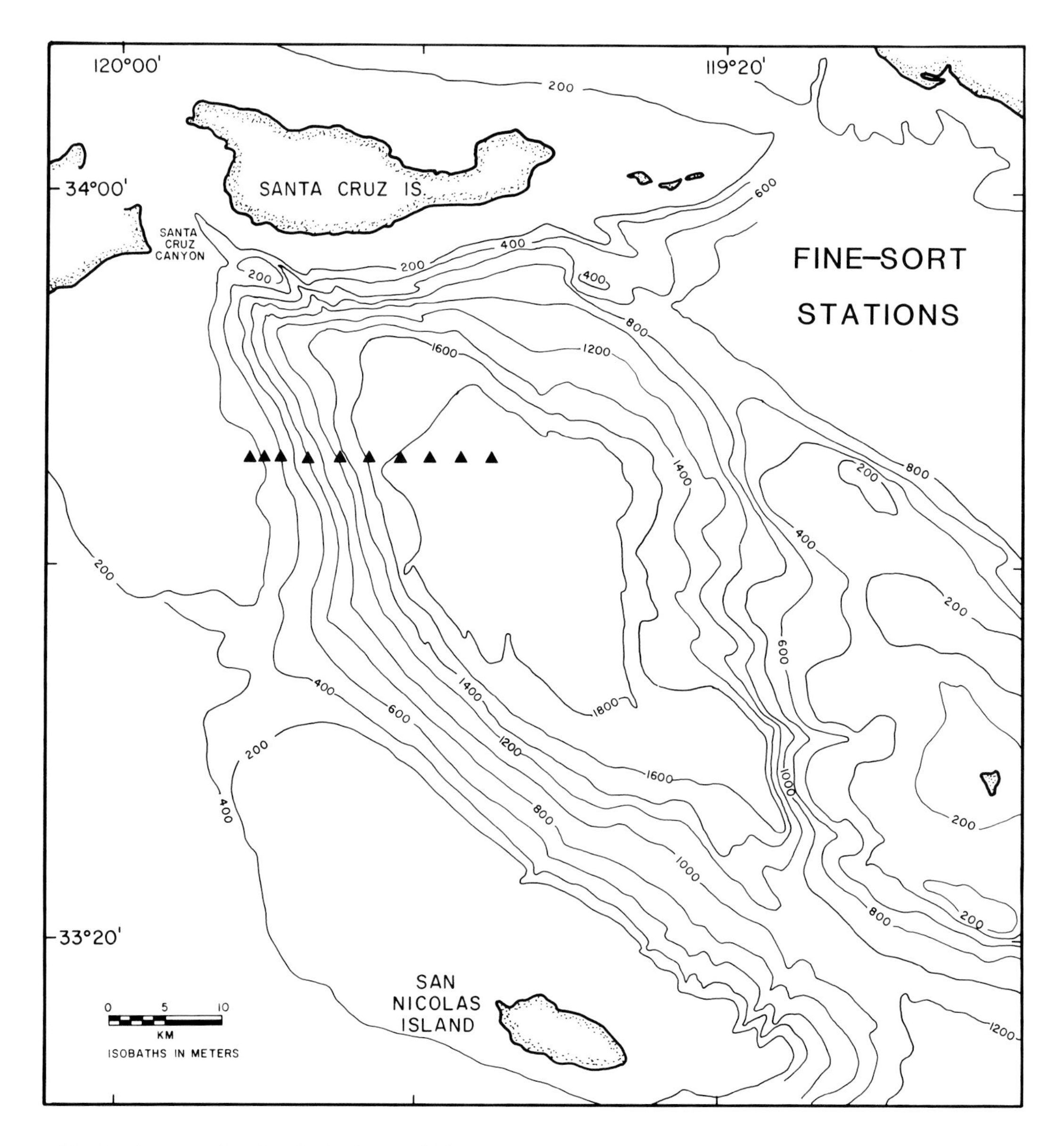

Fig. 6.—Location of stations in the fine-sort biologic transect.

1979). These samples will be referred to as the "fine-sort transect." Correlations of the density of major invertebrate groups (number of individuals per square meter) with water depth and dissolved oxygen content at the 10 stations in the fine-sort transect are shown in Figure 7. In all areas below the ridge crest the population is dominated by polychaete annelids (44 to 74% of the total population density); the density of these increases abruptly at the base of slope due to the occurrence of *Myriochele* (Family Oweniidae), a small tubulous polychaete. Crustaceans are second in abundance on the ridge and slope (ranging from 14 to 35%) and mollusks become more abundant than crustaceans on the basin floor (ranging from 17 to 33% of the total density). Species richness (number of species per sample) also increases with depth (Edwards, 1979). Variations in the standing crop (net weight per square meter) are more complex due to the influence of single, large organisms, but polychaetes still dominate the population (ranging from 17 to 100% of the total sample; Edwards, 1979).

Depth variations in feeding strategies and life habits of all taxa were examined. Suspension feeders, surface deposit feeders, and subsurface deposit feeders were considered separately; all other feeding strategies (e.g., parasites, scavengers, and carnivores) were placed in the single category called "others" (Fig. 8). At the ridge crest, other feeding groups, suspension feeders, and surface deposit feeders are equally important; subsurface deposit feeders are scarce owing to the coarseness and hardness of the substrate. Most of the organisms on the ridge are motile epifauna.

A rapid decrease in all feeding categories occurs from the ridge crest to the boundary between the ridge and upper slope. The slope is characterized by small numbers of suspension feeders and other feeding groups and by a general increase in surface and subsurface deposit feeders relative to the ridge. There is also a slight increase in motile organisms on the slope as well as increasing numbers of infaunal organisms.

The boundary between lower slope and basin floor is marked by complex relationships of feeding strategy and life habit. In general, surface deposit feeders are more common than subsurface deposit feeders, except at one station (AHF 24278, 1,765 m) where the latter compose 73% of the population because of a localized increase in the motile, subsurface deposit-feeding polychaete *Myriochele* sp. (q.v., Fig 7). Motile infaunal species dominate most of the central basin. Polychaetes, the dominant taxon in the basin, show abundance-versus-depth trends like those of the total phyla, except for an increase with depth in subsurface deposit feeders (Edwards, 1979).

In summary, fine-sort data from the 10-station transect taken along the western flank of the Santa Cruz Basin show the benthic macrofauna to be dominated by

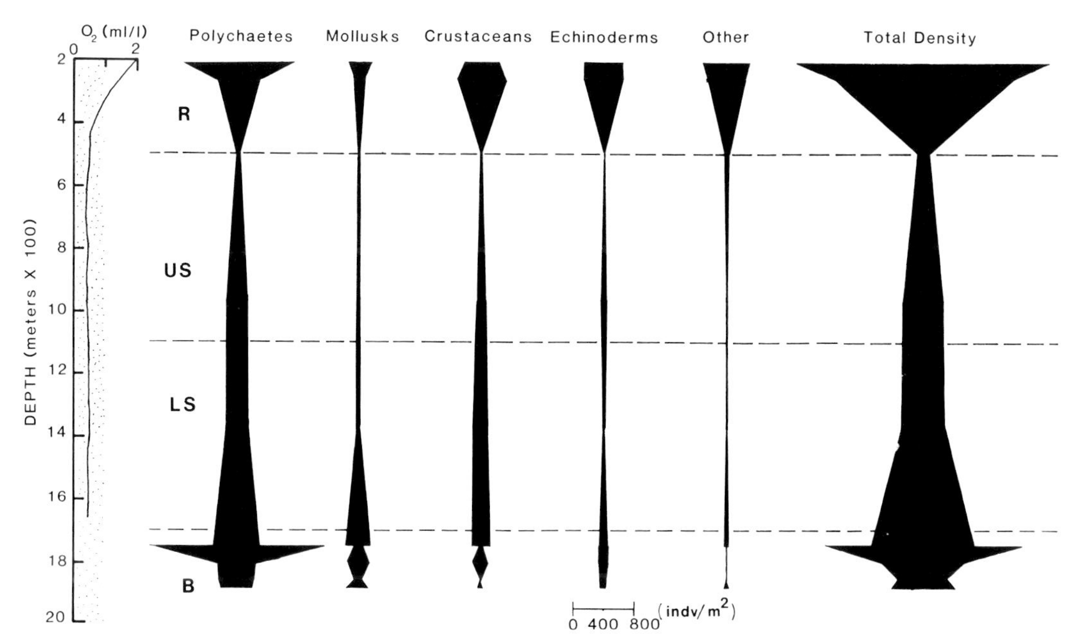

Fig. 7.—Bathymetric distribution of macrofaunal densities (number of individuals per square meter) for major taxa in the fine-sort transect. R: Ridge; US: Upper Slope; LS: Lower Slope; B: Basin Floor.

polychaete annelids. A general decrease in density is noted on the slope, followed by increases on the basin floor proper. The ridge crest is divided equally by other trophic groups (especially carnivores and scavengers), suspension feeders, and surface deposit-feeders, most of which are epifaunal and motile. Relative to the ridge crest, the slope is impoverished in fauna and is dominated by infaunal surface deposit-feeders. The population of the slope is evenly divided between sessile and motile species, with a trend toward increased numbers of motile species at the base. The basin floor is dominated by motile, infaunal, surface, and subsurface deposit-feeding species. The boundary between the slope and the basin floor is a zone with no identifiable patterns.

CHARACTERIZATION OF SUBSTRATE AND EPIBENTHIC FAUNA

Photography is an effective tool for assessing the nature of marine bottom environments (e.g., Hersey, 1967) because it provides an instantaneous record of the sea floor. However, its use is limited to features and fauna at the sediment-water interface (Owen et al., 1967; Fell, 1967) or, at best, to assisting inferences of activity below that interface made on the basis of variations in the sediment surface. The recovery of more than 800 photographs from 117 stations (including both box-core and E.G.&G. stations) affords good coverage of the study area (Fig. 5) and allows a detailed characterization of both the substrate and the epifauna.

Most areas of the ridge top (200 to 500 m) exhibit a high degree of surface roughness due to the presence of outcrops and the coarseness of the substrate. Activity of currents, indicated by well-sorted gravel pavements (Pl. 1A, B), ripple fields, and oriented fauna (e.g., crinoids, Pl. 1A), keeps the ridge top relatively free of sediment except in those regions protected by local topographic irregularities. Total organic-carbon values are low, ranging from 0 to 1.5 percent. Many of the gravel pavements support populations of sponges, hydroids, bryozoans, crinoids, and other characteristic attached and encrusting organisms. Photographs of the sandy areas of the ridge show decapods, regular echi-

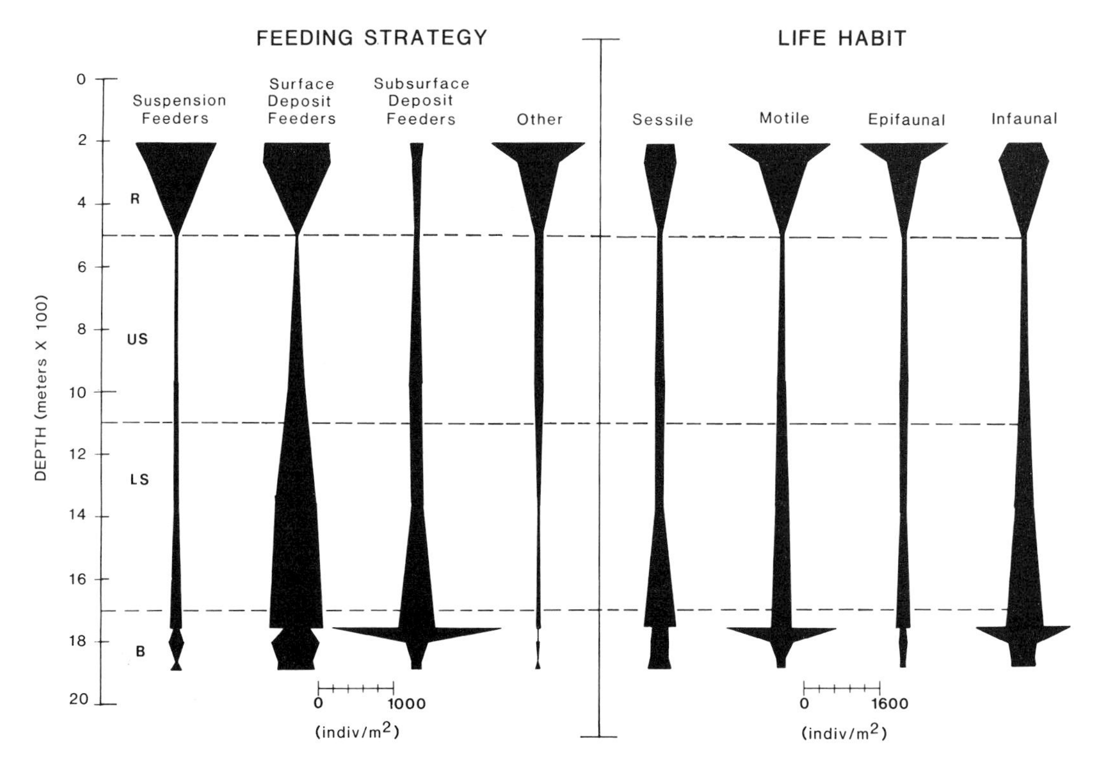

Fig. 8.—Bathymetric distribution of life structures for major taxa in the fine-sort transect. R: Ridge; US: Upper Slope; LS: Lower Slope; B: Basin Floor.

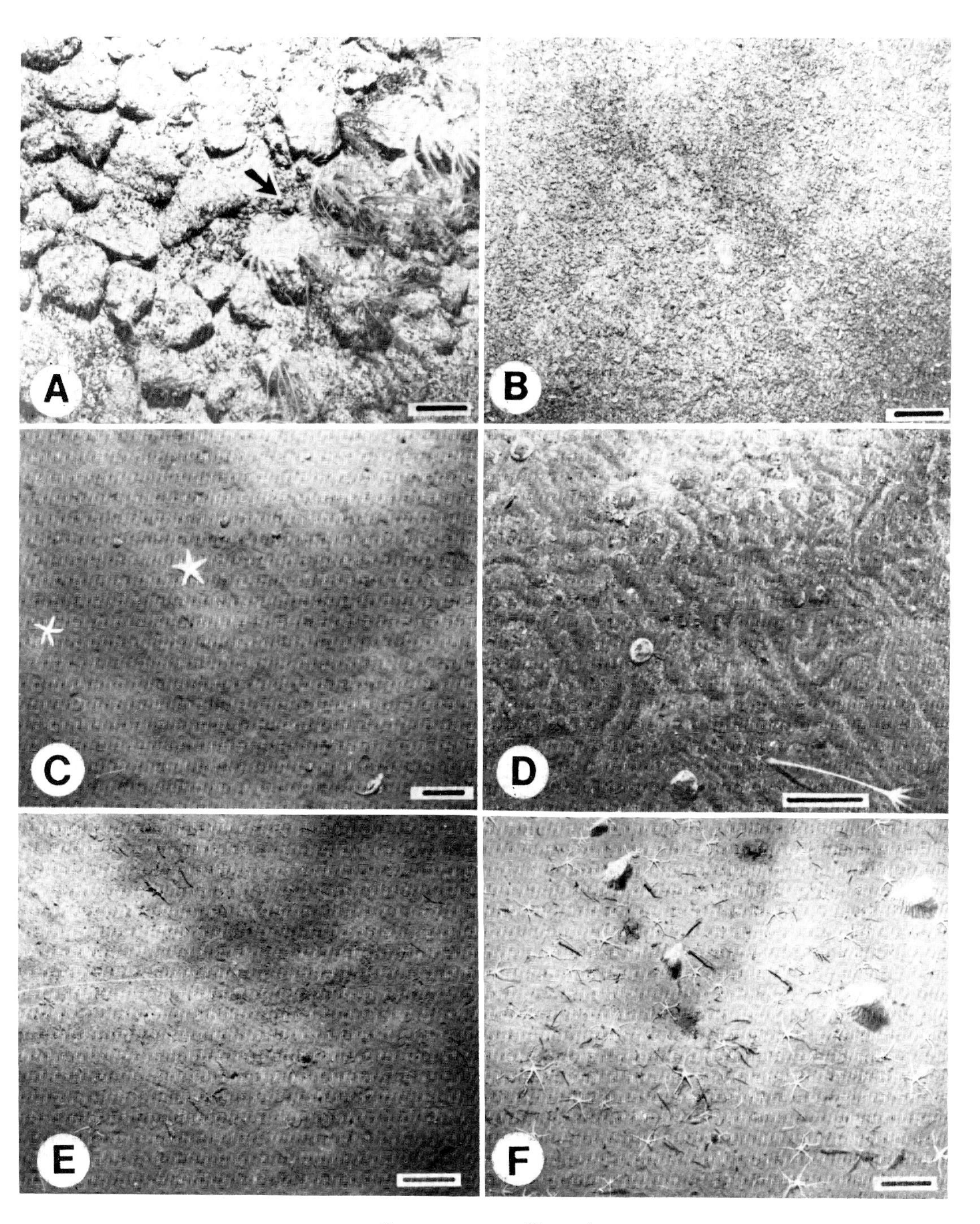

EXPLANATION OF PLATE 1

BOTTOM PHOTOGRAPHS OF SUBSTRATE AND BENTHIC EPIFAUNA. BAR SCALES ARE 20 CM.

FIG. A.— Ridge crest, AHF 24327, at 129 m. Note oriented fauna and coarse substrate.

B.— Ridge crest, gravel pavement, AHF 24326, at 177 m.

C.— Upper slope with pock-marked bottom, AHF 26977, at 615 m.

D.— Upper slope showing *Brissopsis pacifica* plow marks, AHF 24305, at 522 m.

E.— Lower slope showing faunally impoverished surface, AHF 24924, at 1,101 m.

F.— Basin floor, AHF 24280, at 1,808 m. Note faunal densities and oriented pennatulids, indicating current flow.

noids, and gastropods crossing the surface. Pennatulids, asteroids, and actinarians are observed infrequently.

The sediment of the upper slope (500 to 1,100 m), which typically is a poorly sorted, fine sand to coarse silt, is locally rippled and exhibits a very small-scale roughness that is produced by biogenic activity (Pl. 1C, D). Oriented pennatulids are present locally on the upper slope. Considering the low levels of dissolved oxygen in this area, it is surprising that the upper slope supports a rather large epifaunal population which includes asteroids, anemones, regular and irregular echinoids, gastropods, holothurians, ophiuroids, pannatulids, and some demersal fish. It is noteworthy that abundant populations of the irregular echinoid *Brissopsis pacifica* are found in parts of the basin. This organism is particularly effective in mixing sediment at water depths to 600 m (Pl. 1D). On the upper slope, organic-carbon contents increase to an average of 1.5 percent (range = 0 to 4%).

On the lower slope (1,100 to 1,700 m) the sediment is soft, clayey silt with an average organic-carbon content of 3.5 percent (range = 0.4 to 5%). The sea floor in this area is impoverished in fauna compared with other areas (Pl. 1E). Ophiuroids, asteroids, holothurians, anemones, and gastropods are scarce in bottom photographs.

The substrate of the basin floor (below 1,700 m) is a soft, clayey silt to silty clay with organic-carbon contents ranging from 3 to 5 percent. The surface exhibits microtopography that is biologically produced. Epifaunal populations increase markedly on the basin floor; these populations are dominated by tube-dwelling worms and ophiuroids in such large numbers that their arms are commonly in contact (Pl. 1F). Oriented pennatulids occur at the base of the slope (Pl. 1F) and some fish, anemones, asteroids, and holothurians (e.g., *Scotoplanes*) are present on the basin floor.

BIOGENIC STRUCTURES IN BOTTOM PHOTOGRAPHS

In their continuous search for food, animals produce tracks, trails, burrows, and feces as a natural consequence of their passing. It is surprising that much of the sea floor in this bathyal environment lacks traces at a size resolvable with the camera systems used in the study (approximately 0.5 to 1 cm). Because the data are sparse, the following description is necessarily qualitative.

In general, no traces have been noted on the ridge (i.e., to depths of 400 m) because of the rugged nature of the bottom and the coarseness of the sediment. The absence of traces is not restricted to shoal areas, however, and recurs even at the greatest depths of the basin floor.

Tracks and trails.—Echinoderms are the most abundant epifauna visible in the basin and, as in the deep open ocean (Hollister et al., 1975), they are responsible for most of the large tracks and trails viewed on the surface. The most common are ophiuroid traces, which are seen only at depths greater than 700 m. Occasional ophiuroid traces are observed on the basin slope (700 to 1,700 m); these traces increase in abundance near the base of the slope. Although ophiuroid populations are very dense in photographs of the basin floor (cf., Pl. 1F), discrete traces are difficult to resolve in the soupy character of the sediment surface. Curiously, the ophiuroid traces described by Heezen and Hollister (1971) are common and sharply defined, whereas those in the Santa Cruz Basin are generally quite indistinct. Curvilinear imprints of the brittle star's arms are common and in some instances contain impressions of the central disk. Where asteroid populations are dense, the sediment surface has a uniform microscale roughness produced by myriad overlapping imprints.

Recognizable asteroid trails are much less common than imprints of ophiuroids and are found infrequently throughout the upper and lower slope. Shallow-water traces (Pl. 2A, 475 m) are essentially drag marks associated with feeding; these traces reflect the morphology of the animal. A single observation of a deep-water trace (1,360 m) shows a swath of closely spaced, very small depressions produced by the animal's podia (Pl. 2, shown between arrows). This trace is similar to, but much less distinct than, the asteroid "tractor treads" reported at 4,577 m on the continental rise off Great Britain (see Fig. 6.9 of Heezen and Hollister, 1971).

A strikingly similar, though narrower pattern is produced by the regular echinoid *Allocentrotus* (Pl. 2C, between arrows) in 713 m of water, south of Santa Barbara Island. Here, the sediment records the movement of the oral spines and podia as the animal moved across the bottom. Such traces are rare and barely disturb the sediment surface.

In contrast, irregular echinoids, such as *Brissopsis pacifica* (a heart urchin), have a significant effect on the substrate. Plate 1D documents their activity in 522 m of water on the Santa Rosa-Cortes Ridge. These highly active animals are shown plowing along the sediment-water interface, producing distinct furrows with raised, rounded rims. *Brissopsis* is capable of burrowing to shallow depths and producing irregularly shaped tunnels that displace the sediment surface from below. These tunnels collapse following the animal's passage, yielding a highly disturbed surface. In the study area such characteristic features are confined to depths ranging between 440 and 540 m.

Holothurians are a common component of the epifaunal populations in this bathyal environment. Although their traces are reported to be common in the abyss (Heezen and Hollister, 1971; Hollister et al., 1975), they were observed infrequently in the study area. Plate 2D shows the elasipod *Scotoplanes* (arrow) raking the bottom with its podia, possibly in aid of its search for food. Notice that the unidentified holo-

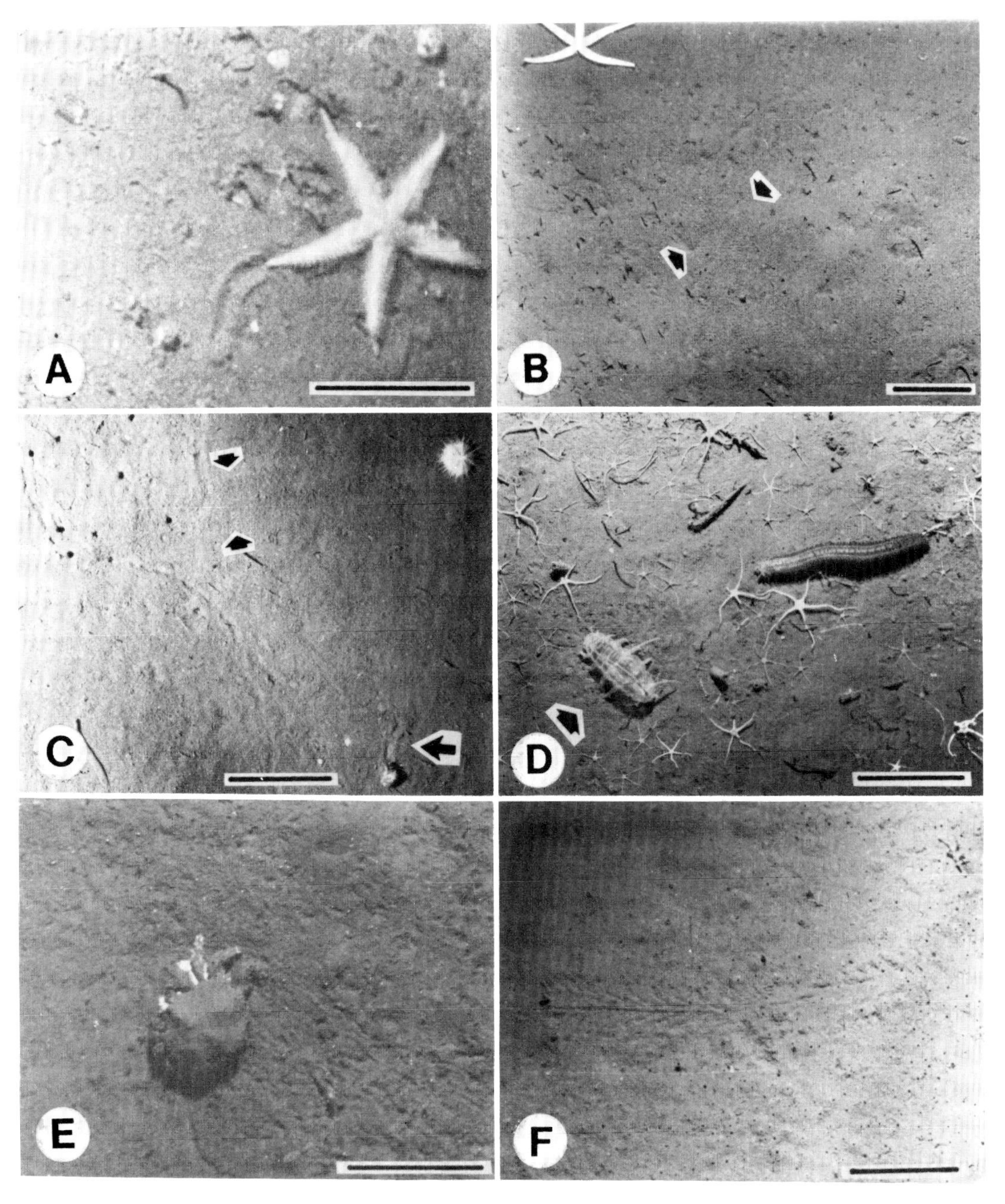

EXPLANATION OF PLATE 2

BOTTOM PHOTOGRAPHS OF TYPICAL TRACKS AND TRAILS. BAR SCALES ARE 20 CM.

FIG. A.— Asteroid and asteroid trace, AHF 25859, at 475 m.

B.— Asteroid and asteroid tracks (between arrows), AHF 26004, at 1,360 m.

C.— Regular echinoid (*Allocentrotus fragilus*) tracks (between arrows) and gastropod trail (single arrow), AHF 26960, at 713 m.

D.— Elasipod holothurian tracks (*Scotoplanes* sp., arrow) in lower left, AHF 24285, at 1,843 m.

E.— Decapod and decapod tracks, AHF 23280, at 270 m.

F.— Unidentified trail, AHF 26008, at 1,408 m.

thuroid to the upper right is leaving no trace on the soft substrate. An area extensively grazed by *Scotoplanes* would be indistinguishable from one heavily traveled by the asteroid shown in Plate 1B.

Gastropods are generally active animals that roam widely across the bottom; their trails are common because their bodies are in constant contact with the substrate. Gastropod trails are rarely observed in depths shallower than 600 m, but they become more numerous with increased depth as the substrate becomes finer. Below 1,500 m, gastropod traces were no longer observed. A typical trace and its maker, *Bathybembix*, is shown in Plate 2C (single arrow, 713 m). The sinuous path produced is a 4-cm-wide, shallow furrow with slightly raised rims, whose shape, relief, and sharpness is controlled by minor variations in the substrate. On close inspection, older traces are recognizable in the same photograph. Exceptionally clear photographs show some gastropod traces to have a medial ridge which may reflect alternating contractions of the two halves of the sole of the gastropod's foot (cf. Fig. 111 of Schäfer, 1972); this configuration suggests rapid movement by means of a pacing action.

Decapod tracks (Pl. 2E, 270 m) are recognized only in the northwestern quadrant of the basin, and they are rare. These tracks occur as a series of closely spaced, elongate grooves that were produced as a leg was dragged forward. The trace outlines are sharp when they are fresh, but their form is gradually blurred with the passage of currents, time, and faunal reworking of the surface.

An enigmatic trace, whose maker is unknown, was encountered in 1,408 m of water southwest of the Santa Cruz-Catalina Ridge. A series of V-shaped depressions connected at the vertices by a medial groove (Pl. 2F; 1,408 m) was seen. The entire trace is approximately 10 cm wide and can be followed for 60 cm before it terminates abruptly.

Depressions.—Many of the faunal groups observed in this study make irregularly shaped depressions on the sea floor during their search for food. Asteroids are capable of producing shallow, circular depressions approximately the size of their body diameter (Pl. 3A). The depressions may have raised rims (Pl. 3A). Asteroid resting traces, which more accurately reflect the shape of the maker (Pl. 3B; 713 m) are also recognizable. Regular echinoids also produce circular depressions, although these are often rimless and generally are uncommon in bottom photographs (Pl. 3C; 466 m). Bottom-dwelling fish, such as representatives of the Scorpaenidae, are occasionally seen in association with irregular, rimless depressions at depths ranging from 700 to 1,300 m (Pl. 3D; 713 m). Stanley (1971) described similar depressions to depths of 914 m on the continental slope southeast of Delaware Bay. Clearly, such large, mobile animals can have a pronounced effect on the sediment—cf., e.g. the escape action of the thornyhead *Sebastolobus* sp. (B.G. Nafpaktitis, personal communication, 1978) that was startled by the box corer in 1,326 m of water (Pl. 3E).

A particularly interesting and diagnostic depression is made by what I term "circle-scribers" (Pl. 3F; 1,756 m), traces of which were found on the lower slope. This is the trace of a sessile, surface deposit-feeding animal, probably a tubulous polychaete. The structures are circular, with a diameter of approximately 14 cm, and have a depressed ring at the outer perimeter. Similar features have been reported from the Somali Basin at 4,853 m (Heezen and Hollister, 1971).

Excrement.—In the marine environment, the majority of feces are small (<1 mm), compact, oval-shaped structures produced by deposit feeders (Moore, 1939). Compared with discrete pellets found within the sediment, feces observable in bottom photographs are large (0.5 to 2 cm diameter). The majority of feces visible in the abyss are recognized to be the product of holothurians (Heezen and Hollister, 1971; Hollister et al., 1975). Many resemble a piece of coiled clothesline; forceably ejected by the holothurian, they take the form of loose strings and knots (Pl. 4A; cf., Fig. 5.14 of Heezen and Hollister, 1971). Clothesline-shaped forms recognized from bathyal depths in this study (Pl. 4B) also are ascribed to holothurians. The strands are typically 0.5 to 1 cm in diameter, and knot clusters are up to 10 cm in length. Tight, planispiral coils (Pl. 4C) recorded in photos are ascribed to an unseen holothurian (cf. Fig. 5.10 of Heezen and Hollister, 1971) defecating in a less forceful manner. The reason that these structures have not been recognized in box cores may be either that they are loosely compacted and easily disaggregated or that their local distribution may preclude sampling with box corers. They occur from 600 m of water to the basin's greatest depths but are most typical of the lower slope and basin floor.

Thin, string-like feces were noted at some ridge and upper slope stations (Pl. 4D, 541 m). These feces were often associated with mounds and probably represent the discarded sediment of an unknown burrowing animal.

Miscellany.—A particularly enigmatic trace was found in bottom photographs in water depths ranging from 700 to 1,000 m (Pl. 5A). These traces were relatively large (0.5 to 1 cm), oval-shaped structures which covered the bottom in amazing profusion (2,000 to 3,000/m^2). Gentle sieving of surface sediment from these areas revealed no fecal or biological material in the proper size class, which attests to their delicate nature. These features may represent manipulation and reworking of the surficial sediment, perhaps by the podia of holothuroids, and the subsequent generation of pseudopellets (cf., Fig. 3.29 of Heezen and Hollister, 1971).

Many burrows were recognized in bottom photographs; these burrows range from single discrete holes (0.5 to 2 cm diameter) at the surface of the substrate, through paired holes, to multiple holes at the

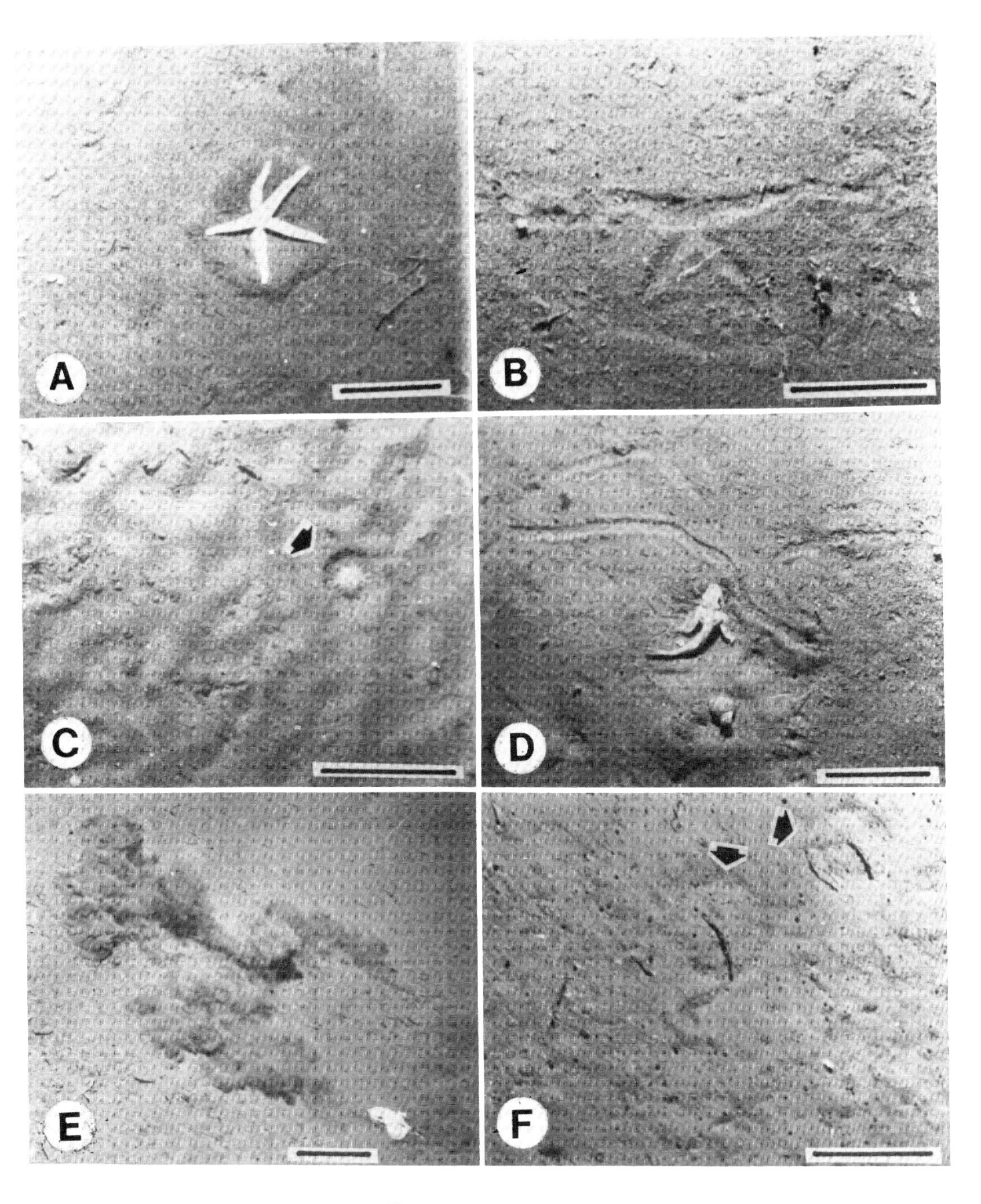

EXPLANATION OF PLATE 3

BOTTOM PHOTOGRAPHS OF TYPICAL BIOGENIC DEPRESSION. BAR SCALES ARE 20 CM.

FIG. A.— Asteroid producing circular depression with raised rim, AHF 26960, at 713 m.
B.— Asteroid resting trace, AHF 26960, at 713 m.
C.— Unidentified regular echinoid producing rimless circular depression, AHF 27167, at 466 m.
D.— Thornyhead fish (Scorpaenidae) producing an irregular depression. Note the gastropod trace nearby, AHF 26960, at 713 m.
E.— Thornyhead fish (*Sebastolobus* sp.) reacting to the descending box corer, AHF 26015, at 1,326 m.
F.— Circle scribers, AHF 24292, at 1,350 m.

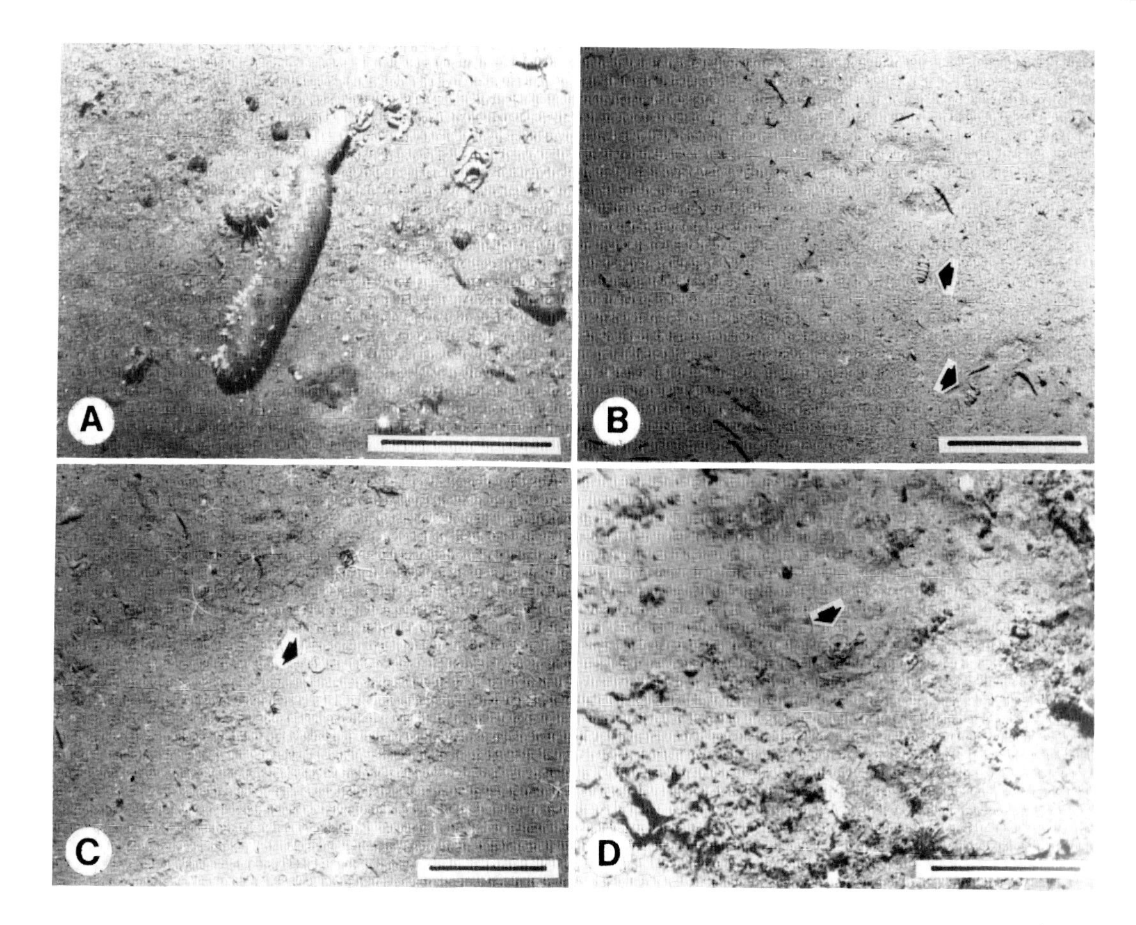

EXPLANATION OF PLATE 4

BOTTOM PHOTOGRAPHS OF EXCREMENT. BAR SCALES ARE 20 CM.

FIG. A.— Unidentified holothurians ejecting (?) feces, AHF 24318, at 407 m.
B.— Clothesline holothurian feces, AHF 26004, at 1,3600 m.
C.— Tight planispiral holothurian feces, AHF 26013, at 1,097 m.
D.— Spaghetti feces associated with mounds, AHF 24315, at 541 m.

apices of mounds, some of which were associated with fecal strings. Such features are distributed randomly throughout the study area at depths greater than 400 m. The makers of these features remain unknown, although some of the burrows are "passively" occupied by ophiuroids (Pl. 5B, 1,097 m). It is difficult to generalize regarding the distributions of these features, with the exception of those in the northwestern quadrant of the basin. Here, between depths of 1,350 to 1,750 m, the bottom is thickly endowed ($2{,}000^{+}/m^2$) with 0.5 to 1.0 cm holes that are at the apices of randomly distributed symmetrical mounds. Most mounds exhibited these openings (Pl. 5C, 1,408 m). Box cores taken in this area showed no surficial evidence of holes; however, the core surface was disturbed and resedimented. Possibly the holes closed at corer impact or while the corer was undergoing vibrations and accelerations during the long ascent to the surface. This region of the Santa Cruz basin is known to have undergone extensive mass movement (Nardin et al., 1979) and these features may be the result of rapid sediment loading and subsequent dewatering or degassing. It is difficult to ascribe them to biological activity, given that the only abundant fauna in this area is the microcrustacean *Apseudes* (G. F. Jones, unpub. data), which does not create such structures.

Perplexing features common to photographs from 200 to 1,200 m have very high point-source reflectivity and have been termed "bright spots" (Pl. 5D, 750 m). These spots are less than 5 mm in diameter, and they have been observed in numbers greater than $3{,}000/m^2$. It is doubtful that they represent benthic foraminifera because individuals of the proper size and number are absent from the bottom samples at these stations (R. G. Douglas, personal communication, 1978). They may be central disks of ophiuroids; however, such large

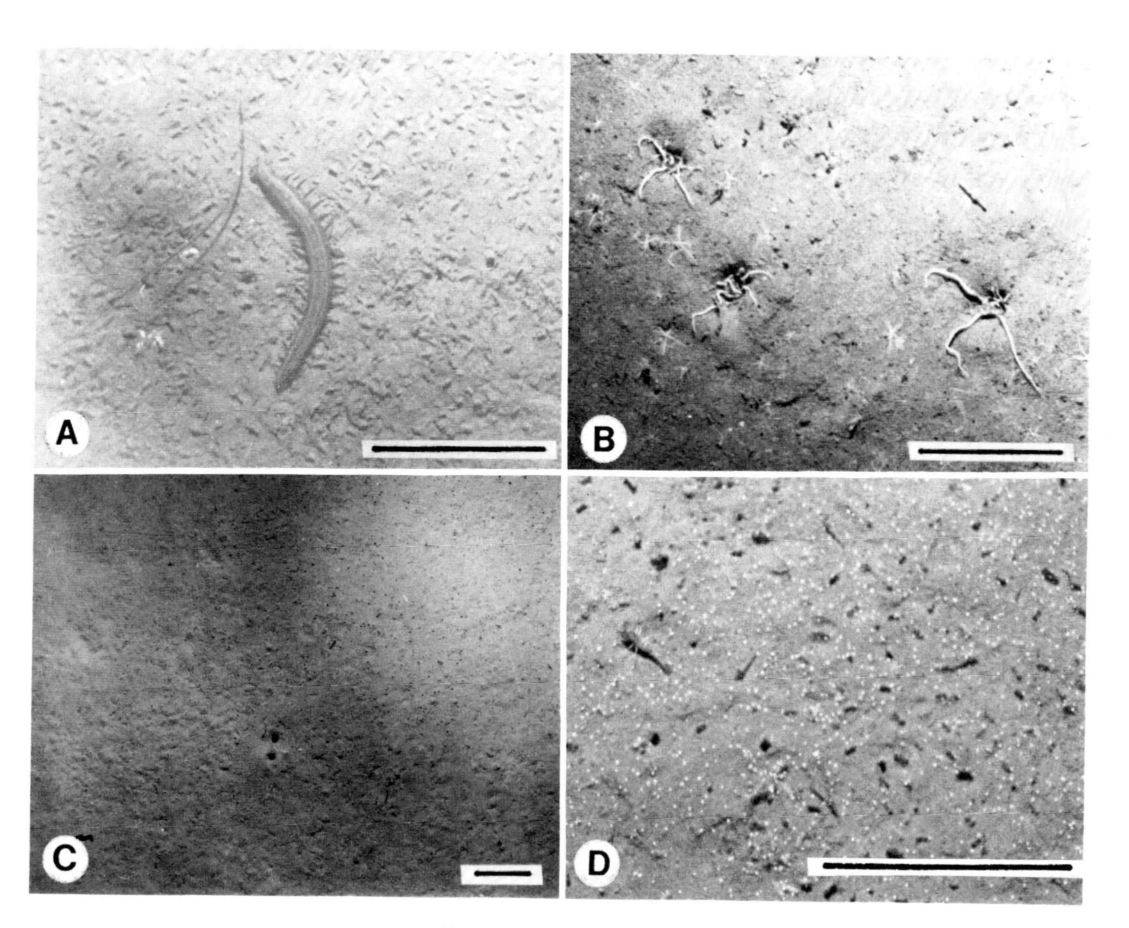

EXPLANATION OF PLATE 5

BOTTOM PHOTOGRAPHS OF BIOGENIC MISCELLANY. BAR SCALES ARE 20 CM.

FIG. A.— Pseudopellets (?) produced by unidentified holothurian (?), AHF 24925, at 748 m.
B.— Ophiuroids passively occupying holes in bottom, AHF 26013, at 1,097 m.
C.— Abundant mounds with apical holes, AHF 26008, at 1,408 m.
D.— Bright spots, AHF 24270, at 750 m. Note association with large pseudopellets (?).

standing crops of ophiuroids are unlikely. Stereopairs of photographs from the stereo-camera system showed bright spots above the substrate, which suggests they may be floccules. Floccules were suggested by Barnes (1970), but he did not observe them in such profusion. If this interpretation is correct, floccules in the study area are being preferentially deposited on the uppermost slope.

BIOGENIC STRUCTURES IN CORES

Visual observation and X-ray radiography of 2-cm-thick longitudinal slabs taken from the box cores show most of the basin sediment to have nondescript bioturbate textures. Figure 9 shows the mottled, bioturbate texture typical of the basin as a whole.

Recognizable biogenic structures are dominantly permanent to semipermanent dwelling tubes or open burrow systems reinforced by mucous secretions. Open burrows with consolidated walls (typically 0.2 to 0.6 cm in diameter) are found less than 10 cm below the sediment-water interface (Fig. 10). From the upper slope to the basin floor, these burrows often occur in horizontally aligned pairs, suggesting a looping pattern.

Sediment collected from central basin areas has well developed, thread-like burrow networks 1 to 2 mm in diameter (Fig. 10). The reticulate pattern is probably produced by foraging polychaetes; it is analogous to burrow patterns recognized by Griggs et al. (1969) at the axis of Cascadia Channel in the northeastern Pacific. The burrow pattern is dense to a depth of 20 cm, beneath which it broadens; the pattern terminates at 30 cm in a distal turbidite layer.

Filled burrow structures are rarely identified except as broad areas of mottled texture. One notable exception was observed in a core from 970 m of water on the

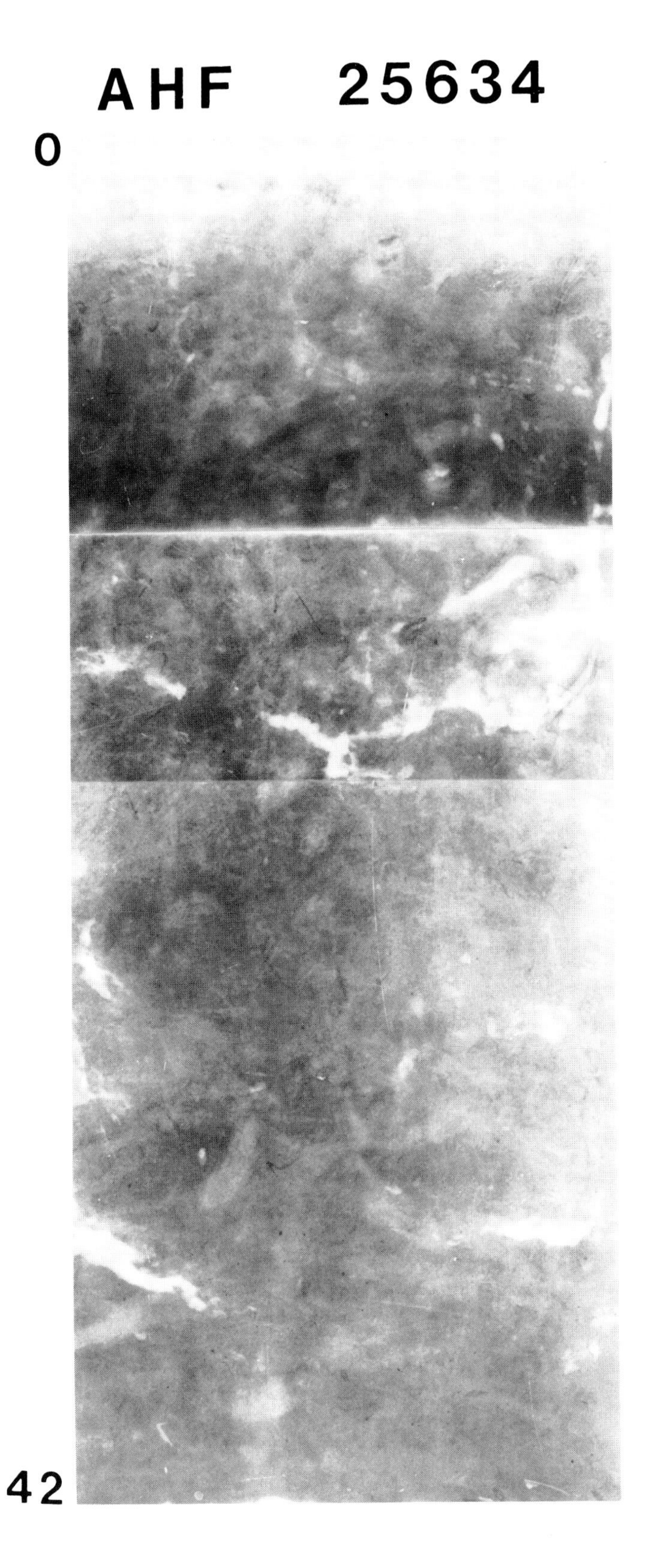
AHF 25634
0
42

Fig. 9.—X-ray radiograph print of box core AHF 25634, at 1,202 m, showing homogeneous nature of sediment. Horizontal lines are splices in the print. Scale in cm.

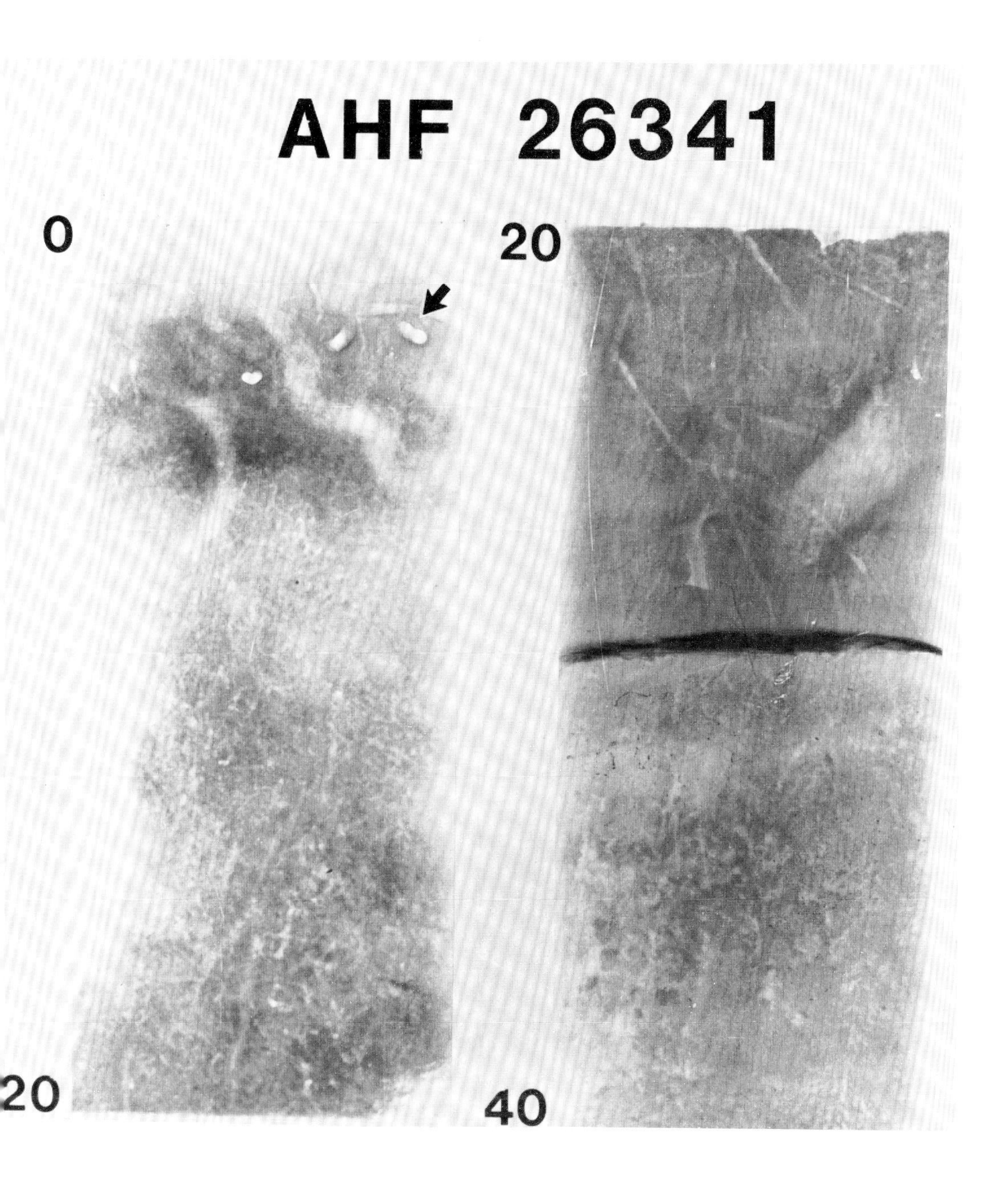

Fig. 10.—X-ray radiograph print of core AHF 26341, at 1,930 m. Note open burrow system at 5 cm (arrow) and threadlike reticulate burrow pattern at 6 to 30 cm, terminating in distal turbidite layer. Scale in cm.

northern slope of the basin (Fig. 11) where a large, horizontal backfill structure was formed. This structure is possibly a feeding trace produced by an irregular echinoid, but it occurs in water considerably deeper than the deepest visual observation of *Brissopsis* (e.g., Pl. 1D, 541 m). Similar structures have been described by Schäfer (1972, see Fig. 183).

Discrete tubes are also recognized in the area. On the ridge crest and upper slope, U-shaped tubes of suspension-feeding polychaetes extend 5 cm into the

Fig. 11.—X-ray radiograph print of box core AHF 26003, at 970 m. Arrow shows backfill structure probably produced by burrowing irregular echinoid (?). Scale in cm.

sediment (Fig. 12A, B). Note that spreiten are not visible in the radiograph (Fig. 12A). Types of tubes which are characteristic of the lower slope and base of slope are shown in Figure 13. A complex of 1-mm-diameter U-shaped to curvilinear tubes of polychaete origin are most common in the upper 5 cm of midslope to basin-floor cores (Fig. 13, #1). Vestiges of buried 1-mm-diameter tubes can be seen down-core. Another type of tube is 6 mm in diameter (Fig. 13, #2) and is inclined to the sediment surface at a 45° angle. This tube, associated with the circle-scribing polychaetes identified in bottom photographs, is restricted to the lower slope. Clustered structures composed of tubes 1 mm in diameter (Fig. 13, #3) are observed infrequently. These clustered structures are probably small, clustered polychaete tubes and are seen to sub-bottom depths of over 20 cm.

DISCUSSION

Life habit and feeding characteristics of benthic macrofauna exercise a strong control over the organism's ability to disrupt physical sedimentary structures or to distort important stratigraphic horizons (such as ash layers or calcium carbonate concentrations). Motile infauna effectively homogenize sediment, whereas sessile species tend to stabilize sediment by constructing dwelling tubes and only disturb the sediment in a narrow zone close to their dwellings. Motile infauna appear to be more important than epifauna in disturbing sediment.

Considering these trends, and given the observed increases in surface and subsurface, motile, infaunal species, an increase in bioturbation would be expected with increasing water depth. But such bioturbation gradients were not definable. Rather, the basin as a whole is characterized either by nonbioturbated, mass transport-produced physical structures or by uniformly homogenized sediment with bioturbate texture. Similar conditions of nondescript mottling have been observed in shallow-water sediment of the Mississippi Delta (Moore and Scruton, 1957) and in deep-sea pelagic carbonates of the western equatorial Pacific (Ekdale and Berger, 1978).

Considering the relatively large faunal densities found in the basin, intense bioturbation is to be expected; nevertheless, the paucity of distinct biogenic structures is, at first, surprising. Surficial sediment (upper 20 to 30 cm) of the slope and basin floor is typically a clayey silt (30 to 50% clay) with a water content of 90 to 130 percent (by weight of dry sediment). These types of sediment behave thixotropically when subjected to remolding at constant water contents (Boswell, 1961; Mitchell, 1976); that is, their shear strengths are reduced dramatically but return to their former levels with time. Infauna, particularly some polychaetes, remold sediment by a bolting action that results from body contractions and expansions (K. Fauchald, personal communication, 1978). The shock transmitted to the sediment can remold it and so cause a reduction in its strength. During the remolded state,

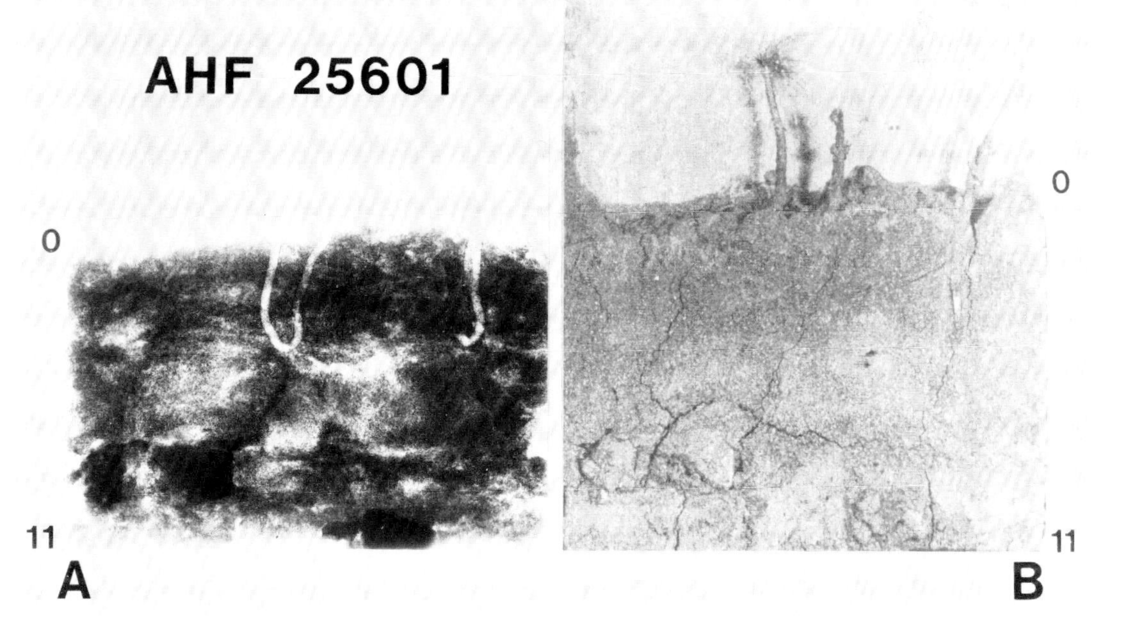

Fig. 12.—A: X-ray radiograph print of U-shaped polychaete tube in box core AHF 25601, at 631 m. Scale in cm. B: Photograph of the slab shown in A. Note polychaete tubes at core surface. Scale in cm.

AHF 24278

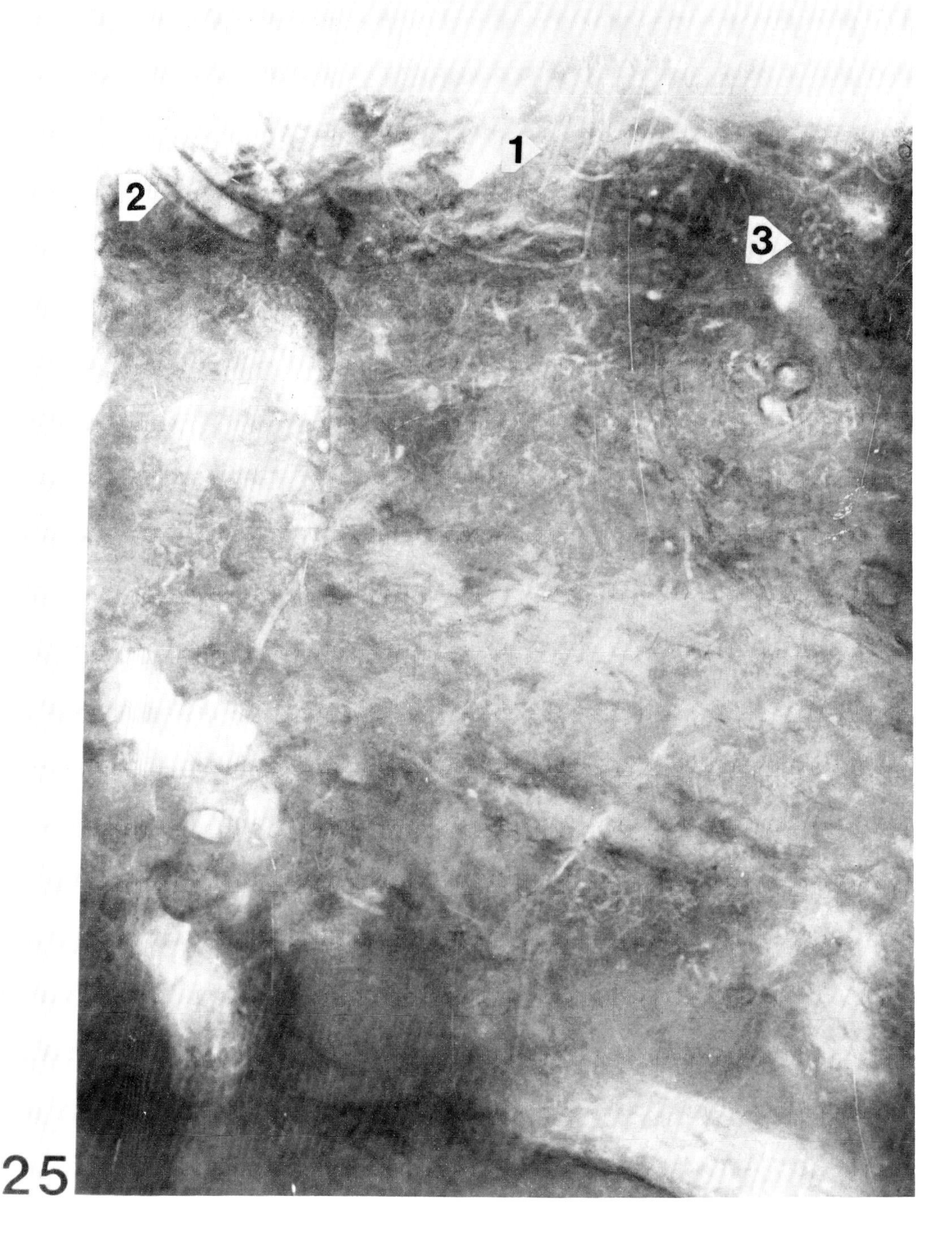

Fig. 13.—X-ray radiograph print of box core AHF 24278, at 1,756 m. 1: Curvilinear polychaete tubes (1 mm diameter). 2: Curvilinear polychaete tube (6 mm diameter). 3: Unidentified 1-mm-diameter burrow cluster. Scale in cm.

the animal moving through the sediment leaves an indistinct burrow. Alternatively, the animal may merely overcome the shear strength of the sediment by force and remold the sediment by passing through it. Continued faunal reworking helps to maintain high water content (Rhoads, 1970) and causes the overprinting of biogenic structures, which results in the observed bioturbate texture.

Structures recognizable on radiographs are essentially shadows of material in the sediment (Bouma, 1969). Uniformity of texture causes a lack of contrasting sediment for radiography and results in a relatively uniform radiograph. Similar observations have been made in deep-sea sediment where diagenetic enhancement allows the visual recognition in split cores of ichnofossils (Ekdale, 1977) which are not seen in radiographs (Ekdale, 1974).

In large part, the potential for preservation of biogenic structures is dependent on the nature of their host sediment (Hallam, 1975). The uniformity of Santa Cruz Basin sediment precludes the development of distinct structures in the uppermost surface sediment (i.e., above 30 cm). Surface traces also have poor preservation potential because the surface sediment is intensely reworked by meiofauna (Cullen, 1973; Chamberlain, 1975) and by surface deposit-feeding macrofauna. Consequently, the individual biogenic structure recorded from this bathyal environment will be biased toward the relatively rare, deep, infaunal burrow where low water content permits plastic dislocation of grains.

CONCLUSIONS

1) The oxygen-deficient bathyal environment of Santa Cruz Basin in the California Continental Borderland supports high benthic densities of macrofauna dominated by polychaete annelids.

2) With increasing water depth, infaunal invertebrates having a deposit-feeding, motile life style become more abundant.

3) The diversity of epibenthic traces is low as a result of the soft, soupy nature of surficial sediment, but some traces can occur in great abundance.

4) Distinct traces within the sediment cannot be recognized readily because of the homogeneity of the sediment. This homogeneity results in a lack of the color contrast necessary to visual observation and a lack of the density contrast that would enable effective X-ray radiography.

5) The sedimentary record from these bathyal environments will contain evidence of small-scale mass movements, and the sediment will have a nondescript bioturbate texture.

6) Preserved trace fossils will be biased toward those formed by deeply burrowing (>20 to 30 cm) infauna and will not represent accurately the original faunal densities or diversities.

ACKNOWLEDGMENTS

I thank Kristian Fauchald and G. F. Jones of the University of Southern California Department of Biological Sciences for providing the biological data used in the life structure analysis; B. E. Thompson and B. J. Balcolm for useful discussions on biological interpretations, and J. K. O'Toole and P. C. Day for assisting in the data collections at sea. I am indebted to the officers, technicians, and crew of the R/V VELERO IV and to the staff of the University of Southern California Marine Support Facility for their support of the field study. M. E. Field and G. W. Hill of the U.S. Geological Survey and A. A. Ekdale (Univ. of Utah) and J. D. Howard (Skidaway Inst. of Oceanography, Savannah) provided helpful and insightful reviews of the manuscript. This research was supported by Bureau of Land Management (BLM) Contract 08550-CT-5-52 and by National Science Foundation (NSF) grant OCE 76-00156. Any use of trade names is for descriptive purposes only and does not imply endorsement by the U.S. Geological Survey.

REFERENCES

Aller, R.C., and Dodge, R.E., 1974, Animal-sediment relations in a tropical lagoon, Discovery Bay, Jamaica: Jour. Marine Res., v. 32, p. 209–232.

Atwater, T., 1970, Implications of plate tectonics for the Cenozoic tectonic evolution of western North America: Geol. Soc. of America Bull., v. 81, p. 3513–3536.

Bainbridge, A., and Show, I. Jr., 1978, Hydrography: Bureau of Land Management Technical Report 1.0, v. II, Washington D.C., Bureau of Land Management.

Barnes, P.J., 1970, Marine geology and oceanography of Santa Cruz basin off southern California [Unpub. Ph.D. dissertation]: Los Angeles, Univ. Southern Calif., 175 p.

Boswell, P.G.H., 1961, Muddy Sediments: Cambridge, W. Heffer and Sons, Ltd., 140 p.

Bouma, A.H., 1969, Methods for the Study of Sedimentary Structures: New York, John Wiley and Sons, Inc., 458 p.

______, and Marshall, N.F., 1964, A method for obtaining and analysing undisturbed oceanic sediment samples: Marine Geology, v. 2, p. 81–99.

CALLAHAN, R.A., AND SHOKES, R., 1978, Southern California Baseline Study and Analysis 1975–1976, v. I, Executive Summary: U.S. Dept. Commerce, Nat'l. Tech. Info. Service, PB-290 739, 35 p.
CHAMBERLAIN, C.K., 1975, Trace fossils in DSDP cores of the Pacific: Jour. Paleontology, v. 49, p. 1074–1096.
COOK, H.E., 1979, Ancient continental slope sequences and their value in understanding modern slope development, *in* Doyle, L.J., and Pilkey, O.H., Jr. (eds.), Geology of Continental Slopes: Soc. Econ. Paleontologists Mineralogists Spec. Pub. No. 27, p. 287–305.
CRIMES, T.P., 1975, The stratigraphical significance of trace fossils, *in* Frey, R.W. (ed.), The Study of Trace Fossils: New York, Springer-Verlag, Inc., p. 109–130.
CROWELL, J.C., 1974, Origin of late Cenozoic basins in southern California, *in* Dickinson, W.R. (ed.), Tectonics and Sedimentation: Soc. Econ. Paleontologists Mineralogists Spec. Pub. 22, p. 190–204.
CULLEN, D.J., 1973, Bioturbation of superficial marine sediments by interstitial meiobenthos: Nature, v. 242, p. 323–324.
DEMARS, K.R., AND TAYLOR, R.J., 1971, Naval seafloor soil sampling and in-place test equipment: A performance evaluation: Naval Civil Eng. Lab. Tech. Report R 730, 59 p.
EDWARDS, B.D., 1979, Animal-sediment relationships in dysaerobic bathyal environments, California Continental Borderland: Unpub. Ph.D. dissertation, Los Angeles, Univ. Southern Calif., 291 p.
EKDALE, A.A., 1974, Geologic history of the abyssal benthos: Evidence from trace fossils in Deep Sea Drilling Project Cores: Unpub. Ph.D. dissertation, Houston, Rice Univ., 156 p.
______, 1977, Abyssal trace fossils in worldwide Deep Sea Drilling Project cores, *in* Crimes, T.P. and Harper, J.C. (eds.), Trace Fossils 2: Geol. Jour. Spec. Issue No. 9, Liverpool, Seel House Press, p. 163–182.
______, AND BERGER, W.H., 1978, Deep-sea ichnofacies: Modern organism traces on and in pelagic carbonates of the western equatorial Pacific: Palaeogeogr., Palaeoclimatol., Palaeoecol., v. 23, p. 263–278.
EMERY, K.O., 1954, Source of water in basins off southern California: Jour. Marine Res., v. 13, p. 1–12.
______, 1960, The Sea Off Southern California: New York, John Wiley and Sons, Inc., 366 p.
______, AND BRAY, E.E., 1962, Radiocarbon dating of California basin sediments: Am. Assoc. Petroleum Geologists Bull., v. 46, p. 1839–1856.
FAUCHALD, K., AND JONES, G.F., 1976, Benthic macrofauna of the southern California Bight: Unpub. Report, Los Angeles, Univ. Southern Calif., 486 p.
FELL, H.B., 1967, Biological applications of sea-floor photography, *in* Hersey, J. B. (ed.), Deep-Sea Photography: Baltimore, The Johns Hopkins Press, p. 207–221.
FIELD, M.E., AND EDWARDS, B.D., 1980, Slopes of the southern California Continental Borderland: a regime of mass transport, *in* Field, M.E., Bouma, A.H., Colburn, I.P., Douglas, R.G., and Ingle, J.C. (eds.), Quaternary Depositional Environments of the Pacific Coast: Los Angeles, Pacific Coast Paleogeogr. Symp. 4, Soc. Econ. Paleontologists Mineralogists Pac. Section, p. 169–184.
GRIGGS, G.B., CAREY, A.G., JR., AND KULM, L. D., 1969, Deep-sea sedimentation and sediment-fauna interaction in Cascadia Channel and on Cascadia Abyssal Plain: Deep-Sea Research, v. 16, p. 157–170.
HALLAM, A., 1975, Preservation of trace fossils, *in* Frey, R.W. (ed.), The Study of Trace Fossils: New York, Springer Verlag, Inc., p. 55–63.
HEEZEN, B.C., AND HOLLISTER, C.D., 1971, The Face of the Deep: New York, Oxford Univ. Press, 659 p.
HERSEY, J.B., 1967, The manipulation of deep-sea cameras, *in* Hersey, J.B. (ed.), Deep Sea Photography: Baltimore, The Johns Hopkins Press, p. 55–67.
HOLLISTER, C.D., HEEZEN, B.C., AND NAFE, K.E., 1975, Animal traces on the deep-sea floor, *in* Frey, R.W. (ed.), The Study of Trace Fossils: New York, Springer-Verlag, Inc., p. 493–510.
JUMARS, P.A., AND FAUCHALD, K., 1977, Between-community contrasts in successful polychaete feeding strategies, *in* Coull, B.C. (ed.), Ecology of Marine Benthos: The Belle W. Baruch Inst. Mar. Biol., Univ. S. Carolina Press, v. 6, p. 1–2.
MITCHELL, J.K., 1976, Fundamentals of Soil Behavior: New York, John Wiley & Sons, Inc., 422 p.
MOORE, D.G., AND SCRUTON, P.C., 1957, Minor internal structures of some recent unconsolidated sediments: Am. Assoc. Petroleum Geol. Bull., v. 41, p. 2723–2751.
MOORE, G.W., 1976, Basin development in the California borderland and in the Basin and Range Province, *in* Howell, D.G., Aspects of the Geologic History of the California Continental Borderland: Am. Assoc. Petroleum Geologists Pacific Section, Misc. Paper No. 24, p. 383–390.
MOORE, H.B., 1939, Faecal pellets in relation to marine deposits, *in* Trask, P. D. (ed.), Recent Marine Sediments: Tulsa, Am. Assoc. Petroleum Geologists Symp., p. 516–524.
NARDIN, T.R., EDWARDS, B.D., AND GORSLINE, D.S., 1979, Santa Cruz Basin, California borderland: dominance of slope processes in basin sedimentation, *in* Doyle, L.J., and Pilkey, O.H., Jr. (eds.), Geology of Continental Slopes: Soc. Econ. Paleontologists Mineralogists Spec. Pub. No. 27, p. 209–221.
OWEN, D.M., SANDERS, H.L., AND HESSLER, R.R., 1967, Bottom photography as a tool for estimating benthic populations, *in* Hersey, J.B. (ed.), Deep-Sea Photography: Baltimore, The John Hopkins Press, p. 229–234.
RHOADS, D.C., 1970, Mass properties, stability and ecology of marine muds related to burrow activity, *in* Crimes, T.P., and Harper, J.C. (eds.), Trace Fossils: Geol. Jour. Spec. Issue No. 3, Liverpool, Seel House Press, p. 391–406.
______, AND MORSE, J.W., 1971, Evolutionary and ecologic significance of oxygen-deficient marine basins: Lethaia, v. 4, p. 413–428.
______, AND YOUNG, D.K., 1970, The influence of deposit-feeding benthos on bottom sediment stability and community trophic structure: Jour. Marine Research, v. 28, p. 150–178.

ROSFELDER, A.M., AND MARSHALL, N.F., 1967, Obtaining large, undisturbed and orientated samples in deep water, *in* Richards, A.F. (ed.), Marine Geotechnique: Urbana, University of Illinois Press, p. 243–263.

SCHÄFER, W., 1972, Ecology and Paleoecology of Marine Environments: Chicago, Univ. of Chicago Press, 568 p.

SHEPARD, F.P., AND EMERY, K.O., 1941, Submarine topography off the California coast—Canyons and tectonic interpretation: Geol. Soc. America Spec. Paper 31, 171 p.

SKOGSBERG, T., 1936, Hydrography of Monterey Bay, California thermal conditions, 1929–1933: Trans. Am. Philosophical Soc. of Philadelphia, New Series, v. 29.

STANLEY, D.J., 1971, Fish-produced markings on the Atlantic outer margin off north-central United States: Jour. Sed. Petrology, v. 41, p. 159–170.

SVERDRUP, H.V., AND FLEMING, R.H., 1941, The waters off the coast of Southern California, March to July, 1937: Bull. Scripps Inst. Oceanography, v. 4, 1936–1941, p. 261–375.

VEDDER, J.G., BEYER, L.A., JUNGER, A., MOORE, G.W., ROBERTS, A.E., TAYLOR, J.C., AND WAGNER, H.C., 1974, Preliminary report on the geology of the continental borderland of southern California: U.S. Geol. Survey Misc. Field Studies Map (MF) 624, 34 p., 9 maps.

YEATS, R.S., COLE, M.R., MERSCHAT, W.R., AND PARSLEY, R.M., 1974, Poway Fan and submarine cone and rifting of the inner southern California Borderland: Geol. Soc. America Bull., v. 85, p. 293–302.

TRACE FOSSILS AND MID-CRETACEOUS ANOXIC EVENTS IN THE ATLANTIC OCEAN

A.A. EKDALE
Department of Geology and Geophysics, University of Utah, Salt Lake City, Utah 84112

ABSTRACT

Patterns and intensity of bioturbation in marine deposits are useful indicators of the response of benthic organisms to fluctuating oxygen levels in bottom waters. Trace fossil assemblages in mid-Cretaceous (Barremian-Albian) Deep Sea Drilling Project (DSDP) core sections from the North and South Atlantic Ocean were examined in order to document the activities of burrowing infauna relative to anoxic episodes in deep Atlantic basins during that time.

Mid-Cretaceous anoxic events typically are represented in DSDP cores by dark, homogeneous or laminated organic muds, which alternate with moderately to heavily burrowed facies containing less organic carbon. Bioturbation intensity and trace fossil diversity appear to correlate inversely with the amount of unoxidized carbon and total sulfur in the sediment, suggesting that the more organic-rich facies were deposited under conditions where oxygen was a limiting factor for benthic infaunal macro-organisms.

The ichnogenus *Chondrites* commonly occurs, sometimes to the exclusion of all other kinds of burrows, in zones immediately above and (or) below unburrowed, laminated (oxygen-depleted) mud. It also occurs in heavily burrowed (oxygenated) limestones containing rich trace fossil faunas, including *Zoophycos*. Therefore, it appears that mid-Cretaceous *Chondrites* were created by animals possessing a broad range of oxygen tolerance. The presence of *Chondrites* alone in an organic-rich deposit probably indicates dysaerobic conditions.

INTRODUCTION

No animal can live under totally anoxic conditions for an extended period of time, so the mere occurrence of burrows in a sediment indicates the presence of at least some oxygen in the bottom environment. However, some bottom-dwelling creatures can tolerate lower oxygen levels than others (Rhoads and Morse, 1971). It is mainly the soft-bodied infauna, such as mud-borrowing polychaetes, that inhabit slightly euxinic sediments. Organisms with calcareous, siliceous, or phosphatic hard parts hardly ever are found living in wholly reducing environments. There are some shelled organisms, such as thalassinidean crustaceans and various bivalves, which may be found inhabiting anaerobic substrates, but even in these cases the organisms require oxygenated bottom water to be pumped through their burrow systems.

Trace fossils may be useful in depositional interpretations of black sedimentary deposits, because if particular kinds of burrows can be related to particular kinds of burrowers with different oxygen threshold tolerances, a trace fossil facies model based on the oxidation potential of the aqueous environment may be established. In other words, a certain type of burrow assemblage may indicate certain chemical conditions of the bottom environment. Such anaerobic ichnofacies models would aid in understanding trends from aerobic to anaerobic conditions (and vice versa) on and in the ancient sea bottom.

Dark, laminated, apparently anaerobic sediments are not widespread in the ocean today. Rather, they occur only in (1) coastal lagoons and intertidal flats, where the rate of sediment supply and especially the rate of supply of organic material is so high that the sediment cannot be completely oxidized; (2) restricted basins, such as the Black Sea or California Borderland basins, where turnover of the bottom water is extremely slow; and (3) areas in the open ocean where the oxygen-minimum zone in the water column impinges upon the ocean bottom. Sediments deposited in these environments may or may not be carbonate-rich, but, depending upon productivity in the overlying water column, they almost always are very high in organic carbon content (often 1% or more). As the dissolved oxygen content of bottom waters approaches zero, the benthic biomass likewise approaches zero.

Rhoads and Morse (1971) studied the distribution of benthic organisms in anaerobic basins of the southern California Borderland, the Gulf of California, and the Black Sea. They found that the lower limit of dissolved oxygen in the water that can be tolerated by multicellular benthic life in these areas is generally about 0.1 ml O_2/l H_2O. "Anaerobic" environments below that oxygen level lack any living thing, either on or in the bottom. "Dysaerobic" conditions (0.1 to 1.0 ml O_2/l H_2O) support a low diversity of small, soft-bodied, infaunal species. "Aerobic" conditions (in excess of 1.0 m. O_2/l H_2O) support a much higher diversity and abundance of benthos, including those organisms which secrete calcareous shells or hard parts.

Patterns and intensity of bioturbation in ancient marine deposits are useful indicators of the response of benthic organisms to fluctuating oxygen levels in the bottom water. For the investigation reported here, trace fossil assemblages in mid-Cretaceous (Barremian-Albian) Deep Sea Drilling Project (DSDP) core sections from the North and South Atlantic were examined in order to document the activities of burrowing infauna relative to anoxic episodes in deep Atlantic basins during that time.

OCEANIC ANOXIC EVENTS

One of the more intriguing results of the DSDP has been the discovery that extensive organic-rich deposits were of widespread occurrence in the World Ocean during various portions of the Cretaceous Period (e.g., see Pl. 1) and, moreover, that these strata apparently were correlative with similar black shale and limestone units known in land-based outcrops on several continents. Interest in these organic-rich, deep-sea deposits has been very high in recent years due to their potential as hydrocarbon source rocks (e.g., see Arthur and Schlanger, 1979).

The organic-rich facies occur in sedimentary cycles which appear to represent fluctuating oxygen levels at the ocean floor. These aerobic-anaerobic cycles are particularly well-developed in two parts of the Cretaceous column, one spanning from the late Barremian to late Albian and the other extending from the late Cenomanian to early Turonian. The magnitude and worldwide extent of these two major episodes of periodic anoxia at the sea floor has led geologists to refer to them as "oceanic anoxic events" (Schlanger and Jenkyns, 1976; Arthur and Schlanger, 1979).

The Barremian-Albian event ("OAE 1" of Arthur and Schlanger, 1979) lasted for about 18 to 20 million years; the Cenomanian-Turonian event ("OAE 2" of Arthur and Schlanger, 1979) was shorter, lasting only 3 to 5 million years or less, but it apparently was considerably greater in stagnation intensity; i.e., in terms of organic richness and carbon accumulation rates. The trace fossil research reported in this paper deals only with the Barremian-Albian event.

The cause(s) and effects of the Cretaceous oceanic anoxic events have been considered by many workers (e.g., Berger and von Rad, 1972; Schlanger and Jenkyns, 1976; Fischer and Arthur, 1977; Hallam, 1977; Ryan and Cita, 1977; Thiede and van Andel, 1977; Roth, 1978; Arthur, 1979; Arthur and Schlanger, 1979; Weissert et al., 1979). Basically, these investigations have outlined three possible causes of extensive deposition of organic-rich, deep-sea sediments during the Cretaceous Period: (1) formation of restricted anoxic basins, (2) expansion of the oxygen-minimum zone in the water column to intersect with vast expanses of the deep-sea floor, and (3) detrital deposition in the deep sea of large volumes of organic debris derived from shallow-water and continental environments.

The first model resembles the situation in the modern Black Sea, where an oxygenated, low-salinity, surface water mass, derived from coastal runoff, overlies a saltier, denser water mass entering from the Mediterranean Sea through the Straits of Bosporus. This denser layer of water stagnates, because the shallow depth of the Bosporus passage acts as a sill, and the resultant density contrast in the water column prevents rapid overturn. Thus, only the top 200 m (or less) of the 2200 m-deep water column in the Black Sea is oxygenated, and so dark-colored, unoxidized sediments accumulate throughout the basin. In the young Atlantic Ocean of mid-Cretaceous age, relatively narrow basins were common, and in some cases bottom circulation may have been impeded by sills. A Black Sea anoxic basin model (Degens and Stoffers, 1976; Ryan and Cita, 1977) may at least partially explain extensive black shale deposition in the Cretaceous Atlantic and possibly also Indian Ocean basins, especially during the Barremian-Albian anoxic event. However, such a model does not adequately account for similar deposits in the Pacific Ocean of the same age.

The second model of anoxic sediment deposition relies on an oxygen-minimum zone considerably thicker than that observed in today's oceans (Schlanger and Jenkyns, 1976; Thiede and van Andel, 1977). Increased latitudinal thermal equitability during the Cretaceous may have caused a general reduction in oceanic circulation, which in turn decelerated the replenishment of oxygen-depleted intermediate water masses and allowed the expansion of the oxygen minimum zone to abnormally great water depths. The impingement of this low-oxygen (or no-oxygen) layer on broad areas of the Cretaceous sea floor, including perhaps the entire floor of some basins in the North and South Atlantic, resulted in widespread accumulation of organic-rich sediment.

The third model, which depends upon periodic rapid influxes of enormous amounts of land-derived organic detritus, implies that black shales and limestones need not have been deposited under truly anaerobic conditions. Aptian-Albian black shales drilled in the Bay of Biscay on DSDP Leg 48 have been interpreted by means of this hypothesis (de Graciansky et al., 1979; Timofeev and Bogolyubova, 1979). However, it seems doubtful that this model can be used to explain adequately the worldwide occurrences of approximately synchronous, organic-rich deposits in mid-Cretaceous sections.

TRACE FOSSIL RECORD

Samples

For this investigation, Barremian-Albian sections of DSDP cores from nine North and South Atlantic drill sites were examined for physical and biogenic sedimentary structures, which were described, photographed, and sampled for chemical analysis (Table 1). The paleodepth and paleogeographic location of each site during the mid-Cretaceous (approximately 110 million years ago) were determined by backtracking the oceanic plate (Fig. 1).

The general stratigraphy of all the cores used in this study is outlined in the appropriate volumes of the

Initial Reports of the Deep Sea Drilling Project. Additional, more precise stratigraphic information for certain samples was provided by P. H. Roth (oral communication, 1980), based on calcareous nannoplankton zonations.

Systematic Ichnology
Ichnogenus CHONDRITES Sternburg, 1833
Chondrites ichnosp.
Pl. 1B

Description.—*Chondrites* is a highly branching burrow system, produced presumably by a deposit-feeding invertebrate. In vertical cross-section in deep-sea cores, *Chondrites* typically appears as a cluster of tiny circular or elongate dots where the slice through the core has truncated the numerous branching tunnels. (For further descriptions of *Chondrites* in DSDP cores, see Warme et al., 1973; Chamberlain, 1975; Ekdale, 1977, 1978.)

Ichnogenus PLANOLITES Nicholson, 1873
Planolites ichnosp.
Pl. 1B

Description.—*Planolites* is a simple, smooth-walled, generally unbranched tunnel produced in a horizontal or sub-horizontal plane presumably by a deposit-feeding invertebrate. In vertical cross-section in deep-sea cores, *Planolites* typically appears as an isolated circular or elliptical dot where the slice has cut transversely through the tunnel. (For further descriptions of *Planolites* in DSDP cores, see Warme et al., 1973; Chamberlain, 1975; Ekdale, 1977, 1978.)

Ichnogenus ZOOPHYCOS Massalongo, 1855
Zoophycos ichnosp.

Description.—*Zoophycos* is a complex burrow system composed of a horizontal or nearly horizontal, sheet-like spreite, which may be flat, curved, inclined or wound in a helix around a vertical axis. In vertical cross-section in deep-sea cores, *Zoophycos* typically appears as a series of chevron-shaped mottles lined up in row where the slice has cut across a portion of the spreite. (For further descriptions of *Zoophycos* in DSDP cores, see Warme et al., 1973; Chamberlain, 1975; Ekdale, 1977, 1978).

SEDIMENT CHEMISTRY

Chemical Analyses

Organic carbon, calcium carbonate, and total sulfur concentrations in each sample were determined. Prepa-

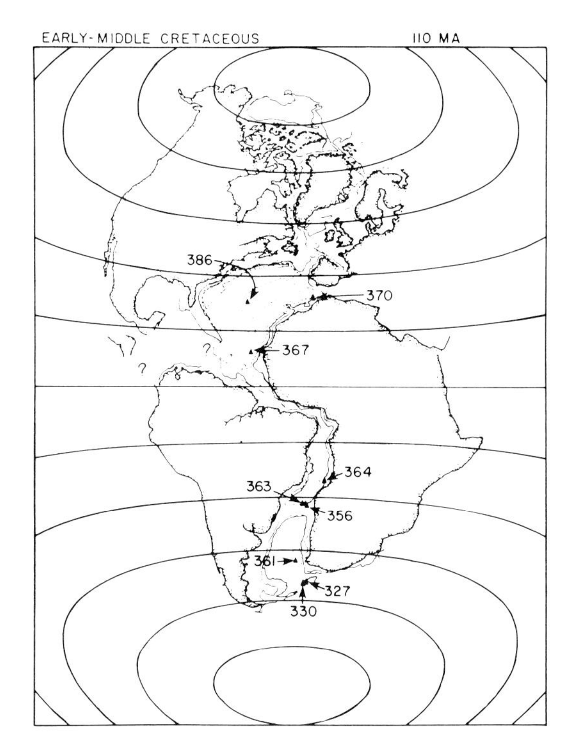

Fig. 1.—Locations of DSDP sites studied for this report, plotted on an Early-Middle Cretaceous (Aptian) paleogeographic map representing the configuration of the Atlantic Ocean approximately 110 million years ago (after Sclater et al., 1977).

TABLE 1.—SAMPLES OF BARREMIAN-ALBIAN SEDIMENT ANALYZED FOR THIS STUDY

DSDP Site	Paleo-Depth (m)	Paleo-Latitude	No. of Samples	$\%C_{Organic}$ Min.	$\%C_{Organic}$ Max.	$\%CaCO_3$ Min.	$\%CaCO_3$ Max.	$\%S_{total}$ Min.	$\%S_{total}$ Max.	NC Zone[1]
327A	Shelf	54°S	21	0.00	8.52	0.17	33.70	0.04	1.98	6–11
330	Shelf	54°S	4	0.51	4.55	0.08	36.20	0.05	2.10	8A
356	3500 (?)	32°S	4	0.23	8.34	4.00	60.83	0.01	3.06	10
361	3000–3500	48°S	13	0.07	4.89	0.00	23.30	0.00	3.62	6–8?
363	Shelf	32°S	4	0.03	0.84	66.00	82.08	0.04	0.19	8–10
364	2950–3100	25°S	23	0.09	15.52	0.07	86.50	0.02	3.30	8–10
367	3600–3850	10°N	30	0.01	3.23	1.00	94.96	0.00	1.33	6–11
370	4150	25°N	42	0.25	4.07	2.25	34.32	0.12	1.87	8–10?
386	1900	23°N	8	0.23	3.64	2.42	31.74	0.22	0.35	7–10?

[1]Calcareous nannofossil zone (P. H. Roth, oral communication, 1980).

A
B

ration of the samples included drying, pulverizing, and random splitting into duplicate subsamples. A LECO automatic carbon determinator analyzed the total carbon content of two duplicate subsamples, one of which had been baked in a muffle furnace at 500°C for 2½ hours prior to analysis to drive off volatile organic compounds. For this study, the total carbon content of the baked subsamples was assumed to represent only carbonate carbon. That value was subtracted from the total carbon content of the unbaked subsample to obtain the percentage of organic carbon in the whole sample. Organic carbon values were assumed to represent a direct, quantitative measure of the degree of oxygenation of the bottom environment; that is, high organic carbon values indicate low oxygenation.

A third split from each sample was analyzed for total sulfur content using a LECO automatic sulfur determinator. No attempt was made to separate sulfur of different oxidation states by wet chemical techniques because most of the samples analyzed contained poorly oxidized sediments. It was assumed that most of the sulfur in the samples was reduced to sulfide.

The results of the chemical analyses and their relation to the distribution of trace fossils in the DSDP cores are shown in Figures 2 through 9. The results from South Atlantic and North Atlantic sites are depicted separately because the nature and intensity of the anoxic events in the two regions were not necessarily the same, perhaps due to differences between the South and North Atlantic in surface productivity, sedimentation rates, and bottom circulation.

For each sample analyzed chemically, morphologically distinct trace fossils were identified and noted, and the intensity of bioturbation was characterized by visual inspection. Unburrowed sediment was categorized as either homogeneous (H) or laminated (L). Burrowed sediment was categorized as sparsely bioturbated (S) if only a very few burrows occurred in an otherwise homogeneous or laminated stratum; moderately bioturbated (M) if burrows were abundant, but

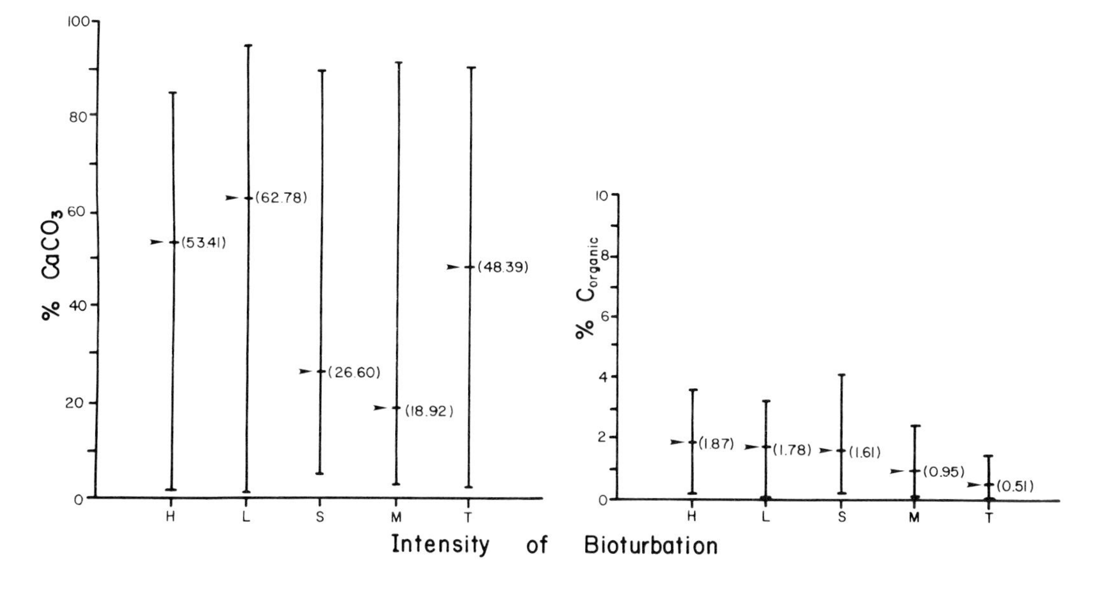

Fig. 2.—Organic carbon and calcium carbonate concentrations at North Atlantic sites plotted against intensity of bioturbation. H, Homogeneous unbioturbated sediment; L, Laminated unbioturbated; S, Sparsely bioturbated; M, Moderately bioturbated; T, Totally bioturbated. Arrows indicate mean values.

EXPLANATION OF PLATE 1

Examples of anaerobic cycles in mid-Cretaceous DSDP cores from the North Atlantic.

FIG. A.— Unbioturbated, laminated limestone ($CaCo_3$ = 86.47%; $C_{organic}$ = 1.63%; S_{total} = 0.61%) grading up into totally bioturbated limestone ($CaCO_3$ = 90.55%; $C_{organic}$ = 0.10%; S_{total} = 0.06%). DSDP Core 367-25-4 (21–40 cm interval). Width of core is 5.5 cm.

B.— Sparsely bioturbated mudstone ($CaCO_3$ = 23.24%; $C_{organic}$ = 2.33%; S_{total} = 0.23%) grading up through moderately bioturbated mudstone with abundant *Chondrites* into totally bioturbated mudstone with abundant *Plantolites* ($CaCO_3$ = 7.912%; $C_{organic}$ = 0.77%; S_{total} = 0.06%). Note the flattened pyrite nodule near the middle of the core. DSDP Core 370-27-1 (51–66 cm interval). Width of core is 5.5 cm.

if a laminated or homogeneous matrix texture was apparent in small, isolated patches; or totally bioturbated (T) if all primary sedimentary features were obscured by burrows or bioturbate textures. The only identifiable ichnogenera in the core sections studied were *Chondrites, Planolites,* and *Zoophycos*.

No direct correlation exists between the intensity of bioturbation and minimum values of $CaCO_3$, organic carbon, or total sulfur (Figs. 2, 3, 8, and 9); the minimum value is near zero in every case. However, for South Atlantic sites there is an inverse relation between maximum values of both organic carbon and sulfur and the intensity of bioturbation. A similar inverse relation is indicated for North Atlantic sites as well, although anomalies exist in the sense that the maximum organic carbon value occurs in sparsely bioturbated (rather than

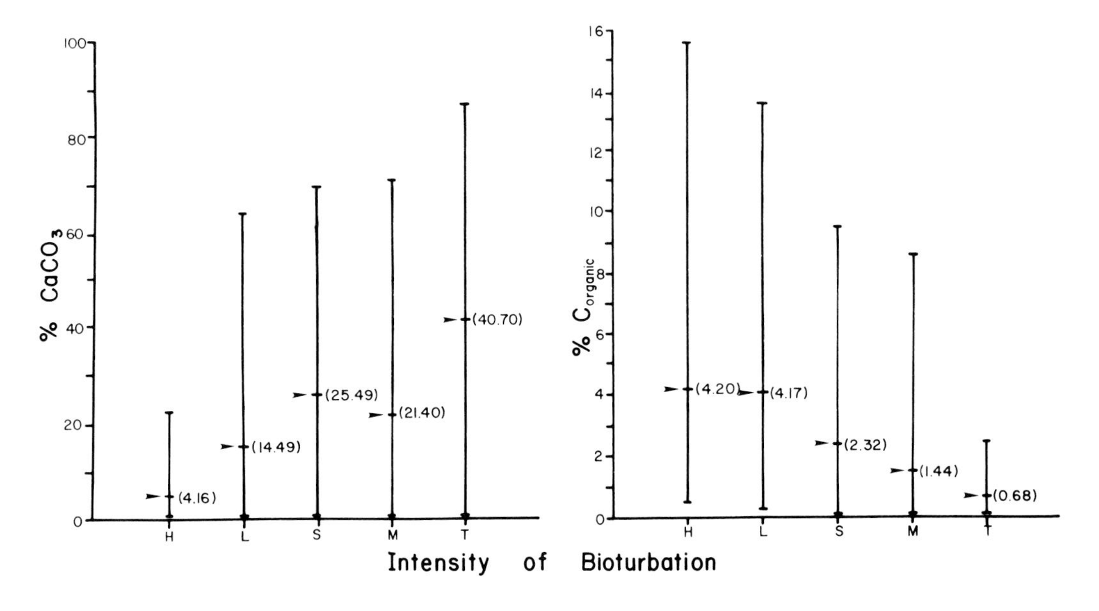

Fig. 3.—Organic carbon and calcium carbonate concentrations at South Atlantic sites plotted against intensity of bioturbation. H, Homogeneous unbioturbated sediment; L, Laminated unbioturbated; S, Sparsely bioturbated; M, Moderately bioturbated; T, Totally bioturbated. Arrows indicate mean values.

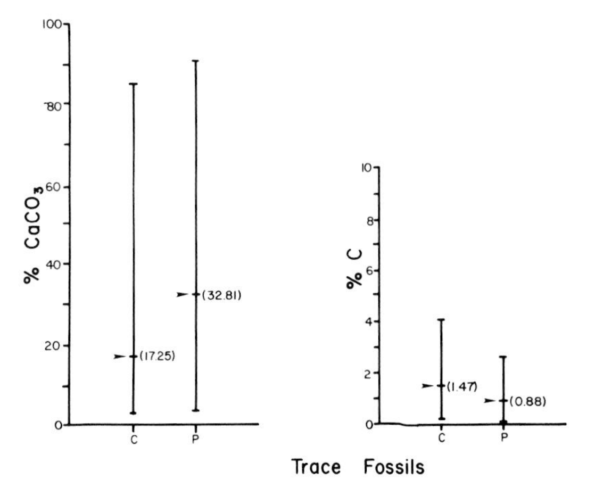

Fig. 4.—Range of organic carbon and calcium carbonate concentrations at North Atlantic sites in which various trace fossils occur. C, *Chondrites*; P, *Planolites*. Arrows indicate mean values.

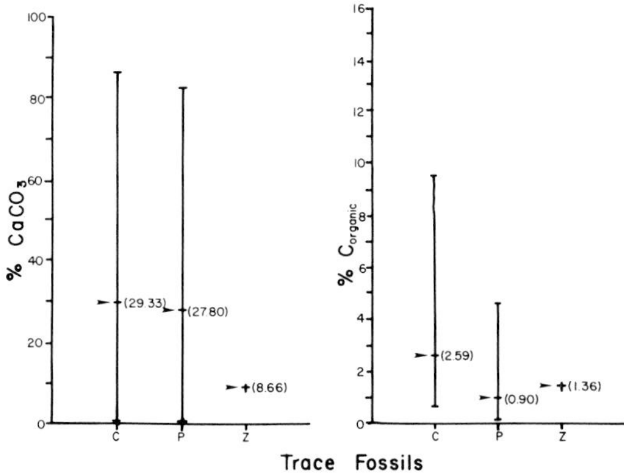

Fig. 5.—Range of organic carbon and calcium carbonate concentrations at South Atlantic sites in which various trace fossils occur. C, *Chondrites*; P, *Planolites*; Z, *Zoophycos* (one sample only). Arrows indicate mean values.

unbioturbated) sediment, and the maximum sulfur value occurs in moderately bioturbated sediment. When mean values are considered, there is a decrease in organic carbon and sulfur content with increasing intensity of bioturbation in virtually all cases.

The relation between bioturbation intensity and $CaCO_3$ content is somewhat confusing. In North Atlantic samples there seems to be no correlation between maximum, minimum, or mean values of $CaCO_3$ content and the intensity of bioturbation. In South Atlantic samples, however, both maximum and (with the exception of an anomaly in sparsely bioturbated sediment) mean values appear to increase with increasing intensity of bioturbation.

Chondrites and *Planolites* were by far the most abundant trace fossils in all the mid-Cretaceous DSDP cores. In fact, the only other identifiable burrow type was *Zoophycos*, which occurred in just one sample of clayey mudstone (8.7% $CaCO_3$; 1.4% $C_{organic}$; 0.1% S_{total}). Both *Chondrites* and *Planolites* were found in sediments with minimum values of $CaCO_3$, organic carbon, and total sulfur that were close to zero (Figs. 4, 5, 8, and 9). In both the North and South Atlantic, the maximum and mean values of organic carbon and sulfur in the deposits in which *Chondrites* occurred are slightly higher than those of the deposits in which *Planolites* occurred. Maximum and mean values of $CaCO_3$ are slightly higher for *Chondrites* in the South Atlantic and for *Planolites* in the North Atlantic.

Although no more than two ichnogenera occur in any sample, it may be instructive to observe the relation between trace fossil diversity and sediment chemistry (Figs. 6, 7, 8, and 9). Once again, minimum values of $CaCO_3$, organic carbon, and sulfur are near zero in every case. Maximum and mean values of organic carbon and sulfur show an inverse relation with trace fossil diversity in the South Atlantic. The North Atlantic samples show a similar inverse relation for organic carbon but not for sulfur content. The $CaCO_3$ content, based on mean values, appears to increase with trace fossil diversity in the South Atlantic but decrease with diversity in the North Atlantic.

Discussion

The amount of unoxidized carbon and total sulfur in the sediment is assumed to be inversely related to the amount of dissolved oxygen available for oxidation reactions in interstitial and bottom waters. Both the intensity of bioturbation and the diversity of biogenic structures decline with increasing percentages of organic carbon and sulfur in a given stratum, thus suggesting that burrowing activity became essentially nil with the onset of oxygen-poor conditions at the sea bottom during various portions of the mid-Cretaceous. These trends are better defined for the South Atlantic sites than for the North Atlantic sites, possibly because the South Atlantic basins may have been more completely restricted by tectonic sills, may have been limited to a narrower water depth range over which the oxygen-minimum zone was juxtaposed, and (or) may have received a greater supply of organic-rich sediment. The relationship between bioturbation and $CaCO_3$ percentage of the sediment is not straightforward because the $CaCO_3$ content is more directly related to fertility in the overlying water column and water depth at the site of deposition (relative to the calcite compensation depth) than it is to the dissolved oxygen content of the bottom water.

All the mid-Cretaceous DSDP core sections examined for this study contain a depauperate trace fossil

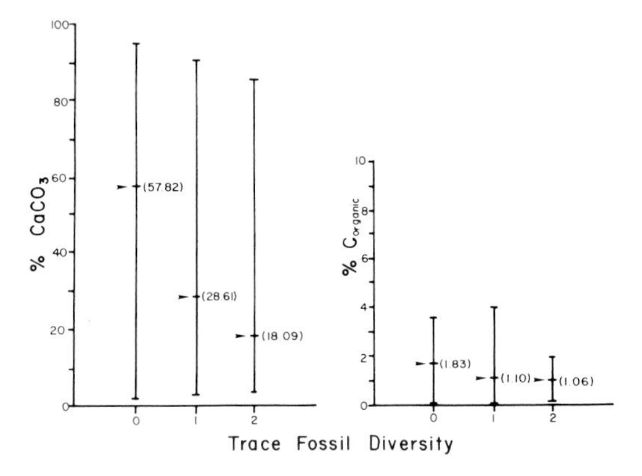

Fig. 6.—Organic carbon and calcium carbonate concentrations at North Atlantic sites plotted against diversity of biogenic structures. O, No burrows in samples; 1, One discernible ichnogenus; 2, Two discernible ichnogenera. Arrows indicate mean value.

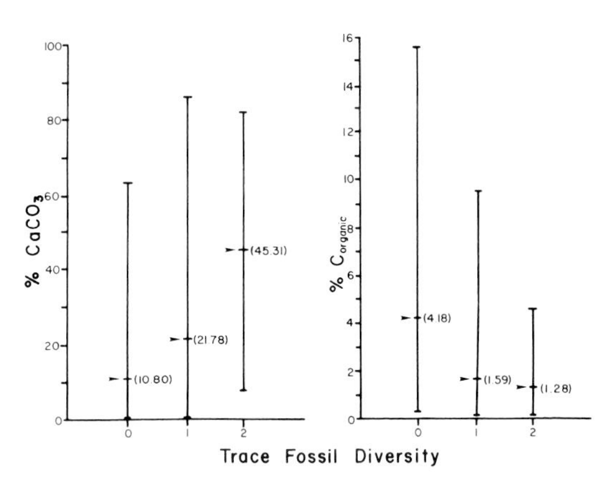

Fig. 7.—Organic carbon and calcium carbonate concentrations at South Atlantic sites plotted against diversity of biogenic structures. O, No burrows in sample; 1, One discernible ichnogenus; 2, Two discernible ichnogenera. Arrows indicate mean values.

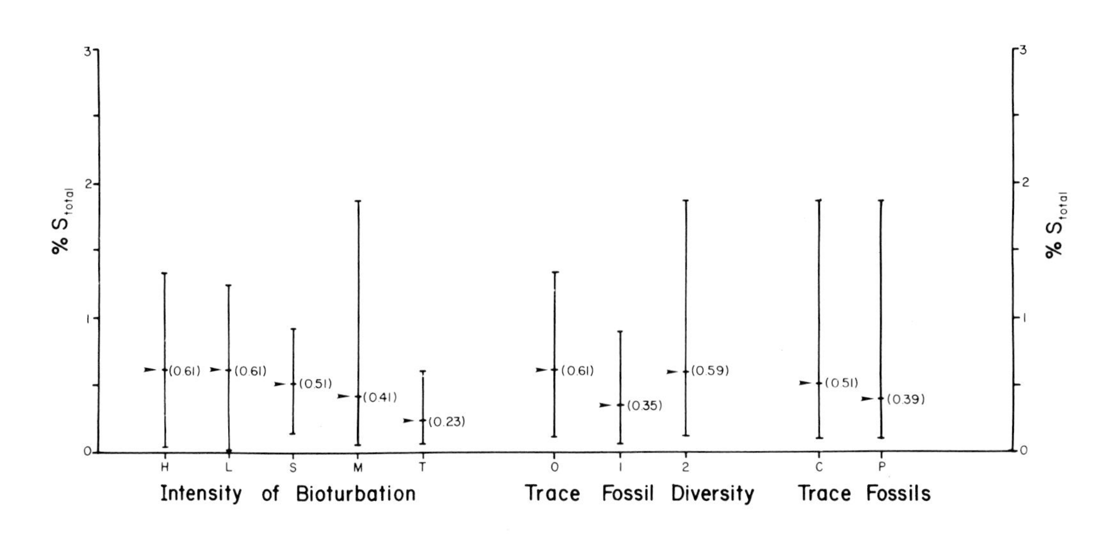

Fig. 8.—Total sulfur concentrations at North Atlantic sites plotted against intensity of bioturbation (H, Homogeneous unbioturbated sediment; L, Laminated unbioturbated; S, Sparsely bioturbated; M, Moderately bioturbated; T, Totally bioturbated), diversity of biogenic structures (O, No burrows; 1, One discernible ichnogenus; 2, Two discernible ichnogenera), and various trace fossils (C, *Chondrites*; P, *Planolites*). Arrows indicate mean values.

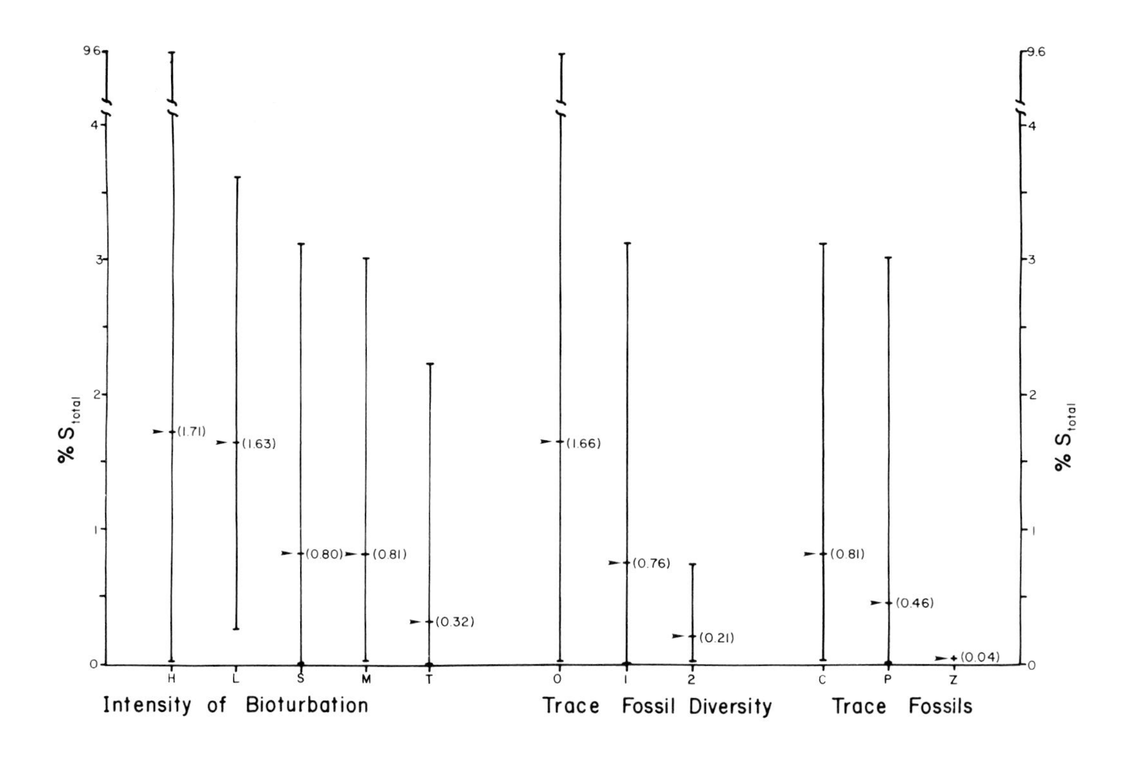

Fig. 9.—Total sulfur concentrations at South Atlantic sites plotted against intensity of bioturbation (H, Homogeneous unbioturbated sediment; L, Laminated unbioturbated; S, Sparsely bioturbated; M, Moderately bioturbated; T, Totally bioturbated), diversity of biogenic structures (O, No burrows; 1, One discernible ichnogenus; 2, Two discernible ichnogenera), and various trace fossils (C, *Chondrites*; P, *Planolites*; Z, *Zoophycos*). Arrows indicate mean values.

fauna. Even in totally bioturbated layers that contain little or no unoxidized carbon, the identifiable trace fossils are limited to two burrow systems, *Chondrites* and *Planolites*, both formed by deposite feeders. Other burrows which are common elsewhere in the deep-sea record, such as *Skolithos*, *Teichichnus*, and *Zoophycos* (e.g., Chamberlain, 1975; Ekdale, 1977; Ekdale and Berger, 1978; Berger et al., 1979) are absent from these cores (with the exception of a single occurrence of *Zoophycos*). This suggests that even the "aerobic" portions of the mid-Cretaceous sedimentary cycles may have had slightly low dissolved oxygen concentrations in the bottom waters relative to the deep-sea norm. In any case, the absence of *Skolithos* and other burrows presumably formed by suspension feeders indicates that sediment-ingesting deposit feeders dominated the infauna.

Chondrites was found in deposits with an organic carbon content as high as 9.49% and a total sulfur content as high as 3.14%; *Planolites* was observed in sediment with an organic carbon content as high as 4.55% and a total sulfur content as high as 3.02%. Both are burrows of deposit-feeding infaunal macro-organisms which apparently could survive (if not thrive) in an oxygen-poor environment. The mean concentrations of both organic carbon and sulfur in *Chondrites*-bearing units are slightly higher than the mean concentrations in *Planolites*-bearing layers, suggesting that *Chondrites*-producing organisms had slightly broader oxygen tolerances. The presence of *Chondrites* alone in an organic rich deposit probably is a good indicator of dysaerobic (i.e., nearly anoxic) conditions at the sea bottom at the time of deposition.

CONCLUSIONS

The following conclusions are derived from this study of trace fossils in mid-Cretaceous DSDP cores from the North and South Atlantic Ocean:

1) Burrows of deposit feeders (e.g., *Chondrites* and *Planolites*) dominate the biogenic structures of organic-rich zones. Burrows of suspension feeders (e.g., *Skolithos*), which may be common in other types of deep-sea deposits, are virtually absent. This observation suggests that the primary food resource at the sea floor during periods of oxygen minima was organic detritus in the sediment and that very little food was available in the water column.

2) Burrows which can be attributed ostensibly to hard-shelled organisms (e.g., molluscs and calcareous arthropods) are absent from organic-rich layers. This observation agrees with those of previous workers that organisms with calcareous, siliceous or phosphatic hard parts typically cannot tolerate oxygen levels as low as the minimum oxygen levels tolerated by many soft-bodied organisms (e.g., various worms).

3) A general decrease in both the intensity of bioturbation and the diversity of biogenic structures coincides with increasing amounts of unoxidized carbon and total sulfur in the sediment. These relationships imply that the abundance and diversity of benthic organisms are drastically reduced when anaerobic conditions prevail at the sea floor.

4) *Chondrites* is the most common trace fossil in organic-rich layers, typically occurring alone in rather high densities. This observation indicates that even though *Chondrites*-producing organisms flourished in oxygenated sediment, they also could tolerate lower oxygen levels in the bottom water and sediment than most, if not all, other burrowers.

ACKNOWLEDGMENTS

This research was supported by National Science Foundation grant no. OCE75-21601 to the author. The DSDP curatorial staff (East Coast Core Repository) assisted with sampling and photography of cores; M. A. Bennett, P. L. Buettner, and L. N. Muller assisted with laboratory analyses; P. H. Roth provided biostratigraphic information; M. A. Arthur, H. A. Curran, C. T. Feazel, and P. H. Roth read various drafts of the manuscript and offered suggestions for its improvement. The help of all these individuals is gratefully acknowledged.

ADDENDUM

Following submission of this manuscript, re-examination of *Chondrites* information in deep-sea sediment and elsewhere in the geologic record leads to a slight reinterpretation of the relation between the distribution of *Chondrites* and oxygen levels in interstitial water. Although *Chondrites*-producers indeed may have possessed extremely broad oxygen tolerances, it appears that *Chondrites*-producers actually preferred oxygen-poor interstitial environments in many (if not most) cases (e.g., see Bromley and Ekdale, 1984; Ekdale et al., 1984).

REFERENCES

ARTHUR, M.A., 1979, North Atlantic Cretaceous black shales: the record at Site 398 and a brief comparison with other occurrences, *in* Sibvet, J.C., Ryan, W.B.F., et al., Init. Rpts. Deep Sea Drilling Project, v. 47 (2): Washington, D.C., U.S. Govt. Printing Office, p. 719–751.

______, AND SCHLANGER, S.O., 1979, Cretaceous "oceanic anoxic events" as causal factors in development of reef-reservoired giant oil fields: Am. Assoc. Petroleum Geologists Bull., v. 63, p. 870–885.

BERGER, W.H., EKDALE, A.A., AND BRYANT, P.F., 1979, Selective preservation of burrows in deep-sea carbonates: Marine Geology, v. 32, p. 205–230.

______, AND VON RAD, U., 1972, Cretaceous and Cenozoic sediments from the Atlantic Ocean, *in* Hayes, D.E., Pimm, A. C., et al., Init. Rpts. Deep Sea Drilling Project, v. 14: Washington, D.C., U.S. Govt. Printing Office, p. 784–954.

BROMLEY, R.G., AND EKDALE, A.A., 1984, *Chondrites*: A trace fossil indicator of anoxia in sediments: Science, v. 224, p. 872–874.

CHAMBERLAIN, C.K., 1975, Trace fossils in DSDP cores of the Pacific: Jour. Paleontology, v. 49, p. 1074–1096.

DE GRACIANSKY, P.C., AUFFRET, G.A., DUPEUBLE, P., MONTADERT, L., AND MULLER, C., 1979, Interpretation of depositional environments of the Aptian/Albian black shales on the north margin of the Bay of Biscay (DSDP Sites 400 and 402), *in* Montadert, L., Roberts, D.G., et al., Init. Rpts. Deep Sea Drilling Project, v. 48: Washington, D.C., U.S. Govt. Printing Office, p. 877–907.

DEGENS, E.T., AND STOFFERS, P., 1976, Stratified waters as a key to the past: Nature, v. 263, p. 22–27.

EKDALE, A.A., 1977, Abyssal trace fossils in worldwide Deep Sea Drilling Project cores, *in* Crimes, T.P., and Harper, J.C. (eds.), Trace Fossils II: Liverpool, Seel House Press, p. 163–182.

______, 1978, Trace fossils in Leg 42A cores, *in* Hsü, K., Montadert, L., et al., Init. Rpts. Deep Sea Drilling Project, v. 42 (I): Washington D.C., U.S. Govt. Printing Office, p. 821–827.

______, AND BERGER, W.H., 1978, Deep-sea ichnofacies: modern organism traces on and in pelagic carbonates of the western equatorial Pacific: Palaeogeogr., Palaeoclimatol., Palaeoecol., v. 23, p. 263–278.

______, BROMLEY, R.G., AND PEMBERTON, S.G., 1984, Ichnology: Trace fossils in sedimentology and stratigraphy: Soc. Econ. Paleontologists Mineralogists Short Course Notes 15, 317 p.

FISCHER, A.G., AND ARTHUR, M.A., 1977, Secular variations in the pelagic realm, *in* Cook, H.E., and Enos, P. (Eds.), Deep-water Carbonate Environments: Tulsa, Soc. Econ. Paleontologists Mineralogists Spec. Pub. 25, p. 19–50.

HALLAM, A., 1977, Anoxic events in the Cretaceous ocean: Nature, v. 268, p. 15–16.

RHOADS, D.C., AND MORSE, J.W., 1971, Evolutionary and ecologic significance of oxygen-deficient marine basins: Lethaia, v. 4, p. 413–428.

ROTH, P.H., 1978, Cretaceous nannoplankton biostratigraphy and oceanography of the northwestern Atlantic Ocean, *in* Benson, W.E., Sheridan, R.E., et al., Init. Rpts. Deep Sea Drilling Project, v. 44: Washington, D.C., U.S. Govt. Printing Office, p. 731–759.

RYAN, W.B.F., AND CITA, M.B., 1977, Ignorance concerning episodes of ocean-wide stagnation: Marine Geology, v. 23, p. 197–215.

SCHLANGER, S.O., AND JENKYNS, H.C., 1976, Cretaceous oceanic anoxic events: cause and consequences: Geologie en Mijnbouw, v. 55, p. 179–184.

SCLATER, J., HELLINGER, E., AND TAPSCOTT, E., 1977, The paleobathymetry of the Atlantic Ocean from the Jurassic to the present: Jour. Geology, v. 85, p. 509–552.

THIEDE, J., AND VAN ANDEL, T.H., 1977, The paleoenvironment of anaerobic sediments in the late Mesozoic south Atlantic Ocean: Earth Planet. Sci. Letters, v. 33, p. 301–309.

TIMOFEEV, P.P., AND BOGOLYOBOVA, L.I., 1979, Black shales of the Bay of Biscay and conditions of their formation, Deep Sea Drilling Project Leg 48, Holes 400A, 402A, *in* Montadert, L., Roberts, D.G., et al., Init. Rpts. Deep Sea Drilling Project, v. 48: Washington, D.C., U.S. Govt. Printing Office, p. 831–853.

WARME, J.E., KENNEDY, W.J., AND SCHNEIDERMANN, N., 1973, Biogenic sedimentary structures (trace fossils) in Leg 15 cores, *in* Edgar, N.T., Saunders, J.B., et al., Init. Rpts. Deep Sea Drilling Project, v. 15: Washington D.C., U.S. Govt. Printing Office, p. 813–831.

WEISSERT, H., MCKENZIE, J., AND HOCHULI, P., 1979, Cyclic anoxic events in the Early Cretaceous Tethys Ocean: Geology, v. 7, p. 147–151.

INDEX OF ICHNOTAXA

This index lists all trace fossil and tracemaker organism scientific names used in this volume. For each paper, only the page number of the formal description or first mention of each taxon is indexed; page numbers for illustrations of the various taxa are in **boldface**. Modern and fossil animal tracemaker names are indicated by (B); names of mat-producing algae are followed by (A), and names of endolithic algae are indicated by (E).